Computer Algebra in Physical Research

N. N. Govorun

IV International Conference on

Computer Algebra in Physical Research

Memorial Volume for N. N. Govorun

Dubna, USSR 22 – 26 May 1990

Editors:

D. V. Shirkov

Laboratory of Theoretical Physics
Joint Institute for Nuclear Research
USSR

V. A. Rostovtsev & V. P. Gerdt

Laboratory of Computing Techniques and Automation
Joint Institute for Nuclear Research
USSR

World Scientific

Singapore • New Jersey • London • Hong kong

Published by

World Scientific Publishing Co Pte Ltd
5 Toh Tuck Link, Singapore 596224
USA office: 27 Warren Street, Suite 401-402, Hackensack, NJ 07601
UK office: 57 Shelton Street, Covent Garden, London WC2H 9HE

British Library Cataloguing-in-Publication Data
A catalogue record for this book is available from the British Library.

IV International Conference on
COMPUTER ALGEBRA IN PHYSICAL RESEARCH

ISBN-13 978-981-02-0687-1
ISBN-10 981-02-0687-9

PREFACE

This volume contains the papers presented at the IV International Conference on Computer Algebra in Physical Research which was held in Dubna, Moscow Region, USSR, 22-26 May, 1990. Previous conferences were held also in Dubna in 1979, 1982 and 1985. The Conference was organized by the Laboratory of Computing Techniques and Automation and the Laboratory of Theoretical Physics of Joint Institute for Nuclear Research under the sponsorship by Soviet-Polish Joint Enterprise "Interface" and Scientific-Technical Center "Eventus".

The Conference was dedicated to the memory of Prof. N. N. Govorun, the Chairman of the organizing committees of all the previous meetings and conferences on computer algebra which were held in Dubna. He had also promoted a given Conference at the early stage of its organization.

This Conference Proceedings includes papers on computer algebra systems, algorithms and their applications to science, especially to physics. Invited reports were presented in the morning sessions and are published in the early part of this volume; contributed papers presented in the afternoon sessions are also included. In addition the list of reports presented at poster sessions is given. There were also demonstrations of various software systems and packages.

The scope of the Conference covers the main directions of computer algebra activities in the USSR with good presentation of research projects and papers of other countries.

It is our pleasure to acknowledge all contributors to these Proceedings. Unfortunately, a number of interesting papers could not be accepted. The talks in the Proceedings are arranged in sections in accordance with their presentation at the seminar.

The Editors

CONTENTS

Computer Algebra Systems and Algorithms

ADVANCES IN INTEGRATING SYMBOLIC, NUMERIC AND GRAPHICS COMPUTING

Paul S. Wang*

Department of Mathematical Sciences, Kent State University, Kent, OH

Abstract

Modern computer workstations are equipped with high resolution graphics displays, sufficient memory, ample processing power, and networking capabilities. The workstations are usually connected on a local area network (LAN) that also links powerful mainframe computers, parallel processors, and super computers. This hardware and operating environment provides a powerful platform to build integrated scientific computing systems that can significantly increase the productivity of contemporary scientists and engineers. When symbolic, numeric, graphics and document processing facilities and techniques are combined in an integrated environment, they reinforce one another so that the whole is bigger than the sum of the parts.

We envision an integrated system consisting of compute servers that are controlled by an overall system-independent user interface (SUI). SUI runs on a user's graphics workstation and each compute server can run on any processor on the LAN. Through SUI, the compute servers interact with one another and with the user. The user interface is window/mouse oriented with graphics and networking capabilities. The compute engines will include existing systems such as Vaxima, Reduce, Maple, Matlab, NAG, LINPACK, DITROFF, and LATEX. An interface architecture and protocol must be designed to accommodate existing as well as future compute engines.

Research in this direction has been going on at Kent and elsewhere. Presented are some recent developments at Kent: symbolic derivation of numerical code for finite element analysis; automatic numeric code generation based on derived formulas; interactive display and manipulation of graphs for mathematical functions; automatic inclusion of mathematical expressions produced symbolically in documents; generation of code for parallel processors and CRAY super computers; architecture, design, protocol and implementation of SUI.

We also describe several other software packages in these directions.

*Work reported herein has been supported in part by the National Science Foundation under Grants CCR-8714836 and EET-8714628.

ON THE DESIGN OF AN ARTIFICIAL INTELLIGENCE ENVIRONMENT FOR COMPUTER ALGEBRA SYSTEMS

J. Calmet and **I.A. Tjandra**

Institut für Algorithmen und Kognitive Systeme, Universität Karlsruhe, FRG

I - Introduction

Knowledge Representation is the field of Artificial Intelligence (AI) that studies how to represent and manipulate knowledge. We adopt the point of view that algebraic algorithms are a special sort of knowledge . This leads to reformulate the concept of a CAS in the framework of AI.

The first task was to design a general hybrid knowledge representation system capable of handling mathematical knowledge. This task has been completed and a system called MANTRA /1/ is available for this purpose.

The second task is to define the concept of Mathematical Knowledge. This study is based on the definition of computing domains as Abstract Computational Structures. The inference procedure which permits the system to ensure that a mathematical operation on a given domain is valid is realized through Learning methods. This approach leads to a feasible solution to the problem of type inference in CA which is known to be undecidable /2/ i.e. not solvable by a universal algorithm.

Such an environment /3/ is well suited to extend the capabilities of CAS. For instance, it is trivial to integrate graphics since the kernel of MANTRA is based upon a graphical representation of knowledge. The design of explanation and tutoring capabilities or the introduction of databases are straightforward since they are simply built upon existing tools of the environment. It becomes thus possible to gather in such an environment the very diverse knowledges available for Feynman integrals for instance.

II - An Hybrid System for Knowledge Representation: MANTRA

MANTRA (Modular Assertional, Semantic Network and Terminological Representation Approach) is a Shell for knowledge system. It is best described in the dissertation of G. Bittencourt /1/ where a complete list of references is given. The main characteristics of MANTRA are:

- the introduction of a multilevel architecture for hybrid systems together with a methodology to define a unified semantics for knowledge representation methods and their interaction,
- the integration of two features which are usually not found in hybrid systems: Inheritance with exceptions and heuristic programming,
- the extension of the Frame terminological language to accept n-valued relations instead of binary roles only,
- the semantic definition of non-monotonic inheritance with exceptions using the four-valued logic approach,
- all algorithms for inference procedures are decidable,
- high interaction among the different knowledge representations covered by the system.

There are six main methods to represent knowledge: Logic, semantic nets, production systems, frames, procedural representation and direct representation. The first four are recognized to be the most important ones and have been implemented in MANTRA. They can be further divided into two categories according to their functions in representing knowledge. Logic, frames and semantic nets are used to express declarative knowledge and are called epistemological methods because they are associated to models in the external world. A production system represents a complete programming paradigm including the procedural aspect and is called a heuristic method because of this feature. One or more inference procedures are associated to each of the epistemological methods.

First order logic offers a very powerful inference procedure allowing to generate all of the entailed (i.e. implicitly represented) knowledge by a given amount of explicitely represented knowledge. But, the problem of verifying if a given element of knowledge is entailed by a certain amount of explicit knowledge is not decidable in first order logic. To avoid this problem we adopt a non-complete inference procedure and a query procedure which always terminates with an answer. They were introduced by Patel-Schneider [/4/]. These procedures are semantically defined through a four-valued logic approach [/5/].

Our frame method is a terminological language as in Brachman[/6/] but it extends its capabilities by accepting n-valued relations. The main inference procedure is the subsumption procedure. A concept or relation subsumes another one if the former includes the latter.

Semantic networks are methods based on the abstraction of nodes and edges. The method adopted in our approach allows to define hierarchies with exception. The inheritance procedure is the main inference procedure available in this method. It allows to verify if a class is a subclass of another one given an explicit hierarchy consisting of either default or exception links. The so-called skeptical inheritance approach [/7/] has been selected within, again, the four-valued semantic approach.

These three epistomological methods are not very powerful when standing alone because of the adoption of the four-valued semantics which weaken the deductive power of the system. But, this choice was mandatory in order to have only decidable inference algorithms and a semantically sound system. The deductive power is however very much increased by the association of the three methods. This association is performed through the definition of hybrid inference procedures allowing to generate new knowledges from those acquired through the three methods separately. This interaction is defined not by the inference procedures themselves but by the semantic specification of the results that these procedures should provide.

The architecture of MANTRA is defined along three levels: the epistomological level, the logical level and the heuristic level as shown in Figure 1.

(i) The epistomological level includes at present, but it is extendable, the three methods described above. Each of the corresponding module is thus constructed around some epistemological notions: predicates, functions and constants in the logic module, concepts and relations in the frame module, classes and hierarchies in the semantic net module. Syntactically, each one of these modules consists of a set of primitives which are used to manipulate the epistomological notions. The four-valued semantics associates to each notion two models. The positive model consists of the domain elements that are known to be instances of the notion while the negative model consists of the elements that are known not to be instances of the notion. This framework models aspects of ignorance, an instance belongs to neither of the two models, and of inconsistancy, an instance belongs to both models, and thus constitutes a semantics for an incomplete inference mechanism.

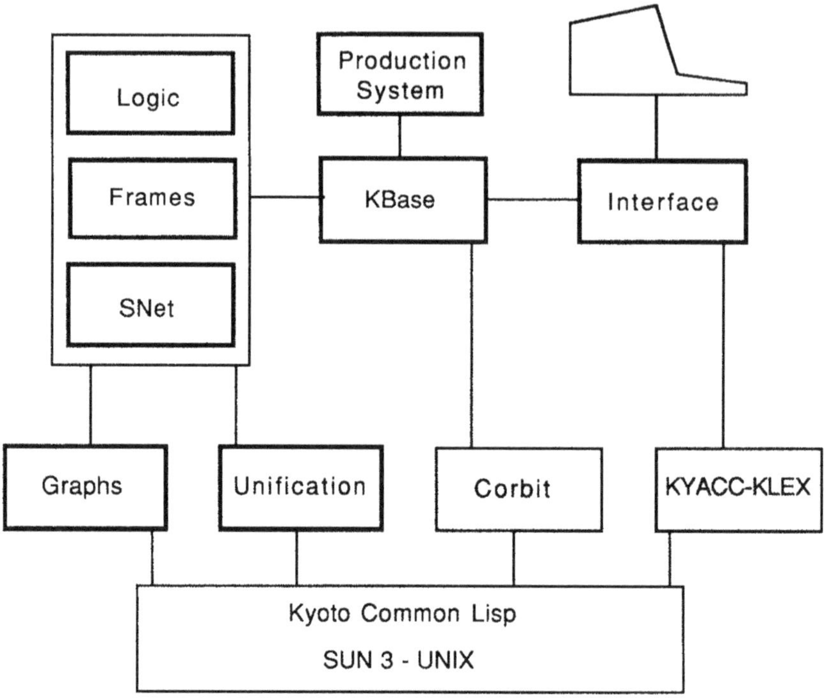

Figure 1 : The MANTRA system and its components

(ii) The second level introduces the concept of knowledge base. Two primitives are used to store facts and to interrogate knowledge bases respectively. There are nine possible interactions among modules: logic-frame, logic-semantic networks (SN), frame-SN, frame-logic, SN-logic, SN-frame, logic-frame-SN, frame-logic-SN and SN-logic-frame. All these interactions have been semantically defined and algorithms for the first three have been proposed and proven sound and complete with respect to the semantic definition.

(iii) The third level is the heuristic level. It consists of primitives allowing the definition of production systems. At this level the conflict resolution and flow control strategies applied during the execution of these production systems can be explicitely chosen through special primitives.

To facilitate the interconnection and communication between the different implemented methods a single data abstraction has been adopted: The directed graph. Directed graphs subsumes several of the most commonly used data structures and they are also suitable for an interactive system because of their inherent graphical character.

The common basis for all the implemented inference procedures is the unification function. A specialized unification package has been implemented in KCL. It is based upon the almost linear algorithm of Martelli and Montanari /8/.

III - Representation of Algebraic Knowledge

We want to redefine algebraic algorithms as a sort of epistemological knowledge. We present here a sketch of the methodology that we propose. It is presented in greater details in /9,10/. We introduce the concept of algebraic knowledge The key and obvious idea is that an algebraic algorithm is valid in some specific mathematical domains. This leads to define a term algebra as a set of well defined objects (domains) together with their specified operations. It is thus conceivable to regard such an object set with its defined operations as a unit of a computational structure /11/ and domains as models of Abstract Computational Structure (ACS).

We define an ACS as a pair (Σ, P) where Σ and P are defined as follows:

- Σ is a signature that consists of :
  - (i) a carrier set that could include other primitive carrier sets,
  - (ii) a set of operations (functions) whose arguments belong to the carrier set.
- P is a set of properties determining the semantics of the operations defined in Σ.

It follows that one may represent an ACS with four different parts:

(i) the ACS head part (Σ-i), (ii) the ACS function part (Σ-ii),

(iii) the ACS initial law part (P) (iv) the ACS learned law part (P).

The head part includes the identifier of the ACS and the carrier sets identifications. An ACS inherits the properties of the primtive carrier sets. The function part consists of a set of descriptions of the function belonging to the ACS. The initial and learned law parts determine the semantics of the functions defined by the user.

The representation in MANTRA is such that the ACS's are joined together into a semantic network where each node corresponds to an ACS and where the links specify which properties are inherited by an ACS. Clearly, this creates a "graph" of the computational domains. We comment only on two issues connected with the decision and learning methods which must be designed in order to get complete sets of laws and to enlarge the semantic net of ACS's. Details may be found in /9,10/.

The first issue deals with the roles of functions according to the concept of term algebra. The term algebra T_{sig} with respect to a signature Σ consists of terms of well-defined strings of elements of the carrier set of Σ and operation symbols of Σ respecting their arity. To determine whether or not two terms of T_{sig} are equal, we analyze syntactically their equality. The most important condition to represent terms of T_{sig} is that every t of T_{sig} be finitely generated otherwise there exists some t of T_{sig} that cannot be represented. In relation with our definition of an ACS we specify the formalism to build parse trees for all elements of T_{sig}. Among the many available formalisms available express terms of T_{sig} we have adopted the formalism of Context Free Grammar (CFG) which provides a powerful formalism to express complex expressions. Moreover, the word problem of CFG is known to be decidable. To represent a CFG we define the concept CFG-descr. The context-free productions rules, as well as laws in the ACS, are represented using first-order logic.

At this stage one must consider the problem of the learned laws which is the second issue. Since using the CFG approach leads to have rewrite rules the problem is thus to find a completion algorithm. Currently available completion algorithms can be used only when strong restrictions are imposed on the reduction system. In practice, we cannot rely on them. We propose a methodology inspired by works in the domain of machine learning /12/. The principle of this method is as follows. Keep the rules defined by users as they are, i.e. do not try to make this rule system complete at the beginning. Instead, the interpretation of each term should be used as a positive training instance from which new knowledge can be acquired incrementally. The set of new acquired rules constitutes the learned rules part of the ACS. A preliminary method to achieve this analysis has been achieved and some algorithms given /10/. We are investigating methods from explanation based generalization /13/ to enhance the efficiency of reduction processes.

IV - Conclusion

To take full advantage of the environment implies to design a CAS tailored to it. Although this is part of our project, the main benefit of the MANTRA environment is certainly to study theoritically problems in Knowledge Engineering. For immediate practical use, we have integrated REDUCE. This allows to add new capablilties to REDUCE such as databases and tutoring capability, but because of the architecture of REDUCE one cannot fully use the capabilities of MANTRA.

References:

1 G. Bittencourt, *An Architecture for Hybrid Knowledge Representation,* Dissertation Thesis, University of Karlsruhe, January 1990.

2 J. Calmet, H. Comon and D. Lugiez, *Type Inference Using Unification in Computer Algebra.* Proc. AAECC-4 Conference, LNCS **307**, pp. 25-28, 1988.

3. J. Calmet, *Intelligent Computer Algebra System : Myth, Fancy or Reality ?* In "Trends in Computer Algebra", R. Janßen Ed., Springer-Verlag LNCS **296**, pp. 2-11, 1988.

4 P.F. Patel-Schneider, *A Decidable First-Order Logic for Knowledge Representation.* Proceedings of IJCAI-9, pp. 455-458, 1985.

5 N.D. Belnap, *A Useful Four-Valued Logic.* In "Modern Uses of Multiple-Valued Logics", J.M. Dunn and G. Epstein Eds., D. Reidel Pub. Co., 1977.

6 R.J. Brachman, R.E. Fikes and H.j. Levesque, *KRYPTON : A Functional Approach to Knowledge Representation.* IEEE Computer, **16 (10)**, pp. 67-73, 1983.

7 J.F. Horty, R.H. Thomason and D.S. Touretzky, *A Skeptical Theory of Inheritance in Nonmonotonic Semantic Nets.* Report CMU-CS-87-175, Carnegie-Mellon Univ., 1987.

8 A. Martelli and U. Montanari, *An Efficient Unification Algorithm.* ACM-TOPLAS, **4(2)**, pp. 258-282, 1982.

9 I.A. Tjandra, J. Calmet and G. Bittencourt, *A Framework for Representing Algebraic Knowledge Using a Hybrid Knowledge Representation System.* Proc. Fourth Int. Symp. on Knowledge Engineering, Barcelona, 1990.

10 I.A. Tjandra and J. Calmet, *Learning Complete Computational Structures.* Preprint submitted to a conference, 1989.

11 B. Listov and J. Guttag, *Abstraction and Specification in Program Development.* The MIT Press, Mc Graw Hill Co., 1986.

12 R.S. Michalski, *A Theory and Methodology of Inductive Learning.* Artificial Intelligence, **20**, pp. 111-161, 1983.

13 T.M. Mitchell, R.M. Keller and S.T. Kedar-Ceballi, *Explanation -Based Generalization : A Unifying View.* Machine Learning, **1**, pp. 47-80, 1986.

SPECIFICATION AND PROGRAMMING METHODOLOGIES FOR AXIOMATIZABLE OBJECTS MANIPULATION: TASSO PROJECT

G. Cioni †, *A. Miola* ‡

† Istituto di Analisi dei Sistemi ed Informatica, C.N.R., Viale Manzoni 30, 00185 Roma, Italy, Tel. and Fax +39 - 6 - 770031

‡ Dipartimento di Informatica e Sistemistica, Università degli Studi "La Sapienza", Via Buonarroti 12, 00185 Roma, Italy, Tel. +39 - 6 - 733412, Fax +39 - 6 - 734616.

1. Introduction

The characteristics and the power of the software systems currently available for mathematical problem solving (DERIVE, MACSYMA, MAPLE, MATHEMATICA, to name a few) are still not sufficiently adequate to answer the quantitative and qualitative needs in different application areas.

The research and development in scientific computing system design has been very intensive in the last few years, also with significant contributions from industrial environments. However the relevant achievements obtained in this field are still based on classical software design methodologies and techniques, such as, for instance, syntactic rewriting rules systems.

It seems necessary to attach the problem from a globally new point of view, by defining new structural characteristics of the desired systems in order to allow the user to work in a much more effective way. In particular the role of a computer should be furtherly enlarged from a pure computing instrument to a tool for supporting the creative and qualitative activity in mathematical problem solving.

Therefore the main subject of the present project is the definition and the application of programming methodologies to offer the effective and complete integration of different computing paradigms (numerical, algebraic and analytical) at a semantic level [MM87]. At the same time the correctness of all the computations (both from the mathematical and the informatic points of view) has to be guaranted. Furthermore the possibility for a qualitative analysis of the results must also be made available by allowing the formal verification of the mathematical properties of the objects to be dealt with.

In particular the project under development on specification and programming methodologies for axiomatizable objects manipulation (TASSO Project) covers the following main items:

Research partially supported by M.U.R.S.T. Projects: "Calcolo Algebrico e Simbolico", "Metodi e Strumenti per l'Elaborazione Non-numerica"; C.N.R. Projects: "Matematica Computazionale", "Sistemi Informatici e Calcolo Parallelo".

- programming methodologies for the abstract specification;
- assiomatic definitions of the mathematical objects, with the consequent use of automated deduction mechanisms, as a new foundamental computing tool;
- algebraic and heuristic methods for applied mathematics, at a very high level of abstraction respect to the domains where the problems are defined.

Therefore the main objective of TASSO is to furnish a specification and implementation language (TASSO-L) to support the design and development of a system for mathematical problem solving.

TASSO is based on the object oriented programming paradigm, and the language Loglan [KSW90] has been initially used as implementation language. On this basis several mathematical objects have been assiomatically specified (numbers, matrices, polynomials, etc.) and one can deal with them on a safe way from the type checking point of view.

At the same time several mechanisms for automated deduction (Resolution, Connection Method, Sequent Calculus) have been made available (TASSO-D) both to the user and to the system itself to check and derive properties of the mathematical objects generated during the computation.

A prototype version of TASSO system runs on a 386-based computer under both MS-DOS and UNIX.

2. Motivations

Let us consider some of the major problems one has to face in using the currently available systems. We do not refer here to any particular system. As a matter of facts the problems we like to address here have to be interpreted as general and common problems in symbolic computation, when a system designed and implemented with inadequate methodological approaches is used.

Let us discuss this matter throughout very simple examples which already bring to unsatisfactory computing situations. In general, one can attempt the following computations:

1)

$$\frac{d}{dx}\log x \longrightarrow \frac{1}{x}$$

2)

$$\sum_{1\leq i\leq 10} i \longrightarrow 55$$

3)

$$\sum_{0\leq i\leq 10} x^i \longrightarrow 1+x+x^2+\cdots+x^{10}$$

4)

$$\int x^n dx \longrightarrow \frac{x^{n+1}}{n+1}$$

Apart from efficiency and performance evaluation, those computations are acceptable as correct from both mathematical and informatic points of view.

However if we attempt a slightly different set of computations, we often obtain results which are definitely inacceptable for several reasons.

Example 1

Step 1:

Let y be $\frac{1}{x}$

Step 2:

Let z be $\frac{d}{dx} \log x$

Step 3:

$$y - z \longrightarrow 0$$

This example is mathematically incorrect, obviously because the function $\frac{1}{x}$ and the function $\frac{d}{dx} \log x$, even with the same analytical expression $\frac{1}{x}$, are defined in two different domains. And this is not taken in any consideration by a system designed simply on the basis of syntactic rewriting rules.

Example 2

Step 1:

Let y be x^i

Step 2:

$$\sum_{0 \leq i \leq 10} y \longrightarrow 11x^i$$

This example is wrong for completely different reasons. The symbol i is used with different meanings in the two successive steps. A system designed without appropriate semantics can often bring to such an unpleasant situation.

Examples 3

3.1)

$$\infty - \infty \longrightarrow 0$$

3.2)

$$\infty + \infty \longrightarrow 2\infty$$

3.3)

$$\infty \cdot 0 \longrightarrow 0$$

This example shows the obvious need for a precise algebra and therefore for an adequate semantics to deal with infinity quantities.

Example 4

$$\sum_{-1\leq i\leq 10} \int x^i dx \longrightarrow ?$$

In general a message like "division by zero" occurs in this case.

This example again presents the case where the symbol i of the integrand function x^i is not bound by the constraint given in the summation. In order to get ride of this computation the sum operator and the integral operator must be interchanged, then the sum must be spleeted into two parts. Then the computation could proceed as it follows:

Step 1:

$$\longrightarrow \int \sum_{-1\leq i\leq 10} x^i dx$$

Step 2:

$$\longrightarrow \int (x^{-1} + \sum_{0\leq i\leq 10} x^i) dx$$

Step 3:

$$\longrightarrow \log x + x + \frac{x^2}{2} + \cdots + \frac{x^{11}}{11}$$

Certainly some mathematical knowledge is need to carry out this computation. But a powerful and flexible system is needed to support the computation.

The TASSO project has been sollicitated by the above considerations on the state of the art on the design and implementation of systems for symbolic computation to be mainly used in mathematical problem solving.

3. Abstract specification of objects and methods

The objects of the TASSO systems are abstract entities susceptible of an axiomatic specification. For example we can consider logic formulas of a given logic calculus or algebraic structures (such as Group, Ring, Field, etc.). For each object the basic definition axioms must be given together with the formal specification of its attributes.

The specification formalism should reproduce the typical hierarchy of the objects. For instance, a Ring must be specified by assuming the specification given for the Group structure. Exactly as it is done in abstract algebra.

However other objects, very common and important in mathematical problem solving, must be provided (e.g. matrices, polynomials). Furthermore, they must be specified in such a way to allow dynamic treatment, as it is often required in real applications.

The specification of an object will be unique, while the instantiation mechanism will allow the use of objects dynamically defined (e.g. matrices over polynomials over ... matrices of integers).

Furthermore for each object the set of admissible computing methods must also be specified.

The specification of a computing method must also be done at the highest possible level of the object hierarchy, in an abstract way. For instance the Euclidean Algorithm must be specified as an attribute of the Euclidean Domain, assumed as a particular object and specified at the appropriate level. Then the instantiation of the object must imply the instantiation of all the related computing methods.

It is very interesting to stress here how the desired specification mechanism allows to exploit the main results on the integration and the amalgamation of numerical and algebraic methods [ML88], [MI90a], [MI90b]. A specification at very high level of abstraction can be given, for instance, for the solution of algebraic equations in abstract domain, so to include both the classical Newton method as well as the Hensel construction for polynomial equations over the integers as special cases.

Furthermore the recent result on the possible unified interpretation of the Hensel construction for algebraic equation solving and of the Buchberger method for the solution of systems of polynomial equations, could support this research line towards the definition of a general method for abstract approximation construction [MM88].

4. The language TASSO-L

Following the previous considerations on objects and methods to be manipulated in the environment system, we can assert that the most natural programming paradigm is the *object-oriented.* According to the literature (e.g. [ST87]), this approach can be well characterized by the following equation

$$\textit{Object-Oriented} = \textit{ADT} + \textit{Inheritance}.$$

In fact all the mathematical objects are well modelled by ADT, their abstract structure is well captured and their implementation is encapsulated at a hidden level respect to the user. Then, mathematical objects represented by ADT can be considered as generic structures in which data, methods and all the possible attributes are specified and localized. In this sense, a strong type checking mechanism, acting on all the defined objects, is the key to support the correctness of the types and subtypes used in all the different steps of the execution.

The inheritance gives the possibility to specialize or extend an already defined ADT. In this way the typical hierarchy of mathematical objects is made correspondent to the tree of the inherited structures.

Furthermore a correct cooperation between abstraction and inheritance allows also to obtain *parametric polymorphism.* Following this approach the common properties of some data structures are defined at the highest possible level and the methods to perform specific operations are dynamically defined only upon the appropriate operands.

An important characteristic to be considered is the necessity to mantain a high degree of correctness of the performed operations, also when dynamic data are defined: i.e. the flexibility of the language respect to user's needs must be balanced by preventing incorrect or ambiguous types definitions. This fundamental objective can be obtained by a correct regulation of the instantiations mechanism of the dynamic data, together with a powerful type checker.

Among the existing object-oriented programming languages we choose the most interesting, correct and flexible features [LMRT90], [RT90] and we have used some of them, namely Eiffel [ME88], Smalltalk [IN78], C++ [ST86] and Loglan [KSW90], as environments for preliminary experimentations.

Among these languages the one we have considered with particular attention is Loglan also because it refers to the theory of algorithmic logic [MS87] as the main theoretical framework where the specification and the correctness verification can be carried out.

In the TASSO-L the objects are realized by *classes* and are connected in a tree structure by a multi-level *prefixing*. The attributes of the objects can be dynamically defined and redefined by *virtual* specifications.

The following schema presents an example of the connections among different objects which directely correspond to mathematical objects. In this schema the attributes defined and made inheritable explicitely are given in every node.

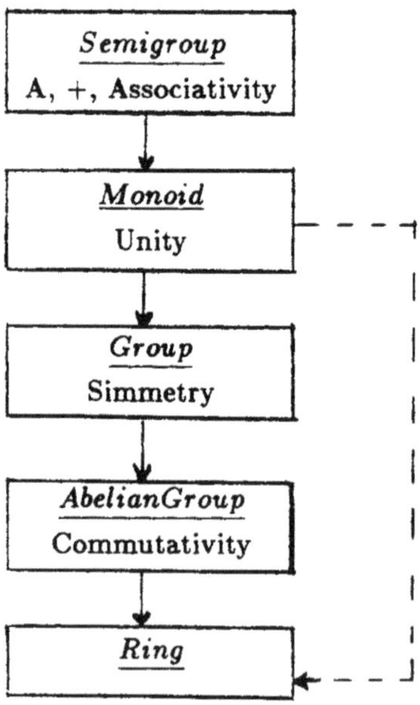

The object Ring will inherit throughout the hierarchical path (the plain lines path), all the specified attributes, (namely the elements A, the operation +, the existence of a unity and the properties of Associativity, Simmetry and Commutativity). In this case strict-inheritance mechanism is used, i.e. the relation between sons and father is always a "is-a" relation.

However the object Ring requires a second operation (i.e. the operation $\times$) which has to be an associative operation. Then a non-strict inheritance is also necessary (given in the schema by the dashed lines) to reuse the basic operation $+$ of the object Semigroup, redefined as $\times$.

5. The automated deduction mechanisms TASSO-D

The language is the main element of the environment and it must incorporate an automated deduction mechanism. This mechanism can be activated directly by the user when he has to accomplish some steps of deduction, but it has also to be automatically used by the system itself as a tool to guarantee the correctness of the objects definitions.

Let us examplify some of the properties treatable by the automated deduction mechanisms.

Properties defined by the user during the specification of the objects. They concern the inheritance relations and can be tested during the compilation phase on the basis of the static tree of the prefixing and of the compatibility rules for the types.

Properties of dynamically created objects. They must be verified after the creation of the objects in order to guarantee the correctness of the operations.

Properties of an object generated by a computation. They must be verified on a user's request and can fall in two cases. In the *generation case* the user could ask for the properties of the object starting from known hypotheses. In the *causes for events case* the user can check the possible hypotheses under which known properties of the object have been obtained.

The TASSO system makes available, in the same environment, different mechanisms which could be extended to higher order logics. These mechanisms offer a good degree of human orientation and in some cases their applicability is increased by the interactive use. The proposed tool just helps the user in his work. He has the knowledge of his application field, can modify the deduction path or stop it or, also, execute it under the control of a specific ad-hoc strategy. Therefore the interaction is useful in this perspective.

We have considered different approaches. The system includes, at the moment, an object-oriented implementation of the *Resolution* with different kinds of *Backtracking* [BC88], the *Connection Method* [BI87], [FT90] and a *Sequent Calculus* [GA86], [BCC90].

The Connection Method is based on the possibility of representing wff as matrices whose elements are (matrices of) atoms, and a validation of a formula is obtained by searching a path of connections between columns of the matrix, which represent the clauses of a NDF of the formula. Then, this method can be defined as an algorithmic mechanism operating upon a single data structure which directly corresponds to the given formula.

Some remarks can be made on the characteristics of this mechanism.

- The transparency of the method derives from the possibility of always working with the same unmodified matrix, created in the input step. In particular it has to be stressed that any kinds of preprocessing to transform it in a normal form can be avoided.
- This mechanism can be considered flexible in the sense that it is extendible to higher order logics and that, from a proof based on this Connection Method, the corresponding proof based on natural Gentzen calculus can be automatically obtained.
- The Connection Method, using a single data structure, in which the formula is completely encapsulated, is particularly suitable of an object-oriented implementation and in this way the uniformity with the entire system can be maintained.

The extension of the deduction capabilities to include the *Abduction Rule* in the *Resolution* approach (as proposed by [KO83]) is known in the literature. Similarly the extension from the verificative case to the generative one (namely to the causes for events) has been proposed for the Connection Method [FT89].

Furthermore the possibility of defining a single method to support different kinds of deductions, namely verificative, generative and causes for events, has been considered. To this purpose a *Sequent Calculus* has been defined because this method results to be intrinsecally suitable by being based on a set of rules easily defined and applied according to the different objective.

Starting from the Gallier's proposal [GA86], we defined a Sequent Calculus with a set of rules for both the decomposition of a given formula in elementary atoms, and for the composition of many sequents into a normal form. We started from the propositional case and we designed the kernel of a tool to be used not only for the verificative approach, but also for the generative and causes for events ones, in the predicative theory. In this case we used some strategies in order to augment the efficiency and to give some halting criteria which take into account the undecidibility of the general problem.

6. Current development

The design of TASSO system is still in progress mainly on the integration of the different moduli, described in the previous sections, and on the user's interface.

In fact the current implementation of TASSO-L and TASSO-D have been widely tested as single moduli and their interaction and cooperation will be verified now on an appropriate set of examples of real applications.

The starting point of this experimentation has been proposed by D. Wang by stating a set of interesting problems together with some solution strategies [WA90]. Those examples came from different areas of applied mathematics, such as irrational expressions simplification, stability of differential equations, linear algebra, limits, geometry reasonings and number theory.

According to the original motivation for the TASSO project (see section 2), all these examples might be impossible to be treated by the existing systems for the mathematical problem solving and can be assumed as a good bed for the test of the system.

References and Bibliography

[BC88] S. BONAMICO, G. CIONI: *Embedding flexible control strategies into object oriented languages*, LNCS 357, Springer Verlag, (1989).

[BCC90] S. BONAMICO, G. CIONI, A. COLAGROSSI: *A Gentzen based deduction method for the propositional theory*, Rapp. IASI 289, (1990).

[BI87] W. BIBEL: *Automated theorem proving*, II Ed., Fried. Vieweg & Sohn, Braunschweig/Wiesbaden (1987).

[CA82] B. BUCHBERGER, G.E. COLLINS, R. LOOS: *Computer Algebra - Symbolic and algebraic computation*, Springer Verlag, (1982).

[CiSa] G. CIONI-A. SALWICKI (Eds): *Advanced Programming Methodologies*, Academic Press (1989).

[FT89] G. FORCELLESE, M. TEMPERINI: *A system for automated deduction based on the Connection Method*, Rapp. IASI 268, (1989).

[FT90] G. FORCELLESE, M. TEMPERINI: *Towards a logic language: an object-oriented implementation of the Connection Method*, LNCS 429, Springer Verlag, (1990).

[GA86] J.H. GALLIER: *Logic for Computer Science*, Harper & Row, Publishers, (1986).

[HH82] G. HUET, J. HULLOT: *Proofs by Induction in Equational Theories with Constructors*, Journal of Computer and System Science, 25, pp 239-266, (1982).

[HO80] G. HUET, D.C. OPPEN: *Equations and Rewrite Rules: A Survey.* In Formal Language Theory: Perspectives and Open Problems, Book, R.V., ed., Academic Press, (1980).

[HU86] G. HUET: *Deduction and Computation*, Foundamental of A.I., Bibel W., Jorrand P. (Eds), LNCS 232, pp. 39-74 (1986).

[IN78] D. INGALLS: *The Smalltalk 76 Programming System Design and Implementation*, Proc. V POPL Conf., pp. 9-16, (1978).

[JSC87] *Rewriting Rules Techniques and Applications*, Special issue of the Journal of Symbolic Computation, n. 1/2, (1987).

[KO83] R. KOWALSKI: *Logic for Problem Solving*, The Computer Science Library, (1983).

[KSW90] A. KRECZMAR, A. SALWICKI, M.WARPECHOWSKI: *LOGLAN '88 - Report on the Programming Language*, LNCS 414, Springer Verlag, (1990).

[LMRT90] C. LIMONGELLI, M. B. MELE, M. REGIO, M. TEMPERINI: *Abstract Specification of Mathematical Structures and Methods*, Proc. DISCO '90, LNCS 429, pp. 61-70, Springer Verlag, (1990).

[ME88] B. MEYER: *Object Oriented Software Constructions*, Prentice Hall, (1988).

[ML88] A. MIOLA, C. LIMONGELLI: *The amalgamation of numeric and algebraic computations in a single integrated computing environment*, Rapp. DIS 18.88, (1988).

[MI90a] A. MIOLA (ed.): *Computing tools for scientific problems solving*, Academic Press, (1990).

[MI90b] A. MIOLA (Ed.): *Design and Implementation of Symbolic Computation Systems*, LNCS 429, Springer Verlag, (1990).

[MM87] G. MASCARI, A. MIOLA: *On the integration of numeric and algebraic computation*, LNCS 307, Springer Verlag, (1987).

[MM88] A. MIOLA, T. MORA: *Constructive lifting in graded structures: a unified view of Buchberger and Hensel methods*, Academic Press JSC, (1988).

[MS87] G. MIRKOWSKA, A. SALWICKI: *Algorithmic logic*, PWN Warszawa & D. Reidel Publishing Company, (1987).

[RT90] M. REGIO, M. TEMPERINI: *A short review on Object Oriented Programming: Methodology and language implementations*, Proc. ACM Sympos. on Personal and Small Computers, (1990).

[ST86] B. STROUSTRUP: *An Overview of C++*, SIGPLAN Notices, Vol. 21, N. 11, pp. 7-18, (1986).

[ST87] B. STROUSTRUP: *What is Object-Oriented Programming*, ECOOP 87, (1987).

[WA90] D. WANG: *Some examples for testing an integrated system*, Rapp DIS, (1990).

On the computer algebra system working as a part of powerful computer system

M.V.MIKHAYLJIK, A.N.SOTNIKOV, A.A.TRUSHINA
Institute of cybernetics problems of the USSR Academy of Sciences

ABSTRACT

Software of the powerful computer system must include a part that can perform symbolic transformation. We describe investigations, carring out by the Institute of cybernetics problems of the USSR Academy of Sciences in the field of computer algebra system.

Analisis of modern approaches for development the analitic manipulations system, working as part of powerful computer system, shows that the users main requests and therefore the programmers efforts are concentrated on one hand at widering of tasks spectrum, which are to be solved with emploiment computer algebra system, and on the other hand on the ensuring "friendly" interface to work with it.

The success of the second way requests intellectualization of means of intercouse with the system, quite simply allowing to enter the new objects into the consideration (for example in mathimatical physics, celestial mechanics ets) and to set the operations for them. To a considerable extent this approach is in the interfase class of "menu".

Reseaches held in the field of analitic manipulations in the Institute of cybernetics problems of the USSR Academy of Sceinces are directed on creating of main components of computer algebra system, functioning within a powerful computer system.

1.Arithmetics manipulations

Rational numbers of any accuracy are one of the main algebraic objects. We have realized algorithms for addition, subtraction, multiplication and divisionof any length numbers. These algorithms are discribed in same works[1,2]. It is known that any positive integer N can be represented in numeration system by base $b=2^{n-1}$ (n is a number of digits in coputer) in a following way:

$$N = n_m b^m + \ldots + n_1 b + b_0, \quad 0 \leqslant n_i < b, \quad n_m \neq 0$$

If the number is negative, the same representation can be used where all n are negative. If $n_i=0$, we assume it also has corresponding positive or negative sign.

Such number representation is designed in computer as one way directed list, elements of which contain value n_i. As the very left machine's digit renains unused, the number's sign will be fixed in it (0 the number is positive, 1 the number is negative). The addition algorithm for numbers with the same sign doesn't require spesial comments, and when the numbers has different signs they are added by digits at first (the digits of any signs are obtained as a result), then digits with wrong signs are corrected. Subtraction is made by changing sign at the subtrahend and following adding it to the minuend. For multiplication on this stage modified for memory diminution "school" algorithm is used. Division is to be executed in accordance with algorithm, stated in Knuth[2].

For the purposes of comfort, beside multiplication and addition of long numbers, multiplication and division of long number on one digit is also to be made, which allows to simplify and speed up fulfillment of operations in simplier cases. Programms of conversion numbers from input form in internal structure (in numeration system by base b) and back are also designed. In the first case the line is translated at first to representation by base 10^k and then using Gorner scheme and already existing operations of long numbers multiplication and addition in system by base b is translated in internal structure. In the second case the program of long number division on is repeatedly used and division remainders will be digits in the representation in numeric system by base 10^k.

At present we work on representation and realization of operations with the Poisson series in computer.

2.Boolean minimizations

In Boolean algebra we consider the problem of logical formula minimization and obtaining its representation in different basises. It is known that Boolean function may be set by the table, in which for every string of variables values the function value in this string is indicated. Or it may be set as a formula using variable sighs and signs of basis elementary functions (such as for example conjunction, disjunction, negation ets). Within elaborated programming complex a user may set any function $f(x_1, \ldots, x_n)$ by the table, then obtain its representation as a formula (at present as an perfect disjunction normal form) and finally simplify this formula.

Possibility of such transformation is very important for ingeneers designing logical schemes, because they often know the accordance between significance of input values to output values beforehand. This significance can be done by the table and the scheme can be designed in accordance with a formula using elements that realize basic functions.

To simplify the formula user has a possibility to obtain shortened disjunction normal function (d.n.f.) and d.n.f. closed to minimal from free d.n.f. In the first case complex the Blak algorithm is used. It includes two transformations:

$$xK' \vee \neg xK'' = xK' \vee \neg xK'' \vee K'K''$$

$$K' \vee K'K'' = K'$$

Using the first one and then the other transformation guarantees obtaining of shortened d.n.f. To obtain pseudominimum d.n.f. the algorithm of gradient descent is used. It concludes consecutive choose maximal intervals covering the great possible number of the rest uncovering single vertexes from shotened d.n.f.

3.MENY system

All eleborated programs and also any other program productions, created by users can be connected to the MENU system. The MENU system allows users to create their own spread tree without using programming languages, secures all oportunities for tree walking, connecting users programs to its windows and their starting. Each vertex of the menu tree contains at most 8 windows.

Eleborater get such edit opportunities for menu tree, window rename, a new window creation, window deletion and changing of window function. When a window is created there is some function (call of subroutine or going to a submenu) to be put in accordance to it. Any time one can change window function (instead of subroutine call get going to a submenu or otherwise) or change function substance. For example subroutine can be changed in case of the subroutine call. With the MENU help you can delete earlier created window if it is necessary.But if there is submenu, which is connected with it, it can be deleted or written in the archives. A great number of different submenu can be kept in the archives and the system provides scanning the archives, and if it is necessary connected a suitable submenu to the current window. So with the MENU's help one can connect subnemu created by different programmers into one common menu for the whole system. It can be connected to all user's programs and through it one can get the system with the convenient user's interface from a total combination separeted modules.

The MENU system can also be used in other fields differing from computer algebra.

All programs are realized on the programming C language, so they are called earsely transfered.

References

1. Computer Algebra. Symbolic and Algebraic Computation (ed by B.Buchberger), Moscow, "Mir", 1986.
2. D.Knuth. The art of computer programming, v.2, Moscow, "Mir", 1977. 3. Discrete mathematics and mathematical problems of cybernetics, Moscow, "Nauka",1974.

PROJECT OF A TOOL FOR CREATING OF COMPUTER ALGEBRA SYSTEMS

A.V.ASTRELIN, E.V.PANKRAT'EV, AND A.F.SLEPUKHIN

Department of Mechanics and Mathematics, Moscow State University, 117234, Moscow, U.S.S.R.

Abstract. *The paper contains a description of a project of a tool for implementation of computer algebra systems. The tool consists of a kernel and a flexible shell that provides possibilities to design object-oriented systems of different kinds.*

Many computer algebra systems were created abroad in the last years. Some of them such as REDUCE, MACSYMA, MAPLE are well known in our country. Increasing interest to computer algebra caused similar home-made investigations, the most known of which are algorithmic language "ANALITIK" and system "ALCOR", and also a number of specialized systems[4,5].

Particular computer algebra investigations take place on mathematical department of MGU for some years. For their further development we have to design a program enviroment that is able to unit them in integral system supporting their interaction and to the own development. Before that the most part of our computer algebra programs were written in REFAL language, which had good opportunities for design of complex data structures but was replaced by "C" because of low speed and noneffective memory usage. But after refusing of REFAL as the base language it became problematic to create universal system and the folloing work was creation of separate packages.

After reading of the descriptions of some computer algebra system[1-3], we saw that most of them (may be except SCRATCHPAD) are large libraries of functions, working with data types from some fixed set and language allowing to use this library and to create new functions.

Such structure of system seems to be not enough flexible. For example, even if we are able to create new type, it is impossible to expand the existing operations over it. More acceptable were abilities of system SCRATCHPAD and language SMALLTALK to create new types dynamically. And we formulated the basic requirements to the future system:

- possibility to create new types and expand existing operations over them;
- presence of an extensible language with possibility to change syntax while system work.

These requirements imply that the future system should consist of a kernel and dynamically attached shell. Besides that we want not to bound application area of our system with the computer algebra, but strive to make it possibly more universal. Therefore in process of system development we moved from kernel to shell all functions except dynamic memory and system work management. Also functions and data became to have similar form.

In fact, the system kernel became a tool for supporting of object-oriented languages. We begin to consider computer algebra system as a language, allowing to create new data types and extend its library.

Now the kernel is an emulator of some virtual processor.

Basic distinguishing property of this processor is a specific memory structure, unit of that is so-called object of language. In system work there are automatic creation and deleting of objects and garbage collection. Every object has a type which, on one hand, infly on the work of processor with this object, and, on another hand, functions to work with this type are setting while the type creation. The program module in this processor is the object too, and it is possible to call another objects. Interface of the different modules carries out through global symbols. Every global symbol has its own number that doesn't change during system work and consists a reference to some object. For more effective work and for implementation of the machine-dependent operations there is a possibility to execute the fragments of real processor programs.

Now let's consider the desired form of shell. One of its main parts should be the supporting of polymorphic functions, i.e., fuctions depending of types of its arguments. These functions may be dynamically defined for the new argument types when new modules are loading. At first we wanted to include polymorphic functions in the kernel, but later we refused this idea for better performance. Another important part is defining of new types by the setting of functions for work with them. These two principal system properties should be reflected in ways to declare types and objects in user language that is the third part of shell. We are going to implement the flexible syntax, at least to allow to use constants of new data type after creation of this type. Besides the shell contains a library of different functions. Thus there are several levels of work with the system:

- interactive work at simple programming in system;
- creating of new functions and extending of library;
- introducing of new data types and operations over them (and expanding of the existing operations);
- changing and developing of user language;
- creating of new shell over the kernel.

After implementation of system on the level of this description its further development is unpredictable and depends of wishes and possibilities of users.

The work on the system was began a year ago. In this work we meet some problems. The main of them are following ones:

- parameter passing and return of values (variables and arrays have no fixed addresses but sometimes there is neccessity to change values of parameters which may be either variables or fields of complex data);

- change of elements of such complex structures as nested arrays (every such change may cause change of whole structure);

- dynamic loading of module in active system (in this module, as a rule functions, variables and types defined in other modules are used and system should provide access to them).

Now we have selected some ways to solve these problems, but each of them needs further studying. Unfortunately, even the structure of the kernel depends on the solution of these problems (in process of the search of optimal solutions we have to rewrite the system from very begining several times), therefore work on the kernel will be finished only when there will exist sufficienly developed computer algebra system on its base. However we hope that computer algebra is not the unique domain of owr system and its kernel may be used in other problems that needed object-oriented approach.

REFERENCES

1. K. Futi, H. Sudzuki, "Yazyki programmirowaniya i ckhemotekhnika SBIS (Programming languages and scheme-technique VLSI)," MIR, Moskwa, 1988.
2. R.D. Jenks, *A Primer: 11 Keys to New SCRATCHPAD*, Lect. Notes Comput. Sci. **174** (1984), 127-147.
3. "MARPLE. Reference Manual. 5th ed," WATCOM Publications Ltd, Canada, 1988.
4. M.V. Grosheva, G.B. Efimov, *O sistemakh analiticheskikh wychislenij na EWM. (On computer algebra systems)*, in "Pakety prikladnykh programm. Analiticheskie preobrazowanija," Nauka, Moskwa, 1988, pp. 5-30.
5. M.V. Grosheva, *Prikladnye i ekspluatacionnye wozmozhnosti sistem analiticheskikh wychislenij. (Applying possibilitie of computer algebra systems)*, in "Pakety prikladnykh programm. Analiticheskie preobrazowanija," Nauka, Moskwa, 1988, pp. 30-38.

EXPERIMENTAL SOFTWARE TOOL FOR DEVELOPING COMPUTER ALGEBRA SYSTEMS FOR PERSONAL COMPUTERS

M.M.Nisheva-Pavlova
Department of Mathematics, University of Sofia
Anton Ivanov Str. 5, 1126 Sofia, Bulgaria

The paper examines some problems of the application of computer algebra systems (CAS) in research, design and education. Some difficulties concerning the application of existing CAS in these fields are discussed and a project of an intelligent software tool for developing CAS for "Pravetz-16" compatible microcomputers is presented. The design and the implementation of the experimental version of this instrument computer algebra system are briefly described.

In the last years computer algebra systems (CAS) have been widely used in many fields of science and engineering. These systems make easy the solution of different types of theoretical and applied problems connected with the execution of some complicated and labour-consuming transformations of algebraic expressions. Computer algebra systems have been successfully used in the automation of scientific computation and design and in the last years there is a trend to application of CAS in teaching mathematics in the secondary schools and universities[1,3].

The application of CAS in research, design and education is connected with several difficulties. The first one is that most of the existing CAS can not be used directly for implementation of special purpose program systems. The second one is that some specialities of the design and implementation of the most widely used CAS don't allow them to be integrated with other program systems. Because of that it is advisable to create instrument CAS intended for developing special purpose program systems.

These are the main reasons of the project of an experimental intelligent software tool for developing special purpose computer algebra systems for "Pravetz-16" compatible microcomputers being under development at the Institute of Mathematics of the Bulgarian Academy of Sciences and the Faculty of Mathematics and Computer Science of the Sofia University.

With respect to its design this instrument system will be a simplified version of an intelligent computer algebra system[2,4] with incomplete knowledge base including some types of knowledge about the mathematical objects and their properties and the available algorithms of algebraic transformations. The general structure of the system is shown in figure 1.

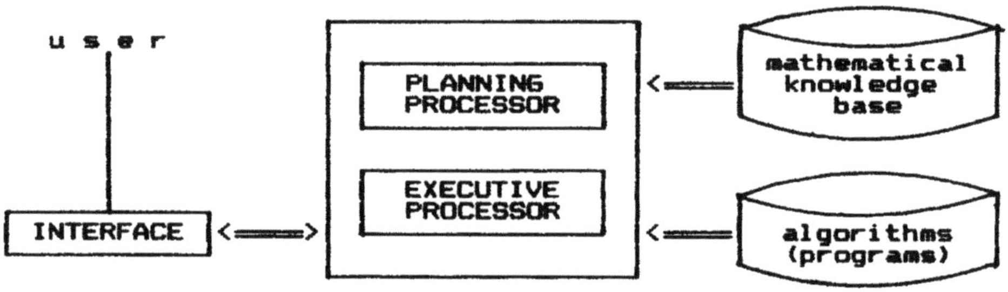

Fig.1. General structure of the system

The interface module will be organized as a hierarchical menu system. In this simple and convenient way the user will be able to define his query to the system, i.e. to determine the mathematical problem he wants to be solved.

The planning processor will analyze the user queries and on the base of the system knowledge about the problem types and the available algorithms for their solving it will create the plan of system's operations. This plan is a sequence of algorithms (programs) that must be performed for the solution of the given problem.

The main function of the executive processor will be to organize the execution of the plan of operations created by the planning processor, i.e. to organize the execution of the necessary "elementary" transformations of the expressions that are given by the user or obtained during the system's performance. It will also realize the basic mathematical facilities of the system including the following operations:

- infinite precision integer and rational arithmetic,
- polynomial arithmetic, opening the brackets and reducing the like terms in algebraic expressions,
- formal differentiation,
- definition of substitution rules in algebraic expressions,
- automatic and user-drived algebraic simplification,
- partial factorization of algebraic expressions,
- computation of algebraic expressions,
- rational function manipulation,
- elementary transcendental function manipulation,
- some simple graphic operations.

The knowledge base will include some types of knowledge about the classes of mathematical problems that can be solved by the system, about the main properties of the different objects and the available algorithms of algebraic transformations. The form of knowledge representation is not finally specified.

The user's work with the software tool will be supported by some proper means for filling the system knowledge base with information about new types of mathematical problems and the algorithms of their solution, about new mathematical object types and their properties and about new symbolic manipulation rules and algorithms. In such a way the system could be extended and used as a software tool for developing special purpose program systems.

An experimental version of this software tool is under development using the PROLOG language. The programs implemented till now constitute the kernel of the executive processor of the system realizing its basic functions: infinite integer and rational arithmetic, univariate polynomial arithmetic, algebraic simplification (opening the brackets and reducing the like terms), partial factorization of simple algebraic expressions.

These programs can be considered as an extension of the PROLOG language with several predicates for algebraic manipulation. This extension forms the first stage of the development of a PROLOG-based intelligent problem solving environment for elementary algebra.

REFERENCES

1. K. Aspetsberger and B. Kutzler, Symbolic Computation - a New Chance for Education. Proceedings of the IFIP TC 3 1st European Conference on Computers in Education (Lausanne, Switzerland, July 24-29, 1988), pp. 331-336.
2. J. Calmet, Intelligent Computer Algebra System: Myth, Fancy or Reality? LNCS Vol. 296 (Springer-Verlag, 1987), pp. 2-11.
3. J.Moses, Algebraic Computation for the Masses. SIGSAM Bulletin, Vol. 15, No. 3 (1981), pp. 5-8.
4. V. Tomov and A. Gerov, Intelligent Program Systems. Proc. of the First International Seminar on Problems and Applications of Artificial Intelligence (Varna, Bulgaria, September 21-25, 1987), pp. 58-66 (in Russian).

MAS
Modula–2 Algebra System

Heinz Kredel
University of Passau,
Innstraße 33, D-8390 Passau, FRG *

Version 0.30

The first Version of the MAS computer algebra system is available for Atari ST / GEM–TOS, Commodore Amiga / Amiga–DOS and IBM–PC / MS–DOS (or compatible) computers. It is available on 3 1/2 inch disks from the author, it will be also available via electronic mail from the author or from USENET.

MAS is an experimental computer algebra system combining Modula–2 program development, a LISP interpreter with a Modula–2 like language, generic function support and access to the comprehensive ALDES/SAC–2 algebraic algorithm libraries.

Current implementations run on an Atari 1040ST / GEM–TOS, Commodore Amiga / Amiga–DOS and IBM–PC / MS–DOS, further implementations are planed on VAX / VMS, IBM–370 / VM–CMS, Sun / Unix. It is completely written in the programming language Modula–2.

Overview

MAS is a computer algebra system in contrast to symbol manipulation systems. The MAS data objects are not considered as symbolic expressions, but they are considered to be elements of algebraic structures. So working with MAS usually requires that you first have to decide in which algebraic structure you want to compute. Data objects can then be converted from their external representation (character sequences) to their internal representation (lists over numbers). These

*E-mail: kredel at unipas.uucp

internal objects can then be used as inputs to programs which are implementations of algebraic functions.

The main concept is to develop algorithms using a standard Modula-2 programming environment and to use a LISP-like interpreter for interactive calculation.

Modula-2 was choosen because it is relatively easy to convert algorithms coded in the ALDES/SAC-2 computer algebra system to Modula-2. By this approach the comprehensive and reliable ALDES/SAC-2 libraries could be reused (650 original + 450 new programms). LISP was choosen because it seemed to be most flexible and well understood to provide a basis for interactive usage and experimenting with generic functions.

To combine these two worlds, the LISP interpreter has been written itself in Modula-2. Using Modula-2 procedure types, the compiled procedures are made accessible from the LISP interpreter. To add a new compiled procedure one just uses an ordinary Modula-2 compiler and then relinks the whole system. This guarantees maximum efficiency of the developed algorithms.

To avoid communication with the system in pure LISP syntax, the MAS language has been provided. The MAS parser generates LISP S-expressions, which are then evaluated by the LISP interpreter. Although one can write complete programs in the MAS language, its primary usage is as data generation language and tayloring the access to the compiled procedures.

Contrary to the design of other systems the MAS interpreter has very limited knowledge of data structures. He only knows atoms (integers with absolute value less than beta = 2**29), symbols and lists over atoms and symbols. So whatever data types are provided by the compiled algorithms (arbitrary precision integers, rational numbers, floating point numbers, strings, polynomials etc.) the user is responsible for their correct handling.

This concept assures maximum flexibility for experimenting with new data types and algorithms. However the interactive usage of these data structures is somewhat cumbersome and error prone.

Further the LISP interpreter has been extended to allow generic lambda constructs, type definitions and some kind of signature definitions. The type declaration concept equips the interpreter with the knowledge that some variables and objects have types. Applying generic functions to these typed variables or objects causes the interpreter to look for an executable function, which can work with the specific object types.

The type definition requires the specification of read and write programs for that type. The interpreter will then be able to call these algorithms for input and output of typed objects.

Finally the portability of the software should be very high, since all storage and input/output management is handled by only two library modules MASSTOR and MASBIOS. They have been first implemented for an Atari 1040 computer running TOS and GEM and we plan to implement them also on other computers. Currently available libraries:

MAS Storage Management System

MAS Basic Input/Output System

MAS LISP Interpreter System

MAS Parser System

ALDES/SAC-2 Integer Arithmetic System,

ALDES/SAC-2 Rational Number System,

ALDES/SAC-2 Integer Factorization System,

MAS Arbitrary Precision Floating Point System,

ALDES/SAC-2 Polynomial System,

ALDES/SAC-2 GCD and Resultant System,

ALDES/SAC-2 Polynomial Factorization System,

ALDES/SAC-2 Algebraic Number System,

ALDES/SAC-2 Real Root Isolation System,

DIP Distributive Polynomial System,

DIP Buchberger Algorithm and Groebner Base System,

DIP Ideal Real Root Isolation System,

DIP Ideal Decomposition System,

and further algorithms for Computer Algebra.

MAS is an open system. Beside the ALDES/SAC-2 and the MAS libraries other libraries can be connected to the interpreter. In a first test we used a numerical library developed at the Technische Hochschule Aachen by Prof. G. Engeln-Muellges and Prof. F. Reutter. It was not only possible to call numerical programs from MAS, but moreover it was also possible to call MAS routines from the numerical programs.

Disclaimer

MAS can be copied and distributed freely for any non-commercial purposes. MAS can only be incorporated into commercial software with the permission of all Authors:

> MAS, 1989, H. Kredel,
>
> ALDES/SAC-2, 1982, G.E.Collins and R.Loos.

I make no warranty and disclaim any usefulness of the programm. Use this program at your own risk.

Passau, 5. November 1989 Heinz Kredel

FORM, a program for the manipulation of very large formulae

J.A.M.Vermaseren

CERN/NIKHEF-H Amsterdam

Most computer algebra programs provide the user with a mathematics workbench. They give the user control over massive amounts of built in knowledge and much of this is used automatically. This is what most users need. There are however problems for which this approach turns out to be "sub-optimal". The design of Form had some of these problems in mind. Hence Form can be considered complementary to a number of these computer algebra packages. Throughout this talk the term computer algebra is used for systems that have much built in knowledge, much of which is applied automatically, while the term symbolic manipulation indicates systems that are mainly rewrite systems.

Many high level operations in computer algebra can only be implemented efficiently if the whole expression to which they are applied resides inside the physical memory. These operations need simultaneous input from more than one term. A good example of such an operation is factorization in which all terms in an entire expression have to "conspire" in order to make the expression factorizable. Such an operation will be very time consuming once the expression doesn't fit inside the physical memory. I will call operations like the above "non-local operations". Local operations take their input from only one term at a time. This means that one can process formulae sequentially and there is much less need to have the whole formula inside the physical memory. Alas it isn't possible to construct a symbolic manipulation program from local operations only. The operation that brings expressions to a unique form is necessarily non-linear as it has to do some sorting. On the other hand it is possible to build a symbolic manipulation program in which the only non-local operation is a variation of sorting. This can be done very efficiently, even when the expressions are disk based. What is even more interesting is that many algorithms that are implemented in a non-local way in computer algebra programs can in principle be rewritten as a sequence of local operations and sort operations. It is not clear how much this influences their performance as not much research has been done in this direction.

A system that has its expressions disk based will use its memory in an entirely different way than a memory based system. The main memory use of Form concerns various types of buffers to not let the disks slow down the execution significantly. The advantage of this is that even on small computers Form can deal with rather large formulae, although very small buffer sizes may make execution somewhat slower. This means that even inside an MS-DOS computer with 640 Kbytes formulae of Mega bytes can be manipulated. On bigger computers FORM will usually run in a space that is smaller than 4 Mbytes and a large part of that can even be swapped to disk without influencing the performance noticeably. It is even possible to control the size of the buffers by means of a special setup file. At times this can be very handy. On such systems the available diskspace is usually the factor that limits the size of the expressions.

To design a program like Form that must deal with big expressions and be very fast involves some tradeoffs. Of course the definition of local operations needs the definition of a "term". For most large scale formula crunching the addition (and subtraction) and the multiplication (and division) have a special role. The addition separates the terms and the multiplication separates subterms. Both operations are associative and the addition is commutative. The multiplication isn't necessarily commutative. This scheme looses some generality and is less elegant than a completely flexible setup, but it is a minor loss. Other operations can be built in by using functions and defining rules for the interactions of the arguments of the functions. This is of course slower than addition and multiplication.

In the beginning most of the syntax of Form was a modernized version of the language of Schoonschip [1]. This was however abandoned when the Schoonschip syntax prevented the use of modern language constructs. An example of this is the introduction of the semicolon to indicate the end of a statement. Before this was done it was impossible to use indentation. The statements in Form still use the Schoonschip construct of a keyword, possibly followed by subkeys, after which come either the body of the statement or a number of parameters. This construction is used in many languages. Because Form knows several types of variables and more are planned for the future the decision has been taken to make it a strong typing language. Hence all symbols, vectors, indices, functions, sets and expressions have to be declared properly. In the declaration it is possible to assign some properties to these objects. An example is the power restriction on symbols. This can be very useful for power series expansions, as terms that contain such an object to an "illegal" power are set to zero automatically at a very early stage. Sets are a hybrid between arrays and real mathematical sets. They are mainly used for pattern matching in which a potential match can be restricted to an element of a set. Functions can be declared to be commuting functions or non commuting functions. The order of non commuting functions will always be respected.

The pattern matching of Form is very extensive. Much attention has been given to wildcarding of more than one object and to very special wildcard modes. Yet the scheme isn't complete as some types of wildcarding might require very large search times and could introduce ambiguities. In future versions most of those modes will also be included anyway. Some examples of wildcarding that are allowed are:

```
id  F(x?,??)*g(??)*x1^n1?*x2^n1?*p?.q?^n2? = ...
```

The language of Form knows several types of expressions. Expressions are the formulae that are being manipulated. There are local and global expressions. Global expressions can be kept after a program has terminated. After the proper command they are stored inside a special file and take the status of "stored expressions". The other possible status is that of an active expression. Active expressions are the expressions that are being manipulated. In principle all operations act on all active expressions, although there is a way to exclude expressions from such activity temporarily.

Form statements are grouped in modules. Each module is translated only when its turn has come. This means that at the module level Form acts as an interpreter. The statements inside the module are compiled, so that the execution of a module will proceede rapidly. This hybrid fashion is dictated by some considerations of flexibility. Internally the variables aren't stored by name, but rather by a number in a table. It is possible for Form to "forget" names as only names that have been declared as global variables cannot be removed from the tables. This loosing of variables and expressions is a special command. Hence the fresh translation before a module is executed. This compilation takes rarely a measurable amount of time anyway.

After a module has been executed the expressions are brought to normal form by sorting them and adding the coefficients of terms that are otherwise identical. This means that the sorting is only executed when the user asks for it. Such a feature provides great flexibility and can speed up execution considerably. An example of such a case would be a sequence of substitutions in which each substitution affects only a small fraction of the terms. In such a case sorting after each substitution would be a waste of time. For small expressions one would hardly notice the difference, but for expressions with more than 10^5 terms this can become annoying.

As in most modern languages Form is equiped with a preprocessor that can treat the input on a character level before this input is offered to the compiler part of Form. This preprocessor is rather flexible as it contains preprocessor variables, loops, conditions, macro's (named procedures) and a little calculator that allows constructs of the type:

```
#define MAX "10"
Local F0 = 1;
Local F1 = 1;
#do i = 2,'MAX'
  .sort
  drop F{'i'-2};
  skip F{'i'-1};
  Local F'i' = F{'i'-1}+F{'i'-2};
  print;
#enddo
.end
```

Here the define assigns the string 10 to the preprocessor variable MAX. Preprocessor variables have to be enclosed by quotes when they are used. This allows them to be concatenated to form new strings. The do loop parameter i is also a preprocessor variable. It gets assigned a string that can be interpreted numerically. The curly brackets invoke the calculator. The string between them will be interpreted as a numerical expression, be evaluated and then it is reconverted to a string. This means that the first pass in the loop generates the statements

```
  .sort
  drop F0;
  skip F1;
```

```
Local F2 = F1+F0;
print;
```

This can be very useful when setting up recursions.

The substitutions are in principle a one shot deal. Their pattern is matched, possibly using some rules that are given in a subkey, the matching object is taken out, and only after all matching due to this statement and possible follow up statements that are indicated as such is finished are all the right hand sides inserted. After this the statement is only used again for the other terms that haven't had their turn yet or the statement is reused when it has been placed inside a loop:

```
id  x = x+1;
```

will not result in an infinite loop.

```
repeat;
  id  x = x+1;
endrepeat;
```

would result in an infinite loop, as the repeat loop will execute the statements inside the loop until they have no more effect. In the above example execution would come to a halt, unless the expressions don't contain the object x. Repeat loops, and if statements can be nested, but their "body" cannot contain and end-of-module instruction.

The arithmetic of Form is in principle over the rational numbers, but it can also be switched to integer arithmetic modulus a finite number. Fractions in such a mode may cause problems if this number isn't a prime number. If the number isn't too large it is possible to print all coefficients as powers of a (user provided) generator. Currently no floating point numbers have been built in although the internal notations leave room for it. In principle symbolic manipulation has not much need for floating point numbers as all transcendental numbers are usually given names like "pi". This is unlike computer algebra where the system may try to evaluate expressions when the user asks for it.

Because High Energy Physics has been one of the main reasons to develop symbolic manipulation and nowadays it cannot do without, Form would not be complete without efficient Dirac algebra algorithms. Both 4-dimensional and n-dimensional algorithms have been built in. The n-dimensional algorithms use only the anticommutation rules for the evaluation of traces. The 4-dimensional algorithms are faster as they use some identities that are specific to 4 dimensions. Of course the Chisholm identities have been built in. An example of a typical nontrivial computation in High Energy theory is the amplitude for the reaction 2 gluons $\rightarrow W^+W^-$. This reaction needs a quark loop. To compute such a reaction in the most general case (massive quarks, offshell vector particles) is quite some task. One obtains integrals with four powers of the loop momentum in the numerator and these should be expressed in terms of integrals with no powers of this momentum in the numerator. Details of the method used are given in ref [2]. Below is the processing of one of the diagrams.

```
    #include declare.h
    .global
    G   fermi = ( -(g_(1,Q)+m1)*g_(1,e1)*g7_(1)
          *(g_(1,Q)+g_(1,p1)+m2)*g_(1,e2)
          *(g_(1,Q)+g_(1,p1)+g_(1,p2)+m2)*g_(1,e3)
          *(g_(1,Q)+g_(1,p1)+g_(1,p2)+g_(1,p3)+m2)*g_(1,e4)*g7_(1)
*-- Plus the reversed string to eliminate Levi-Civita tensors
          )/N1/N2/N3/N4;
    trace4,1;
        id  p1.e1 = 0;
        al  p2.e2 = 0;
        al  p3.e3 = 0;
    .sort
Time =      16.00 sec     Generated terms =        1886
            fermi         Terms in output =         339
                          Bytes used      =       18882
        id  Q.p1 = N2/2 - N1/2 + s1.p1;
        id  Q.p2 = N3/2 - N2/2 + s1.p2;
        id  Q.p3 = N4/2 - N3/2 + s1.p3;
        id  Q.Q = N1 + m1^2;
    .sort
Time =      20.35 sec     Generated terms =         716
            fermi         Terms in output =         527
                          Bytes used      =       27284
    #call reduce4{NN|N1|N2|N3|N4|S|p1|p2|p3|p4|v|Delta3|D1234}
    .sort
Time =      60.73 sec     Generated terms =        2582
            fermi         Terms in output =        1946
                          Bytes used      =      111348
    #call reduce3{NN|N1|N2|N3|S123|p1|p2|p34|mu|nu|Delta123|C123|0|0}
    #call reduce3{NN|N1|N2|N4|S124|p1|p23|p4|mu|nu|Delta124|C124|0|0}
    #call reduce3{NN|N1|N3|N4|S134|p12|p3|p4|mu|nu|Delta134|C134|0|0}
    #call reduce3{NN|N2|N3|N4|S234|p2|p3|p14|mu|nu|Delta234|C234|1|p1}
        id  QQPP = QQPP*2/3*(1+epsilon/6);
    .sort
Time =     143.98 sec     Generated terms =        4069
            fermi         Terms in output =        3351
                          Bytes used      =      284206
    #call reduce2{NN|N1|N2|+|p1|s1|s2|B12|0|0|1|1}
    #call reduce2{NN|N1|N3|+|p12|s1|s3|B13|0|0|1|1}
    #call reduce2{NN|N1|N4|-|p4|s1|s4|B14|0|0|1|1}
    #call reduce2{NN|N2|N3|+|p2|s2|s3|B23|1|p1|0|1}
    #call reduce2{NN|N2|N4|+|p23|s2|s4|B24|1|p1|1|1}
    #call reduce2{NN|N3|N4|+|p3|s3|s4|B34|1|p12|0|1}
    .sort
Time =     227.58 sec     Generated terms =        4181
            fermi         Terms in output =        3656
                          Bytes used      =      330898

*-- Here are some more statements for some simplification
```

```
    print +f;
    .store
Time =      428.99 sec    Generated terms =        3478
             fermi        Terms in output =        3466
                          Bytes used      =      362086
   fermi =
      + A1 * ( 16*delta(p1,p2,p34,e3)*e_(v,p1,p2,e4)
        *p1.e2*p12.p12^-1*p12.e1*Delta3^-1*Delta123^-1 + .....
       etc.
    save fermi.sav;
    .end
```

The file declare.h contains a list of declarations that can be used in the various programs that are used to come to a final solution of the problem. Then the amplitude is entered and the polarizations of the vector particles are kept in terms of polarization vectors, so that the diagram can be extended if needed. Next some simplifications are performed to keep things as simple as possible. Then come the calls to some subroutines, called procedures. These procedures can either be provided inside the program, or they can be present in the current directory. In such procedures one can solve a problem once in a general way. Because of the complexity of the problem the answer is in a somewhat abstract form. It contains still many Gram-determinants which may or may not take a simple form when they are written out. This depends on the problem. In this case it may be better to rewrite all those objects into symbols that are computed in a Fortran program. If the problem is sufficiently simple (like photon-photon scattering) one would substitute these objects and obtain eventually a formula that fits on less than half a page. The above program ran on an Atari-ST home computer with only one Mbyte of memory.

Form can be used also for many problems that have no connection whatsoever to the perturbation expansions of field theory. Below is a small example of the expansion of the Campbell-Hausdorf series.

```
    #define MAX "6"
    S   i;
    F   A,B;
    #do k = 1,'MAX'
      S  x'k'(:'k');
      F  C'k';
    #enddo
    .global
    #do k = 2,'MAX'
    G   F'k' = sump_(i,0,'k',x'k'*A/i) * sump_(i,0,'k',x'k'*B/i)
             - sump_(i,0,'k',x'k'*C1/i) + x'k'^'k'*C'k' ;
    #do j = 2,'k'
      id  C{'j'-1} = C{'j'-1}+x'k'*C'j';
    #enddo
    id  x'k'^'k' = 1;
    id  x'k' = 0;
    id  C1 = A+B;
    #do j = 2,'k'-1
        .sort
```

```
    id  C'j' = F'j';
#enddo
print;
.store
#enddo
save chseries.sav;
.end
```

This program makes an intensive use of the preprocessor features. There are loops and the preprocessor calculator facilitates the setting up of some recursions. The program results in a successive calculation of the terms in the Campbell- Hausdorf series as shown below:

```
F2 = 1/2*A*B - 1/2*B*A;
F3 = 1/12*A*A*B - 1/6*A*B*A + 1/12*A*B*B + 1/12*B*A*A
     - 1/6*B*A*B + 1/12*B*B*A;
F4 = 1/24*A*A*B*B - 1/12*A*B*A*B + 1/12*B*A*B*A - 1/24*B*B*A*A;
etc.
```

There exists a closed form for this expansion due to Dynkin (see [3]) but this formula still contains many summations and isn't necessarily more convenient. By changing the value of the preprocessor variable MAX one can run the expansion as far as the computer will allow. In practise this means that a mainframe or large workstation can obtain F12 in less than 20 minutes.

The program that expresses the expansion in terms of commutators is a little more complicated. To make it really efficient it was necessary to build in some extra patterns. These will become available in version 2. It takes less than an hour to express F12 in terms of commutators and the expression thus obtained is shorter than the expression coming from the Dynkin formula, due to the application of relations between the commutators. Below are for instance the results for F6 and F8:

```
F6 =
     - 1/720*C(A,A,B,B) + 1/240*C(A,B,A,B)
     + 1/1440*C(A,B,B,B)+ 1/1440*C(B,A,A,A);
F8 =
     - 5/24192*C(A,A,A,B,B,B)  + 1/2520*C(A,A,B,A,B,B)
     + 1/20160*C(A,A,B,B,B,B)  + 1/15120*C(A,B,A,A,B,B)
     - 1/2016*C(A,B,A,B,A,B)   + 1/20160*C(A,B,A,B,B,B)
     + 1/20160*C(A,B,B,A,A,B)  - 1/10080*C(A,B,B,A,B,B)
     - 1/60480*C(A,B,B,B,B,B)  - 1/60480*C(B,A,A,A,A,A)
     - 1/10080*C(B,A,A,B,A,A)  - 1/20160*C(B,A,B,A,A,A)
     + 1/20160*C(B,B,A,A,A,A) ;
```

These were separate programs. They took 12 sec and 129 sec respectively on an Atari–ST computer. The function C is defined by

$$
\begin{aligned}
C(x_1, x_2, \ldots, x_n) &= x_1 C(x_2, \cdots, x_n) - C(x_2, \cdots, x_n) x_1 \\
C &= AB - BA
\end{aligned}
$$

This looks very much like the Ad operator in ref [3] The program that produced this expression is rather short but uses very high level wildcarding. Some essential statements in the program are:

```
repeat;
    id  C(??)*E?AB = E*C(..) - C(E,..);
    id  B*A = A*B - C;
endrepeat;

id  C(??,B,A) = C(..,A,B);
repeat   id,disorder,C(??)*C(???)=C(...)*C(..)-C(C(...),..);
repeat   id  C(?,C(a?,??),???) =
             C(.,a,C(..),...)-C(.,C(..),a,...);
repeat   id  C(?,C,??) = C(.,A,B,..)-C(.,B,A,..);
```

The wildcarding like `C(??)` means that any argument field will be picked up. This becomes even wilder in `C(?,C(a?,??),???)` which means: C with a number of arguments, then an argument that must be C with at least one argument, and then any number of extra arguments. In the right hand side the dots match the question marks. The disorder option will cause the substitution to be made only if the objects are not properly ordered, due to the fact that they are non commuting. In the end there are relations between the commutators. These can be derived systematically and each even power introduces more of them. Due to these relations the obtained expressions aren't unique and it isn't easy to set up the relations such that the result is minimal.

Form 1.0 is nowadays available on many different computers. Even though the system still contains occasional errors the situation doesn't seem to be worse than with other systems. Due to the policy decision to make version 1.0 available for free it has already a large number of users. At the moment version 2 is being prepared. It will have a large number of new features, many of which will have their main use outside the perturbation expansions of High Energy Physics. In addition some research is being done to determine what is the best way to parallellize Form. It seems to be quite feasible.

References

[1] Schoonschip was written by M. Veltman. It is described in: H. Strubbe, Comp. Phys. Comm. 8 (1974) 1

[2] G. J. van Oldenborgh and J. A. M. Vermaseren, Zeitschrift für Physik C46 (1990) 425

[3] M. Postnikov, Groupes et algèbres de Lie (Editions MIR, Moscou, 1985)

ALGEBRAIC PROGRAM DIRAC ON IBM PC

A.G.Grozin
Institute of Nuclear Physics, Novosibirsk, USSR
H.Perlt
K.Marx University, Leipzig, GDR

The main design goal of the project DIRAC [1-4] was to obtain a simple, time- and especially storage- efficient, computer-independent program for standard Feynman graph and similar calculations. The program was coded in Pascal.

The work at the project started in 1982 on the computer Odra-1305 compatible with the ICL-1900 series. The version 1.0 contained the basic polynomial algebra, and 2.0 - also the tensor algebra. They had an assembly-like input language. It was replaced by a more convenient one in the version 2.1. It was also transfered to ES computers (IBM-360, 370) and SM-Electronica series (PDP-11). This version [1,2] was distributed to several organizations and used for solving some physical problems [5-7]. An experimental version 3.0 [3] contained a partial implementation of the Dirac matrix algebra. It has not been completed, and was not distributed.

Recently the program was adapted for IBM PC compatible personal computers [4]. The input language was somewhat extended, and some service capabilities were added. This version is called 2.2. It uses Turbo Pascal. The source code is about 3800 lines (123K) long. A version 2.15, intermediate between 2.1 and 2.2, is now available on IBM-360, 370, PDP-11, ICL-1900 compatible computers. It contains, in addition to [1], conditional statements, constructs // in declarations, flag state checks, multiplication of a tensor by a polynomial, more strict error checks. The flags SOUND (sound signal when DIRAC is ready to accept a command) and TIME (printing execution time of every command) are IBM PC (version 2.2) specific. File manipulations (commands <, >, and HALT) are implemented as in the PDP-11 version [1]. Capital and small letters are not distinguished in input; output is in capital. The symbol @ is replaced by # on IBM PC.

A detailed description of data structures and algorithms used is contained in [1,3]. A user's guide to the input language can be found in [4]. It's more formal description is also given in [1] (except new features of the version 2.2). Here we shall focus on the comparison of DIRAC with other algebraic systems available on IBM PC compatible computers.

Any algebraic program should be based on an efficient implementation of the polynomial algebra. In DIRAC polynomials are in the expanded lexically ordered representation. Polynomials may depend on a bounded number of scalars, standard bound is 8. This bound may be easely changed by changing a constant and recompiling the program. The bound should not be too large (say, more then 16), because in the opposite case the storage would be used inefficiently. Powers of scalars occupy 1 byte, and therefore cannot exceed 255. Coefficients of polynomials are rational numbers. Their numerators and denominators are 4-byte integers, and therefore they cannot exceed 2147483647.

Algorithms of polynomial operations are described in [1,3]. A good implementation of polynomial multiplication is the most important for the overall efficiency. Multiplication is performed by joining of the ordered streams obtained by multiplying every term of the first polynomial by the second one. The streams are organized in a linear list. Every time after advancing the current stream it moves to it's correct new position by looking through this list. If polynomals are nearly dense, then the number of comparisons needed is of the order 1, and the multiplication time is proportional to nm, where n and m are the numbers of terms in the first polynomial and in the second one. But if they are sparse, then the number of comparisons is of the order of m, and the time is proportional to nm^2. This algorithm can be easely improved, if the streams were organized as a balanced tree instead of a linear list. The multiplication time then would be proportional to nm ln m. However, procedures dealing with balanced trees are rather complicated, and the factor in this estimate is large. The algorithm behavior would worsen for dense polynomials and for all not too lengthy ones. Therefore the simple procedure described above was retained. Knowing it's behavior, a user can economize time, if he multiplies a longer polynomial by a shorter one, and not vice versa.

Raising a polynomial to an integer power is implemented by a special algorithm [1,3], more efficient than sequential multiplication.

A practical efficiency comparison of the polynomial algebra in DIRAC on IBM PC/AT with REDUCE 3.2 (an efficient Japanese implementation [8]) and muMATH [9] was performed. To that end, the calculation time (without printing) of the expressions $(1+x_1+...x_k)^n$ was measured as a function of n and k. DIRAC and REDUCE have built-in tools for time measurements; in the case of muMATH the time was measured by watch. The results are presented on the figure. The stars on the ends of lines mean that they can't be continued due to the lack of storage.

muMATH turned out to be the least efficient system. It is seen to prefer powers of the form 2^n. DIRAC is more efficient then REDUCE for polynomials up to 4 variables. At a larger number of variables asymptotically more efficient algoritms of REDUCE work better.

The main advantage of DIRAC is, however, a much more economical use of storage. On a computer with 640K operative memory, DIRAC can process polynomials with about 20000 terms in total. The Japanese REDUCE can process about 600 terms in polynomials of 8 variables. True, at a lower number of variables this number would be higher for REDUCE and not for DIRAC. Recompiling DIRAC, one can obtain a variant for 4 variables, that would process 20% more terms. Hence, DIRAC on PC can be used to solve much more large-scale problems than REDUCE, if it's possibilities are enough for these problems.

Tensors are built of vectors with indices, the metric tensor, and the unit antisymmetric tensor. The space dimension may be arbitrary, in particular given by an analytic expression. All sqares and scalar products of vectors must be expressed via scalars. The number of indices is bounded, the standard bound is 8. It can be easely changed by

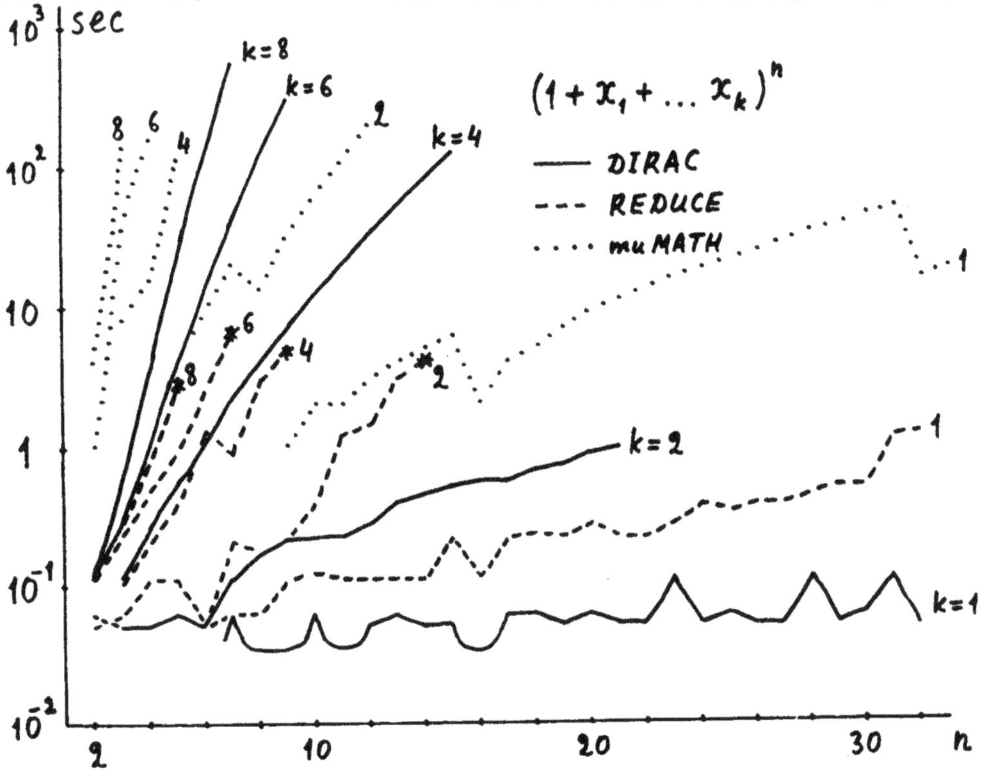

changing a constant and recompiling the program. It should not be too large (say, not greater than 20), because otherwise the storage would be used ineconomically. If the unit antisymmetric tensor is used in calculations, the number of it's indices is subtracted from the bound. The number of vectors is bounded too, the standard bound is 8. Increasing this bound influences only the size of the scalar product table.

As a tensor algebra test, the expressions

$$\overline{\frac{\partial}{\partial x_{i_1}} \cdots \frac{\partial}{\partial x_{i_n}} \frac{1}{x^2}}^{\,2}$$

were calculated (the bar means averaging over x directions, n is even) This example closely models typical calculations in the problem of higher power corrections in QCD [7], in the theory of orientational phase transitions, and in other areas. It contains the typical set of tensor operations: multiplication with contraction over repeated indices, substitutions (used for averaging), and differentiation over a vector. The program for n=4 is

```
VECTOR X;INDEX I1,I2,I3,I4;SCALAR Y,Z;TENSOR A,B,C;POLY O;
DIM(4)  (X.X=Y)  A=(Z)  COM 'Z=1/Y'
COM 'DIFFERENTIATION'
B=DIF Z:#A  C=(-2*Z^2*X.I1)  B=B*C  A=DIF X.I1:A  A=A+B
B=DIF Z:#A  C=(-2*Z^2*X.I2)  B=B*C  A=DIF X.I2:A  A=A+B
B=DIF Z:#A  C=(-2*Z^2*X.I3)  B=B*C  A=DIF X.I3:A  A=A+B
B=DIF Z:#A  C=(-2*Z^2*X.I4)  B=B*C  A=DIF X.I4:A  A=A+B
COM 'AVERAGING - RECURRENT RELATIONS'
B=(Y/5*I1.I2*X.I3*X.I4+Y/5*I1.I3*X.I2*X.I4+Y/5*I1.I4*X.I2*X.I3)
A=SUB I1,I2,I3,I4:X.I1*X.I2*X.I3*X.I4=B:A
B=(Y/3*I1.I2)
A=SUB I1,I2:X.I1*X.I2=B:A
COM 'MULTIPLICATION WITH CONTRACTION'
O=1  A=SUB Y*Z=O:A  B=+#A  A=A*B  WRITE A  END
```

The calculation time for n=4 is 0.4 sec, for n=6 - 5.4 sec, for n=8 - 390.8 sec. REDUCE solved this problem 12.1 sec for n=4. It could not calculate the 6-th derivative due to the lack of storage. It did not solved the problem for n=6 even on a mainframe computer using 4M. muMATH with the SYMBOL-HIT package [10] couldn't calculate the 3-d derivative.

Hence, DIRAC is several tens times more efficient than REDUCE in tensor calculations actively using multiplication with contraction, substitutions with formal indices, and differentiation over vectors. These operations are implemented in DIRAC as special procedures. Moreover, sufficiently difficult tensor calculations can't be performed in

REDUCE on PC due to the lack of storage. muMATH with SYMBOL-HIT turns out to be the least time- and space- efficient system.

Files can be used as subroutines without parameters and local variables. For example, this file tries to divide A by X-1:

```
B=1  B=SUB X=B:#A
IFO B  B=(X+1)  A=SUB X=B:A
    B=(Y)  A=A*B  B=1  A=SUB X*Y=B:A  B=(X-1)  A=SUB X=B:A
ELSE TEXT 'FAILED TO DIVIDE'
ENDIF  HALT
```

The following file calculates 1-loop Feynman integrals. It integrates the tensor A over vector P; S=1/(P.P+1); M, N, L, R are indices.

```
C=(INT)  A=A*C
COM 'INTEGRALS WITH 4 P'
B=(PI^2/3360*M.N*L.R+PI^2/3360*M.L*N.R+PI^2/3360*M.R*N.L)
A=SUB M,N,L,R:INT*S^8*P.M*P.N*P.L*P.R=B:A
B=(PI^2/1440*M.N*L.R+PI^2/1440*M.L*N.R+PI^2/1440*M.R*N.L)
A=SUB M,N,L,R:INT*S^7*P.M*P.N*P.L*P.R=B:A
COM 'INTEGRALS WITH 2 P'
B=(PI^2/120*M.N)  A=SUB M,N:INT*S^6*P.M*P.N=B:A
B=(PI^2/24*M.N)   A=SUB M,N:INT*S^5*P.M*P.N=B:A
COM 'INTEGRALS WITHOUT P'
C=(PI^2/6)  A=SUB INT*S^4=C:A
C=(PI^2/2)  A=SUB INT*S^3=C:A
HALT
```

This table can be easely extended. Such a table integration was used in [5] to calculate 1-loop corrections in O(5) Skyrm model.

We are grateful to V.P.Gerdt for interest to the work and support

References

1. A.G.Grozin. Preprint INP 83-117, Novosibirsk, 1983.
2. A.G.Grozin. Proc. 3-d Conf. "Dialogue man-computer", Serpukhov, 1984, p. 205.
3. A.G.Grozin. Proc. Int. Conf. on computer algebra and it's applications in theoretical physics. JINR D11-85-791, Dubna, 1985, p.76.
4. A.G.Grozin, H.Perlt. Preprint JINR E11-89-537, Dubna, 1989.
5. Yu.N.Kafiev. Theor. Math. Phys. 61, 267, 1984.
6. V.N.Baier, A.G.Grozin. Zeit. fur Phys. C29, 161, 1985.
7. A.G.Grozin, Yu.F.Pinelis. Phys. Lett. 166B, 429, 1986.
8. T.Yamamoto, Y.Aoki. Proc. Eurocal-87, Lecture Notes in Computer Science 378, Springer Verlag, 1987, p.134.
9. D.Stoutemyer. muMATH User's Guide. Softwarehouse, Honolulu, 1985.
10. A.Turski, P.Krawczyk. SYMBOL-HIT Reference Manual. Warsaw University, 1987.

SYMBOLIC COMPUTATION SYSTEM ON PERSONAL COMPUTERS FOR BOOLEAN ALGEBRA

S.V.Kolyada
Glushkov Institute of Cybernetics
Glushkov avenue, 20, 252207 Kiev, USSR

ABSTRACT

Boolean algebra as pure and applied mathematical theory is used to model processes of situation analysis and decision making, for instance, to support computer aided design of integrated circuits, to automate theorem proving and so on. Such massive application of boolean algebra requires developing automation tools to carry out evaluation and formula transformation in boolean algebra. The symbolic computation system for boolean algebra APAL-PC implemented for IBM PC personal computers is described in the paper.

Historically algebra of logic was invented by G.Boole [1] and A.De Morgan [2]. These papers were dedicated to algebraization of Aristotel logic. Suggested there language of algebra of logic (which is called boolean algebra too) permits to write particular formulae (propositions) and formally prove their truth or falsity using algebraic tools for formula transforming.

Boolean algebra (BA) has various applications in different mathematical theories [3], in computer technology, informatics, linguistics and biology; this theory is used especially intensively in design of discrete devices (including computer hardware) - in the theory of switching circuits [4], automata theory [5] and so on. Discrete device design process includes functional design step where the designer needs to write a mathematical model of the device under development as the system of boolean equations. Then this system can be considered as a subject for various formula transformations. Discrete device design features as massive activity, so it is important to have automation facilities for evaluating and transforming boolean formulae.

Let's consider massive applied problems in BA here since automation of the operations required just for these problems promises highest whole gain. They include normal forms construction [6-9], solving logical equations [9], proving boolean functions equivalence, proving tautology or contradiction for boolean function, simplification of boolean formulae (this problem is similar in many aspects to the appropriate problem for other mathematical theories [10]) and so on.

Analysis of these and other problems demonstrates the requirement to have the following set of facilities to make analytical transforming boolean formulae by computer:

- set of functional bases to write formulae ({ 0,1,&,v,~ } and { 0, 1,+,* } at least);
- set of systems of equations;
- set of operators to evaluate boolean functions using the truth tables, to expand and to factor out formulae, to substitute into formulae for transferring boolean formulae from one functional basis to another, to order formulae, to apply equations, to test formulae for graphical identity, to prove tautology and contradiction of boolean formulae as well as equivalence of boolean functions.

Some operations over boolean functions (evaluation using the truth tables, normal form constructing, minimizing and so on) have been implemented so far. However the author isn't familiar with symbolic computation systems supporting sufficiently rich sets of operations over boolean formulae like the set indicated previously. We will describe the program system APAL-PC intended to support symbolic computations in BA by computer.

The APAL-PC system consists of the following components:

- a set of functional bases for writing boolean formulae;
- a set of user-defined substitutions of formulae for variables;
- a set of systems of equations for each functional basis;
- a set of operators to process formulae;
- a set of operators to save and restore results of intermediate formula transformations.

Functional_bases. There are two functional bases { 0,1,&,v, ,~ } and { 0,1,+,* } in APAL-PC to write formulae, which can consist of constants, variables and operation symbols. User can transfer boolean formulae from one basis to another.

Substitutions. User can define, redefine and cancel substitutions of formulae for variables in his APAL-PC-program by operators _subst and can_subst appropriately.

Systems_of_equations. There are two systems of equations for each functional basis in APAL-PC: system of equations which contain the same variables in their left parts and system of different equations (there are no commutative equations). This partition was made due to the need to test identity of subformulae related to the same variables. For instance, these systems of equations for the basis { 0,1,+,* } are as follows:

- $\{ x + x = 0,\ x * x = x,\ x * (x * y) = x * y,\ x + (x + y) = y \}$;
- $\{ 0 * x = 0,\ x * 0 = 0,\ 1 * x = x,\ x * 1 = x,\ 0 + x = x,$
  $x + 0 = x,\ x * (y + z) = x * y + x * z,\ (x + y) * z =$
  $x * z + y * z \}$.

These systems of equations are used for boolean formula transformation by applying equations to given formula according to some strategy. User can easily add his own systems of equations, if desired.

Operators_to_process_formulae:

- copy_f to copy formulae;
- eval_f to evaluate formulae;
- subst_f to substitute formulae for variables into given formula;

- group_f to shift associative parentheses to the right or to the left;
- compare_f to compare two given boolean formulae according to the order embedded into APAL-PC;
- order_f to order formulae;
- expand_f to expand formulae;
- factor_f to factor out formulae;
- chbasis_f to transfer formulae from one functional basis to another equivalently;
- equiv_f to prove the equivalence of two given boolean formulae;
- ident_f to test graphical identity of two given formulae;
- apply_f and other operators to apply one equation or system of equations to given formula.

All previously listed operators have several modes of execution, this feature creates wide range of facilities to process formulae. For example, ordering can be made by order_f using the six modes: by nonincreasing or nondecreasing directly, along with substituting or simplifying.

Saving and restoring results. There are operators save_f and restore_f in APAL-PC to save and restore results of formula transformations for given number of steps.

Debugging tools. There is operator test_f in APAL-PC to test correctness of formula writing. Each operator of APAL-PC includes facilities to print messages if incorrect operator parameters are given.

The experience of design of similar system APAL-ES [11] implemented in System/370 environment was taken in consideration during developing APAL-PC. APAL-PC has been implemented for IBM PC personal computers on the base of the programming language C and universal tools for formula processing designed at Glushkov Institute of Cybernetics [12]. APAL-PC is the library of C-functions and macrodefinitions.

Conclusion. The symbolic computation system for BA APAL-PC is described in this paper. APAL-PC allows to program and execute effectively various operations for boolean formula processing. APAL-PC is under system debugging now. Main lines of APAL-PC development are designing versions of this system for computer aided design of integrated circuits and for automated instruction on BA according to university courses.

REFERENCES

[1]. Boole G. The mathematical analysis of logic, being an essay toward a calculus of deductive reasoning. Cambridge and London, 1847.

[2]. De Morgan A. Formal logic: or, the calculus of inference, necessary and probable. London, 1847.

[3]. Sikorski R. Boolean algebras. Springer-Verlag, 1964.

[4]. Shannon C. The synthesis of two-terminal switching circuits. - Bell System Techn. Journ., 1949, v.28, #1.

[5]. Glushkov V.M. Synthesis of digital automata. - Phismathghiz, Moscow, 1962. - 462 p. (in Russian).

[6]. Discrete mathematics and mathematical problems of cybernetics/ Ed. Yablonsky S.V., Lupanov O.B. - Nauka, Moscow, 1974. (in Russian).

[7]. McCluskey E. Minimization of Boolean Functions. - Bell System Techn. Journ., 1956, v.35, pp.1417-1444.

[8]. Quine W.V. On cores and prime implicants of truth functions. - Amer. Math. Monthly, 1959, v.66, #9.

[9]. Zakrevsky A.D. Logical equations. - Nauka i tekhnika, Minsk, 1975. (in Russian).

[10]. Computer Algebra: Symbolic and Algebraic Computation / Ed. B.Buchberger, G.E.Collins, R.Loos. - Springer-Verlag, 1982, 1983.

[11]. Kapitonova Yu.V., Kolyada S.V. Symbolic computation system for boolean algebra. In: New generation computer design: architecture, programming, intellectualization. Novosibirsk, 1986, p.110-117. (in Russian).

[12]. Kapitonova Yu.V., Letichevsky A.A., Konozenko S.V. Algebraic programming system APS-1. - Cybernetics, 1989, #1.

HECAS-2: NEW VERSION OF THE COMPUTER ALGEBRA SYSTEM FOR HIGH ENERGY PHYSICS

S.N.Grudtsin, V.N.Larin
Institute for High Energy Physics, Protvino, 142284, USSR

ABSTRACT

The general structure of the system is presented and some of the new possibilities are considered. The trends of future system development are briefly discussed.

The possibilities and the input language of the previous HECAS version were considered in reference/1/. Here we describe briefly the system structure and the essential modifications distinguishing HECAS-2 from it.

The system is written in FORTRAN extended by the package working with symbols and the package of a dynamic storage allocation. The language choice was dictated by several reasons, among them is including the system into the INTERCOMP integrated package/2/ written in FORTRAN too. It includes a numeric algorithms library, graphics, DBMS, editor and text processor. So, the integration idea of the computer algebra with the numeric algorithms is realized.

The source memory needed for the complete compiled code is 300K bytes for the 32-bits computers. The high machine independence of the system ensured its adaptation on various computers: ICL, VAX, EC, PC.

1. The General Structure of the System

The system has block-modular arrangement. The extended scheme of the system structure contains two blocks (kernel and block of independent modules-BIM) and two modules (interface with dynamic storage and module of matrix processing). It is represented at fig.1.

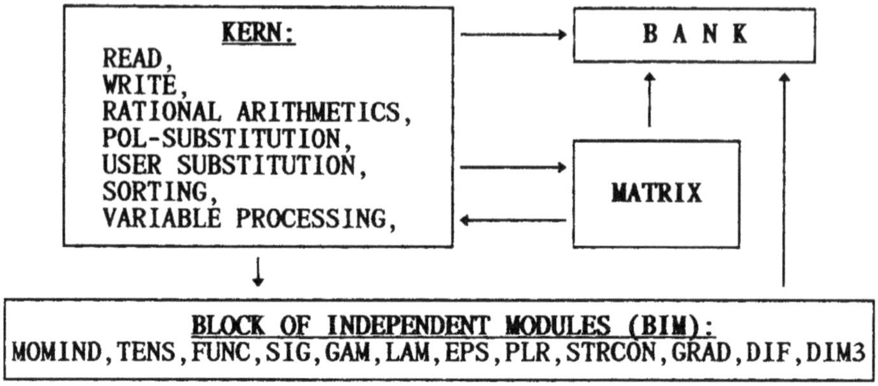

Fig.1. The structure of the CAS HECAS.

The data structures have two forms of the representation: static - FORTRAN arrays, and dynamic - the chains of the "work space banks"/3/. The first is used for data processing, the second one - for compact data storing. Such an organization solves the alternative: compact storage - effective processing of the "large" expressions (for "short" ones it is not critical). The large expressions have, as a rule, polynomial structure which can be considered as a chain of the addendums (monomials), each being a list of the multipliers with numeric coefficient. Therefore the inner representation of a polynomial is a chain of banks, each storing the multipliers of a monomial, coefficient and a link to the next one.

The interface with dynamic storage BANK contains two basic routines, one packing information for compact store in the main memory, and another unpacking it for further processing. In both cases one monomial is transferred at one call. Such an organization permits to choose the most rational manner of data packing for various computers and problems.

The kernel of system KERN (together with BANK module) is the minimal HECAS configuration. We consider the functions of two modules only: POL-SUBSTITUTION and SORTING (the functions of others follow from the names to some extent). The POL-SUBST. module executes the main operation under expressions - forms the current monomial for processing. For this purpose the polynomials contained in the processed expression are substituted term by term. Since after processing the monomial is returned into module may again contain the polynomials, this procedure has a recursive character. The important function of this module first introduced into HECAS-2 is processing connected polynomials at the depth specified.

The monomial formed in the POL-SUBST. is transferred to the SORTING module which sorts the objects in classes and invokes BIM routines to process the objects of the specific classes. After processing all object classes, they go into the output monomial in turn, i.e. it becomes ordered in classes (within classes ordering is already executed by the special BIM modules).

The block of the independent modules is the unification of the modules processing the objects of individual classes and performing some operations. Any of these modules (or even all) can be excluded from the working version.

MOMIND - processing vectors and indices; TENS - processing tensor objects containing calls of the modules of the concrete tensors, e.g. the unit totally antisymmetric tensors (EPS module), polarization tensors (PLR module). The FUNC module processes the functions; SIG, GAM, LAM process σ-, γ- and λ-matrices, respectively; STRCON processes the structural constants of SU(3) group. The DIF module executes scalar differentiation, and GRAD module calculates the covariant derivatives. The DIM3 module reduces 4-dimensional expressions to 3-dimensional ones.

The MATRIX module processes the matrices. It has an original status in the system hierarchy because it not is only called by the kernel routines but itself calls the POL-SUBST. module substituting polynomials. This is related to the fact that the matrix elements are generally polynomials and therefore the main purpose of this module is forming (according to certain rules) expressions from them for further processing by the BIM modules. Therefore the MATRIX module can be regarded as a superstructure. Such a principle of constructing the two-level CAS with underlying level of processing the polynomial structures and having the algorithms forming them from the matrix elements at a higher level can be extended for the various mathematical objects.

2. HECAS-2: New Possibilities

Now we consider the modifications which necessitated our announcing about the creation of a new version. These are related to the modifications of the input language syntax and the mechanism of the forming and realizing the substitutions.

So, the input language syntax became harder. Complicated objects (functions, tensors, etc.) are not allowed now to be recorded with blanks between their names and the names of their "arguments". Therefor special delimiters become obligatory. For the functions these are parentheses and commas, for the tensors these are points, etc. For example, F^2(X,Y), T.I.J, TF.I.J.(X). The hard syntax permitted to raise the adequacy of the system response to user's errors. At the same time, the certain language constructions were simplified. For instance, now one can set substitutions without pointing to the command "$LET", the functions of the trace calculation from γ- and λ-matrices (GSP and LSP, accordingly) were introduced. They can replace a cumbersome construction by "spinor brackets". So, the expression U2.?A_!A_!B_U1.?A, where ! is γ-matrix, and U2.?A, U1.?A are "spinor brackets", can be written as GSP(!A_!B) too.

In the previous version a part of the substitutions was set by direct links, and another part was placed into the tables individual for every object class (the multitabulated mode). These possibilities remained, but the main regime was added under which almost all substitutions are placed into one table (unitabulated mode). Now the last substitution is put into the table head, but not into the end as before, therefore the algorithm of the substitution realization will try to use it first. This can be avoided by use the instruction $HECAS1 transferring the system into the mode with "old" mechanism of substitutions. Moreover, the user can flexiblely manage the substitutions with the help of three new commands-keys: $UNISUB - unitabulated ↔ multitabulated mode switch; $LASTSUB - the switch pointing to the head or end of the table the next substitution to be placed; $ADDSUB - the switch regulating the filling of the table by the substitutions with identical left-hand sides.

Now we dwell briefly on some of other innovations.

1. The "functional" substitution SUB(X^2=Y,P.Q=Z,...,POL) was introduced. The "local" substitutions pointed in its "body" are applied only to the POL expression to be cancelled at once.

2. The function X^Y with X, Y arbitrary expressions (including rational numbers, e.g., X^(1/3), (1/2)^Y) was introduced.

3. The possibility appeared to work in n-dimensional space.

4. The derivative with respect to vector was introduced, e.g. D'(P.I,EX), with P vector, I index, and EX arbitrary expression.

5. The operation of a covariant differentiation was introduced. The covariant derivatives in the system language are presented by the tensors with names G1′, G2′,...,G9′ (the number points to the quantity of the indices equal to the quantity of ∇-operators). To canonize the expressions including these objects, the rules of the rearrangement of the indices are used, which correspond to the commutation relations, set by the command $GRCOMM_n. For n=0 the indices are not rearranged, for n=1 they are rearranged with the zero commutator, and for n=2 the indices are rearranged with the commutator corresponding by the Riemannian geometry:

$$\nabla_b \nabla_a [\Phi_{ijk..}] = \nabla_a \nabla_b [\Phi_{ijk..}] + R^x_{iab} \Phi_{xjk..} + R^x_{jab} \Phi_{ixk..} + R^x_{kab} \Phi_{ijx..} + \ldots,$$

where R is the Riemannian curvature tensor (RI4), $\Phi_{ijk..}$ is the expression with 4-indices. The other command, $BIANCI, switches application of the Ricci and Bianchi identities.

6. The services were extended. So, the new command $PSIZE; gives information on the size of any expression.

3. Prospects

Now we will discuss briefly the trends of the future system development. A main problem will be to remove the restrictions still inherent in HECAS-2. One of them is related to the static nature of the data structures used to store the names and attributes of the objects. Therefore change over from static to dynamic structures is the major problem everywhere, where it does not lead to an essential deterioration of the system efficiency. In HECAS-2 all substitution tables were made dynamical, what removed the limitation on the number of the substitutions.

Other direction is a development of the two-level structure, namely creation of the apparatus to work with tensor components. Extension of the system feasibilities depends on the user needs.

4. References

1. S.N.Grudtsin,V.N.Larin, IHEP preprint 87-26,(Serpukhov,1987).
2. S.N.Grudtsin, IHEP preprint 81-118, (Serpukhov, 1981).
3. S.N.Grudtsin, IHEP preprint 81-119, (Serpukhov, 1981).

INTEGRATED SYSTEM INTERCOMP AND COMPUTER LANGUAGE FOR PHYSICISTS

S.N.Grudtsin
Institute for High Energy Physics, Protvino, 142284, USSR

ABSTRACT

This paper contains a description of general approach to physical software integration, development history and modern stage of INTERCOMP system, containing a large set of language and program means for description and computer analysis of physical models. The system has a high level interpreted language and includes a powerful analytic calculation subsystem, numeric algorithms library, relational DBMS, graphic package, editor and text processor.

1. Software Integration

The main feature of modern development of physical software is the understanding of the importance of software unification and integration/1/. During passed years many particular systems and packages have been developed for different scopes of application, such as numeric algorithm libraries, systems for data bases management, analytic calculations, graphics, text processing etc. So, many physicists have possibility to derive the analytic form of their formula, extract the necessary information from data bases, carry out a numeric calculation, plot graphics of the final results. But a great problem is to use the incompatible systems and packages at different stages of processing, and this requires from user high qualification and large efforts. He has to know a number of different languages, to realize the output-input conversion for all stages of successive processing. And still, many desired things are found to be impossible because of the successive character of processing.

In order to eliminate these problems, at IHEP there has been designed, developed and put in use the integrated system INTERCOMP. It has a high-level interpreted language, containing all necessary means for numeric analysis, analytic calculations, graphics, data base access and text processing. This system allows user to code all usual stages of processing as one program in one language, using cross-calls of different subsystems. The INTERCOMP language is the integrated one, it involves many types of objects and provides large-scale programming. It has a resident compiler and interpreter, written in FORTRAN, and all its intrinsic possibilities are based on FORTRAN packages and library subroutines.

2. INTERCOMP Development

The history of INTERCOMP development originates from the first version of interpreted language with FORTRAN-written compiler and interpreter of 1979/2/. It was initially intended for use in data bases as a powerful language for data selection and on-line processing including graphic output. Then the system was supplemented with various numeric algorithms and special functions written in FORTRAN and fixed into it. They provide numeric integration, minimization, fit of functions, matrix computations which allow a user to write most of relatively simple algorithms in a compact form, familiar to physicist and do it very quickly.

In later years the system was supplemented with special tools for direct access to relational data bases and text processing. So, the system became a convenient and powerful tool for physicists, allowing wide possibilities for model calculations, numeric analysis and data representation, and also to describe all stages of processing as one source in one language.

A further natural step was to supplement the system with tool for analytic calculations to achieve some kind of completeness for scope of physical computation. Although it had to be one of the system modula it was designed as the independent system with its own language, and then included in the main system. A difference of languages for numeric and analytic stages seems to be necessary because of an essential difference in their logics and data structures. So, the first language is the algorithmic one with all standard operations and control constructions (IF, GOTO, LOOPS, USER SUBROUTINES), and the second one is closer to an editor for a special form of mathematical expressions. The former is the "main" language, and its special relations with latter one makes it possible to code complex algorithms with analytic operations (similar to REDUCE language).

3. The Computer Algebra System HECAS

The language of analytic modulus realization had to be FORTRAN, to make it compatible with all others INTERCOMP components. Initially we tried to use the SAC-1 package as the ready analytic modulus, but it was found to be very poor for high energy physics requirements. So, after the study of the existing analytic systems (most written in LISP) we decided to develop a new FORTRAN-written system of the modern level of efficiency and language possibilities in accordance with following requirements:

1. Efficiency - as of SCHOONSCHIP system/3/ (the pattern of efficiency, having a high productivity and the relatively small memory size).
2. Language - as or more convenient than REDUCE/4/.
3. Modular structure of system - for future introduction of new kinds of objects following modern physics development.
4. Independence of data and algorithms.

The main part (nucleus) of the system was completed in 1982, then it was supplemented with the basic general-purpose algorithms and special ones for High Energy Physics. The first publication on HECAS was in 1983/5/, after it had solved the large field theory problem with γ-matrices trace computation. During the language design much attention was paid to the simplicity of all notations, to readability of program text and to unification of the language logics. So, this language appeared to be very convenient and easy to study, and acquired a lot of users among the IHEP physicists. The total size of HECAS codes is 12000 FORTRAN cards (excluding one short accembler routine). This system is highly machine independent and is implemented at ICL, VAX, PC and EC computers. The basic description of the HECAS language was published in 1987 /6/.

There are 2 ways to supplement a system with new algorithms:

1) to implement this algorithm in HECAS+INTERCOMP language;
2) to develop the equivalent FORTRAN subprogram and include it into the system as a new modulus.

The first is easy for any user to realize. The second is possible for advanced ones only, but provides much more efficient realization. Worth to note that the second way is really acceptable for the user, who knows FORTRAN and the modular structure of HECAS. It is not so hard, as it seems!

4. HECAS-INTERCOMP Connection

The HECAS language has not means for large-scale programming such as "IF" and "GOTO" operations, "LOOPS", macro-definitions and numeric calculations. They are not necessary in many simple cases when the standard set of HECAS tools is sufficient. But the user needs them in more complex cases.

Most of such means are present in the INTERCOMP language, and the user can write his algorithm in INTERCOMP, using the HECAS language insertions ("calls" for HECAS) as simple algorithm operators. The necessary connection between HECAS and INTERCOMP is provided by means of "associated variables" and a special evaluation mechanism.

The associated variables are introduced as ordinary INTERCOMP variables and denote the "numeric values" of appropriate HECAS objects (of $VAR or $MOM type). The name of associated variable is the HECAS one prefixed by "&" sign. Any assignment of the numeric value to such a variable causes the substitution of this value in HECAS expressions (it is equivalent to value substitution defined by $LET-construction).

The special INTERCOMP function VAL("EXPR") gives the numeric value of HECAS expression "EXPR" for current numeric values of its constituents (if any value used is not defined the error is produced). It makes it possible to use the analytic expressions as the "variable" part of numeric algorithms, and to use all powerful tools of INTERCOMP to control the analytic processing and further numeric analysis.

So, HECAS is the natural part of the INTERCOMP system, dynamically and flexibly connected with other software.

5. INTERCOMP as the Integrated System

After the appearance of HECAS-subsystem, the list of INTERCOMP tools became close to completeness for physical purposes. Now this list contains:

1. High level interpreted language with all standard tools.
2. A large collection of numeric algorithms for physical analysis (special functions, linear algebra, function integration, minimization and approximation, polynomial roots etc.).
3. The special means to deal with graphics and histograms (definition, filling, representation, transformation, etc.) and graphical primitives for pictures formation/7/.
4. "Text processing" - high-quality output of scientific texts, formulas and graphics.
5. "Text editing" - the language means for complex text generation and transformation.
6. The tools for relational data base management - structure definition, tables creation, data filling, search and restoring. It provides direct access to data elements from the language level through the special associated variables of INTERCOMP - the images of data elements in base records. It makes the relational data bases to be the regular tool of programming, on same level with arrays and data structures.
7. The powerful analytic calculation subsystem/6/.

It makes it possible for user to describe the complete solution of his problem as one compact source, using the single language. Such a description can easily be understood, changed and used by other users. And the accumulation of these descriptions leads to the "Library of Physical Models".

Then a further natural step is to develop the "Bank of Physical Models" as a system of particular algorithms and data, having the complex structure of mutual relations.

6. References

1. J.Calmet, J.A.van Hulzen, in Computer Algebra. Symbolic and and Algebraic Computation, ed. B.Buchberger, G.E.Collins, R.Loos, (Springer-Werlag, Wien New York, 1983).
2. S.N.Grudtsin, IHEP preprint 81-118, (Serpukhov, 1981).
3. H.S.Strubbe, Comp. Phys. Comm. 8 (1974) 1.
4. A. C. Hearn, REDUCE Users Manual. Version 3.1, (RAND Corporation, Santa Monica, 1984).
5. S.N.Grudtsin, V.N.Larin, in Proccedings of Third Conference "Dialogue Man-Computer", (Serpukhov, 1984), p.201.
6. S.N.Grudtsin,V.N.Larin,IHEP preprint 87-26,(Serpukhov,1987).
7. S.N.Grudtsin, IHEP preprint 83-134, (Serpukhov, 1983).

COMPUTER INTERACTIVE SYSTEM FOR CALCULATION OF PARTICLE COLLISION CHARACTERISTICS AT HIGH ENERGIES

E.Boos, M.Dubinin, V.Edneral, V.Ilyin,
A.Kryukov, A.Pukhov, V.Savrin, S.Shichanin
INP, Moscow St. Univ.119899 Moscow, USSR

ABSTRACT

Problem of theoretical symbolic calculations automation for physics on future colliders is considered. Design and applications of system CompHEP created for this purpose are discussed.

1. General purposes

Theoretical foundation of the collider physical program consists of detailed computations of possible collision and decay processes necessary for the identification of the interesting phenomena and separation them from the background. Almost in all cases complicated computer calculations were performed in the framework of various colliders physical programs. Many problems of collider physics could not be solved without the strong computer support. We think that nowadays the problem has several features which facilitate the creation of special program packages with high level of service.

Let us list these features.

1. There is a large number of physical field theory models (Standard Model with extensions, compositeness models, SUSY, GUT, ...). New models are created by theoreticians continuously. A great number of model parameters are usually not fixed.
So the system should contain standard set of models and a *possibility to change them or to* create *new ones should be available.*
2. It is desirable to make computations symbolically as far as possible to use intermediate results for theoretical analysis (approximations, estimates, special cases etc.). However the final results (distributions, histograms etc.) can be obtained in many cases only with the help of numeric calculations.
So developed symbolic-numeric interface is necessary.
3. Diagrams, final distributions and histograms must be

represented graphically.
So the system must include a graphical interface.
4. A great number of Feynman diagrams appears for each physically interesting process. Backgrounds complicate the situation significantly.
A large number of rather simple similar calculations must be done for physical phenomena analysis.
5. The tree level of perturbation theory is necessary in almost all cases. One can consider the limited number of exclusive subprocesses types: 2⟶2, 2⟶3, 2⟶4, 1⟶2, ..., 1⟶5.
For this reason (intermediate) results obtained have some typical analytical structure.

Our purpose is the elaboration of the universal interactive system with simple user interface. Such system gives the possibility to make fast calculation of any physical process interesting for the user in any experimental environment on the basis of particle interaction lagrangians built in the system or interaction lagrangians given by the user.

2. CompHEP system

We have presented[1] several months ago the interactive system CompHEP for solution of "symbolic" part of above-stated problem. The following possibilities are realized now (version 2.0.):

- *choice and modifications of the physical model;*
- *selection of the process (and subprocess);*
- *helicity states for massless fermions;*
- *definition of composite objects (hadrons, ...);*
- *consideration of inclusive processes;*
- *graphical generation of Feynman diagrams;*
- *generation of a formula corresponding to diagram according to Feynman rules (in REDUCE codes);*
- *symbolic calculation of squared matrix elements.*

Note that another approach based on numeric calculation from the very beginning is under development in KEK (Tsukuba, Japan)[2].

3. Physical testing of CompHEP

We calculated the following processes using the field theory models built in CompHEP (QED, QCD, QED+QCD+effective 4-fermion interaction, Standard Model) in order to be sure that we have correct interaction lagrangians in lepton, quark, gauge boson and higgs sectors. The results were compared with the results obtained previously by other authors.

QED: $e^-e^-\to e^-e^-$ $\quad \gamma e^-\to\gamma e^-$ $\quad e^+e^-\to e^+e^-$ $\quad e^+e^-\to\gamma\gamma$;

QCD, 2⟶2, massless quarks[3]:

$gg \to gg$ $\quad q\bar{q} \to gg$ $\quad gg \to q\bar{q}$ $\quad gq \to gq$;

QCD, 2⟶2 and 2⟶3, heavy quarks[4]:

$q\bar{q} \to Q\bar{Q}$ $\quad gg \to Q\bar{Q}$ $\quad qQ \to qQ$ $\quad q\bar{q} \to Q\bar{Q}g$ $\quad qg \to Q\bar{Q}q$ $\quad gg \to Q\bar{Q}g$;

Standard Model:

$\mu \to e\nu\nu$ (effective 4-fermion interaction)

$e\nu \to e\nu$ $\quad ee \to \mu\mu$;

processes for LEP physical program[5]:

$H \to \mu\mu$ $\quad H \to WW$ $\quad H \to Z\mu\mu$ $\quad H \to W^+\bar{u}d$

$q\bar{q}, e^+e^- \to ZZ$ $\quad q\bar{q}, e^+e^- \to W^+W^-$

$e^+e^- \to ZH$ $\quad t\bar{t} \to Hg$ $\quad H \to b\bar{t}g$

$W \to \gamma\bar{u}d$, $t \to Wb\gamma$ (radiative amplitude zeros[6])

4. Nearest future developments

The nearest future development of CompHEP is seen by us in the following way:

- development of CompHEP facilities for polarization effects (massive spinor and vector particles);
- creation of a module for generation of (optimized) FORTRAN function for cross-sections (for future implementation in LUND like packages);
- creation of a module for rewriting the symbolic answer in various kinematic variables (s,t,u or θ,E for 2⟶2 etc.);
- application of improved Born approximation substituting functional form factors to the lagrangians (development of "Edit Model" option).

5. References

1. E.E.Boos et al., *Preprint* **89-63/140**, (INP MSU, Moscow, 1989);
   E.E.Boos et al., in *Proc. of Int. Workshop on Software Engineering, Artificial Intelligence and Expert System for High Energy and Nuclear Physics*, (Lyon, France, March 1990).
2. H.Tanaka, T.Kaneko, Y.Shimizu, *KEK Preprint* **90-29**, (KEK, Tsukuba, May 1990);
   T.Kaneko, in *Proc. of Int. Workshop on Software Engineering, Artificial Intelligence and Expert System for High Energy and Nuclear Physics*, (Lyon, France, March 1990).
3. J.F.Owens et al., *Phys. Rev.* **D18** (1978) 1501.
   B.Combridge et al., *Phys.Lett.*, **70B** (1977) 234.
4. E.Eichten et al., *Supercollider physics, Rev. Mod. Phys.* **56** (1984) 579.
5. *Physics at LEP-CERN Yellow Rep.* **86-02** (1986).
6. K.O.Mikaelian, M.A.Samuel and D.Sahdev, *Phys.Rev. Lett.* **43** (1979) 746.

USER-FRIENDLY FEATURES OF ORTOCARTAN

Andrzej Krasiński
N. Copernicus Astronomical Center, Polish Academy of Sciences,
Bartycka 18, 00 716 Warszawa, Poland

1. The purpose of this article.

The users of computer-algebra systems often report troubles that seem to have been prevented in ORTOCARTAN[(1),(2)]. This note will describe the algorithms of algebraic simplification, differentiation, substitutions and handling arrays designed for ORTOCARTAN that provide ways around the difficulties. It is not claimed that our algorithms are the best theoretically possible, but they were tested in practice, with positive effect, and may inspire the designers of other systems to make similar improvements.

2. General features of the system.

The system ORTOCARTAN was originally implemented in University of Texas LISP 4.2 on CDC Cyber 73 computers, and all pieces of LISP code quoted below refer to that version. The system performs a calculation specific to gravitation theory that will not be described here (see references quoted in Ref. 1).

Four kinds of variables exist in the system: 1. Constants; 2. Coordinates (the independent variables for differentiation); 3. Functions, they depend on specified variables but are otherwise undetermined; 4. Symbols that represent explicit algebraic expressions. The assignment of variables to these classes is done by the user on input. For each function only those arguments are specified on which it depends directly (the details are described in the user's manual, available separately). For example, for the composite function $f\ (g(x,y),\ u(x,y),\ z)$ one declares f as a function of g, u and z, then g and u as functions of x and y. The arguments of each function (only the direct ones) are stored on the property-list of the function-name under the indicator DEPS. Using again the last example, (GET (QUOTE F) (QUOTE DEPS)) would yield the value (G U Z), and (GET (QUOTE G) (QUOTE DEPS)) would yield (X Y).

A symbol has the expression it represents stored on its property list under the indicator CONTENT. In algebraic calculations, the symbol-names only appear, while differentiations enter the CONTENTs of the expression.

3. Algebraic simplification.

We apparently managed to avoid repeated simplification, the frequently reported problem[(2)]. A thorough simplification, on all levels of the expressions processed, is done only on the user's input in order to standardize the expressions, by the function SIMPLIFY. If its argument is an atom, then the atom is returned as the value of SIMPLIFY. Otherwise, the function calls MAP to scan the argument (now known to be a list) and to apply SIMPLIFY to each element of the list. In this way the simplification is transferred to the deep levels of the list, even though the reader of this text does not yet know what it means to SIMPLIFY an expression. This is just one of many examples of power and elegance provided by recursion. After this stage, SIMPLIFY inspects the first element of the list, assuming it to be a name of a function. Depending on the name, it calls one of the specialized functions: SMINUS, SPLUS, SLOG, STIMES, SEXPT, SEXP or DERIV (if the name is none of the seven, then the expression is returned as it stands, but with the arguments of the unknown function simplified internally). These specialized functions, their names assumed self-explanatory, each perform only top-level simplification characteristic to the respective algebraic operation. For example, STIMES only looks for opportunities to add up exponents of the same base, removing the factor if the exponents add up to zero, or to expand the product by the distributivity rule if a factor is a sum, etc. Whenever a specialized function enters, its argument (-s) is (are) guaranteed to be internally simplified. This rule applies throughout the whole calculation, and in fact SIMPLIFY is never called again because from the moment when the whole input is standardized, the kind of simplification needed is under control, and it must occur only on top level.

All sums and products are ordered according to an unambiguous algorithm. The ordering is based on priorities of all atoms. A priority is a number stored on the atom's property list under the indicator PRIOR. The relative priorities of lists are determined by comparing the priorities of their CAR-s (should they be identical, the CDR-s are compared in the same way). The priorities of two objects can be equal only if the objects are EQUAL. This method may be inapplicable to noncommutative algebras, but in ordinary calculus it has the advantage that the ordering of the output depends only on the ordering of the input (determined by the user), and does not depend on the history of the session. Consequently, it will be identical in every run of the system if the input was not reordered.

Unfortunately, the conventional way of ordering polynomials is not the most natural in LISP, and so the output may look disordered to the user. Nevertheless, ordering in ORTOCARTAN follows a strict algorithm and is unique.

4. Differentiation.

All derivatives to be calculated are assumed to be total derivatives. In the example from the previous section, the total derivative of f with respect to x is $\frac{Df}{\partial x} = \frac{\partial f}{\partial g}\frac{\partial g}{\partial x} + \frac{\partial f}{\partial u}\frac{\partial u}{\partial x}$. The differentiation penetrates all expressions recursively down to the level where either direct dependence on the variable is spotted or it is found out that the expression does not depend on the variable. Therefore it does not matter how composite the function to be differentiated is. Let us quote an actual example[(3)]. Suppose t, x, y and z are the coordinates, a and b are constants, $V = (z + b)\ S(t,Z)$ where S is a function, and $Z = \frac{1}{2}\,(a - x^2 - y^2 - z^2)/(z + b)$ is one of the SYMBOLS. Then

$$\frac{\partial V}{\partial z} = S - Z\,\frac{\partial S}{\partial Z} - z\,\frac{\partial S}{\partial Z}\,.$$

ORTOCARTAN had no difficulty to process this expression (but a substitution was necessary to collapse the coefficient of $\frac{\partial S}{\partial Z}$ in the second term back to Z).

The results of differentiation of atomic variables (functions and symbols) are stored on the property lists of the appropriate atoms with the indicator DERIVS, as lists of pairs of the form ((<variable 1><value of the derivative>) (<variable 2> <value of the derivative>) ...). Thus each derivative is calculated at most once; when needed again, it is retrieved from the property list.

This method of coding dependencies of functions on their arguments has one disadvantage: we found it impossible to shuffle the names of arguments of a functional expression at later stages of the calculation. Hence, definite integrals cannot be properly differentiated because the integrand may depend on a different variable after the differentiation than before (but we never found any user in need of doing such an operation).

Algebraic (top-level only) simplification is automatically called on the result of differentiation on every level of recursion.

4. Substitutions.

In many LISP systems, a system function is available that performs substitutions. These system functions are usually rather inefficient because they copy the original expression on all levels even when no substitution occurred. We designed our own substituting procedure that avoids copying as much as possible, and after actually performing the substitution, automatically calls the algebraic simplification to do its work, but only upward from the level where the substitution was done. When no substitution occurred, the simplification does not enter and no part of the expression is copied.

The rule that simplification should immediately follow every actual substitution is important both for the efficiency of the program and for the user's convenience. In some systems[2] simplification is called only at the end of a series of substitutions, then it must go through the whole structure which might explode to impossible size through the unsimplified consecutive modifications. To force simplification into the intermediate expressions may require programming skill inaccessible to nonexpert users. Moreover, substitutions that are not immediately followed by simplifications may lead to ambiguities. The user who decides to insert one more substitution, assumes the substitution to occur within the last expression he has seen. This will not necessarily be the case when simplification is withheld and enters after the (currently) last substitution.

In ORTOCARTAN, the substitutions are performed by the functions SUBSTITUTE and EXCHANGE. SUBSTITUTE has merely two tasks: to provide the initial values to the variables referenced by EXCHANGE, and to print a message to the user in case a substitution was tried unsuccessfully (the messages can be suppressed). The main work is done by EXCHANGE. The key point of the algorithm is this: the variable A is initialized in SUBSTITUTE with the value NIL. This variable is nonlocal to EXCHANGE, and when a substitution is actually done, the value of A is reset to T. Whenever EXCHANGE calls itself recursively, it verifies the current value of A afterwards. This value being T is a signal to begin copying the structure in which EXCHANGE operates and calling the simplifying functions. The procedure is quite simple except in case of functions with many arguments (e. g. sums and products) where the decision to start copying and simplifying may have to be taken in the middle of scanning a list.

Contrary to some reports[2], there are no difficulties to replace *parts* of sums or products by new expressions. The algorithm is almost identical to the one that finds the difference of two sets and adds new elements to the resulting set - such programs are often exercises in LISP courses. However, this task is so easy in ORTOCARTAN just because of the ordering procedure mentioned before. If a sum is a subset of a larger sum, then its ordering is identical in both places.

ORTOCARTAN can do pattern-matching. A special class of variables called MARKERS may be defined by the user. A marker (necessarily an atom) may represent any expression and may be used to conveniently abbreviate the lists of substitutions. For example, suppose the variable r represents $(x^2 + y^2 + z^2)^{1/2}$. Such r could be defined as one of the SYMBOLS, then, however, the derivatives of r would not be very compact (e.g. $r,_x = x\,(x^2 + y^2 + z^2)^{-1/2}$) and substitutions would have to be used to collapse such expressions to smaller size ($r,_x = x/r$). It is simpler to define r as a function and replace its derivatives by $r,_x = x/r$, etc. Instead of defining these three substitutions, the user may define M to be a marker, and then request just one substitution, $r,_M = M/r$ which will be equivalent to the three. Other situations when MARKERS are convenient are e.g.: $\tan(M) = \sin(M) / \cos(M)$ (the equation will then be applied with any argument) or $a^M = 0$ (this implies neglecting all powers of the parameter a and is equivalent to making a linear approximation in a).

From the user's point of view, substitutions by pattern-matching do not differ from ordinary substitutions. The presence of the argument MARKERS is an alarm-bell for the system, and the special processing of the MARKERS is done automatically whenever any one of them is spotted.

Since no reports are known to this author about other working systems in which pattern-matching would be included, the algorithm will not be disclosed in order to preserve this (possibly) unique piece of glory for ORTOCARTAN for some longer time.

5. Manipulating the tensors.

Tensors with many indices are handled by system functions in U. T. LISP. In order to retrieve or insert any element of such a tensor, its position must be specified by a list of indices. When done carelessly, these operations may easily overuse the available space and force too many entries of the garbage collector, making the system rather inefficient. The apparently most natural way to retrieve the component $(i\ j\ k\ l)$ of a tensor of rank 4 is to call the system function (GETEL <tensor-name> (LIST I J K L)). Then, however, each entry of GETEL creates a new 4-element list that instantly becomes garbage, and the number of calls to GETEL in a single run of ORTOCARTAN may go into tens of thousands. Moreover, these operations are usually done in sequences in which whole tensors are scanned. Again, the apparently most natural way to do this is to define an analog of the ALGOL statement DO running through all values of every index. Then, each value of each index becomes a newly generated number, and this causes again too many entries of the garbage collector.

We solved these problems as follows. We defined a global list of all values of indices, named LIND, with the value (0 1 2 3) (this is the range of indices for all tensors processed in gravitation theory, it may be changed by modifying just one line of the code). Then, (MAPC LIND (FUNCTION (LAMBDA (I) ...))) does the same work as <FOR I = 1 TO 3 STEP 1 DO ...>, but uses only a single copy of each number. Similarly, instead of generating new lists of indices for each call to GETEL, we defined a few global lists (one was not enough), among them LIJKL with the initial value (NIL NIL NIL NIL). Its elements are then consecutively replaced with the appropriate numbers by the function RPLACA before each call to GETEL or SETEL. This method had a dramatic effect on the efficiency, even though the code of the system became longer by about 10%. The execution time went down by the factor 2, and the number of garbage collections by the factor 10.

It is often claimed that with personal computers, efficiency is not important because one can leave the program to run all night long. However, errors in the input data, unavoidable now as ever before, usually force the user to run the program several times on the same problem (*especially* when the problem is complicated and requires a long calculation). Everyone who knows the pain of waiting for the results will agree that it makes a great difference whether the answer is known after 5 minutes or after 2 hours (this is the real order of magnitude of speed difference between fast and slow systems). Hence, efficiency of a system does really make it more user-friendly.

References

1. A. Krasiński, in: *International Conference on Computer Algebra and its Applications in Theoretical Physics*. Edited by N. N. Govorun. Joint Institute of Nuclear Research 1986, p. 50.

2. References to the trouble-causing systems will be omitted in order to avoid negative advertising.

3. A. Krasiński, *J. Math. Phys.* **30**, 433 (1989).

EXCALC - A Package for Calculations in Modern Differential Geometry

Eberhard Schrüfer
Gesellschaft für Mathematik und Datenverarbeitung mbH,
Institut F1, Postfach 1240, D-5205 St. Augustin 1
F.R.G.

May 24, 1990

Abstract

The capabilities and the design philosophy of the differential geometry package EXCALC are described. Examples from physical Field Theories and from the Cartan-Kähler theory of partial differential equations are used to illustrate the capabilities of this system. The current status and projected future developments of the implementation are discussed.

1 Introduction

Ever since Einstein formulated his theory of general relativity, the belief that differential geometry is the proper mathematical setting for formulating the laws of nature has grown. Given the success of differential geometry in physics, it is only natural to ask for computer algebra programs that can manipulate the objects of differential geometry directly. There are several programs available: CTENSOR, ITENSOR (MACSYMA), SHEEP, STENSOR, MathTensor (Mathematica), EXCALC, SUPERGRAV (REDUCE) to name just a few. These packages have very different degrees of sophistication. In almost any general purpose Computer Algebra System it is possible to patch together some tools with standard language constructs like loops, pattern matching rules, noncommutative variables, etc. However, programs constructed this way are hard to use, very specialized, and in general do not perform very well.

In particular, in order to get mathematicians and theoretical physicists interested in using computer algebra as a standard tool for their daily work, a program must have an interface appearing to be as natural as a standard textbook. Input must be accomplishable by only a few keystrokes or mouse clicks, and output should resemble usual mathematical notation. Furthermore, results must be real objects that can be manipulated later in any reasonable context and are not just printed. In particular, this requires that the package does not corrupt the imbedding computer algebra system. The programming constructs provided need to be very high level and flexible for ease of programming and novice use.

A program trying to realize these goals is EXCALC, whose version 2 is distributed with REDUCE 3.3. It implements a very important part of differential geometry, the EXterior CALCulus. The exterior calculus has proven to be extremely useful in such areas as electrodynamics, general relativity, supersymmetric theories and the study of differential equations.

2 Capabilities of EXCALC version 2

The basic objects EXCALC handles are exterior forms and vectors.

The following operations are supported:

EXCALC symbol	Meaning	Textbook symbol
^	exterior multiplication	$\wedge$
d	exterior differentiation	d
@	partial derivative and coordinate vector	∂
#	Hodge duality operator	$*$
_|	inner product	$\rfloor$ or i
|_	Lie derivative	$\pounds$
vardf	variational derivative	δ

Metric spaces can be easily defined by providing a set of basis 1-forms together with a metric. The induced connection forms are available through a simple command.
To get a feeling of how natural the syntax of EXCALC is, we will look at a concrete example. It is concerned with checking a solution [2] of Einstein's equation on a Riemannian manifold. The mathematical setting for this is in the next section.

2.1 EXCALC 2 example for verifying an Einstein space

2.1.1 The mathematical setting

The geometrical arena of interest here is Riemannian geometry. It can be described by a bilinear relation between the basis 1-forms e^i, the metric tensor g

$$g = g_{i\,j} \cdot e^i \otimes e^j$$

where the usual convention is adopted that there is an implicit sum over indices with equal names in contragredient positions.
If the matrix $g_{i\,j}$ is a diagonal matrix with only 1's and -1's as elements, one says that the 1-forms e^i constitute an orthonormal frame (more precisely, a coframe) for the metric space. On the other hand, if the basis 1-forms e^i are simply the coordinate differentials, the e^i's are said to be a natural or coordinate frame. As will be seen below, it is much more efficient to work with frames for which the coefficient matrix $g_{i\,j}$ is constant.
Since the metric components $g_{i\,j}$ form a nondegenerate symmetric matrix, it is clear that orthonormal frames always exist since a symmetric matrix can always be diagonalized.
In Riemannian geometry, given the coframe and the metric, there are uniquely determined connection forms ω^i_k, which play a similar role to the Christoffel symbols. The connection forms can be obtained by differentiation. Using a constant metric, for example, they can be calculated by:

$$d\,e^i = -1/2 \cdot c_{k\,l}{}^i \cdot e^k \wedge e^l$$

$$\omega_{k\,l} = 1/2 \cdot (c_{k\,l\,m} + c_{k\,m\,l} - c_{l\,m\,k}) \cdot e^m$$

Most important is the curvature 2-form R obtained by

$$R^i{}_j = d\omega^i{}_j + \omega^i{}_k \wedge \omega^k_j$$

Einstein's field equations take the following simple form:

$$*(e^i \wedge e^k \wedge e^l) \wedge R_{k\,l} = -16 \cdot \pi \cdot g \cdot *T^i$$

where T^i are the energy-momentum forms, the sources of the field and $*$ is the Hodge-duality operator. The Einstein 3-forms (the left-hand side of the equation above) therefore must be identically zero for fields in a vacuum. The following code should be self-explanatory. Nevertheless, we note that the statement *fdomain* declares the dependence of a variable, *pform* declares a variable or an indexed quantity to be an exterior form of the indicated valence, and *riemannconx* calculates the Riemannian part of the connection and stores it under the name given by the argument. A covariant index is denoted by a preceding minus sign of the index. The summation convention is understood, and there is no need for explicit raising or lowering of indices as this is performed by EXCALC automatically. Indexed objects can be declared completely *symmetric* or *antisymmetric*. For more details see the EXCALC manual [1].

```
fdomain al=al(u4),bet=bet(u4);

let sin(u1)**2 + cos(u1)**2 = 1;

pform al=0,bet=0;

coframe e(u1) =  d u1, e(u2) = d u2, e(u3) = d u3, e(u4) = d u4
 with metric g = u4**2/4*(al*(cos(u1)*e(u2) + sin(u1)*sinh(u2)*e(u3))**2
                    +  1/al*(-sin(u1)*e(u2) + cos(u1)*sinh(u2)*e(u3))**2
                    +  bet*(e(u1) + cosh(u2)*e(u3))**2)
                  + 1/bet*e(u4)**2;

riemannconx om;

pform r(k,l)=2;

antisymmetric r;

r(-k,-l) := d om(-k,-l) - om(-m,-k) ^ om(m,-l)$

let bet = ((a**4 - u4**4)*(b**4 - u4**4))**(1/2)/u4**4,
     al = (a**4 - u4**4)**(1/2)/(b**4 - u4**4)**(1/2);

pform einstein(k)=3;

einstein(k) := #(e(k)^e(l)^e(m))^r(-l,-m);

        U1
EINSTEIN   := 0

        U2
EINSTEIN   := 0

        U3
EINSTEIN   := 0

        U4
EINSTEIN   := 0
```

The same calculation in an orthonormal frame:

```
fdomain al=al(u4),bet=bet(u4);

let sin(u1)**2 + cos(u1)**2 = 1;

pform al=0,bet=0;

coframe e(1) = u4/2*al*(cos(u1)*d u2 + sin(u1)*sinh(u2)*d u3),
        e(2) = u4/(2*al)*(-sin(u1)*d u2 + cos(u1)*sinh(u2)*d u3),
        e(3) = u4/2*bet*(d u1 + cosh(u2)*d u3),
        e(4) = 1/bet*d u4
  with metric g = e(1)*e(1) + e(2)*e(2) + e(3)*e(3) + e(4)*e(4);

riemannconx om;
```

```
pform r(k,l)=2;

antisymmetric r;

r(k,l) := d om(k,l) + om(k,-m) ^ om(m,l)$

r(4,3); %to display just this particular component of the curvature.

    1  2             4                    3
 2*E ^E *( - @  AL*AL  + @  AL - @  BET*AL *BET)
              U4          U4      U4
---------------------------------------------------
                           3
                         AL *U4

     3  4                          2
   E ^E *(@      BET*U4*BET + @  BET *U4 + 3*@  BET*BET)
          U4 U4               U4              U4
 + -------------------------------------------------------
                                 U4


let bet = ((a**4 - u4**4)*(b**4 - u4**4))**(1/4)/u4**2,
    al = (a**4 - u4**4)**(1/4)/(b**4 - u4**4)**(1/4);

pform einstein(k)=3;

einstein(k) := #(e(k)^e(l)^e(m))^r(-l,-m);

        1
EINSTEIN  := 0

        2
EINSTEIN  := 0

        3
EINSTEIN  := 0

        4
EINSTEIN  := 0
```

3 Features of EXCALC version 3

3.1 Principal goals

Version 2 of EXCALC has been very successfully applied to relativity (especially to the so-called quadratic Poincare Gauge Theory), to cosmology, to the study of differential equations by Cartan's theory and to dynamical systems.
To broaden the area of applications for EXCALC to unified theories, supersymmetric theories, strings, etc. and to ease calculations in the Cartan-Kähler formulation of differential equations, new mathematical objects and operations need to be included. Among them are:

- Spinor-valued forms

- Forms with values in Lie algebras
- Indicial simplification
- Multiple spaces
- Clifford algebra
- Graded manifolds and the whole machinery on them
- General tensor and spinor operations
- Exterior systems and ideals
- Mappings

However, it becomes clear from the outset by looking at a typical formula

$$\mathcal{L} = -\frac{1}{2} R^{ab}_{\mu\nu} V^c_\rho V^d_\sigma \varepsilon_{abcd} \varepsilon^{\mu\nu\rho\sigma} + \alpha \mathcal{D}_\mu \bar{\Psi}_\nu \gamma_5 \gamma_\rho \Psi_\sigma \varepsilon^{\mu\nu\rho\sigma} + \ldots$$

that an even **better interface** allowing all the symbols to be introduced in a *painless* way and permitting a high (text processor-like) quality of display **is indispensable.** With the availability of workstations and PC's with bitmapped screens, such an interface can now be realized. For its implementation, the X11 window system [4] has been chosen here because it is a de facto standard, and the resulting interface should be quite portable. It uses the XView toolkit [5] to manage the window accessories like scrollbars, menus, panels etc.

3.2 The X interface

The following design decisions have been made (influenced by the GI/S interface [3] and some equation editors of textprocessing systems).

- Freely arrangeable multiple windows with formula and text display in near typesetter quality.
- Input from remapable keyboard or through mouse click on *compose windows.*
- Utilization of special keys for entering sub- and superscripts, tildes, bars, dots, daggers etc.
- Cut and paste operations between all dynamic windows.
- Special compose window for selecting nonkeyboard symbols (e.g. Greek symbols, math symbols - as an alternative to keyboard remapping) and structures (matrices, complicated 2-dimensional operators).
- Input editor acting on incrementally parsed input.
- Mathematical operations on selected parts of an expression.

3.3 Implementation status

- Symbols:
  All symbols available from the interface are accepted.
  Input is now case sensitive!
  Symbols can be 'composed' (by the compose key to put, for example, a bar on top of it).

- Operators:

EXCALC symbol	Meaning
$\wedge$	exterior multiplication
d	exterior differentiation
∂	partial derivative
∂_{x^a}	and coordinate vector
*	Hodge duality operator
$i_X\omega$	inner product
$\$_X\omega$	Lie derivative
$\delta_X\Lambda$	variational derivative
$\otimes$	tensor product
$\nabla_X Y$	covariant derivative
$[F, G]$	commutator
$\in$	declarations

- Indices:
Indices are introduced when working with the X interface by using the cursor keys (if not otherwise customized).
Groups of indices can be *typed* (i.e. they can have the property of being labels, spinoral, tensorial, etc.), and they can be given individual ranges (e.g. {0,1}, {t,r,θ,ϕ}, etc.). Mixed symmetries on indices can be declared.
There are several operations on indices like dotting for spinors.

- Declarations:
The syntax of declaration is not yet fixed and will probably use the $\in$ symbol. Currently, a simple extension of the EXCALC 2 declarations is used. For example:
pform {x,y,z}={1,spinor},g(a)={k,LieAlgebra:S}, {s0,s1,s2,s3}={0,odd};
would declare x,y,z to be spinor-valued 1-forms, g to be an indexed k-form with values in the Lie algebra S, and s0,s1,s2,s3 to be odd elements of a graded manifold.

4 Examples of EXCALC version 3

4.1 Variation of a local supersymmetric Lagrangian in 3 dimensions

4.1.1 Mathematical setting

The following example is taken from a work by Howe and Tucker. They consider a Lagrangian 3-form built up from the following fields:

- The coframe field e^a
- The connection forms $\omega^{a\,b}$
- A spinor-valued 1-form χ (Rarita-Schwinger field)
- A SL(2,R) valued 1-form $\gamma = \gamma_a e^a$

We have the following defining relations for the torsion, curvature and covariant derivative of the spinor field:

$$T^a = d\,e^a + \omega^a{}_b \wedge e^b$$
$$= d\,e^a + \epsilon^a_{\ b\,c} e^c \wedge \omega^b$$
$$R^{a\,b} = d\,\omega^{a\,b} + \omega^a{}_c \wedge \omega^{c\,b}$$
$$F^a = -\frac{1}{2}\epsilon^a_{\ b\,c} R^{b\,c}$$
$$D\,\chi = d\,\chi + \frac{1}{2}\gamma_a \omega^a \wedge \chi$$

where ϵ is the total antisymmetric symbol and

$$\omega^a = -\frac{1}{2}\epsilon^a_{\ b\,c}\omega^{b\,c}$$

The Lagrangian in question is given by:

$$\Lambda = e^a \wedge F_a - \frac{1}{6}\epsilon_{a\,b\,c}e^a \wedge e^b \wedge e^c + \frac{i}{2}\bar{\chi} \wedge D\,\chi + \frac{i}{4}\bar{\chi} \wedge \gamma \wedge \chi$$

We now demonstrate the X interface and the built-in variational derivative of EXCALC 3 by calculating the field equations resulting from the variation of this Lagrangian.

Sample session:

REDUCE development version ...

on DIV;

$$F^a := d\omega^a - \frac{1}{2}\cdot\varepsilon^a_{\ bc}\cdot\omega^b \wedge \omega^c;$$

$$F^a := d\omega^a - \frac{1}{2}\cdot\omega^b \wedge \omega^c \cdot \varepsilon^a_{\ bc}$$

$$D\ \chi := d\chi + \frac{1}{2}\cdot\gamma_a\cdot\omega^a \wedge \chi;$$

$$D\ \chi := d\chi + \frac{1}{2}\cdot\gamma_a\cdot\omega^a \wedge \chi$$

$$\Lambda := e_a \wedge F^a - \frac{1}{6}\cdot\varepsilon_{abc}\cdot e^a \wedge e^b \wedge e^c + \frac{i}{2}\cdot\bar{\chi} \wedge D\ \chi + \frac{i}{4}\cdot\bar{\chi} \wedge \gamma \wedge \chi;$$

$$\Lambda := e_a \wedge d\omega^a - \frac{1}{2}\cdot e_a \wedge \omega^b \wedge \omega^c \cdot \varepsilon^a_{\ bc} - \frac{1}{6}\cdot e^a \wedge e^b \wedge e^c \cdot \varepsilon_{abc} + \frac{1}{2}\cdot\bar{\chi} \wedge d\chi\cdot i - \frac{1}{4}\cdot\bar{\chi} \wedge \chi \wedge \gamma\cdot i + \frac{1}{4}\cdot\gamma_a\cdot\omega^a \wedge \bar{\chi} \wedge \chi\cdot i$$

$$\delta_{e_a}\Lambda;$$

$$d\omega^a - \frac{1}{2}\cdot e^b \wedge e^c \cdot \varepsilon^a_{\ bc} - \frac{1}{4}\cdot\gamma^a\cdot\chi \wedge \chi\cdot i - \frac{1}{2}\cdot\omega^b \wedge \omega^c \cdot \varepsilon^a_{\ bc}$$

$$\delta_{\omega^a}\Lambda;$$

$$de_a + e^c \wedge \omega^b \cdot \varepsilon_{abc} - \frac{1}{4}\cdot\gamma_a\cdot\chi \wedge \chi\cdot i$$

$$\delta_{\chi}\Lambda;$$

$$i\cdot\left(d\chi + \frac{1}{2}\cdot\gamma_a\cdot\omega^a \wedge \chi - \frac{1}{2}\cdot\chi \wedge \gamma\right)$$

4.2 Calculation of generalized characteristics of differential equations

4.2.1 Mathematical setting

In the Cartan-Kähler theory, differential equations are represented by systems of exterior equations $\Im$. A standard way to construct such a system is via the contact ideal C and a suitable map i. The pullback of the contact ideal under i yields an exterior system representing the original differential equations. EXCALC 3 allows for the definition of mappings and derives from them the actions of pullback and pushforward.

The calculation of the classical Cauchy characteristic vector fields $Char$ amounts in this formulation to the requirements that any vector X of $Char$ must annihilate the system of 1-forms $\Im$ and must also yield zero modulo $\Im$ when contracted with the exterior differential closure of $\Im$.

$$Char = \{X \mid i_X\Im, i_X d\Im mod\Im\}$$

This constitutes a system of linear equations for the coefficients of the vector X. One way to implement this would be to write a blackbox function CreateCharacteristics, for example, but this would be unnatural and not extensible. Instead, EXCALC 3 uses literally the same construct as above and therewith allows variations in the definition of the characteristic vector fields, which are needed to solve Cauchy's problem on a wider class of differential equations, for example.

For the Cauchy characteristic vector fields to give rise to a solution they have to be transversal to the sum of the pushforward of the Cauchy data and the so-called focal vectors. The focal vectors form the kernel of the composition of the pushforward of the natural projection of the corresponding jets and the pushforward of the map i. This complicated statement can be quite naturally expressed in EXCALC 3 by a two line statement as can be seen from session log at the end of this paragraph.

The Cauchy characteristic vectors above rarely exist. However, it can happen that when the exterior derivative of a subset J of $\Im$ is taken instead of the whole $\Im$, sets of vectors called J characteristics [7] do exist and there is some theory how to calculate the subsets J.

$$JChar = \{X \mid i_X\Im, i_X dJ mod\Im\}$$

For certain exterior systems, the subsets J can be found from a conformal tensor constructed from the exterior product of the forms in $\Im$ and the exterior derivative of linear combinations of the forms in $\Im$ [7]. This again can be literally used in an EXCALC calculation, as shown in the window dump.

In the following sample session the following is demonstrated: First an exterior system for a Monge Ampere equation is generated from the contact ideal C via the map i. Then Cauchy characteristics $Char$ are sought, but none exist. The focal vectors V are calculated next. Then a subset J of exterior equations is computed. Finally, the J characteristic vectors $JChar$ are constructed.

Sample session:

REDUCE development version ..

$\mathrm{PFORM}\ \{x,y,z,p,q,r,s,t,u,v\}=0;$

$\text{mapping } i\!:\ (x,y,z,p,q,u,v) \longrightarrow (x=x,y=y,z=z,p=p,q=q,r=x+v\cdot u^{2},s=v\cdot u,t=v);$

$C := \{dz-p\cdot dx-q\cdot dy,\ dp-r\cdot dx-s\cdot dy,\ dq-s\cdot dx-t\cdot dy\}\$$

$\mathfrak{I} := i^{*}(C);$

$\mathfrak{I} := \{-dx\cdot p-dy\cdot q+dz,\ dp-dx\cdot(u^{2}\cdot v+x)-dy\cdot u\cdot v,\ dq-dx\cdot u\cdot v-dy\cdot v\}$

$X := Cx\cdot\partial_{x}+Cy\cdot\partial_{y}+Cz\cdot\partial_{z}+Cp\cdot\partial_{p}+Cq\cdot\partial_{q}+Cu\cdot\partial_{u}+Cv\cdot\partial_{v}\$$

$\mathrm{Char} := \{X \mid (i_{X}\mathfrak{I},\ i_{X}d\mathfrak{I} \bmod \mathfrak{I})\};$

$\mathrm{Char} := \{0\}$

$\text{mapping } \pi\!:\ (x,y,z,p,q,r,s,t) \longrightarrow (x=x,y=y,z=z,p=p,q=q);$

$V := \{X \mid \pi_{*}(i_{*}(X))\};$

$V := \{\partial_{v},\partial_{u}\}$

$\Phi := \alpha\cdot(C1 := \mathrm{FIRST}(\mathfrak{I}))+\beta\cdot(C2 := \mathrm{SECOND}(\mathfrak{I}))+\gamma\cdot(C3 := \mathrm{THIRD}(\mathfrak{I}))\$$

$\Psi := \delta\cdot C1+\varepsilon\cdot C2+\chi\cdot C3\$$

$d\Phi\wedge d\Psi\wedge C1\wedge C2\wedge C3;$

$-2\cdot dp\wedge dq\wedge du\wedge dv\wedge dx\wedge dy\wedge dz\cdot v\cdot(u^{2}\cdot\beta\cdot\varepsilon+u\cdot\beta\cdot\chi+u\cdot\varepsilon\cdot\gamma+\chi\cdot\gamma)$

$\text{let } \alpha=1,\beta=0,\gamma=0,\delta=0,\varepsilon=1,\chi=-u;$

$J := \{\Phi,\Psi\};$

$J := \{-dx\cdot p-dy\cdot q+dz,\ dp-dq\cdot u-dx\cdot x\}$

$\mathrm{JChar} := \{X \mid (i_{X}\mathfrak{I},\ i_{X}dJ \bmod \mathfrak{I})\};$

$\mathrm{JChar} := \{\partial_{v},\ \frac{(\partial_{p}\cdot x+\partial_{x}-\partial_{y}\cdot u+\partial_{z}\cdot p-\partial_{z}\cdot q\cdot u)}{x}\}$

4.3 Conclusions

It is hoped that with the availability of version 3 of EXCALC an attractive workbench is availbale for mathematicians and physicists. It will probably available through the REDUCE network library later this year.

References

[1] Schrüfer, E. (1987), *EXCALC User's Manual*, RAND Corp. (distributed with REDUCE 3.3)
Schrüfer, E., Hehl, F.W.H., McCrea, J.D. (1987), *Exterior Calculus on the Computer: The REDUCE-Package EXCALC Applied to General Relativity and to the Poincare Gauge Theory,*

[2] Nielsen, B., Pedersen, H. (1988), *A Note on Einstein Metrics*, SIGSAM Bulletin, vol. 22, no. 1
Schrüfer, E., (1988) *A comment on "A Note on Einstein Metrics"*, SIGSAM Bulletin, vol. 22, no.3

[3] Young, D.A., Wang, P.S. (1987) *GI/S: A Graphical User Interface for Symbolic Computing Systems*, Journ. Symb. Comp., vol. 4, no. 3

[4] The X Window System is a trademark of MIT

[5] XView is a trademark of Sun Microsystems Inc.
Heller, D. (1989) *XView Programming Manual* O'Reilly & Associates, Inc.

[6] Howe, P.S., Tucker, R.W. (1978) *Local supersymmetry in (2+1) dimensions*, J.Math., Phys., 19, 4

[7] Gardner, R.B. (1970), *A differential Geometric Approach to Characteristics*, Proceedings of Symposia in Pure Mathematics, vol. XVI

INFORMATION SYSTEM "ALGEBRAIC COMPUTING SYSTEMS" FOR PERSONAL COMPUTERS

S.A.Pritomanov
Department of Physics, Chemical and technological institute, Gagarin av. 8, Dniepropetrovsk, 320640 USSR

ABSTRACT

We elaborated the information system **Algebraic Computing Systems** *(ACS) for the aim of collection and systematization of all information about computer algebra and algebraic computing systems. The system intend for active users of algebraic computing systems which interesting the problems in this branch of science, having own literature databases and taking part in different conferences for analitical computation.*

1.General information about system

The system **ACS** was written on internal language **FRED** integrated packets **FRAMEWORK**-II and **FRAMEWORK**-III for personal computers **IBM PC/XT,AT**. Minimal RAM for **FRAMEWORK**-II packet is 512K and 640K for **FRAMEWORK**-III. After starting Framework you have to load the main block of information system and give command to begin working. The main block take about 100K of RAM. All remainder of memory (including expanded or extended memory) the system use for working databases which overlay loading by user. There are English and Russian versions of system. All information may be receive on display or printer.

User may easy correct and and modify the information system with own requirements in different branches of physics and mathematics. At present time the system orientate on the information about using the computer algebra in general relativity and gravitation.

The **ACS** system contain:

- the information about algebraic computing systems and its technical features;
- comparative features of algebraic computing systems;
- special features of algebraic computing systems for personal computers;
- description of REDUCE-3.2 and REDUCE-3.3 systems for personal computers IBM PC/XT,AT and compatible;
- different databases for users information.

2.Structure of ACS.

The **ACS** consist of many level menus by means of which user may recieved all interesting information.

Main system menu:

ACS

```
├──→Algebraic Computing System in GRG
├──→Active users
├──→Literature
├──→Conferences
├──→Organizations
├──→REDUCE for PC
└──→Quit
```

Second level menu:

Algebraic computing system in GRG
- Large systems
- Small systems
- Quit

Active users
- Soviet
- Foreign
- Quit

Literature
- Foreign Journals
- Soviet Journals
- Books
- Review

Conferences
- Soviet conferences
- Foreign conferences
- Future conferences
- Quit

REDUCE for PC
- USA Version 3.2
- USA Version 3.3
- Japan version 3.2
- Japan version 3.3
- Quit

Third level menu:

Large Systems
- **REDUCE**
  - Information about system
  - Programs for GRG
  - Programs for Newman-Penrose
  - Exterior calculations
- **MACSYMA**
- **SHEEP**
- **CAMAL**
- Comparison of systems
- Quit

Small systems
- **SAC**
  - Information about system
  - Purpose
- **ALTRAN**
- **ORTOCARTAN**
- **POLYNOM**
- **GRATOS**
- **FORMAC**
- **SYMBAL**
- Comparison of systems

USA Version 3.2, 3.3.
- Structure of library
- List of files
  - **REDUCE**
  - **FASL**
  - **RFASL**
  - **RUTILS**
- Creation freeze files
- Run REDUCE

Japan Version 3.2, 3.3.
- Structure of library
- List of files
  - **REDUCEJ**
  - **OVR**
  - **XQT**
  - **EXAMPLES**
- Run REDUCE
- Quit

The menu structure is defined for **ACS** and may be modify only by means of FRAMEWORK packet.

3. Databases.

The **ACS** system allow to create, correct and sort the databases:

- -active users database in Soviet Union - **ACSSPSOV**;
- -active foreign users - **ACSSPFOR**;
- -computer algebra reviews database - **ACSREVIE**;
- -Soviet journals in computer science database - **ACSSOVIE**;
- -foreign journals in computer science database - **ACSFOREI**;
- -database for books in computer algebra - **ACSBOOKS**;
- -Soviet conferences on computer algebra database - **ACSSOVCO**;
- -international conferences on computer algebra database - **ACSFORCO**;
- -future conferences on computer algebra - **ACSCONFE**;
- -database of Soviet and foreign departments in which the computer algebra system using for physical research - **ACSDEPAR**.

It is necessary for normal work of **ACS** that all these databases to be in the default directory. The user load database from the suitable menu and remove it from memory under exiting of this menu. All new information the system automatically saves on disk.

After loading the database the menu for work with database is appear:

Main Menu

- **Database Maintenance**
  - → **Add**
  - → **Edit**
  - → **View**
  - → **Delete**
  - → **Help**
  - → **Quit**
- **Filter Database**
  - → **Filter**
  - → **Un-Filter**
  - → **Help**
  - → **Quit**
- **Print**
- **Help**
  - →**Main Menu**
  - →**Maintenance**
  - →**Filter**
  - →**General Info**
  - →**Sign Off**
- **Quit**

Maintenance menu allow add, edit, view and delete the database fields.

Filter menu allow to filter database by several fields simultaneously.

Command **Print** print full or filtered database.

Menu **Help** give the information about working of suitable menu.

Command **Quit** save the database on disk, clear the memory and return user to main menu.

In conclusion remind that we assume that user have an elementary experience of working with **FRAMEWORK**-II or **FRAMEWORK**-III packets. Author constantly modify the information system **ACS** and include new information in the databases

ON A COMPACTNESS OF TWO DIMENSIONAL OUTPUT

A.A.Markov

Republic Engineering Center of Automation Design. Politechnical Inst., 603000, Gorky, Dalnya, 15. USSR

ABSTRACT

The style of formulas representation as formulas systems is proposed. The mathematical model and optimizing problems are discussed for the problem of two dimensional output' compactness.

1. Introduction.

The important aspect of working out of computer algebra systems (CAS) is the ensuring of adequate service, particularly, the ensuring of two dimensional (many-level) formulas' output, that is a formulas representation in natural mathematical notation. The algorithm for translation to such representation of formulas has been made long ago (see, for example, [1,2]). Most of CAS ensure two dimensional form of formulas. In review [3] the outputs'forms of 10 different CAS are demonstrated and more wide problem about easy-readable output is discussed.

The following problem, after creation of algorithm to translate the formulas into two dimensional form, is the problem of formulas' *transfer* from line to another. Now this problem is usually solved (for example, in REDUCE) either by the change to an one dimensional (FORTRAN-similar) outputs' form, or by non compactness transfer, which increases the real size of the formulas.

Formula F in a many-level notation (two dimensional formula=TFD) is characterized by *length* L(F) and *height* H(F) - the number of levels. The length of identifier (of the variable or function, which hasn't special mathematical notation) and of constant is a number of characters in it. Then we define, for example,

$$L(\ x/y\ ) = \max \{\ L(x)\ ,\ L(y)\ \}\ ,$$

$$L(\ x+y\ ) = L(x) + L(y) + 1 \quad , \quad L(\ x^y\ ) = L(x) + L(y)\ ,$$

etc. for all functions, which are used in the formulas. Consider the example of lengths' definition for formula with more difficult representation. Let integral in TFD is printed as

```
X2
 /
 !   F(X) DX        ,
 /
X1
```

so $L = \max \{\ 4 + L(X2),\ L(XI)\ \} + L(F(X)) + L(X) + 1$. The height is defined by analogy.

With the *system of re-designations*, let's represent the TFD F by formulas' system $S=\{F_i\}$, which are length-bounded with L_0=const and, may be, which are height-bounded (H_0). In that case, when L_0 isn't more then the pages' width (display width), the problem of transfer is irrelevant. (More precisely, it's relevant for infinite-length constants and for such a occasions like as "length of identifier is more then L_0". We'll not consider this cases). We'll solve the optimization problem of two dimensional outputs' compactness under this restrictions: the minimization of the number of formulas in the system (N-Problem) and the minimization of the sum of formulas' heights (Σ-Problem). Let $L_0=17$ (the height isn't bounded) for formula F with $L(F)=20$, $H(F)=6$

$$F=\frac{a+b+c+d+e+f+g}{m} - \frac{a_i^b}{m^n},$$

then, for the following representations

$$S=\left\{ \begin{array}{l} F=\frac{a+b+c+d+x}{m} - \frac{a_i^b}{m^n} \\ \\ x=e+f+g \end{array} \right\}, \quad S_1=\left\{ \begin{array}{l} F=\frac{a+b+c+d+e+x}{m} - y \\ x=f+g \\ y=\frac{a_i^b}{m^n} \end{array} \right\}$$

which are satisfied for the restriction, the system S is more compact then S_1 in the sense of both N- and Σ-Problems.

In Section 2 the mathematical model of compactness problem is described. The Sections 3 and 4 are intended for N-Problem and Σ-Problem under one and two restrictions respectively.

2. Decompositions of a Rooted Trees.

Natural representation of formulas are the rooted trees, so below we use the corresponding terminology. Describing model is more wide, then it's necessary for compactness outputs' problem. Let K is the class of rooted, directed from a root trees, the terminal vertices of which are labeled with weight $\bar{w}(v)= (w_1,\ldots,w_r)$, $w_i=$ const $\in Z^+$, and internal vertices - with the vectors of functional symbols $\bar{m}(v) = (m_1, \ldots ,m_r), m_i \in M$. Let the functions, which are correspond to symbols in M, are integer-valued, non-decreasing, symmetric (in the sense, that they are the same value on any permutation of their arguments) and defined for any number of arguments. The weight of internal vertices of the tree is recurrence calculated as $\bar{w}(v)=(m_1(w_1(v_1),\ldots,w_1(v_k)),\ldots,m_r(w_r(v_1),\ldots,w_r(v_k)))$, where $v_1, \ldots ,v_k$ - the sons of v. The trees' weight is equal to weight of its' root.

The result of decomposition of tree T in vertex v is the tree T^v - rooted subtree with root v and the tree T_v - result of substitution of T^v by terminal vertex with weight $(w_1^0,\ldots,w_r^0)$. Let the root v^* is a greatest element of T. *Decomposition* R(W) *of tree* T(V,E) *in class* K *with set* W $\subseteq$ V ($v^* \in$ W) *is representation* T *as the aggregate of trees* $\{ T\langle M(v,W)\rangle \in K,\ v \in W \}$, *where*

$M(v,W) = \{ x \in V \ :\ x \leqslant v \ \&\ v > y > x \Rightarrow y \notin W \}$,

and $T\langle X\rangle$ - *subgraph, generated with vertices set* X .

For the fixed $\bar{s}=(s_1, \ldots, s_r)$, $s_i \in Z^+$, the set of permissible decompositions P consists of $R=\{T_1,\ldots,T_k\}$ such that $w_i(T_j) \leqslant s_i$ for $i=\overline{1,r}$ and $j=\overline{1,k}$. Let the functional $F(R) = f(w_i(T_1), \ldots, w_i(T_k))$ is given. Optimization problem $A(F,T,\bar{s})$ of decomposition of T is $F(R) \rightarrow \min$ for $R \in P$.

Obviously, optimization problem of compactness of two dimensional output is particular case of $A(F,T,\bar{s})$. Below, $N(R)$ - number of parts of decomposition ($A(N,T,\bar{s})$ correspond to N-Problem), and $\Sigma(R)$ is the sum of weights of trees, which are included to decomposition R ($A(\Sigma,T,\bar{s})$ corresponds to Σ-Problem).

3. Minimization under "Length-Bound" Restriction.

In this section we suppose r=I, that is only one function is defined on the trees' vertices. All formulated statements were proved in [4]. Designate the set of sons of v as $\varepsilon(v)$. Let $V_{or}=\{v\in V : (w(T^v)>s)\&(w(T^u)\leqslant s)$ for all $u\in \varepsilon(v)\}$ and for $v\in V_{kp}:\varepsilon(v)=\{v_1,\ldots,v_n\}, w(v_1)\geqslant\ldots\geqslant w(v_n)$, and $\varepsilon'(v)=\{v_1,\ldots,v_k\}$, where $w(1,\ldots,1,w(v_{k+1}),\ldots,w(v_n))\leqslant s < w(1,\ldots,1,w(v_k),\ldots,w(v_n))$. Let's formulate the greedy algorithm G for building $W \in P$: if $V_{or}=\emptyset$, then $W=\{v^*\}$, else $W(T)= W_1(T) \cup W(T_{W_1})$, where $W_1(T)= \bigcup_{v\in V_{or}} \varepsilon'(v)$. The complexity of G are $O(n\cdot\log(n))$ with memory and $O(n)$ with time for T with n vertices.

Theorem I. *Algorithm* G *solves the problem* $A(N,T,s)$.

Theorem 2. *Problem* $A(\Sigma,T,s)$ *is* NP-*hard*.

The NP-hardness of Σ-Problem under two restrictions immediately follows from theorem 2.

4. Optimization under Two Restrictions.

Let two functions are defined on the vertices of a tree,and $\bar{s}=(L_0,H_0)$.Statements formulated in this section were proved in [5].

Let's designate optimum for $A(N,T,\bar{s})$ and $A(\Sigma,T,\bar{s})$ as N_0 и Σ_0 respeotively; the results of $\mathfrak{A}$ (see[5], certain generalization of G with the same complexity) – as $N_{\mathfrak{A}}$ and $\Sigma_{\mathfrak{A}}$.

Theorem 3. $N_{\mathfrak{A}}(T) \leqslant 2\cdot N_0(T) - 1$.

Theorem 4. $\Sigma_{\mathfrak{A}}(T) \leqslant 2\cdot H_0\cdot\Sigma_0(T) - H_0$.

Theorem 5. *Problem* $A(N,T,\bar{s})$ *is solvable in polynomial-bounded time.*

The idea of automatons over trees with finite delay k is a little different with the standard definition. The delay means that labeling of trees' vertices with output symbols is performed after k times since it was read by automaton.

All greedy algorithms solving $A(N,T,\bar{s})$ are in the class of such automatons with k=1.

Theorem 6. *There is no automaton over trees with finite delay k solving* $A(N,T,\bar{s})$.

Certain precise extreme estimations of the number of formulas in system are given in [4,5], like following :

Theorem 7. $N_0(T) \leqslant \frac{1 + L_0}{3\cdot L_0}\cdot n + 1$ *for* $L_0 \geqslant 3$ *for any* (Σ,M)*-tree* T *with* n *vertices.*

5. References

1. Y.Syret, in *Proc.* SYMSAM-2 , ed. S.R.Petrick, (New York: ACM, 1971), p.90.
2. W.A.Martin, in *Proc.* SYMSAM-2 , ed. S.R.Petrick, (New York: ACM, 1971), p.78.
3. D.Y.Y.Yun, R.D.Stoutemayer, in *Encyclopedia of Comput. Sci. and Techn.*, **15**, eds. J.Belzer, A.Holzman, A.Kent, (New York – Basel: Marcel Dekker 1980), p.235.
4. A.A.Markov, *Discrete Math.*, **1**, 2 (1989), p.3. (Russian).
5. A.A.Markov, *Discrete Math.*, 1991 (Russian, to apper).

REDUCE as a Numerical Tool

John Fitch
School of Mathematical Sciences
University of Bath
England

Abstract

The use of computers for algebraic computations are well known, and are established as algebraic tools for physical sciences. In this paper concentration will be on the assistance that algebraic calculation can make for numerical calculations. This assistance includes analysis of problems, preparation of programs, and code generation for numerical programs and supercomputer architectures. There are also that start of a number of truely cooperative algebraic-numeric compuational systems. Particular emphasis is made of the current cooperative project between NAG Ltd and the University of Bath.

1 Introduction

The relationship between algebraic calculations and numeric ones has often been seen as a conflict. In this paper I wish to continue the start I made in [7] to show that the correct connection is a complementary one, and increasingly a friendly one. Of course there are calculations where the final result is an algebraic one, as typified by the determination of the Einstein equations from a description of the space metric. In many more styles of calculation the ultimate answer is a number. This paper considers the use of an algebraic system to start numerical calculations, and to find the correct solution, and the use of algebra to generate numeric programs. It ends with the consideration of the coming combined algebraic-numeric systems.

There is a long tradition of numeric calculation, but the less common algebraic method can often be a useful addition to the toolkit of the physical scientist. One should use the best tool for the job.

2 Planetary Models and Power Series

A favourite illustration of mine comes from planetary structure theory. In considering models of the structure of terrestrial planets for investigations of their internal structure, cosmological effects and mountain creation, Lyttleton and Fitch [13] were led to investigate multi-zoned spherically symmetric planets. A planet such as the Earth has two substantial zones, the core and the mantle, as well as a small crust. The structure can be easily seen to be represented by the equations of hydrostatic equilibrium relating the pressure p at a radius r, the mass m inside the radius r and the density ρ at a radial distance.

$$\frac{dm}{dr} = 4\pi r^2 \rho$$
$$\frac{dp}{dr} = -\frac{Gm\rho}{r^2}$$

To complete the equations we need a relationship between the pressure and the density. There is no theoretical relationship known, but by empirical investigations of the seismic data the formula

$$p = \frac{a}{b}\left[\left(\frac{\rho}{\rho_0}\right)^b - 1\right]$$

has a very close fit in each zone with different values of a, b and ρ_0. The boundary conditions are that at the centre of the planet the mass is zero and the pressure is finite, at the surface the mass must be the total mass of the planet and the pressure must be zero. In addition there must be continuity of m and p at the internal boundaries.

There is an immediate form of solution, which is to integrate from the initial values at the centre of the planet with a guessed value for $p(0)$ until the pressure is zero. One can iterate to find a model with the correct total mass. The problem with this solution strategy is that there is an apparent singularity at the origin. One way to avoid this is to start at a small distance ϵ from the centre of the planet. The initial values for p, ρ and m can be determined by calculating a power series expansion about the origin.

There are a number of ways in which an algebra system can do this expansion. The conceptually simple method is to construct a power series to some small degree, and then to substitute them into the equations, and then solve the equations formed by equating coefficients. A more certain way is to use the method of repeated approximation, also known as Newton's method.

In the past I have described this method in detail, including in [6]. The basic idea is to start with assuming that the small quantities are ignorable, and from that generate a zeroth approximation to the solution. We then assume that we have the solution correct to the nth order in the small quantity, and that the true solution is given by a small correction to the approximation. The details are not relevent here. The methods are more easily available now as truncated power series.

2.1 Truncated Power Series

The general techniques of truncated series are easy to program in a straightforward fashion. However there are better ways of achieving the calculation of power series, with guaranteed accuracy of truncation. The method has been described by Norman [15]; here we give a short introduction to the idea.

If a quantity A is represented by a power series

$$A = \sum a_i x^i$$

then we can consider the quantities a_i as the power series. If A is calculated by the addition of two quantities, then there is a natural relationship

$$A = B + C \quad \longrightarrow \quad a_i = b_i + c_i$$

Similarly the operations of multiplication, differentiation, and integration are given by the forms

$$A = B \times C \quad \longrightarrow \quad a_i = b_k c_{i-k}$$
$$A = \frac{dB}{dx} \quad \longrightarrow \quad a_i = (i+1)b_{i+1}$$
$$A = \int B dx \quad \longrightarrow \quad a_i = \frac{b_{i-1}}{i}$$

Similar results apply for reciprocal, and by consideration of the differential equations that define *sine* and *cosine* it is possible to construct similar recurrence relationships. These recurrences can be used to implement a power series package, which with the addition of memo functions can be made into a very efficient mechanism. A package doing precisely this is available for REDUCE, written by Barnes and Padget [1], following the Norman model. They compile algebraic expressions into a nest of recurrence relations of the form above, and provide automatic memory of previously calculated values. By making the internal data structure into a mode, they also implement combination of series calculated independently without the error which simpler techniques can introduce. It is possible to extend the series to have an additional power of the series variable multiplying the whole series, and this provides for simple poles and other more complex series.

3 Duffing Equation and Generation of FORTRAN

The example of power series expansion uses computer algebra as part of the algorithm. Another activity with which computer algebra is well suited is the writing of the actual FORTRAN code. At the simplest level this involves the printing of algebraic expressions in FORTRAN notation, and all major algebra systems provide this facility. As a demonstration of this we will consider another series expansion problem, the quasi-linear differential equation of Duffing

$$y'' + y = \epsilon y^3$$

where ϵ is a small parameter, and we require a periodic solution. It is well known (see for example Nayfeh [14]) that in order to produce such a solution it is necessary to introduce a stretched time. Rather that give the whole details of the solution, the reader is referred elsewhere, such as Barton and Fitch [2], or Fitch, Norman and Moore [8]. It is sufficient here to give the program in REDUCE, with the resulting approximation printed in a form ready for presenting to the FORTRAN compiler.

```
COMMENT A Program for Duffing's Equation y''+y=ey^3;
ORDER A;
FOR ALL B,C LET COS(B)*COS(C) = (COS(B-C)+COS(B-C))/2;
FOR ALL B    LET COS(B)**2      = (1+COS(2*B))/2;
Y := A * COS(X);
C := 1;
OPERATOR CS;
FOR N:=1:4 DO <<
    LET E^(N+1) = 0;
    R := E*Y^3 - C^2*DF(Y,X,2) - Y;
COMMENT Get the cos(x) term;
    COSTERM := R - SUB(COS(X)=0,R);
    R := SUB(COS(X)=0,R);
    C := C - COSTERM/(2*A);
COMMENT Perform the integration;
    FOR ALL J LET COS(J*X) = CS(J*X)/(1-J^2);
    TEMP := R;
    FOR ALL J CLEAR COS(J*X);
    FOR ALL B LET CS(B) = COS(B);
    Y := Y + TEMP;
    FOR ALL B CLEAR CS(B);
>>;
ON FORT;
Y := Y;
C := C;
```

which produces the output

```
      y=-(a*(1440.*a**8*cos(4.*x)*e**4+200295.*a**8*cos(3.*
     . x)*e**4+1346800.*a**8*cos(2.*x)*e**4+1188000.*a**8*e
     . **4+123552.*a**6*cos(4.*x)*e**3+146880.*a**6*cos(3.*
     . x)*e**3+336960.*a**6*cos(2.*x)*e**3+842400.*a**6*e**
```

```
. 3+34560.*a**4*cos(4.*x)*e**2-662400.*a**4*cos(2.*x)*
. e**2+2505600.*a**4*e**2-1036800.*a**2*cos(2.*x)*e+
. 3110400.*a**2*e-2073600.*cos(x)))/2073600.
c=-(19589.*a**8*cos(x)*e**4-5600.*a**6*cos(x)*e**3-
. 10560.*a**4*cos(x)*e**2+23040.*a**2*cos(x)*e-11520.)/
. 11520.
```

This was produced on an HLH Orion 1/05, running Cambridge LISP and REDUCE 3.3; some other implementations of REDUCE may print the answer in upper case letters. This answer is syntactically correct, but can be criticized on a number of counts, such as re-evaluations. Also what this has given is only a collection of assignment statements, and a numerical program is considerably more than that. In the next section we will consider the production of a whole program.

Another case where I have found this simple writing of FORTRAN to be effective is in coordinate transformations. Some years ago I was involved with some calculations based on separation of variables of Navier-Stokes equations, where the solution methods were numerical equating of coefficients, which is fundamentally a matrix inversion. We used computer algebra to construct the FORTRAN program, passing all the tedious work of expressing the equations in our strange coordinate system to the automated algebra.

4 Generating whole Programs

Printing in FORTRAN is a great benefit, but it does not address the real problems of automatic generation of programs. A short look at the example of output given above clearly shows three things;

1. It only generates assignment statements, which are only a small part of the eventual program;
2. There is unnecessary recalculation, for example in constructing the powers;
3. There is no provision for efficient use of vector processors.

It order to be a real general help it is necessary to address these problems. The first two need immediate attention, while the third can wait for such machines to gain wide availability. Still some indication of progress in that direction would be of comfort. We now briefly consider these three aspects of program generation, with the GENTRAN system, and its associated optimisations.

5 GENTRAN

GENTRAN [10] is a component of an algebra system which addresses the issue of complete programs, and of program segments. It exists in a number of forms, in particular for MACSYMA and for REDUCE. It is the second version I describe.

In a sense GENTRAN provides a separate language dedicated to the production of FORTRAN (or RATFOR, C or PASCAL). It is either embedded in an algebra system, or calls upon it. In its simplest form it duplicates the facilities of ON FORT, but in a more general fashion. If we have calculated a polynomial F and we wish to generate the program to calculate F/F' then we can write the REDUCE program segment

```
GENTRAN Q := EVAL(F)/(EVAL(DF(F,X))$
```

The EVAL function in GENTRAN calls on REDUCE to perform the evaluation. If F were the polynomial $2x^2 - 5x + 6$ then the answer printed would be

```
Q=(2.0*X**2-(5.0*X)+6.0)/(4.0*X-5.0)
```

In this the keyword GENTRAN switches to the new language, and EVAL calls REDUCE from it.

The power of GENTRAN lies in its ability to translate entire program segments, expressed as REDUCE blocks, and to provide declarations both at the user's direction and automatically.

```
GENTRAN <<
    DECLARE << A-H, O-Z : IMPLICIT REAL*8;
               M(4,4)   : INTEGER >>;
    FOR I:= 1:4 DO
        FOR J := 1:4 DO
            IF I=J THEN M(I,J):=1
                   ELSE M(I,J):=0;
    DECLARE I,J : INTEGER;
    .....
>>
```

which translates into

```
      IMPLICIT REAL*8 (A-H, O-Z)
      INTEGER M(4,4),I,J
      DO 25001 I=1,4
        DO 25002 J=1,4
          IF (.NOT. I .EQ. J) GOTO 25003
            M(I,J)=1.0
            GOTO 25004
25003     CONTINUE
            M(I,J)=0.0
25004     CONTINUE
25002   CONTINUE
25001 CONTINUE
    .....
>>
```

with similar code for the other potential target languages. This reduces the difficulty of writing FORTRAN, and moves the concentration of the programmer to REDUCE.

GENTRAN also deals with segmentation of long expressions, creating temporary variables as necessary.

```
VAL := (A+B+C+D)^2$
MAXEXPPRINTLEN!* := 24$
GENTRAN VALUE :=: VAL$
      T0=A**2+2.0*A*B
      T0=T0+2.0*A*C+2.0*A*D
      T0=T0+B**2+2.0*B*C
      T0=T0+2.0*B*D+C**2
      VALUE=T0+2.0*C*D+D**2
```

Template processing is used when the whole FORTRAN program is written, but it needs the insertion of a number of expressions. The user supplies the program, with indications as to when to switch to REDUCE. For example to write a function to calculate the determinant of a 3 by 3 matrix directly one could prepare a template which says

```
      REAL*8 FUNCTION DET(M)
      REAL*8 M(3,3)
;BEGIN;
      OPERATOR M$
      MATRIX MM(3,3)$
      MM := MAT((M(1,1),M(1,2),M(1,3)),
                (M(2,1),M(2,2),M(2.3)),
                (M(3,1),M(3,2),M(3.3)))$
      GENTRAN DET :=: DET(MM)$
;END;
      RETURN
      END
;END;
```

From these examples I hope it is possible to comprehend the extent to which GENTRAN can form part of a united numerical-algebraic approach to problem solving. GENTRAN has a large and detailed manual [9] from which these examples have been taken. The manual gives all the details necessary to construct complex programs.

6 Optimisation and Vectorisation

The second criticism of the simple printing in FORTRAN syntax was the number of times there were recalculations. This problem of representing algebraic expressions efficiently is an old one [12]. Recently significant progress has been made in this area by van Hulzen and his associates [16, 17], but it is beyond the intentions of this paper to describe their work.

The third aspect which needs attention, namely vectorisation, has been much less well served, although recently two papers have appeared on this topic. One of these considers the generation of high quality code for a CRAY [18], and can be viewed as a progress report rather than a final solution. The other [11] is a very different approach, but again must be seen as a preliminary report. It is to be hoped that this aspect of the algebraic-numeric interface will receive more attention in the next few years.

7 The IRENA Interface

Previously in this paper I have considered computer algebra as a preprocessor to numerical computations, at various degrees of sophistication. It is now time to consider progress towards a real combined system. IRENA is the first attempt at a true integrated environment. It is largely the work of Dewar [3, 5], with support from NAG Ltd. In its present form IRENA is an interface between REDUCE and the NAG subroutine library. The aim is to provide the user with an easy access to the NAG library.

In the NAG library there is a routine for numerical quadrature called D01AJF. It takes twelve parameters to calculate $\int_B^A F(x)dx$. These are given in full in figure 1. Three of these parameters are the limits A and B, provided as real variables, and the real-valued function of one real argument F, and clearly it is essential that they be provided by the user. There are 2 reals for error control, EPSABS and EPSREL, which only sophisticated users would want to change from the recommended default values. There are a number of workspace variables, and parameters for returning the result, and numerical error, together with an indication of success or failure.

There are three classes of parameter. The first can be called the mathematical ones, which describe the problem (A, B and F in our case). The second class are algorithmic, being concerned with details of how the problem is to be solved; we can place EPSREL and EPSABS in this class. The last class are wholly concerned with the implementation, and include the workspace arrays. In IRENA the principle

Parameters for D01AJF		
2 reals	A, B	limits of integration
function	F	user defined
2 reals	EPSABS, EPSREL	error control
2 arrays	IW, W	workspace
2 integers	LIW, LW	dimensions of workspace
2 reals	RESULT, ABSERR	returned values
1 integer	IFAIL	diagnostic

IW(1) contains number of subdivisions used

Figure 1: Description of D01AJF

is that the user must specify the first class, and may specify the second. There is no need for any consideration of the third. So for example to integrate $(\log x - 1)^{-10}$ the interaction would go

```
2: do1ajf(a=0, b=1, f(x)=(log(x)-1)^(-10));
93: result;
0.09892913264061787
94: abserr;
2.003058472163449e-13
95: divisions;
4
```

where a number of internal steps have been omitted. The **result** and **abserr** variables are used to return the answer and error estimate. The somewhat obscure use of **IW(1)** has been packaged to the variable **divisions**. An experienced NAG Library user will note that there is no apparent check on the the value of **IFAIL**, which the documentation stresses must be done after any use of a NAG routine. In fact this check is made internally. If an attempt is made to perform an integral with a singularity for example, IRENA reports the appropriate error message from the NAG manual.

```
97: d01ajf(a=0,b=1,f(x)=1/x);
 ** ABNORMAL EXIT from NAG Library routine D01AJF: IFAIL =      1
 ** NAG soft failure - control returned

    The maximum number of subdivisions allowed with the given workspace has
    been reached without the accuracy requirements being achieved.  Look  at
    the integrand in order to determine the  integration  difficulties.  If
    the position of a local difficulty within the interval can be determined
    (e.g.        a         singularity         of          the          integrand
    or its  derivative,  a peak, a  discontinuity...)  one
    will probably gain from splitting up the  interval  at  this  point  and
    calling   the   integrator  on  the  subranges.  If  necessary,  another
    integrator which  is  designed  for  handling  the  type  of  difficulty
    involved,  must  be  used.  Alternatively consider relaxing the accuracy
    requirements       specified       by        EPSABS       and       EPSREL,
    or increasing the amount of workspace.

188:
```

So it can be seen that IRENA provides a simple user interface to all the numerical library.

A second example should re-enforce this feeling. Suppose we wish to find

$$\min\ \left((x_1 + 10x_2)^2 + 5(x_3 - x_4)^2 + (x_2 - 2x_3)^4 + 10(x_1 - x_4)^4\right)$$

subject to the conditions:

$$1 \leq x_1 \leq 3,\ -2 \leq x_2 \leq 0,\ 1 \leq x_4 \leq 3$$

starting from the initial guess $(3, -1, 0, 1)$. It is decided to secect the routine E04JAF which uses a quasi-Newton algorithm, and is designed to be used with continuous functions with continuous first and second derivatives.

```
4: e04jaf(bounds=rec[ 1:3, -2:0, *:*, 1:3],
4:     vec x {3,-1,0,1},
4:     funct1(x1,x2,x3,x4)=(x1 + 10*x2)^2 + 5*(x3 - x4)^2
4:                                         + (x2 - 2*x3)^4 + 10*(x1 - x4)^4);
 ** ABNORMAL EXIT from NAG Library routine E04JAF: IFAIL =     5
 ** NAG soft failure - control returned

   There is some doubt about whether the point X found by //E04JAF//  is  a
   minimum.  The  degree  of  confidence  in  the result decreases as IFAIL
   increases. Thus, when IFAIL = 5 it is probable that the final X gives  a
   good  estimate  of  the  position of a minimum, but when IFAIL = 8 it is
   very unlikely that the routine has found a minimum.
{X F}

93: x;
[          1          ]
[                     ]
[ - 0.0852325899880082]
[                     ]
[   .40930359107539   ]
[                     ]
[          1          ]

94: f;
2.433787512120732
```

The internal mechanism used by IRENA is based on matching the input syntax with a description of the NAG routines, including which parameters are necessary, and what algorithm parameters need to be provided. From these two fragments of code are produced, on a simple C interface, and the other a FORTRAN function, for which GENTRAN is used.

IRENA is now capable of reading the machine versions of the NAG library in order to determine the parameters classes automatically, and to identify the necessary error messages. While a little hand assistance is still necessary, this area is progressing [4]. IRENA also uses the GENTRAN/SCOPE optimiser, as experiments have shown that it is frequently more efficient overall, even for a single use of the code!

Work is now underway for automatic routine selection, so the interface for the first example will eventually be

```
integrate(a=1, b=1, f(x)=(log(x)-1)^(-10));
```

When we have this, REDUCE will truely be a numerical tool.

8 Conclusions

In this paper I have explored a number of aspects of the ways in which computer algebra systems can be used in conjunction with FORTRAN, or other similar languages, to provide a powerful tool for researchers. The cooperation ranges from simple preprocessing to the sophisticated integrated system which is signposted by IRENA. I hope that all users will be able to find some hints here to help them with their own applications.

References

[1] A. Barnes and J. A. Padget. A univariate power series package for REDUCE. In *Proceedings of ISSAC 90, Tokyo*, 1990.

[2] D. Barton and J. P. Fitch. The application of symbolic algebra system to physics. *Reports on Progress in Physics*, 35:235–314, 1972.

[3] M. C. Dewar. IRENA — an integrated symbolic and numerical computation environment. In *Proc. of ISSAC '89*, pages 171–179. ACM Press, New York, 1989.

[4] M. C. Dewar. PhD thesis, Univ. of Bath, England, 1990.

[5] M. C. Dewar and M. G. Richardson. Reconciling symbolic and numeric computation in a practical setting. In *Proc. of DISCO '90, Capri*, 1990.

[6] J. P. Fitch. Applying computer algebra. In *Proc. of International Conference on Computer Algebra and its Applications in Theoretical Physics*, pages 262–275, 1985.

[7] J. P. Fitch. The Symbolic–Numeric Interface. *Computer Physics Communications*, 1990.

[8] J. P. Fitch, A. C. Norman, and P. M. A. Moore. The automatic derivation of periodic solutions to a class of weakly nonlinear differential equations. In *Proceedings of the 1981 ACM Symposium on Symbolic and Algebraic Computation*, pages 239–244, 1981.

[9] Barbara L. Gates. Gentran user's manual - REDUCE version. Memorandum INF-85-11, Twente University of Technology, Department of Computer Science, The Netherlands, June 1985.

[10] Barbara L. Gates. A numerical code generation facility for REDUCE. In *Proc. SYMSAC '86*, pages 94–99, July 1986.

[11] V. V. Goldman and J. A. van Hulzen. Automatic code vectorization of arithmetic expressions by bottom-up structure recognition. In J. Della Dora and J. Fitch, editors, *Computer Algebra and Parallelism*, pages 119–132. Academic Press, 1989.

[12] A. C. Hearn. The problem of substitution. Technical Report AI-70, Stanford Artificial Intelligence Report, 1968.

[13] R. A. Lyttleton and J. P. Fitch. Cosmologically changing G and the structure of the earth. *Monthly Notices of the Royal Astronomical Society*, 180:471–477, 1977.

[14] A. A. Nayfeh. *Perturbation Methods*. John Wiley & Sons, 1973.

[15] A. C. Norman. Computing with formal power series. *Transactions on Mathematical Software*, 1:346–356, 1975.

[16] J. A. van Hulzen. Breuer's grow factor algorithm in computer algebra. Memorandum 332, Department of Applied Mathematics, Twente University of Technology, The Netherlands, April 1981.

[17] J. A. van Hulzen, B. J. A. Hulshof, B. L. Gates, and M. C. Van Heerwaarden. A code optimization package for REDUCE. In *Proc. of ISSAC '89*, pages 163–170. ACM Press, New York, 1989.

[18] Sanjiva Weerawarana and Paul. S. Wang. GENCRAY: A portable code generator for Cray Fortran. In *Proc. of ISSAC '89*, pages 186–191. ACM Press, New York, 1989.

SYMBOLIC-NUMERIC INTERFACE FOR PERSONAL COMPUTERS

KORNAUKHOV M.A.

Department of Computational Mathematics and Cybernetics, Moscow State University, Moscow, SU-119899, USSR

ABSTRACT

There is a number of problems that require both symbolic and numeric computations. And usually numeric results are computed with formulae derived by symbolic systems. Symbolic-numeric interface discussed was constructed specially to make this process more convenient.

1. Introduction.

The idea of the system that helps to combine both symbolic and numeric computations in a single program is clearly realized now[3] . Such facilities, but in very rudimentary form, are included in almost every Computer Algebra (CA) system - the examples are REDUCE and MAXIMA. But existing features are inconvenient because in order to incorporate some formulae, derived by symbolic system, into his program, one must take a number of steps - generate source code (usually in FORTRAN) from the derived formula, then compile it and link into his program. And the presented system was designed in order to avoid such inconveniences. In was constructed for those people who deal mainly with numeric computations and don't want to learn another more programming language - the CA system's language - but still want to use symbolic computations in their work. There is only one restriction (of course avoidable but very useful) - language for numeric programs must support some features of object-oriented programming.

2. Key Ideas of Implementation

2.1 The Main Requirements.

Symbolic-numeric interface of this kind must meet following requirements:

- It must provide complete language independence - i.e. the user of such system (if he don't want to program in CA system directly) must be able to use the main features of CA system without any notion of CA system's language;
- Interface must provide an ability to use all basic features of CA system involved;
- Interface must resolve all problems concerned with data transformations.

2.2 The Solution - Object-Oriented Programming.

Object-oriented programming (OOP) technique helps to resolve all these problems. The new class type *formula* incorporates all the needed features and hides undesirable details in a very elegant manner. And OOP technique gives us an ability to implement even a so called *dynamic loading* facility - it is almost invisible for the user of the interface. In other words, when you want to get an executable code from the results of symbolic computations, you must first of all call one of *formula*'s methods to generate source program from the result of computations, compile and load this program. Then you call some other method to set the values of variables, involved in computations, (the list of such variables is also available). And after that you call the third *formula*'s method to get the results of computations. It's rather simple and all this is done during the user's program execution. And computations are quick - because the resulting formulae are compiled, not interpreted.

2.3 On the Choice of Systems Involved.

This work was done for single-user/single-process operation environment on iAPX 80x86 based machines - for MSDOS environment. We choose MuMath-83[*,2] as basic symbolic system because it is the only system which is rather compact (it takes about 180 K of RAM to run while REDUCE 3.2[4] takes 450-500 K). And we used TURBO-Pascal[**,4] as complimentary numeric system. We chose it because it is the only system for these machines which is widely spread and supports object-oriented programming. As the operational environment is single-process, the interface package uses common memory area mechanism an input/output redirection to organize interconnection between CA system and numeric one.

3. Example.

The idea of this example was taken from [3]. It implements Newton's method for finding the root of equation **F(x) = 0**. Iteration formula is:

---

* MuMath is a registered trademark of Soft Warehouse Inc.

** TURBO-Pascal is a registered trademark of Borland Inc.

$$x_{n+1} = x_n - F(x_n)/F'(x_n).$$

F' is calculated using CA system, and both F and F' are compiled and loaded using *dynamic loading* facility.

3.1 Description of Procedures Used.

Formula.Init(VarName : String) - object initialization (must be called explicitly). It assigns internal CA system variable named VarName to object;
Formula.Assign(Value : String) - assign string representation of expression to the internal CA system variable, associated with this object;
Formula.MakeExecutable - *dynamic loading* facility;
Formula.SetVarValue - set value for variable in the compiled expression;
Formula.Call(Result) - Boolean function that returns TRUE iff no error occurs during computations of compile expression; *Result* is the result of these computations;
Formula.Done - shut-down object;
Dif(Source, VarName, Order, Destination) - perform symbolic derivation of *Source* with the respect to *VarName* *Order* times. Result is placed into *Destination* (of course, all is done by the CA system and source and result of derivation are the CA system variables associated with *Source* and *Destination* respectively). *Dif* returns *TRUE* iff the process ends correctly.

3.2 Program Itself.

```
Program Newton;
Uses   MuSimp;
Const NIterations = 20;
        EPS = 1E-06;
Var     F, Df : Formula;
        B : Boolean; I : Byte;
        S : String; X0, X1 : Real;
Function RecRel (X : Real) : Real;
Var R1, R2 : Real;
Begin
        F.SetVarValue('X', X); Df.SetVarValue('X', X);
        If F.Call(R1) And Df.Call(R2) Then RecRel := X - R1/R2
                                      Else RecRel := 0;
End;
Begin
        ReadLn(S)                      {Read string representation formula F }
        F.Init('F');                   { Object initializations }
        Df.Init('DF');
        F.Assign(S);                   { Assign value to formula }
        If Not Dif(F, 'X', 1, Df) Then Halt;
        If Not F.MakeExecutable Then Halt;
        If Not Df.MakeExecutable Then Halt;
        B := False;
        X0 := 0.6;
        I := 1;
        While (I <= NIterations) And Not B Do Begin
                X1 := RecRel(X0);
                B := Abs(X1-X0) < EPS;
                X0 := X1; Inc(I);
                WriteLn(I,'-th iteration; ',X0);
        End;
        F.Done;
        Df.Done;                       { Objects shut-down }
End.
```

3.3 How it works.

This program implements trivial Newton's algorithm. First fo all it reads string representation of F from standard input. After that it performs initializations of object for F *(F)* F' *(Df)* and assigns input value to *F*. The next step is symbolic derivation - via *Dif* function. Then we try to compile and load expressions for both F and F'. If all is correct, *RecRel* function, that implements recurrent relation of Newton's method, is called repeatedly, until the method converges. In *RecRel*, *SetVarValue* is called to set new value to variable in F (F is assumed to depend only on x i.e. F = F(X)), and than *Call* is called to get results of the expression evaluation.

4. Concluding remarks.

This model of symbolic-numeric interface represents a new type of such systems (on classification of tools for these interfaces see [3]). And we hope that this system is much more convenient than ANS[3] because it is completely constructed using only one programming language.

5. References.

1. *TURBO-Pascal User's Guide version 5.5*, (Borland International, U.S.A., 1989).
2. *MuMATH-83 User's Manual*, (Soft Warehouse, U.S.A., 1983).
3. Suzuki,M., Sasaki,T., Sato,M., Fukui,Y.,in *Proceedings of European Conference on Computer Algebra EUROCAL'87 (Lecture Notes in Computer Science)*, Vol 378.
4. Yamamoto,T., Aoki, Y., in *Proceedings of European Conference on Computer Algebra EUROCAL'87 (Lecture Notes in Computer Science)*, Vol 378.

Current Trends in Source-Code Optimization

J.A. van Hulzen

Twente University, Department of Computer Science
P.O. Box 217, 7500 AE Enschede, The Netherlands

ABSTRACT

The design of SCOPE, our Source-Code Optimization PackagE for REDUCE, is discussed. The present hardware and software engineering developments demand modifications in the original design. We shortly mention present and future work directed towards implementing such modifications.

1. Introduction

The computer landscape is changing rapidly. The quickly increasing computing power of desk-top tools will enormeously stimulate the use of computer algebra systems in routine computations and will indespensably lead to a demand for program generation tools based on these sytems.

One obvious reason is that increase in computer power naturally leads to a comparable increase in complexity and size of research problems. But complex problems invariantly lead to complex programs, which cannot realistically be coded by hand and thus require facilities for automatic mathematical processing and automatic code generation. As an example we mention the class of non-linear and optimization problems which make use of gradient methods and which require the construction of Jacobians or Hessians for their solution. Such problems range from structural engineering problems to econometric modelling. Although numerical libraries, such as NAG[1] and ISML[2] offer efficient solution engines, additional problem dependent procedures must be coded by the user. This coding is tedious and error prone and -moreover- it is required for every problem instance. This is where automatic generation becomes essential. Such a program generator ought to accept the problem definition on a relatively abstract level, and then ought to apply algebraic and symbolic processing to finally produce an auxiliary procedure to be used in conjunction with the numerical program library. We have made such a program generator, called GENJAC, for the construction of Jacobians and Hessians [9]. GENJAC delivers the auxiliary procedure in a NAG-library compatible fashion. Other examples are FINGER, a package to assist in finite element analysis [17], and a program generator, described by Miller, for the efficient evaluation of Fourier series [15].

Another important reason is that the present state of computer network facilities makes desk-top access possible to a great variety of architectures, ranging from the more traditional von Neumann-like types to vector- and parallel architectures. So a (computer algebra supporting) workstation can be considered as a front-end for an arbitrary machine. This complicates our former view on program generation aspects of the symbolic-numeric interface [12]. The program generator needs a problem decomposition strategy, simply to be able to make an adequate map on the target machine possible. However, such a strategy is usually related to a specific class of problems and to the characteristics of the target architecture. Hence it is required -in view of the problem definition- to find a consistent task distribution for the target architecture. Once it is known if some form of decomposition is fruitful we will be

1) NAG FORTRAN Library Manual Mark 11, Numerical Algorithm Group, Oxford, 1984.
2) International Mathematical & Statistics Library, Inc. (ISML).

confronted with another aspect of the generation process : The decomposition ought to be translated into a syntactically (and of course also semantically) correct program in a specific target language, possibly specially designed to be used on the target machine. A powerful tool for such a task was GENTRAN [3,4]. The present version allows to translate REDUCE (or MACSYMA) commands, including function definitions, into equivalent and syntactically correct programs, coded in RATFOR, FORTRAN-77 or C. The REDUCE version also allows to produce PASCAL code. It is in use for both GENJAC and FINGER. But how useful will GENTRAN be in future? At present it lacks features to assist in generating programs for alternative architectures. This has lead to the development of GENCRAY [18] and GENW2 [16]. Both packages are used for the production of finite element code, to be executed on a CRAY- or a Warp-machine, respectively. The output of GENCRAY is FORTRAN-77 or CRAY FORTRAN-77, while GENW2 is a W2-code generator. W2 is a PASCAL-like high level programming language for the Warp array. These variations on the GENTRAN-line form an answer to the question how to use a workstation as a front-end. But is it an adequate answer? How many substitutions for XXX in GENXXX will be needed? Or do we need another approach to GENTRAN?

A third relevant aspect, attendant to automatic program generation, is the problem of automatic source-code optimization. This is a crucial aspect because code generated from symbolic computations often tends to be made up of lenghty expressions, to be grouped together in blocks of straightline code in a program for numerical purposes. The main objective of source-code optimization so far has been to minimize the arithmetic complexity of such blocks of code. That means minimizing the number of elementary arithmetic operations. However, such a strategy is based on program execution on von Neumann types of architectures. SCOPE, our Source-Code Optimization package for REDUCE [7,13], now available, like GENTRAN, through the REDUCE network library, was especially designed for this purpose.

But if a computer algebra system is used as a front-end facility for program generation the question what source code optimization is does not allow a straightforward answer for different reasons. One reason is the target architecture and another the complier in use for the target language. And these two reasons are of course related. As stated above, the definition of a complex problem leads via a possible decomposition to a translation into a target language. Such a raw program can be subjected to transformations leading to fine grained improvements and thus to more efficient code. These improvements, for instance inside in parallel executable problem solution components, may be based on the present possiblities of SCOPE. However it is in fact prescribed by the compiler in use for the target language what the overall effect will be of heuristic strategies applied on sets of arithmetic defining expressions with the purpose to minimize the number of elementary arithmetic operations. Introduction of many short arithmetic assignment statements, defining how to assign values to common sub-expressions representing variable names can lead to many time consuming store-operations. Explicit application of an addition chain algorithm to redefine the treatement of integer-valued exponents can interfere with actions automatically taken by an optimizing compiler [14]. Hence a general approach to source to source mappings can hardly be qualified as efficient and the program generation process has to made quite subtile when a computer algebra system is used as front-end.

In summary it can be stated that the need for program generation tools is created by the increasing problem size. It is complicated by the variety of target architectures and thus of target languages. It is further confused by efficiency issues related to often not explicitly known features and facilities offered by computer manifacturers. We conclude that a complete automatization of the program generation task is propably not yet possible, and thus that user intervention is still required. Furthermore we believe that attempts to parameterize the characteristics of hardware concepts and of optimizing compilers may assist in reducing user tasks beyond problem formulation.

We shortly indicated the context of our present work on source-code optimization and related topics. SCOPE, as it is operational now, is completely coded in RLISP. It requires some GENTRAN facilities to interface with

REDUCE. Its input, in its most simple form, is a set of syntactically correct REDUCE assignment statements. Its output is the optimized version of the input, given in the syntax of one of GENTRAN's target languages. It is possible to combine this standard output with a list of declarations, defining the type of the input names and of the system generated common sub-expression names. The output is in fact produced as a side-effect of a SCOPE application. At present we are working on a new version of SCOPE, taking into account the design considerations for a future REDUCE 4 version [6]. An implication of this desire is to realize a functional behaviour of the future version, for instance by making both GENTRAN and SCOPE completely independent of each other. The earlier indicated variety of target machines has some implications, like our wish to make our program generation tools more powerful. We shortly discuss the present features in the next section, before outlining our present work and indicating future developments in the following sections.

2. SCOPE's present status

In [10,11,13] we described the overall optimization-strategy of SCOPE as a composite function $R^{-1} \circ T \circ R$. The function R defines how to store the input E_0 in an expression data base D_0. This D_0 is formed by two matrix-structures and a function table. The incidence matrices represent E_0, a set of arithmetic expressions, in a two-dimensional structure where the rows represent expresssion or sub-expression references and the columns represent indentifier references such as variable and function names. The function names are taken from the function table, consisting of a list of pairs of function applications occurring in E_0, and system selected names functioning as their placeholders during the optimization process. Arguments of functions are similarly entered in the matrix-structures when ever relevant. A given sub-expression will be entered in one of two types of incidence matrices, one for sums and one for products, depending on the nature of the arithmetic operation at the top level of the expression. The two matrices are correlated by auxillary predecessor-successor information at the row level for every sub-expression reference. The actual entries in the matrices are either multiplicative numerical coefficients for the sums matrix or powers for the products matrix. The inverse function R^{-1} defines the output-production. The function T defines the optimization process itself. It essentially consists of a heuristic remodelling of the (extendable) matrices in combination with storing information required for a fast retrieval and correct insertion of the detected cse's in the output. This is accomplished by an iteratively applied search, resulting in a stepwise reduction of the arithmetic complexity of the input set, using an extended version of Breuer's grow factor algorithm [1,10,11]. It is applied until no further profit is gained, i.e. until the reduction in arithmetic complexity stops. Before producing output, a finishing touch can be performed to further reduce the arithmetic complexity with some locally applied techniques. The overall process can be summarized as follows:

$$R : E_0 \rightarrow (D_0, \text{profit}_0)$$

$$T_\beta : (D_i, \text{profit}_i) \rightarrow (D_{i+1}, \text{profit}_{i+1}),\ i = 0,\ldots,\lambda-1.$$

$$F : (D_\lambda, \text{profit}_\lambda) \rightarrow D_\lambda$$

$$R^{-1} : D_\lambda \rightarrow E_\lambda$$

D_0 is created as a result of an R-application performed on input E_0. The termination condition depends on some profit criterium related to the arithmetic complexity of the latest version of the input, D_i. Hence we assume $\text{profit}_i = \textit{true}$ for $i = 0,\ldots,\lambda-1$ and $\text{profit}_\lambda = \textit{false}$. The function T is composite as well, and defined by $T = F \circ T_\beta^{\lambda}$. T_β defines one iteration step, i.e one application of the extended version of Breuer's algorithm. The function F defines a finishing touch, resulting in the final version D_λ of D_0, used to produce the output E_λ. We omit a further discussion of the different algorithms used for optimization; this can be found in [10,11,12,13], for instance.

This view on SCOPE as a composite function operating on E_0 and leading to E_λ is a somewhat idealized picture of reality, as shown by the following two examples.

Example 2.1

Part of a REDUCE session is shown. F denotes an expression, which is needed in some calculation in combination with some of its partial derivatives with respect to A and B. The FORTRAN code required to carry out such a computation can easily be made using the **optimize**-command. The assignment symbol :=: results in an evaluation of the righthandside, before the assignment statement is further processed. The list of assignment statements, is only shown directly after the **optimize**-command itself if the SCOPE-switch **acinfo** is on, a default situation. This list defines the actual input for SCOPE. In addition tables of elementary operations are shown before and after the optimization process, giving the arithmetic complexity of the sets E_0 and E_λ, respectively. The SCOPE-switch **primat** is also turned on, leading to a schematic visualization of the matrices, being part of the expression data base. Both the initial and final state are presented.

```
load scope$

f:=sin(a*x + 2*b)^2*cos(a*x + 2*b)$

on primat,fort$

optimize {f:=:f,fa:=:df(f,a),fb:=:df(f,b),faa:=:df(f,a,2),
          fab:=:df(f,a,b),fba:=:df(f,b,a),fbb:=:df(f,b,2)}
         iname s;



                      2
F := SIN(A*X + 2*B) *COS(A*X + 2*B)

                         2                   2
FA :=  - (SIN(A*X + 2*B)  - 2*COS(A*X + 2*B) )*SIN(A*X + 2*B)*X

                            2                   2
FB :=  - 2*(SIN(A*X + 2*B)  - 2*COS(A*X + 2*B) )*SIN(A*X + 2*B)

                             2                   2                   2
FAA :=  - (7*SIN(A*X + 2*B)  - 2*COS(A*X + 2*B) )*COS(A*X + 2*B)*X

                               2                   2
FAB :=  - 2*(7*SIN(A*X + 2*B)  - 2*COS(A*X + 2*B) )*COS(A*X + 2*B)*X

                               2                   2
FBA :=  - 2*(7*SIN(A*X + 2*B)  - 2*COS(A*X + 2*B) )*COS(A*X + 2*B)*X

                               2                   2
FBB :=  - 4*(7*SIN(A*X + 2*B)  - 2*COS(A*X + 2*B) )*COS(A*X + 2*B)


%%%%%%%%%%%%%%%%%%%%%%%%%%%%%%%%%%%%%%%%%%%%%%%%%%%%%%%%%%%%%%%%%%%%%%%%%%%%%%
%%% The matrices shown form slightly modified output.                      %%%
%%%%%%%%%%%%%%%%%%%%%%%%%%%%%%%%%%%%%%%%%%%%%%%%%%%%%%%%%%%%%%%%%%%%%%%%%%%%%%
```

```
Sums matrix :

   |  B|exp|Father
------------------
  1|  2| 1 | S1
  3|   | 1 | FA
  5|   | 1 | row 4
  8|   | 1 | FB
 10|   | 1 | row 9
 13|   | 1 | FAA
 15|   | 1 | row 14
 18|   | 1 | FAB
 20|   | 1 | row 19
 23|   | 1 | FBA
 25|   | 1 | row 24
 28|   | 1 | FBB
 30|   | 1 | row 29
------------------

Function table :
----------------
 S0 | SIN(S1)
 S2 | COS(S1)
----------------
```

```
Products matrix:

   | S2 S0  X  A|coeff|Father
-----------------------------
  0|  1  2      |  1  | F
  2|         1  1|  1  | row 1
  4|     1   1   | -1  | row 3
  6|     2      |  1  | row 5
  7|  2         | -2  | row 5
  9|     1      | -1  | row 8
 11|     2      |  2  | row 10
 12|  2         | -4  | row 10
 14|  1      2  | -1  | row 13
 16|     2      |  7  | row 15
 17|  2         | -2  | row 15
 19|  1      1  | -1  | row 18
 21|     2      | 14  | row 20
 22|  2         | -4  | row 20
 24|  1      1  | -1  | row 23
 26|     2      | 14  | row 25
 27|  2         | -4  | row 25
 29|  1         | -1  | row 28
 31|     2      | 28  | row 30
 32|  2         | -8  | row 30
-----------------------------
```

```
Number of operations in the input is:

Number of (+,-)-operations : 26
Number of (*)-operations : 66
Number of integer exponentiations : 14
Number of other operations : 20

%%%%%%%%%%%%%%%%%%%%%%%%%%%%%%%%%%%%%%%%%%%%%%%%%%%%%%%%%%%%%%%%%%%%%%%%%%
%%% This is the code we need.                                          %%%
%%%%%%%%%%%%%%%%%%%%%%%%%%%%%%%%%%%%%%%%%%%%%%%%%%%%%%%%%%%%%%%%%%%%%%%%%%


      S1=2*B+X*A
      S0=SIN(S1)
      S4=S0*S0
      S2=COS(S1)
      F=S4*S2
      S12=2*S2*S2
      S10=(-S12+S4)*S0
      FA=-S10*X
      FB=-2*S10
      S7=-S12+7*S4
      S9=S7*S2*X
      FAA=-S9*X
      S11=2*S9
      FAB=-S11
      FBA=-S11
      FBB=-4*S7*S2


%%%%%%%%%%%%%%%%%%%%%%%%%%%%%%%%%%%%%%%%%%%%%%%%%%%%%%%%%%%%%%%%%%%%%%%%%%
%%% It is all we get if we replace the command on primat$ by the       %%%
%%% command off acinfo$.                                               %%%
%%%%%%%%%%%%%%%%%%%%%%%%%%%%%%%%%%%%%%%%%%%%%%%%%%%%%%%%%%%%%%%%%%%%%%%%%%

Number of operations after optimization is:

Number of (+,-)-operations : 3
Number of (*)-operations : 16
Number of integer exponentiations : 0
Number of other operations : 2
```

```
%%%%%%%%%%%%%%%%%%%%%%%%%%%%%%%%%%%%%%%%%%%%%%%%%%%%%%%%%%%%%%%%%%%%%%%%%%%%
%%% Observe that some of the common sub-expression names lost their      %%%
%%% role during the optimization process: S7 and S8 in an additive        %%%
%%% sense and S6 in a multiplicative sense. The function table is         %%%
%%% omitted, because it was not changed during the process.               %%%
%%%%%%%%%%%%%%%%%%%%%%%%%%%%%%%%%%%%%%%%%%%%%%%%%%%%%%%%%%%%%%%%%%%%%%%%%%%%
```

Sums matrix :

	S12	S11	S9	S10	S8	S7	S6	S4	B	exp	Father
1									2	1	S1
3										1	FA
8				-2						1	FB
13										1	FAA
18		-1								1	FAB
23		-1								1	FBA
28										1	FBB
37	-1							7		1	S7
38	-1							1		1	S8
41			2							1	S11
42							2			1	S12

Products matrix :

	S10	S9	S7	S8	S6	S4	S3	S2	S0	X	A	coeff	Father
0						1		1				1	F
2										1	1	1	row 1
4	1									1		-1	row 3
14		1								1		-1	row 13
29			1					1				-4	row 28
33								1		1		1	S3
34									2			1	S4
36								2				1	S6
39			1				1					1	S9
40				1					1			1	S10

□

Example 2.2

Another form of the **optimize**-command is shown. The generalized form of the REDUCE **structr**-command is applied on a set of assignment statements including a matrix definition. A **gstructr**-application results in a list formed by a mixture of toplevel-subexpressions, in this case denoted by names starting with a V, and arguments of the function **gstructr**. The list given directly after the **optimize**-command itself defines the actual input for SCOPE, implying that the evaluation of the **gstructr**-command is part of a form of preprocessing, which is organized during the parsing of the **optimize**-command. Here the **optimize**-command is extended with a **declare**-option and a first letter rule (**iname** s) for the common subexpression names to be generated.

```
off exp$

load scope,struct$

matrix a(2,2)$
a:=mat((x+y+z,x*y),((x+y)*x*y,(x+2*y+3)^3-x))$

optlang!*:='c$

optimize gstructr <<a; b:=(x+y)^2; c:=(x+y)*(y+z);
                d:=(x+2*y)*(y+z)*(x+z)^2>> name v
         iname s
         declare <<a(2,2),b,c,d:real>>$

A(1,1) := X + Y + Z

A(1,2) := X*Y

V2 := X + Y

A(2,1) := V2*X*Y

                         3
A(2,2) := (X + 2*Y + 3)   - X
```

```
             2
B := V2

V5 := Y + Z

C := V2*V5

                          2
D := (X + 2*Y)*(X + Z) *V5

DOUBLE X,Z,S5,Y,S8,V2,S6,S4,A[3][3],B,V5,C,D;
{
    S5=X+Z;
    A[1][1]=S5+Y;
    S8=Y*X;
    A[1][2]=S8;
    V2=X+Y;
    A[2][1]=S8*V2;
    S6=X+2*Y;
    S4=S6+3;
    A[2][2]=-X+S4*S4*S4;
    B=V2*V2;
    V5=Y+Z;
    C=V5*V2;
    D=S6*S5*S5*V5;
}
```

For shortness we removed the operation counts from the original output. □

Part of the evaluation of an **optimize**-command, in fact part of the actions collected under the name R, is realized with some features of GENTRAN. We use GENTRAN to translate LISP prefixforms -made by REDUCE as the first step of the evaluation of input- into possibly equivalent but different prefixforms, which are then used to built the expression data base D_0. These prefixforms are algebraically evaluated -simplified- when one of the GENTRAN assignment operators :=: or ::=: is used. The input is taken literally if we use the operator :=. Another GENTRAN module is used to create a symbol table. This table is required if an **optimize**-command is extended with a **declare**-option, as done in example 2.2.

The form of the **optimize**-command differs in both examples. The command syntax allows different choices. Therefore, it requires an extension of the top level REDUCE evaluation module. This extension is partly based on the application of GENTRAN functions, for instance for handling declarations. An advantage of this interface between SCOPE and GENTRAN is that SCOPE runs under new REDUCE versions if GENTRAN is adequately maintained. However, a complication of the desire to operate with prefixforms is that direct coefficient domain information is missing -as available in the internal algebraic form, the SQ-form- when building the expression data base D_0. Therefore, up to now, we limited the coefficient domain to $\mathbf{Z}$.

The construction of D_0 -user anonymously built up with vectors using RLISP- requires some initial "cleaning up" activities to guarantee a healthy start of the optimization process. During evaluation of T the data base is continuously modified until some final version D_λ is obtained. Information about these activities can be obtained as a side effect by turning on certain SCOPE-switches. The optimization process $T = F \circ T_\beta{}^\lambda$ is in a sense static. Both F and T_β are very composite, thus defining a number of transformations on some version of the expression data base. Hense an identical stategy is applied for different architectures. Experiments with a CRAY-machine showed for instance, that more time can be spend in executing output of SCOPE than in executing the coresponding input, because the introduction of many short expressions resulted in an inefficient register use.

The activities covered by R^{-1} are as diverse as those collected under the name R. D_λ is used to produce a prefixlist ϕ, consisting of pairs of the form (expression name, expression). This list is finally used to produce output.

However it may be necessary to have it preceded by a list of declarations, either because the **declare**-option in the **optimize**-command was used or since the SCOPE-switch **optdecs** is turned on. The final version of the symbol table -borrowed from GENTRAN- is obtained using the final version of ϕ. It may happen that ϕ contains redundant structures, as is further illustrated in

Example 2.3

Some refinements of ϕ can be found in examle 2.1.
The sub-expression S9 = S7*S2*X has the form S9 = S7*S3 in the products matrix. Since S3 (= S2*X) occured only once as an operand in an expression in ϕ (row 39), it is removed by substituting its value back before output is produced. In addition the pair defining its occurrence is skipped from ϕ. That is the reason that seemingly a not discovered sub-expression occurs in both S9 and FBB.
Part of the preparation of ϕ consists of the application of an addition chain algorithm. So, S6 for instance, defines $S2^2$, as shown in row 36 of the products matrix. Since S6 is only occuring once in an additive sense at the end of the optimization process (see row 42 in the sums matrix) the same mechanism as applied for S3, is also used to remove S6 from ϕ. □

So once ϕ is made a number of different actions, collected in one procedure -anonymous to the user- are performed to obtain the final result. It is simply text collected in a file or displayed on a screen. It has no direct connection with the REDUCE environment, although that can simply be arranged via ϕ.

3. Towards a new version of SCOPE

It may be obvious now that some revision is needed to increase the overall flexibility and robustness of SCOPE. The use of GENTRAN modules is not necessarily in conflict with the desire to improve the usability of the package. It is always possible to simply copy these modules. However the proposed REDUCE 4 strategy is, phrasing it in a few words: *(Algebraic) functions define actions on (lists of) standard quotients resulting in (lists of) standard quotients*. Such an approach does not exclude the use of switches to influence the production (or reduction) of additional information.

Although our expression data base D_i, $i = 1,...,\lambda$ can not be qualified as a list of standard quotients, it is this data structure we need to make explicitly available to obtain a functional behaviour. The new, still experimental version allows to assign such a structure. Another aspect of SCOPE.2 is that all algorithms and transformations can now be applied as an isolated action, operating on an identifiable instance of the expression data base and resulting in a modified version of it, which is again identifiable. Moreover we introduced functions to perform an inverse transformation on an expression data base, leading to a list of assignment statements in prefixform, similar to the list ϕ, mentioned in the previous section.
We intend to reimplement the GENTRAN features, used in a SCOPE setting, such as symbol table handling, in a similar fashion. Special attention is for instance needed for symbol table operations when using SCOPE as development tool for GENJAC. Thus far SCOPE actions were limited to the optimization of blocks of straightline code, i.e. to local activities. But since we have made a module for data dependences analysis operating on blocks of straightline code and on loop structures as well [8], it may be attractive to extend the optimization activities by giving them a more global character. This work in progress requires a more flexible use of the symbol table. The present version of GENJAC relies on the first letter rule of FORTRAN, allowing to make the declarations implicitly. Such extended declaration facilities will increase the usability of SCOPE in other settings as well. A flexible combination of GENTRAN's template processing possibilities and SCOPE's capabilities to reduce the arithmetic

complexity of sets of assignment statements may lead to interesting application packages.

Another aspect of this work, we reported on some time ago [5], is our interest in a possible use of expression data bases, like being used for SCOPE, for other activities. For instance, investigations to obtain information about the profit to be gained by code vectorization. This is an additional reason for increasing the flexibility of the applicability of the basic features of SCOPE. Structure recognition, or AC-unification, is related to symmetry detection (see for instance [17]) and to possibilities to increase the efficiency of differentiation. An interesting application of such facilities is probably work recently reported by Gatermann [2]. We believe such techniques to be variants of the algorithms we implemented, when creating SCOPE. A typical example given by Gatermann is:

$$1 - c.x_1 - x_1.x_2^2 - x_1.x_3^2 = 0$$

$$1 - c.x_2 - x_2.x_1^2 - x_2.x_3^2 = 0$$

$$1 - c.x_3 - x_3.x_1^2 - x_3.x_2^2 = 0$$

Construction of features allowing to gather such structural information as part of a problem solution strategy will also contribute to the set of flexible desk-top program generation tools we believe to be needed.

4. Some conclusions

We suggested that a more flexible use of packages like GENTRAN and SCOPE will become an indispensable future need when using a computer algebra system for the generation of code for numerical purposes. We believe that SCOPE.2, as it is in development now, can largely contribute to the automatization of code generation, but that a similar flexible approach to GENTRAN is also required.

References

[1] Breuer, M.A.: "Generation of optimal code for expressions via factorization", *Comm. ACM* **12,6**, 330-340 (1969).

[2] Gatermann, K.: *Symbolic solution of polynomial equation systems with symmetry*, Preprint SC 90-3, ZIB Berlin (March 1990).

[3] Gates, B.L.: "A numerical code generation facility for REDUCE", *Proceedings SYMSAC '86* (B. W. Char, ed.),94-99. New York: ACM-Press (1986).

[4] Gates, B.L., Wang,P.S.: "A LISP-based RATFOR code generator", *Proceedings 1984 MACSYMA User's Conference* (V.E. Golden, ed.), 319-329. Schenectady, N.Y.: Gen. El. (1984).

[5] Goldman, V.V., van Hulzen, J.A.: "Automatic code vectorization of arithmetic expressions by bottom-up structure recognition", *Computer Algebra and Parallelism* (J.Della Dora and J. Fitch, ed.s), 119-132. London : Acad. Press (1989).

[6] Hearn, A.C.: Private communication.

[7] Hearn, A.C.: *REDUCE user's manual, version 3.3*. Santa Monica, CA: The Rand Corporation (1988).

[8] van Heerwaarden, M.C.: *Data dependences analysis for program generation and optimization*, Masters Thesis, University Twente, Dept. of Computer Science (june 1989).

[9] van den Heuvel, P., van Hulzen, J.A., Goldman, V.V.: "Automatic Generation of FORTRAN-coded Jacobians and Hessians", *Proceedings EUROCAL '87* (J.H. Davenport, ed.), Springer LNCS-series **378**, 120-131. Heidelbeg: Springer Verlag (1989).

[10] van Hulzen, J.A.: "Breuer's grow factor algorithm in computer algebra", *Proceedings SYMSAC '81* (P.S. Wang, ed.), 100-104. New York: ACM-Press (1981).

[11] van Hulzen, J.A.: "Code optimization of multivariate polynomial schemes: A pragmatic approach", *Proceedings EUROCAL '83*, (J.A. van Hulzen, ed.), Springer LNCS series **162**, 286-300. Heidelberg: Springer Verlag (1983).

[12] van Hulzen, J.A.: "Program Generation Aspects of the Symbolic-Numeric Interface", *Proceedings of the JINR III Conference on Computer Algebra and its Applications in Theoretical Physics, Dubna 1985*, 104-113. Dubna (1987).

[13] van Hulzen, J.A., Hulshof, B.J.A., Gates, B.L. and van Heerwaarden, M.C.: "A code optimization package for REDUCE", *Proceedings ISSAC'89* (G. Gonnet, ed.), 163-170. New York: ACM-Press (1989).

[14] Knuth, D.E.: *The art of computer programming, Vol.* **2** (Second Edition). Reading, Mass.: Addison-Wesley (1980).

[15] Miller, B.R.: "A program generator for efficient evaluation of Fourier series", *Proceedings ISSAC '89* (G.H. Gonnet, ed.), 199-206. New York: ACM-Press (1989).

[16] Sharma, N. and Wang, P.S.: "Symbolic derivation and automatic generation of parallel routines for finite element analysis", *Proceedings ISSAC '88* (P. Gianni, ed.), Springer LNCS-series **358**, 33-56. Heidelberg: Springer Verlag (1989).

[17] Wang, P.S.: "FINGER: A symbolic system for automatic generation of numerical programs in finite element analysis", *J. Symbol. Comput.* **2**, 305-316 (1986).

[18] Weerawarana, S. and Wang, P.S.: "GENCRAY: A portable code generator for Cray Fortran", *Proceedings ISSAC'89* (G.H. Gonnet, ed.), 186-191. New York: ACM-Press (1989).

CONSTRUCTION OF ECONOMICAL COMPUTATION FORMULAE IN COMPUTER ALGEBRA SYSTEMS

E.V.ZIMA

Department of Computational Mathematics and Cybernetics, Moscow State University, Moscow, SU-119899, USSR

ABSTRACT

Implementation of various numerical methods often needs organization of computation using complex iterative formulae, i.e. cycles. The time for performing a program on a computer and the necessary storage capacity are largely dependent on form of these formulae. It is desirable to construct computations so that each iterative step might use the results obtained at the previous steps as completely as possible. This implies the association of the given formulae with the recurrent relations, that bring about the same result but economize the arithmetic operations. A special algebraic methods has been created that provides for the automatic construction of such recurrent relations. This method and main ways of its use in the algorithms of cycle optimization, in the systems of symbolic computations and in algorithms of automatic parallel programs construction are explained. All particular problems that appear during the discussion are solved using this method. Many examples and programs are given.

1. Systems of recurrent relations.

Systems of recurrent relations of the type

$$f_{t-1}(x,i)=\begin{cases} a_{t-1}(x), & i=m, \\ f_{t-1}(x,i-1) \ @_t \ f_t \ (x,i-1), & i>m, \end{cases} \quad \textbf{(1)}$$

$$t=1,2....,k; \quad f_k(x,i)=a_k(x),$$

are being considered in [1,2]. Here $@_t \in \{+,\times,\hat{}\}$, $x=(x_1,...,x_s)$ and i, m - integer variables. Such systems are completely determined by the set of functions $a_0(x)$, ..., $a_k(x)$ and the values of operations $@_1$, $@_2$,...,$@_k$. Therefore, for shorting we shall write such systems in linear form: $f_0(x,i)=\{m, @_1, a_0(x), @_2, a_1(x), ..., @_k, a_{k-1}(x), a_k(x)\}$.

Algebraic operations are determined upon (1) type systems. Let it be, e.q., $m=0$. By using system $\{0, \times, x^{-1}, \times, x^2, x^4\}$, which conforms to the expression $x^{2i^{\wedge}2-1}$ and system $\{0, \times, 1, \times, z^2, \times, z^6, z^6\}$, which conforms to the expression $z^{i^{\wedge}3+i}$. It is easy to get the system of the type (1), which will conform to the expression $x^{2i^{\wedge}2-1}z^{i^{\wedge}3+i}$. This is $\{0, \times, x^{-1}, \times, x^2z^2, \times, x^4z^6, z^6\}$ system.

2. Construction of systems of recurrent relations.

A general algorithm of construction of systems of type (1) by the expression which sets the function which is calculated in cycle is worked out. The construction is done in endorder circuit of the binary tree which defines this expression. The possibility of this construction is based on the fact that any leaf node of this tree can be related to the system of recurrent relations of type (1) (that is, the system $\{m ,+ , m ,1\}$ if the leaf node contains variable i, or the system $\{m , + , u\}$ if the leaf node contains other variable u (number u)), and on the usage of operations which the internal nodes of the tree contain to the systems we already got. For example, the result of the application of the algorithm to the expression $f(i)=(2i+1)!+z^{i^{\wedge}2}x^{2i^{\wedge}2+1}(i=0,1,...)$ will be an expression which operands systems of type(1):

$\{0, \times, 1, +, 6, +, 14, 8\} + \{0, \times, x, \times, zx^2, z^2x^4\}$.

The methods of organization of calculations in cycles with the help of such expressions are determined. Program for calculation and output of expressions values $f(i)$ if $i=0,1,...,n$ with the help of the constructed systems of recurrent relations runs as follows:

```
f:=1+x; write(f):
a1:=1; a2:=6; a3:=14; a4:=8;
b1:=x; b2:=z*x^2; b3:=z^2*x^4;
for i:=1 step 1 until n do
  begin
    a1:=a1*a2; a2:=a2+a3; a3:=a3+a4;
    b1:=b1*b2; b2:=b2*b3;
    f:=a1+b1; write(f)
  end
```

3. Generalizations of systems of recurrent relations.

Consider two generalizations of systems of type (1). The first is the result of substitution in (1) $f_k(x,i-1)$ for the function $a_k(x)$, which does not depend on i. We can easily show that for any expression there exists a certain system of recurrent relations. The number of arithmetic operations in computation of the next value $f_0(x,i)$ using the system of recurrent relations is not greater than the number of arithmetic operations in the direct computation of $f_0(x,i)$.

The second generalization is intended for the construction of recurrent relations that define values of functions $s_0(i)=sin(P_n(i))$, $c_0(i)=cos(P_n(i))$, where $P_n(i)$ is a polynomial of i. Suppose $P_n(i)$ is a polynomial - at that case values of $P_n(i)$ for $i = m, m+1, \ldots$ are defines by system $\{m, +, b_0, \ldots, +, b_{n-1}, b_n\}$. Therefore values of functions $s_0(i)$, $c_0(i)$ for $i = m, m+1, \ldots$ are defines by system

$$s_{t-1}(i)=\begin{cases}\sin(b_{t-1}), & i=m,\\ s_{t-1}(i-1)\times c_t(i-1)+c_{t-1}(i-1)\times s_t(i-1), & i>m,\end{cases}$$

$$c_{t-1}(i)=\begin{cases}\cos(b_{t-1}), & i=m,\\ c_{t-1}(i-1)\times c_t(i-1)-s_{t-1}(i-1)\times s_t(i-1), & i>m,\end{cases} \qquad (2)$$

$$t=1,2,\ldots,n;\ s_n(i)=\sin(b_n), c_n(i)=\cos(b_n).$$

Such relations make it possible to exclude standard trigonometric function designators from cycle bodies.

4. Cycles optimization.

Main aspects of the application of the given methods for the optimization of cycle programs are considered. The method of cycle reorganization using the construction of systems of recurrent relations was developed. In order to use this method for the cycles without control variable one should find the elementary recurrent relation inside the cycle body (in the cycle with a control variable this variable defines such relation). The solution of this problem in the limits of considered method is given. The method was developed for the reorganization of computation in functions and procedures which are called from cycle bodies. It can be easily shown that such a well known program optimization methods as code motion and transformation of inductive variables can be obtained as a result of construction of elementary systems of recurrent relations of (1) type (with $k=0$ and $k=1$).

5. Systems for symbolic computations.

The situation when the numerical solution of a problem on computer is precedes by symbolic transformations in system of symbolic computations is typical enough. The formulae which we get in the systems of symbolic computations are usually unwieldy and direct calculations on them are ineffective. Besides, computations with arbitrary precision used in systems of symbolic computations make arithmetic operations very expensive. Hence, the problem of economizing the number of arithmetic operations in this case is still very essential.

For example, consider $f(x)=e^{-x^2/2}sin(x^3)2^{x^3+x}$ where $x = 0, h, 2h, ..., ph$, $h=1/p$ and REDUCE-3.2 as symbolic algebra system. Let $T(p)=Q(p)/R(p)$, where $Q(p)$ stands for time of computations of f(x) using recurrent relations system, and $R(p)$ stands for time of direct computations of $f(x)$. After the following operators : `on bigfloat, numval; precision 20;` we will get the following : T(10) = 0.24, T(20) = 0.16, T(40) = 0.12.

Algorithms of construction of systems of type (1, 2) were realised by Liberman A.A. in REDUCE-3.2 computer algebra system. After recurrent relations are constructed, effective code for cycle can be generated.

6. Automatic parallelization of cycle computations.

Now-a-days much attention is payed to the algorithms of parallelizing computations that are obviously recurrent. At the same time algorithms of parallelizing of arithmetic expressions evaluation do not take into account the fact that these expressions can be in a cycle body [3]. It is shown that method of recurrent relations systems construction can be efficiently used to automatic parallelization of computations in cycles. Systems of recurrent relations have linear structure and can be transformed to "parallel" form more easily then arithmetic expressions that have tree structures. Using methods given it is often possible to reorganize computation effectively and obtain better results in parallelization. Methods of parallel programs construction for different models of parallel architecture were developed.

Now consider the function $f(x)=e^{-x^2/2}sin(x^3)2^{x^3+x}$ for $x=0, h, 2h, ..., nh$. Let S, C, U, V, W be vector registers of the length L (L > 4). The statement S:=SHIFT(V) assigns to a vector S the components of a vector V with a one-component-left shift, i.e.
S[1]:=V[2]; S[2]:=V[3]; ... S[L-1]:=V[L]; S[L]:=V[L].
So the algorithm of computation of f(x) using systems of recurrent relations can

look like:

```
V[1]:=sin(0); V[2]:=sin(h^3) e^(-h^2/2) 2^(h^3+h) ;

V[3]:=sin(6h^3) e^(-h^2) 2^(h^3) ;  V[4]:=sin(6h^3) 2^(6h^3) ;
V[5]:=V[6]:= ... :=V[L]:=0;

U[1]:=cos(0); U[2]:=cos(h^3) e^(-h^2/2) 2^(h^3+h) ;

U[3]:=cos(6h^3) e^(-h^2) 2^(6h^3) ; U[4]:=cos(6h^3) 2^(6h^3) ;
U[5]:=U[6]:= ... :=U[L]:=1; write( V[1] );

for i:=1 step 1 until n do
  begin
    S:=SHIFT( V );  C:=SHIFT( U );
    W:=V*C+U*S;
    U:=U*C-V*S;
    V:=W;   write( V[1] )
  end
```

Here each iteration consists of 6 vector arithmetic operations and 5 vector assignments. You can see that number of vector operations remains the same in the case of calculation of values

$$sin(P_n(i))\ exp(Q_m(i))\ 2\hat{\ }R_t(i)$$

(P, Q, R are polynomials of *i* with power *n*, *m* and *t*), if

$$max(n, m, t) < L-1.$$

7. References

1. Zima E.V., *Automatic Construction of Systems of Recurrence Relations.* (in *Journal of Computational Mathematics and Mathematical Physics.*), Vol. 24, N 6, (1984) pp. 193-197.
2. Zima E.V., *Transformations of Expressions Associated with Systems of Recursion Relations.* (Moscow University Computatational Mathematics and Cybernetics, 1985, N 1, pp. 60-66.)
3. Evans, D.J.(ed.), *Parallel Processing Systems.* (Cambridge University press, 1982).

OPTIMIZING TRANSFORMATIONS OF PROGRAMS IN SYSTEMS OF SYMBOLIC COMPUTATIONS

LIBERMAN A.A.
Department of Computational Mathematics and Cybernetics, Moscow State University, Moscow, USSR, SU-119899

ABSTRACT

The author takes a look at the problem of the economically organized calculations and substantiates it's urgency for the Sistems of Symbolic Computations. One of the possible ways of it's solution is suggested for the case of iterational calculations in loop. The method is based on making the most of results of previous steps. Two families of formulas are examined. They are presented in the form of Systems of Recurrent Relations. Different operations on Systems of Recurrent Relations are implemented. They give an opportunity to construct new Systems by means of reference ones and signs of operations. According to the suggested algorithm a tree corresponding to the formula is built and System of Recurrent Relations is constructed by means of tree traversal.On the base of this System a new more economically organized loop is put down. Also, the article contains some information about realization of this algorithm in System of Symbolic Computations REDUCE and results of it's testing.

1. Optimization in symbolic computations.

Systems of Symbolic Computations (SSC) are oriented mainly on the operations with algebraic expressions and on the identical algebraic transformations which have analitical expressions not numbers as results. However, formulas serving as output of computer algebra algorithms are often used later as input for numerical procedures. The problem is that Lisp, the most widely applicated language in the SSC, isn't accomodated to use in numerical methods because of it's slow arithmetic. That's why economically organized computation is almost a must. Optimization is especially important in case of iterational calculations in loop.

Possible approach puts the Systems of Recurrent Relations[1](SRR) in accordance with reference formulas. Optimisation is achieved by means of making the most of the results of previous steps.

This idea may be quite fruitful when applicated in the SSC. There is a great variety of formulas to be under review. It is important that in addition to the arithmetical operations formulas can include names of elementary functions and special operations, inherent in the Systems of Symbolic Computations as well.

2. The basis of loop optimization.

Let $z=(z_1,..,z_s)$ and i be a set of variables. It is required to calculate values of the function $f(z,i)$ by $i=m,m+1,.n$. Providing this function can be presented in form of recurrent formula

$$f(z,i)=A(f(z,i-1),...f(z,i-j),z,i), \qquad (1)$$

where $f(z,m)=u_0(z),...,f(z,m+j-1)=u_{j-1}(z)$, and it is easier to calculate the right side of Eq. 1 by certain $f(z,i-1),...f(z,i-j)$ than to calculate directly $f(z,i)$ for each i.

3. Recurrent relations.

From the great variety of functions which can be recorded in the form of Eq. 1 we chose two families to have a good look at. One of them was first mentioned in [1] and looks as follows:

$$f(z,i)=\begin{cases} u_0(z), & i=m \\ f(z,i-1)\circ_0 f_1(z,i), & i>m, \end{cases}$$

$$\dots$$

$$f_{k-1}(z,i)=\begin{cases} u_{k-1}(z), & i=m \\ f_{k-1}(z,i-1)\circ_{k-1} u_k(z), & i>m, \end{cases} \tag{2}$$

where $u_0(z),\dots,u_k(z)$ is a set of functions, $\circ_0,\dots,\circ_{k-1}$ is a set of signs of operations, where $\circ_i\in\{+,*,**\}$. Such systems are presented in linear form:

$$f(z,i);\{\circ_0,u_0(z),\dots,\circ_{k-1},u_{k-1}(z),u_k(z)\}. \tag{3}$$

Eq. 3 is called "the simplest" if all the signs of operations $\circ=+$, or $\circ=*$, or $\circ=**$ for $j=1,\dots,k$. In case the system is not "the simplest", it is called "mixed". We shall write down "the simplest" system as follows:

$$f(z,i):\{\circ,u_0(z),U_1(z),\dots,u_{k-1}(z),u_k(z)\}. \tag{4}$$

Now let us determine another family. Let $f(z,i)=func(q(z,i))$, where $func\in\{sin,cos\}$, and

$$q(z,i);\{+,u_o(z),\dots,u_k(z)\}. \tag{5}$$

We can put down the recurrent relation for $f(z,i)=sin(q(z,i))$, making use of equality for sinus of sum:

$$f(z,i)=\begin{cases} sin(u_0(z)), & i=m \\ f(z,i-1)*g_1(z,i)+f_1(z,i)*g(z,i-1), & i>m, \end{cases}$$

$$g(z,i)=\begin{cases} cos(u_0(z)), & i=m \\ g(z,i-1)*g_1(z,i)-f_1(z,i)*f(z,i-1), & i>m, \end{cases}$$

$$\dots$$

$$g_{k-1}(z,i)=\begin{cases} cos(u_{k-1}(z)), & i=m \\ g_{k-1}(z,i-1)*cos(u_k(z))-f_{k-1}(z,i-1)*sin(u_k(z)), & i>m. \end{cases} \tag{6}$$

The same goes for $f(z,i)=cos(q(z,i))$. If multiplication of $f(z,i)=func(q(z,i))$ by "the simplest" $*$-system

$$p(z,i):\{*,v_0(z),\dots,v_1(z)\} \tag{7}$$

takes place, we still keep within the limits of the family. A linear form for this family looks as follows:

$$p(z,i)*func(q(z,i)):\{func,j,(x_1,\dots,x_k),(v_{k+1}(z),\dots,v_1(z))\} \tag{8}$$

where $func\in\{sin,cos\}$, $p(z,i)$ is defined by Eq. 7, $q(z,i)$ - by Eq. 5

$$x_i=\begin{cases} (v_i(z),u_i(z)), & i=1,\dots,min(1,k) \\ (u_i(z)), & i=min(1,k)+1,\dots,k \end{cases} \quad \text{and } 1=\begin{cases} 0, & 1\le k \\ 1, & 1>k, \end{cases}$$

For these two families different operations (addition, multiplication, etc.) are introduced. These operations make it possible to unite simple systems into more complicated ones.

There are the leaves of the tree to start with when constructing SSR. The simplest SSR corresponding to variables and numbres are put in accordance with leaves. For numbers and variables except loop parameter it is true

$$g(z,i)=a=\begin{cases} a, & i=m \\ g(z,i-1), & i>m, \end{cases} \tag{9}$$

where $a \in \{z_1,\dots,z_s\}$ or a is a number. SRR $\{+,a\}$ correspond to such

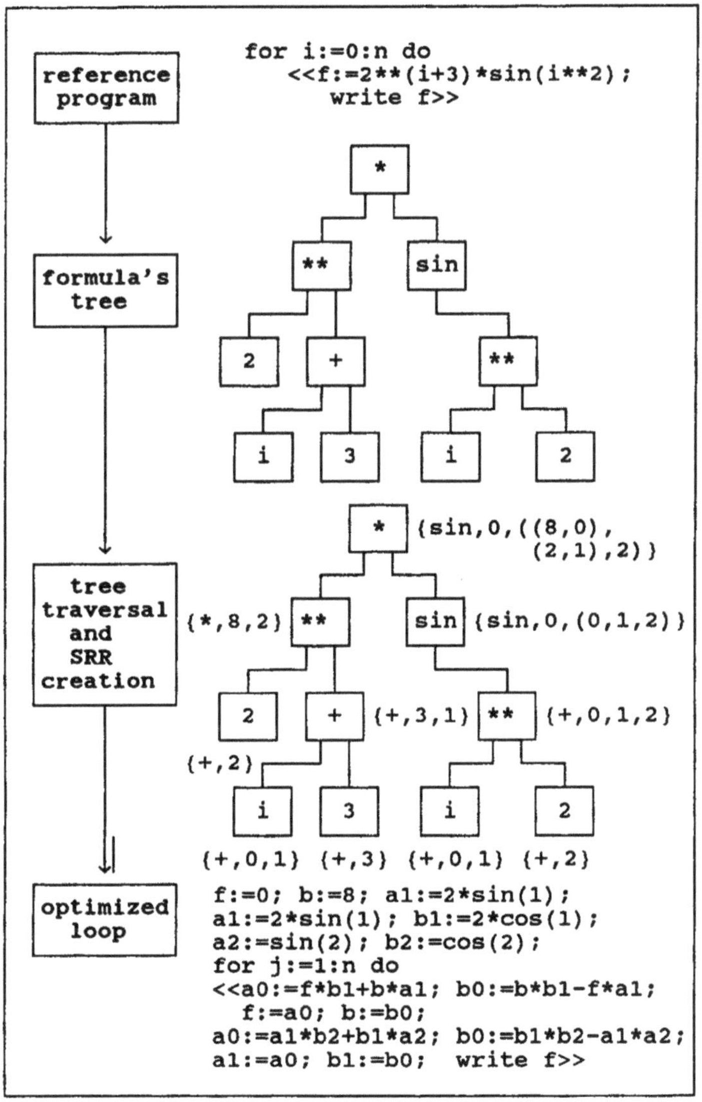

Fig. 1 the formula determined by $f(z,i)$. Variables $i,z_1,\dots,z_s$ and numbers are situated in the leaves of the tree, signs of opeation - in the nodes, a sign of the external operation - in the root.

constructions. For the loop parameter it is true:

$$g(z,i)=i=\begin{cases} m, & i=m \\ g(z,i-1)+1, & i>m. \end{cases} \tag{10}$$

The linear form is $\{+,m,1\}$.

Let's examine two nodes, presenting two SRR with common father node. Suppose that in father node the sign of operationis situated. Let's try to apply the operation to the daughter nodes' systems. If it turns out well, the obtained system is ascribed to the father node. Now let us perform tree traversal from bottom to top, trying to apply the above-stated procedure until it isn't possible any more.

The wanted loop optimization is realized later by means of systems, corresponding to the eldest nodes, for which the foregoing procedure was possible.

This algorithm is illustrated by Fig. 1.

5. Realization.

The suggested algorithm was assumed as a basis for the package for the SSC REDUCE. It's work can be divided into two stadiums:

5.1. Construction of SRR for the $f(z,i)$ or in case it's impossible for some subexpressions of the formula defining $f(z,i)$. This stadium is initialized through running of the procedure *rrscons* with parameters *u,j,m,* where *u* - reference formula, *j* - loop parameter name, *m* - value of the lower bound of the loop parameter(number or variable or expression). As an outcome information the SRR corresponding to the reference function or formula, put down in standard prefix form inherent in LISP language, whose operands are the SRR, are typed out in the output file.

5.2. Generation of program's fragment based on the obtained SRR. This piece of program has to produce the same result as the reference loop, but more economically. The new loop is typed out into the user-defined file and can be used later for calculations with different values of variables and limits of the loop parameter. To generate the fragment of the program one has to call the procedure *genrec* with parameters *u,j,m,n,rab,fil,* where *u,j,m* - the same as in 1, *n* - value of the upper bound of the loop parameter (a number or a variable or an expression), *rab* - the name of a variable, to which the result will be assigned, *fil* - the user-defined name of the file to type out a new loop.

5.3. Tab. 1 illustrates how the program is applied to different calculation formulas.

Tab. 1

reference formula	N	v	t,ms	t1,ms	t2,ms
$a*j^{5}-j-8*j^{3}*z+z+23$	100	10	3790	29000	54700
$e^{-(j*h)^{2}/2}*\sin(j*h)^{3}*$ $*2^{(j*h)^{3}+j*h}$	10	10	8120	25530	170100
$\log(2^{-j^{5}+a*j+j^{2}-z})+j^{4}+$ $+z*j+a*j^{3}$	5	20	7310	19250	103200
$2^{-j*h}/(5-3*\cos(j*h))$	30	20	4500	99350	221030

N - number of steps; v - precision, t - execution time of *genrec* procedure, t1 - time of calculation of function values by means of SRR, t2 - time of direct calculation of formula.

6. References.

1. E.V. Zima, *Automatic Construction of Systems of Recurrent Relations*, (in *Computat. Math. and Math. Phis.* Vol. 24, N 6, 1984, pp 193-197).

ON THE AUTOMATIC CONSTRUCTION OF REPRESENTATIONS OF NON-COMMUTATIVE ALGEBRAS

Gilles Labonté
Royal Military College of Canada
Kingston, Ontario K7K 5L0

The algorithm of "vector enumeration" serves to construct finite dimensional matrix representations of algebras. The only input it requires is a "presentation" for this algebra, that is, a set of equations for its generators. We give a brief description of this construction method.

INTRODUCTION

Motivation: The importance of non-commutative algebra does not have to be demonstrated to Theoretical Physicists and Quantum Chemists. After all, the whole theory of Quantum Mechanics is founded on this mathematical structure.

In practical applications, it is often necessary, or at least very useful, to have available explicit realizations of the representations of algebras. One only has to recall, for example, the importance of the spherical harmonics $\{Y_{l,m}\}$.

Background: This algorithm was devised while we were studying the Lorentz covariant equations used in Quantum Mechanics for the description of relativistic particles of higher spins/1/. These are differential equations of the form $(-i\Gamma\cdot\partial+m)\psi=0$, in which the 4 constant matrices Γ_μ, together with the generators of the Lie algebra of the Lorentz group, satisfy a set of about 24 algebraic equations. It is important to know the possible matrix representations for these Γ's because much of the properties of these equations, such as how waves propagate, are determined by them. The algebraic manipulations required to construct such representations are rather large, they entail very careful and long verifications. Having developed a systematic calculation method, which amounted essentially to a formal computer algorithm, we later reformulated it in a completely general context in references./2,3/

Statement of the problem: The problem solved is the following. Let an algebra E be defined by a set of equations as

$$P_i(X_1, X_2, \ldots, X_n) = 0, \quad i = 1 \quad \text{to } m$$

in which the P_i's are polynomials i.e. linear combinations with coefficients in a certain field K, of monomials in the symbols X_j, for $j = 1$ to n. Find a set of matrices $\{M_j\}$ such that the above equations are satisfied when each symbol X_j is replaced by the corresponding matrix M_j.

Hereafter, R will denote, the set of the polynomials P_j, and we shall call these the "relators" (a term borrowed from "coset enumeration").

Example: Consider the ring which is generated by 4 elements: G_μ, $\mu = 0, 1, 2, 3$, satisfying the 40 equations:

$$G_\mu G_\rho G_\nu \ + \ G_\nu G_\rho G_\mu \ - \ g_{\mu\rho} G_\nu \ - \ g_{\nu\rho} G_\mu \ = \ 0 ,$$

$g_{\mu\nu}$ being the constant Minkowski tensor. This is the Kemmer[4] ring, which is used in the description of relativistic particles of spin 1. Problem: find 4 matrices G_μ with such properties.

The general problem, can be recognized as a form of the "word problem", which was first formulated as such in 1910, by Dehn[5], for group theory, and which is today well studied in computer science. Even earlier statements of the problem can also be found. A particular constructive solution, for the group theory version of it, has been repeatedly rediscovered by Frobenius[6], Dehn[5], Schreier[7], Todd and Coxeter[8], and is now known as the "coset enumeration method". The "vector enumeration method", which we have proposed, is based on the same principles: it is a sort of more general version, applicable to any algebraic structure.

As is well known, it is in general "undecidable" whether a given presentation leads to a finite dimensional representation or not. However, it is possible to prove that vector enumeration will succeed whenever one does exist (see reference 2).

VECTOR ENUMERATION

A helpful introduction to the description of the algorithm of vector enumeration is provided by the following considerations.

Trivial vector enumeration: Let us be given a basis $B_0 = \{v_i : i = 1 \text{ to } k\}$ for a vector space V, and n linear operators $\{M_i : i = 1 \text{ to } n\}$ on this space, which generate the finite algebra E_M.

We call "vector enumeration" the following process by which another basis B for V is constructed.

- Start with a certain vector v in V (which will be called "the generating vector") and denote it as w_1. Set $B = \{w_1\}$.

- Then, consider $M_1(w_1)$; if it is linearly independent of w_1, call it w_2 and add it to B; similarly, if $M_2(w_1)$ is linearly independent of $\{w_1, w_2\}$ call it w_3 and add it to B, etc.... Then, repeat this same process of operating with each M_i

but this time on the vector w_2 , and then on w_3 , etc... This sequence of mappings will obviously terminate since there is a finite number of linear operators M and V is finite dimensional.

Remarks: The vectors of the basis generated by this process all have natural expressions as words with the letters of the alphabet $A = \{M_i : i = 1 \text{ to } n\}$ and a common suffix "v" .

Different choices of generating vectors can produce different representations of E_M .

"Vector enumeration" becomes "coset enumeration" (see Todd and Coxeter[8]) when E is a group and the generating vector H is a subgroup of E ; the basis B is then the set of the cosets of H .

Enumerating vectors starting from a presentation. Let us now consider the less trivial use of vector enumeration, by which a vector space basis will be constructed, and the action of the operators M_i determined, when all that is known is a set of relators for the latter operators.

Since there are in general, many different representations for a same algebra, which one is to be constructed has somehow to be specified. Thus, for example, in coset enumeration, one specifies which subgroup H of the group G , is to be used by giving a presentation for it. We shall similarly assume hereafter that a set of vector relations is given, to characterize the generating vector "v" , as

$$Q_i v = 0 \quad \text{for} \quad i = 1 \quad \text{to} \quad q$$

in which the Q's are polynomials in the generators.

Order relation: During the construction, equations as

$$\textstyle\sum_i c_i \, W_i v \quad = \quad 0 \, ,$$

where the W's are words in the generators of the algebra, will be repeatedly deduced. Each such relation allows for the expression of one vector Wv as a linear combination of the others as:

$$Wv = P$$

Wv will then be said "reducible". The criterion for the singling out of Wv amongst the others vectors, will be that it is the largest one. The order relation on the words is taken to be the following degree order: $W_1 v > W_2 v$ if W_1 is longer than W_2 or, when they have the same length, if $W_1 > W_2$ with respect to some alphabetical order defined on the alphabet $A = \{X_i : i = 1 \text{ to } n\}$.

Here is how the linear operators are defined. If X is the first letter of W , i.e. $W =$ concatenation (X, W') and the above equation holds, then the linear operator M will map $W'v$ as:

$$M(W'v) = P \ .$$

When a relation as above is found during the construction process, the data will be "made compatible" with it by replacing, everywhere it appears, KWv by the linear combination $K(P)$.

The construction: The representation vector space is that of all the possible linear combinations of words that remain non null, while they are compatible with all the vector relations : $Q_i v = 0$, and $P_i W v = 0$ for each P_i in R, and for each vector Wv. Here is how the algorithm constructs a basis for this space.

B is set $= \{v\}$; each word that appears in one of the $Q_i v$, and its suffixes, are added to B. The data is then made compatible with the relations $Q_i v = 0$, through a finite number of simple substitutions.

Each relator P_i is then successively applied on v. All the new vectors introduced are added to the set B. However, the linear relations $P_i v = 0$ imply the reducibility of some vectors, which are then removed from B.

This process is repeated with the next smallest word in B, in the role of v, etc... The cycle of applying relators on a word "w" of B, establishing the necessary adjustments to the data to keep it compatible with the linear relations $P_i(w) = 0$ obtained, then goes on... until the construction stops. This happens when all the relators have been applied to all the words in B.

Example: Here is a simple example, which can be dealt with by hand. Let the generators of an algebra be $\{A, B\}$ and the relators be

$$R = \{AA - e, \quad BB - e, \quad ABAB + 2AB + e\}$$

Let the set of relators for the generating vector be empty, which means that the representation constructed will be the regular representation. Here is how the construction will proceed:

Applying all relators on v yields:

$$AAv = v, \qquad BBv = v, \qquad ABABv = -2ABv - v$$

The new vectors: Av, Bv, ABv and $BABv$ had to be introduced.

Applying all relators on Av yields:

$$0 = 0, \qquad BBAv = Av, \qquad ABABAv = -2ABAv - Av$$

The new vectors introduced are BAv, $ABAv$ and $BABAv$.

Applying all relators on Bv yields:

$AABv = Bv,$ $\qquad 0 = 0,$ $\qquad ABAv = -2Av - Bv$.

The word $BABAv$ then also becomes reducible, and the equation for $ABABAv$ becomes trivial. No new vectors needed to be introduced. Applying all relators on ABv yields :

$0 = 0,$ $\qquad BBABv = ABv,$ $\qquad 0 = 0,$

with no new vectors introduced.

Finally, applying all relators on BAv yields

$$BAv = -ABv - 2v$$

which leads to what is known in coset enumeration as a collapse, that is a cascade of simplifications of the equations already accumulated. This often implies the reducibility of many of the words until then considered irreducible. In the present case, $BABv = -2Bv - Av$ follows and the construction is complete.

CONCLUSION

This algorithm is easy to implement. We have written a version of it for micro-computers; it is in muLISP and uses the field Q of the rational numbers for the coefficients.

Here are some statistics. On an IBM, Model 60, it constructs the 126-dimensional representation of the Kemmer ring, mentioned above, in 53 secs. Construction times for the irreducible representations of the Lie algebra of SU(2) are:

D(2): .5 secs D(15): 3.5 secs D(50): 17 secs.

For the Lie algebra of SL(2,C), the covering group of the Lorentz group they are:

D(1/2,1/2): 1.4 secs D(2,2): = 20 secs D(4,7/2): 118 secs.

Many improvements to the algorithm are still possible. Copies of the program are available without cost, on request.

REFERENCES

1. Labonté, G.: J. Comp. Phys., 1987, **69**, 2, p. 420.
2. Labonté, G.: J.Symb.Comp., 1990, **9**, p.27.
3. Labonté, G.: "On vector enumeration", preprint (1989)
4. Kemmer, N.: Proc. Roy. Soc. London, 1939, **173a**, p. 91.
5. Dehn, M.: Math. Annalen, 1910, **69**, p. 137.
6. Frobenius, F.G.: "Über endliche Gruppen", Sitzungs. d. K. Pruss. Adak. Wiss., Berlin, pp 81-112, (1895)
7. Schreier, O.: Abh. Math., Sem. Hamburg Univ., 1927, **5**, p.161.
8. Todd J.A. and Coxeter H.S.M.: Proc. Edinb. Math. Soc., 1936, **2**, 5, p.26.

Implementational Aspects for Non-Commutative Domains

J. Apel

Department of Informatics
University of Leipzig
GDR-7010

U. Klaus*

Laboratory of Computing Techniques and Automation
Joined Institute for Nuclear Research
Head Post Office, P.O. Box 79, Moscow, USSR

0. Introduction

In this paper we discuss fundamental problems connected with the implementation of non-commutative domains.
In general the representation of such algebraic structures can be difficult since the existence of canonical simplifiers for the elements and algorithms for the basic operations are not ensured.
Therefore, we restrict ourself to special classes of non-commutative algebras, modules and skew fields such that the existence of suitable data structures for their elements and algorithms for computing the basic operations +, -, * and $()^{-1}$ are ensured. In the first section of this article we introduce these structures and give references how its elements can be simplified and how the operations can be performed.
While the first section stresses pure algebraic aspects of computability the second section deals with the coding of the elements using integers and there arithmetic. This is an attempt to avoid time expansive symbol manipulations.
In the last section we describe a small experimental computer algebra system NCD (Non-Commutative Domains) especially designed for the computation in the above domains at an IBM-XT/AT.

---

*Permanent address:
Department of Informatics, University of Leipzig, GDR-7010

1. Algebraic background

Let Γ be the free non-commutative monoid generated by the variables $x_1,\ldots,x_n$ and $<$ a well-ordering on Γ which is (strong) compatible with the monoid operation.
Starting point for our non-commutative computations is the monoid algebra $R\langle\Gamma\rangle$ over the field R of rational numbers. Any element f of $R\langle\Gamma\rangle$ can be uniquely represented in the form

$$f = \sum_{i=1}^{m} \alpha_i \gamma_i \quad \text{with } \alpha_i \in R\setminus\{0\} \text{ and } \gamma_i \in \Gamma \text{ such that } \gamma_m < \ldots < \gamma_2 < \gamma_1.$$

We call γ_1 the *head* of f and denote it by H(f). The ordering $<$ induces a partial ordering $<$ on A by

$$f < g \iff H(f) < H(g).$$

Now we consider an algebra A which is a quotient algebra of $R\langle\Gamma\rangle$ after a two-sided ideal I. Since the elements of A are residue classes modulo I we are faced with the problem of canonical simplification, i.e. the distinction of one representant of each class. A natural choise for the canonical representant of an element f is to take its smallest representant with respect to the partial ordering $<$.
When we are given a finite Gröbner basis/1/ (with respect to $<$) G of I we can compute this minimal representant by reducing an arbitrary representant of f with respect to G. Unfortunately, the existence of finite Gröbner bases of ideals of $R\langle\Gamma\rangle$ is not ensured. But one can try to find a finite Gröbner bases of I using MORA's algorithm/2/ which is successful if it terminates.
Calculation and simplification in the algebra A is much simpler than in its quotient skew field Q(A) provided it exists. All computations in Q(A) are connected with syzygy computation. A useful tool is again the theory of Gröbner bases.
We restrict the class of our algebras A by the demand that the ideal I contains an element of the form

$$x_i x_j + \alpha_{ij} x_j x_i + r_{ij} \qquad \alpha_{ij} \in R\setminus\{0\},\ r_{ij} < x_j x_i$$

for any pair (x_i, x_j) of generators of Γ.
The algebras A with I satisfying this condition are members of the class of G-algebras introduced in /3/ and generalized in /4/. In /3/ and /4/ it is shown how Gröbner bases of one- and two-sided ideals of a G-algebra can be defined and constructed. Furthermore, /3/ deals with the computation of syzygy modules using Gröbner bases and its application to calculation and simplification in the quotient skew-field Q(A) which exists if and only if A has no zero-divisors.
Of course, the theory of Gröbner bases is not only useful for

calculations in skew-fields but also for various ideal operations such as e.g. intersection, decision of ideal membership and elimination of variables. The theory of Gröbner bases is also interesting for modules in connection with similar problems as stated for ideals and also with the computation of higher syzygy modules.
One of the main applications of /4/ was to show how finite dimensional modules over a G-algebra A can be considered as ideals of another G-algebra A' using NAGATA's principle of idealization. All computations of modules can be pushed back to these of ideals this way.

2. Data structures

This section is devoted to the data representation of elements and operations of non-commutative algebras, modules and skew-fields on computers. All these representations are based at these of $R\langle\Gamma\rangle$.
We represent an element of a quotient algebra $A=R\langle\Gamma\rangle/I$ by its canonical representant which can be computed using BUCHBERGER's algorithm. The algorithms for the operations +,-,* are also taken from $R\langle\Gamma\rangle$ and extended by a final computation of the canonical representant of the result.
The elements of an n-dimensional module M over a G-algebra A can be represented by arrays of n elements of A. The basis operations can be easy defined using these of A. In our system NCD we used a second possibility the idealization of M using NAGATA's principle, i.e. we represent a G-algebra A' rather than the module M. A' is the algebra which one gets if the monoid algebra of the monoid $\mathbb{N}^n$ of n-tuples of natural numbers over the algebra A is factored after the ideal generated by all relations $m\circ m'$ where m and m' are non-zero elements of $\mathbb{N}^n$ and $\circ$ is the componentwize addition in $\mathbb{N}^n$.
If the G-algebra A has no zero-divisors then there exists its quotient skew field Q(A). The elements of Q(A) are represented as pairs of elements of A. Algorithms for the operations and canonical simplification in Q(A) based on the theory of Gröbner bases are given in /3/.
The only remaining problem is the representation of the algebra $R\langle\Gamma\rangle$. The elements are represented as lists of monomials which consist of a coefficient from R and a non-commutative power product from Γ. There are different possibilities for the representation of the power products. The usual way is to take a list of the basis elements

$x_1,\dots,x_n$ of Γ. In this case multiplication can be performed by concatenating the lists and checking whether one power product is part of another is done by pattern matching. A second possibility codes the power products by three intergers and transforms the multiplication and the check whether one product is part of another to simple integer arithmetics. Our system NCD contains both possibilities.

We want to give a survey about the coding of the well-ordered free non-commutative monoid $(\Gamma,<)$ generated by $x_1,\dots,x_n$ using integer arithmetics. We require the ordering $<$ to satisfy the additional condition that $(\mathbb{N}^n,\lhd)$ is well-ordered monoid where $\lhd$ is the ordering in $\mathbb{N}^n$ induced by the natural homomorphism φ from Γ onto $\mathbb{N}^n$. Furthermore, the ordering $<$ shall order elements of Γ which have the same image under the homomorphism φ lexicographically w.r.t. $x_1<x_2<\dots<x_n$. The ordering $\lhd$ can be described by an n-dimensional quadratic matrix $\mathfrak{A}$ with elements from $\mathbb{N}$.

Each element $t=x_{i_1}x_{i_2}\dots x_{i_m}$ of Γ will be represented by a triple $t^{\tau}=(\text{length } t, \text{ noncom } t, \text{ com } t)$ of natural numbers. The mappings length, noncom and com are defined as follows:

$$\text{length } t = m$$

$$\text{noncom } t = \sum_{j=1}^{m} (i_j-1)\, n^{m-j}$$

$$\text{com } t = t^{\varphi}\mathfrak{A}\mathfrak{d}^{T} \quad \text{with } \mathfrak{d} = (d^{n-1},\dots,d^2,d,1) \text{ for some fixed non-zero natural number } d$$

The mapping τ restricted to the subset of elements t of Γ satisfying

$$(\text{length } t)*|\mathfrak{A}| < d \quad \text{where } |\mathfrak{A}| \text{ is the maximal entry of } \mathfrak{A}$$

is injective and, therefore, suitable for coding the elements of this subset of Γ.

The following relations show that $(t\circ t')^{\tau}$ can be computed only using simple integer arithmetics from t^{τ} and t'^{τ}.

$$\text{length } t\circ t' = \text{length } t + \text{length } t'$$

$$\text{noncom } t\circ t' = n^{\text{length } t'} * \text{noncom } t + \text{noncom } t'$$

$$\text{com } t\circ t' = \text{com } t + \text{com } t'.$$

The ordering $<$ of Γ can be computed using the relation

$$t < t' \quad \leftrightarrow \quad (\text{com } t, \text{ noncom } t) <_{lex} (\text{com } t', \text{ noncom } t').$$

Finally, the subword problem for elements of Γ, i.e. to check for given elements $t,t'\in\Gamma$ whether there exist $s,s'\in\Gamma$ such that $t=st's'$, can be easy solved using the coding τ using the relation

$$\exists\, s \in \Gamma : t = st' \quad \leftrightarrow \quad \text{noncom } t - \text{noncom } t' \equiv 0 \bmod n^{\text{lenght } t'}.$$

If we compute in a G-algebra we have to solve a second subword problem we have to find an element $s\in\Gamma$ for given $t,t'\in\Gamma$ such that $t^{\varphi}=s^{\varphi}\circ t'^{\varphi}$.

We decode com t - com t' and check whether it is the commutative part of an element s of Γ. Decoding requires computations of residues modulo d and the solution of a system of equations what can be done once for all at definition time of the $R\langle\Gamma\rangle$. The existence check for s restrict the subset of addmissible elements $t\in\Gamma$ to these satisfying

$$2(\text{length } t)*|\mathfrak{A}| < d.$$

We want to close this section with a remark to the size of the region of addmissible elements of Γ.
We have the bounds

$$\text{noncom } t < n^{\text{length } t} \quad \text{and} \quad \text{com } t < d^n.$$

If we assume $|\mathfrak{A}|=1$ (which is true for all important orderings such as total degree ordering, lexicographical ordering and reverse lexicographical ordering) we get

$$\text{com } t < (2 \text{ length } t)^n .$$

3. The program NCD

We designed a small computer algebra system (Non-Commutative Domains) where we realized the data representations from the preceding section. The decision to write a new system rather then to implement the data types in one of the common computer algebra systems is based on the following facts.
While facilities for non-commutative multiplication are included in most computer algebra systems of the second generation such as REDUCE 3/5/ the treatment of non-commutative fractions is impossible. Therefore, we should have used a system of the third generation such as SCRATCHPAD/6/. But our intention was to have a system for a rather small personal computer whère such systems are not available.
NCD consists of a small kernel written in assembler language as described in /7/ which provides a host language for computer algebra applications. This concept makes NCD both fast and easy portable.
NCD is designed especially for calculations in non-commutative rings, modules and skew-fields.
In the present version it only provides the possibility of declaration of such domains and the calculation of the basic operations $+,-,*,()^{-1}$. Particularly, various ideal computations based on Gröbner bases will be added.
By use of the arithmetic coprocessor intel 80187 we got fast procedures for the above described data type of non-commutative polynomials. Using two eight-byte-integers and one one-byte-integer we

can for instance represent power products up to degree 31 for a maximal number of 4 generators of Γ or up to degree 11 for not more than 14 generators.

We want to demonstrate the results we obtained with our experimental system NCD by help of the following example.

Example: We computed the nth-power (n=3,...,8) of the expression x+y+z in the algebra SO(3) with the defining relations yx=xy-z, zx=xz+y, zy=yz-x using the following three different methods.

a) REDUCE 3

b) NCD with symbol manipulation

c) NCD with arithmetic coprocessor 80287

Expression	number of terms of the result	Computing time (in seconds) using method a)	b)	c)
$(x+y+z)^3$	17	0.9	3.5	2.4
$(x+y+z)^4$	34	3.5	17.5	8.8
$(x+y+z)^5$	52	18	72	29
$(x+y+z)^6$	83	102	220	82
$(x+y+z)^7$	113	†	630	222
$(x+y+z)^8$	164	†	†	540

All calculations were performed at an IBM-AT-286 (6 MHz, 640 kByte).

4. References

1. Buchberger B., Gröbner bases: an algorithmic method in polynomial ideal theory. Chapter 6 in: Recent trends in multidimensional systems. (ed.: N.K.Bose), D.Reidel Publ.Comp.,1985.
2. Mora T., Gröbner bases for non-commutative polynomial rings. In L.N.C.S. 229, pp. 353-362, 1986.
3. Apel J., Gröbner bases of non-commutative algebras and its application. (in German) PhD thesis, University of Leipzig, 1988.
4. Apel J., A relationship between Gröbner bases of ideals and modules of G-algebras. to appear in Proc. of Malcev-conference, Novosibirsk, 1989.
5. Hearn A., REDUCE user's manual. The Rand Corporation, Santa Monica, 1987.
6. Davenport J.H., Trager B.M., Scratchpad's view of algebra. Bath Computer Science Technical Report 90-31, University of Bath, 1990.
7. Klaus U., A virtual mashine for a computer algebra system. (in German) PhD thesis, University of Leipzig, 1988.

† computation interrupted without success after 15 minutes

Size Efficient Parallel Algebraic Circuits for Partial Derivatives*

Erich Kaltofen[1] and *Michael F. Singer*[2]

[1]Department of Computer Science, Rensselaer Polytechnic Institute,
Troy, NY 12180-3590; `kaltofen@cs.rpi.edu`

[2]Department of Mathematics, North Carolina State University,
Raleigh, NC 27695-8205; `singer@matagh.ncsu.edu`

ABSTRACT. Given an algebraic circuit or straight-line program of depth d that computes multivariate rational function in l arithmetic operations (additions, multiplications, and divisions), we construct circuits that compute

(1) the first k partial derivatives in a single variable with depth $O(\log(k)(d + \log(k)))$ using $O(k\log(k)\log(\log k)d)$ arithmetic operations;

(2) all first partial derivatives in depth $O(d)$ using no more than $4l$ arithmetic operations.

Our first result is based on Taylor series expansion and essentially parallelizes the Leibniz formula. Our second result parallelizes a construction by Baur and Strassen. A crucial ingredient to the parallel solution is the fact that bounded fan-in computation graphs can be transformed to those where the fan-out is bounded as well while increasing the depth by no more than a constant factor.

1. Introduction

Transformations on structured computations of rational multivariate functions are a commonplace tool for sequential and parallel algebraic algorithm design and complexity analysis. One of the first results is Brent's (1974) transformation of an expression tree to one of logarithmic depth. That transformation generically produces parallel code for expression evaluation and has since then been studied extensively (Miller and Reif 1989), (Kosaraju and Delcher 1988), and (Cole and Vishkin 1988). Surprisingly, the tree contraction technique introduced by Miller and Reif can be generalized to reduce the depth of an algebraic computation graph or a straight-line program (introduced by Strassen (1972) as a model for algebraic computation) to poly-logarithmic in size and degree (Valiant et al 1984), (Miller et al 1986), and (Kaltofen 1988). Another transformation, inspired by the problem of reducing the inversion of a matrix to the computation of its determinant, was found by Baur and Strassen (1983) who show that if one is given a multivariate rational function by a straight-line program, then one can simultaneously compute all first order partial derivatives with only a constant factor growth in the length of the computation. The motivation of this article is to investigate how one might apply derivatives to a straight-line computation of small depth, such as the ones produced by the parallelization methods mentioned before, without substantially increasing the depth while keeping the size minimal.

Baur and Strassen's (loc. cit.) clever construction does not increase the size by more than a constant factor, but it can produce a straight-line program of depth as large as the number

*This material is based on work supported by the National Science Foundation under Grant No. CCR-87-05363 and under Grant No. CDA-8805910 (first author).

of instructions in the straight-line program. In this paper we give a modified transformation that increases the length of the program for all first order partial derivatives of the final function computed by a factor of no more than 4 compared to the original program. Much more significantly our construction retains the depth of the original program within a constant factor. Our construction makes use of another computation graph transformation by Hoover, Klawe, and Pippenger (1984) who show how a bounded fan-in, unbounded fan-out circuit can be transformed into one with bounded fan-out as well by increasing both size and depth by a constant factor only. Their construction essentially adds specially balanced trees of duplication nodes behind nodes of high fan-out. One key ingredient in our transformation is the observation that the circuit which accumulates all partial derivatives from a coarse view-point is a mirror image of the circuit for computing the function itself. Therefore, data flows into dual nodes of high fan-in whose preimage counterparts have high fan out. These nodes only need to perform additions, so we can do these additions in the nodes that in Hoover's et al. construction were just duplication nodes, and therefore do not increase the length at all.

Our second result deals with computing the k-th order derivative in a single variable. Our earlier solution (Kaltofen 1987) made use of the Leibniz formula, which not only increases the length of the program by a factor of $O(k^2)$, but also determines the i-th derivative from the $(i-1)$-st one and therefore multiplies the depth by a factor of $O(k \log(k))$. Our new solution is based on the simple observation that higher derivatives can be read off from the coefficients of the power series expansion in the variable to differentiated by (see (Aho et al. 1975)). In our case, we need to compute Taylor series expansions in a single variable of multivariate rational functions given by straight-line programs. This problem, however, has been studied thoroughly (Strassen 1973) and (Kaltofen 1988). Indeed, by pushing the divisions back to a final single one, we can compute all derivatives in a single variable up to order k by a program whose length has increased by a factor of order $O(k \log(k) \log(\log k))$ and whose depth by roughly a factor of order $O(\log(k))$.

A straight-line program of length l is a sequence of assignment

$$v_i \leftarrow v_i' \circ_i v_i'', \quad 1 \le i \le l,$$

where v_i is a new program variable,

$$v_i', v_i'' \in \mathsf{K} \cup \bigcup_{j=1}^{n} \{x_j\} \cup \bigcup_{k=1}^{i-1} \{v_k\},$$

and the operation $\circ_i$ is either $+$, $-$, $\times$, or $\div$. In this paper we assume that K is an arbitrary field. Every v_i therefore computes a rational function in $\mathsf{K}(x_1, \ldots, x_n)$ and it is not permitted to divide by a v_i'' that is identically 0. All our transformation necessarily must preserve this condition. The depth of each variable is defined recursively as

$$\operatorname{depth}(v_i) = 1 + \max\{\operatorname{depth}(v_i'), \operatorname{depth}(v_i'')\},$$

where depth of operands $v_i', v_i'' \in \mathsf{K} \cup \{x_1, \ldots, x_n\}$ is defined to be zero. The depth of a program is the maximum of the depths of the variables. The depth is a measure of the time it takes to evaluate a straight-line program in parallel. Given l processors, each of which evaluates a variable as soon as the values of their operands are known, the process of evaluating the entire program takes the depth of the straight-line program many rounds.

By $M(n)$ we denote the number of arithmetic operations it takes to multiply two univariate polynomials over K of degree no more than n. By the result in (Cantor and Kaltofen 1987) there is a straight-line program without divisions of length $M(n) = O(n \log(n) \log(\log n))$ and depth $O(\log(n))$ that can compute the coefficients of the product polynomial for this problem.

2. Differential Fields

The results in this paper deal with partial derivatives on multivariate rational functions. If we have a subfield of the complex numbers as the coefficient field, partial derivatives can be defined in an analytic setting by considering the functions as multivariate analytic complex functions. In particular, we obtain the Laurent series representation of such functions, as well as rules on differentiation such as the chain rule. However, we wish to include coefficient fields of finite characteristic in our results as well as use derivatives from a purely algebraic point of view with no analytic interpretation. Therefore, we will appeal to the algebraic theory of differential fields (Kaplansky 1957). Since we only need the very basics of that theory and since we wish to make this article somewhat self-contained, we discuss the necessary definitions and lemmas in this section.

Let R be a commutative ring (with unity). A *derivative* ∂ on R is a map from R into itself such that

$$\forall r, s \in R\colon \partial(r+s) = \partial(r) + \partial(s), \quad \partial(rs) = \partial(r)s + r\partial(s).$$

A *differential ring* is a ring together with a family of derivatives $\{\partial_i\}_{i \in I}$. It is surprising that most of the properties of function fields with derivatives can be modelled by this definition. We shall state the ones we need.

Let 1 be the unit in R. Then

$$\partial(1) = \partial(1^2) = 2 \cdot 1 \cdot \partial(1) = 0.$$

We define as the *ring of constants* of R

$$C_R = \{c \mid c \in R \text{ and } \forall i \in I\colon \partial_i(c) = 0\},$$

which thus is a subring of R.

Let $r \in R$ posses a multiplicative inverse r^{-1}. Then

$$0 = \partial(rr^{-1}) = \partial(r)r^{-1} + r\partial(r^{-1}), \quad \text{hence } \partial(r^{-1}) = -\partial(r)r^{-2}.$$

Therefore, $\partial(r^n) = nr^{n-1}\partial(r)$ for all such r and all $n \in \mathsf{Z}$.

Lemma 2.1. *Let the differential ring be an integral domain. Then any derivative ∂ extends uniquely to the quotient field of R,* QF(R).

Proof. The embedding of R into QF(R) requires us to define $\partial([r/1]) = [\partial(r), 1]$, where $r \in R$ and $[r/s]$ denotes the equivalence class for the pair r/s. Since for $r \neq 0$ the class $[1/r]$ is the multiplicative inverse of $[r/1]$, we must have $\partial([1/r]) = -[\partial(r)/1]\,[1/r^2]$. Hence we must define

$$\forall r, s \in R, s \neq 0\colon \partial([r/s]) = [(\partial(r)s - r\partial(s))/s^2].$$

One now can show easily that this definition is independent of the choice of representative of $[r/s]$, and in fact constitutes a derivative on QF(R). □

We now extend derivatives to series rings constructed from R. The first would be the ring of polynomials over R, $R[x]$. It turns out that one can give $\partial(x)$ any value in $R[x]$ and then uniquely extend a derivative on R to one on $R[x]$. If R is an integral domain, the same holds true for $R(x)$, the quotient field of $R[x]$. In our applications, R will always be a field K, possibly of positive characteristic.

It is useful to embed $\mathsf{K}(x)$ into the field of *extended formal power series*,

$$\mathsf{K}((x)) := \bigcup_{k \geq 0} \frac{1}{x^k} \mathsf{K}[[x]],$$

where $K[[x]]$ is the domain of formal power series over K. The embedding ι requires the identification of polynomial inverses with their power series, that is

$$\iota\left(\frac{1}{a_0 + a_1x + \cdots + a_dx^d}\right) = b_0 + b_1x + b_2x^2 + \cdots + b_ix^i + \cdots, \quad a_0 \neq 0, b_0 = \frac{1}{a_0},$$

where with $a_l = 0$ for $l > d$ we have

$$\forall i \geq 1: b_i = \frac{1}{a_0}(-b_0a_i - b_1a_{i-1} - \cdots - b_{i-1}a_1).$$

We first discuss how to computed derivatives in $K((x))$.

Lemma 2.2. *For any* $g \in K((x))$

$$\partial\left(\sum_{i=-k}^{+\infty} a_ix^i\right) = -\frac{ka_{-k}g}{x^{k+1}} + \sum_{i=-k}^{+\infty}\left(\partial(a_i) + (i+1)a_{i+1}g\right)x^i, \quad a_i \in K, k \geq 0,$$

defines a derivative on $K((x))$ *with* $\partial(x) = g$. □

Note that this lemma allows us the transfer of a derivative ∂' on $K(x)$ to a derivative ∂ on $K((x))$ by defining $\partial(x) = \iota(\partial'(x))$. These derivatives agree on K.

An important special case is that where $\partial(x) = 1$ and K is a subfield of the constant field, i.e., $\partial(a) = 0$ for all $a \in K$. This is because for $K = C$, the field of complex numbers, this derivative agrees with the analytically defined one. In fact, one can by purely algebraic means establish the construction of multivariate MacLaurin series expansions.

Lemma 2.3. *Let* $K(x_1, \ldots, x_n)$ *be endowed with a family of standard partial derivatives* $\partial_{x_1}, \ldots, \partial_{x_n}$ *such that*

$$\forall i: \partial_{x_i}(K) = \{0\},\ \partial_{x_i}(x_i) = 1,\ \partial_{x_i}(x_j) = 0 \text{ for } j \neq i.$$

Consider the multivariate power series expansion for f/g *where* $f, g \in K[x_1, \ldots, x_n]$, $g(0, \ldots, 0) \neq 0$,

$$\frac{f(x_1, \ldots, x_n)}{g(x_1, \ldots, x_n)} = \sum_{i_1 \geq 0, \ldots, i_n \geq 0} b_{i_1,\ldots,i_n} x_1^{i_1} \cdots x_n^{i_n}.$$

Then

$$\forall i_1 \geq 0, \ldots, i_n \geq 0: \left(\partial_{x_1}^{i_1}\partial_{x_2}^{i_2} \cdots \partial_{x_n}^{i_n}\frac{f}{g}\right)(0, \ldots, 0) = (i_1!) \cdots (i_n!)\, b_{i_1,\ldots,i_n},$$

where $\partial_{x_j}^i$ *is applying* ∂_{x_j} *a sequence of* i *times.*

Proof. First we note that the family of derivatives is also valid in $K((x_1, \ldots, x_n))$. Hence by lemma 2.2 we have

$$\partial_{x_1}^{i_1}\partial_{x_2}^{i_2} \cdots \partial_{x_n}^{i_n}\left(\sum_{i_1 \geq 0, \ldots, i_n \geq 0} b_{i_1,\ldots,i_n} x_1^{i_1} \cdots x_n^{i_n}\right)$$
$$= (i_1!) \cdots (i_n!) b_{i_1,\ldots,i_n} + \sum_{l_1 > i_1, \ldots, l_n > i_n} l_1^{\underline{i_1}} \cdots l_n^{\underline{i_n}} b_{l_1,\ldots,l_n} x_1^{l_1 - i_1} \cdots x_n^{l_n - i_n},$$

where $l^{\underline{i}}$ denotes $l(l-1)\cdots(l-i+1)$. Plugging in 0 for x_j proves the lemma. □

We finally prove the chain rule for rational functions.

Lemma 2.4. *Let the derivatives $\partial_{x_1}, \ldots, \partial_{x_n}$ on $\mathsf{K}[x_1, \ldots, x_n]$ be as in lemma 2.3. Similary, define the standard derivatives $\partial_{y_1}, \ldots, \partial_{y_m}$ on $\mathsf{K}[y_1, \ldots, y_m]$. Let $f \in \mathsf{K}(y_1, \ldots, y_m)$ and let $g_1, \ldots, g_m \in \mathsf{K}(x_1, \ldots, x_n)$. Assume that the denominator of f does not become zero in $\mathsf{K}(x_1, \ldots, x_n)$ by setting y_j to $g_j(x_1, \ldots, x_n)$ for all $1 \leq j \leq m$. Then*

$$\partial_{x_i}\Big(f(g_1, \ldots, g_m)\Big) = \sum_{j=1}^{m} (\partial_{y_j} f)(g_1, \ldots, g_m)\, \partial_{x_i}(g_j)$$

Note that by the quotient rule in the proof of lemma 2.1 no zero division occurs in any of the $\partial_{y_j}(f)(g_1, \ldots, g_m)$.

Proof. For $h \in \mathsf{K}(x_1, \ldots, x_n)$ consider

$$\bar{h}(z_1, \ldots, z_n) = h(X_1 + z_1, \ldots, X_n + z_n) \in \Big(\mathsf{K}(X_1, \ldots, X_n)\Big)(z_1, \ldots, z_n).$$

We first note that

$$(\partial_{z_i}\bar{h})(0, \ldots, 0) = (\partial_{x_i} h)(X_1, \ldots, X_n).$$

Hence by lemma 2.3 we have

$$h(x_1 + z_1, \ldots, x_n + z_n) = h(x_1, \ldots, x_n) + \sum_{i=1}^{n} (\partial_{x_i} h) z_i + \cdots.$$

We now apply this expansion to both f and all g_j and obtain

$$f(y_1 + u_1, \ldots, y_m + u_m) = f(y_1, \ldots, y_m) + \sum_{j=1}^{m} (\partial_{y_j} f)\, u_j + \cdots$$

and for $j = 1, \ldots, n$,

$$g_j(x_1 + z_1, \ldots, x_n + z_n) = g_j(x_1, \ldots, x_n) + \sum_{i=1}^{n} (\partial_{x_i} g_j)\, z_i + \cdots.$$

Now we substitute g_j for y_j and $\sum_{i=1}^{n} (\partial_{x_i} g_j)\, z_i + \cdots$ for u_j in the expansion for f. This substitution is valid because for $z_1 = \cdots = z_n = 0$ no zero division occurs by assumption. Considering $h = f(g_1, \ldots, g_m) \in \mathsf{K}[x_1, \ldots, x_n]$ we need to collect the coefficient of z_i in

$$h(x_1 + z_1, \ldots, x_n + z_1) = f(g_1(x_1 + z_1, \ldots, x_n + z_n), \ldots, g_m(x_1 + z_1, \ldots, x_n + z_n))$$

to obtain $\partial_{x_i} h$. By the above substitution this coefficient is

$$\sum_{j=1}^{m} (\partial_{y_j} f)(g_1, \ldots, g_m)\, \partial_{x_i}(g_j),$$

which proves the lemma. □

3. Higher Order Partial Derivatives in a Single Variable

Given $f \in \mathsf{K}(x_1, \ldots, x_n)$ by a straight-line program P, at issue is to construct a straight-line program Q that computes $\partial^k_{x_1}(f)$. Our first solution to this problem used the Leibniz formula for the higher order product rule (Kaltofen 1987). We briefly explain the approach. Let $w \leftarrow u \circ v$ be an instruction in P. The new program Q introduces variables $u^{(j)}$, $v^{(j)}$, $w^{(j)}$, $0 \le j \le k$, such that $w^{(j)}$ computes $\partial^j_{x_1}(w)$. Clearly, we then must have $w^{(0)} \leftarrow u^{(0)} \circ v^{(0)}$ and for $j \ge 1$ and

$$\text{for } \circ = +\text{: } w^{(j)} = u^{(j)} + v^{(j)},$$

$$\text{for } \circ = \times\text{: } w^{(j)} = \sum_{m=0}^{j} \binom{j}{m} u^{(j-m)} v^{(m)},$$

$$\text{and for } \circ = \div\text{: } w^{(j)} = \left(u^{(j)} - \sum_{m=1}^{j} \binom{j}{m} w^{(j-m)} v^{(m)} \right) \bigg/ v^{(0)}.$$

Therefore, $w^{(j)}$ can be computed from

$$u^{(0)}, \ldots, u^{(j)}, v^{(0)}, \ldots, v^{(j)}, w^{(0)}, \ldots, w^{(j-1)}$$

in $\mathrm{O}(j)$ assignments, compiling to a total of $\mathrm{O}(k^2)$ assignments for $w^{(k)}$.

For $k = 1$, this approach is the best we know. In particular, the quotient rule given above only requires 3 additional instructions, leading to a program Q of length no more than $4l$, where l is the length of the original program P. For large k, this approach is, however, not very efficient. Not only does it experience quadratic growth in size, but also leads a program Q of depth $\mathrm{O}(k \log(k)\, d)$, where d is the depth of the original program P. Our new approach is based on reading the derivatives off a Taylor series expansion (see lemma 2.3) and reduces both size and depth of Q. We have the following theorem.

Theorem 3.1. *Let $f \in \mathsf{K}(x_1, \ldots, x_n)$ be given by a straight-line program of length l and depth d. Then f and all derivatives $\partial_{x_1}(f), \partial^2_{x_1}(f), \ldots, \partial^k_{x_1}(f)$, $k \ge 1$, can be computed by a straight-line program Q of length $\mathrm{O}(\mathrm{M}(k)\, l)$ and depth $\mathrm{O}(\log(k)(d + \log(k)))$.*

Proof. Consider the straight-line program P_y that computes $f(x_1 + y, x_2, \ldots, x_n) \in \mathsf{K}(x_1, \ldots, x_n, y)$. By the algorithm Taylor Series Coefficients (Kaltofen 1988, Section 7) we can transform P_y into a straight-line program Q that computes the coefficients $b_j(x_1, \ldots, x_n) \in \mathsf{K}(x_1, \ldots, x_n)$, $0 \le j \le k$, of the expansion

$$f(x_1 + y, x_2, \ldots, x_n) = \sum_{j=0}^{\infty} b_j(x_1, \ldots, x_n) y^j.$$

By the argument given in the proof of lemma 2.4 we then get

$$\partial^j_{x_1}(f) = j!\, b_j(x_1, \ldots, x_n).$$

The program Q essentially keeps for each program variable w the truncated power series coefficients in y in additional variables. Clearly, if we have the assignment $w \leftarrow u \circ v$, then we can compute the coefficients of w from those of u and v. More precisely, let

$$u = \sum_{j=0}^{k} u^{(j)} y^j + \mathrm{O}(y^{k+1}),\ v = \sum_{j=0}^{k} v^{(j)} y^j + \mathrm{O}(y^{k+1}),\ \text{and } w = \sum_{j=0}^{k} w^{(j)} y^j + \mathrm{O}(y^{k+1}).$$

We will have $w^{(0)} \leftarrow u^{(0)} \circ v^{(0)}$ and for $j \geq 1$ and

$$\text{for } \circ = +\text{: } w^{(j)} = u^{(j)} + v^{(j)},$$

$$\text{and for } \circ = \times\text{: } w^{(j)} = \sum_{m=0}^{j} u^{(j-m)} v^{(m)}.$$

For $\circ = \div$ we are reqired to perform power series inversion

$$\left(\sum_{j=0}^{k} v^{(j)} y^j + O(y^{k+1}) \right)^{-1},$$

which in this case can be done since $v^{(0)} = v \neq 0$, because the original program P divides by v. By polynomial multiplication both the case $\circ = \times$ and the case $\circ = \div$ can be done in $O(M(k))$ instructions (Sieveking 1972), (Kung 1974). However, not for any field K can $O(\log(k))$ depth for power series inversion be easily accomplished while asymptotically retaining the instruction count.

We finally show how all but a division in the last instruction can be replaced by additions and multiplications. The idea is already discussed in (Strassen 1973) and introduces for each variable w in P two variables $w^{(1)}$ and $w^{(2)}$ such that both will compute polynomials in $K[x_1, \ldots, x_n]$ and $w = w^{(1)}/w^{(2)}$. Given an instruction $w \leftarrow u \circ v$ we now can compute, e.g.,

$$\text{for } \circ = +\text{: } w^{(1)} = u^{(1)} v^{(2)} + v^{(1)} u^{(2)}, \; w^{(2)} = u^{(2)} v^{(2)}.$$

This transformation does not asymptotically increase either length of depth. Now we apply our transformation to the new program with one division. By using Newton iteration to resolve the final truncated power series inversion we will then add $O(M(k))$ length and $O((\log k)^2)$ depth to the program Q, whose preceding $O(M(k)\, l)$ instructions perform $O(l)$ truncated power series additions and multiplications at a depth of $O(\log(k)\, d)$. □

4. Multiple First Order Derivatives

Given be a straight-line program of length l and depth d that computes $f \in K(x_1, \ldots, x_n)$. In the previous section we developed an algorithm that allows to compute all

$$\partial^k_{x_1}(f), \partial^k_{x_2}(f), \ldots, \partial^k_{x_n}(f)$$

by a straight-line program of length $O\big(nl\, M(k)\big)$ and depth $O\big(\log(k)\,(d + \log(k)\big)$. In this section we treat the case $k = 1$ more closely. We will show that one can then remove the factor n from the length while retaining depth $O(d)$. Our construction is based on the result by Baur and Strassen (1983) who establish a transformation to compute all first order partial derivatives with an increase in length by only a scalar multiple. However, their transformation is not depth preserving. By analyzing their construction more closely, we can prove the following theorem.

Theorem 4.1. *Let $f \in K(x_1, \ldots, x_n)$ be computed by a straight-line program P of length l and depth d. Then f and all derivatives $\partial_{x_1}(f), \partial_{x_2}(f), \ldots, \partial_{x_n}(f)$ can be computed by a straight-line program Q of length no more than $4\,l$ and depth $O(d \log(t))$, where t is the maximum number of times any variable is used as an operand in P, i.e., the maximum "fan-out" in the program P.*

Proof. Let $v_i \leftarrow v_{I_1(i)} \circ_i v_{I_2(i)}$, $n+1 \le i \le n+l$, be the $(i-n)$-th instruction in the program P. Here the function I_1 retrieves the index of the left operand of right-hand side expression, and the function I_2 the right operand. We set $v_i := x_i$ for $1 \le i \le n$, hence have a range for the operand indices of $1 \le I_1(i), I_2(i) < i$. If the left or right operands are scalars, no such indexing will be needed. For $i > n$ the symbol v_i stands, strictly speaking, for a program variable in P, or a node in the computation DAG for f. However, we also use it to identify with it the rational function in $\mathsf{K}(x_1, \ldots, x_n)$ that is computed in this variable. Baur and Strassen's construction proceeds by viewing f as a sequence of functions

$$g_i(y_1, \ldots, y_i) \in \mathsf{K}(y_1, \ldots, y_i), \quad n+l \ge i \ge n.$$

We will have

$$g_i(v_1, \ldots, v_i) = f(x_1, \ldots, x_n) \text{ for all } n \le i \le n+l.$$

The interpretation of g_i is the function that gets computed in v_l if we cut off the instructions for $v_{n+1}, \ldots, v_i$ in the program P and set $v_1, \ldots, v_i$ to the variables $y_1, \ldots, y_i$ whenever the occur in the truncated program. In particular, we want to have

$$g_n(x_1, \ldots, x_n) = f(x_1, \ldots, x_n).$$

Now let

$$h_i(y_{I_1(i)}, y_{I_2(i)}) = y_{I_1(i)} \circ_i y_{I_2(i)} \in \mathsf{K}(y_1, \ldots, y_{i-1}), \quad n+l \ge i \ge n+1,$$

denote the rational function that gets formally computed by the $(i-n)$-th instruction in P. The functions g_i are therefore inductively defined to be related by

$$g_{i-1}(y_1, \ldots, y_{i-1}) := g_i\Big(y_1, \ldots, y_{i-1}, h_i(y_{I_1(i)}, y_{I_2(i)})\Big), \tag{$*$}$$

where initially $g_{n+l}(y_1, \ldots, y_{n+l}) := y_{n+l}$. The goal is to compute

$$\partial_{y_1}(g_n), \partial_{y_2}(g_n), \ldots, \partial_{y_n}(g_n).$$

This is done by using the inductive definition of g_i and the chain rule (lemma 2.4). We first have

$$\partial_{y_j}(g_{n+l}) = 0 \text{ for all } 1 \le j \le n+l-1, \quad \partial_{y_{n+l}}(g_{n+l}) = 1.$$

Now lets assume that at level $n+l-i$ we already compute the derivatives

$$\partial_{y_1}(g_i), \partial_{y_2}(g_i), \ldots, \partial_{y_i}(g_i).$$

From $(*)$ we get by the chain rule for $j_1 := I_1(i)$ and $j_2 := I_2(i)$ that

$$(\partial_{y_k} g_{i-1})(y_1, \ldots, y_{i-1}) = (\partial_{y_k} g_i)(y_1, \ldots, y_{i-1}, h_i(y_{j_1}, y_{j_2})) \text{ for } 1 \le k \le i-1, k \ne j_1, k \ne j_2$$

and

$$\begin{aligned}(\partial_{y_j} g_{i-1})(y_1, \ldots, y_{i-1}) &= (\partial_{y_j} g_i)(y_1, \ldots, y_{i-1}, h_i(y_{j_1}, y_{j_2})) \\ &\quad + (\partial_{y_i} g_i)(y_1, \ldots, y_{i-1}, h_i(y_{j_1}, y_{j_2}))(\partial_{y_j} h_i)(y_{j_1}, y_{j_2})\end{aligned}$$

for $j = j_1$ or $j = j_2$.

The dynamics of these rules are displayed in figure 1. The substitution $h_i(y_{j_1}, y_{j_2})$ for y_i is accomplished by connecting the corresponding node to the nodes for y_{j_1} and y_{j_2} and performing the operation $\circ_i$ in the node. Then the derivatives $\partial_{y_j}(g_{i-1})$ are computed from those of $\partial_{y_j}(g_i)$ plus a value derived from

$$\overline{\partial_{y_i}(g_i)} = (\partial_{y_i} g_i)(y_1, \ldots, y_{i-1}, h_i(y_{j_1}, y_{j_2}))$$

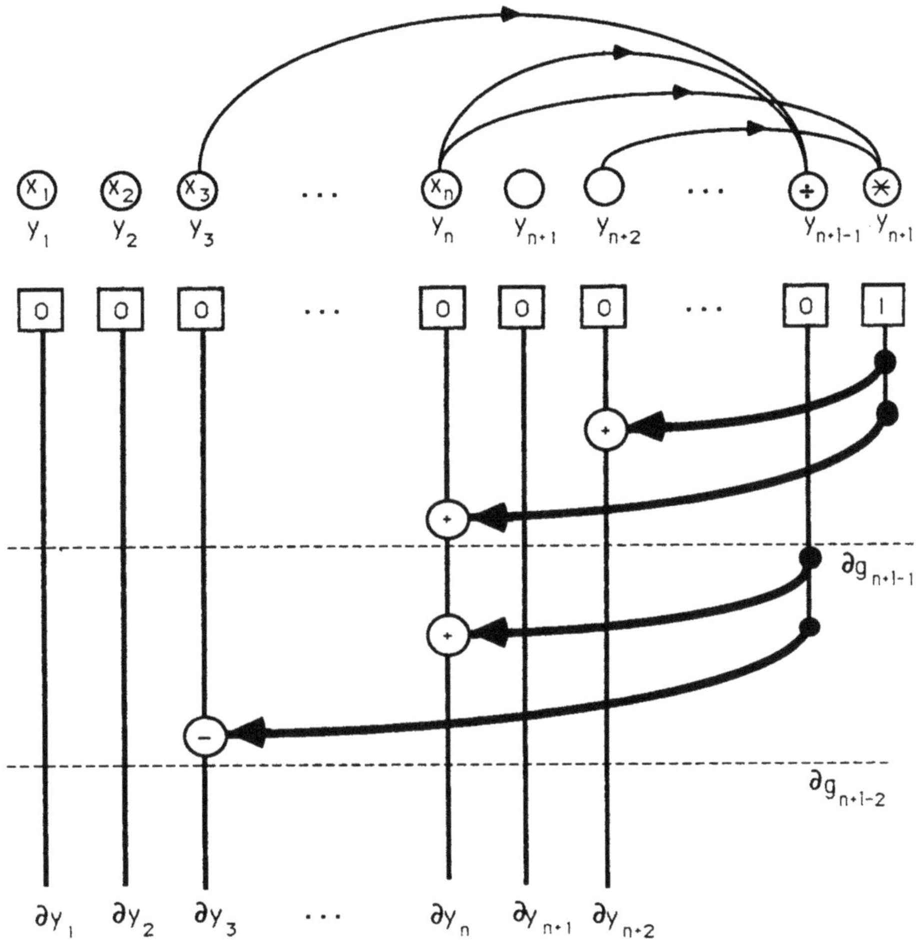

Figure 1: Coarse View of the Baur and Strassen Construction

and the derivatives of $h_i(y_{j_1}, y_{j_2})$. The latter are solely dependent on the nodes corresponding to the variables y_{j_1} and y_{j_2} and require constant work. In figure 1, this is indicated by a thick connection from the line for ∂_{y_i} to ∂_{y_j}. Let us for a moment consider the operation $\circ_i$ with the most costly work, namely division. For

$$h_i(y_{j_1}, y_{j_2}) = \frac{y_{j_1}}{y_{j_2}} \text{ we have } \partial_{y_{j_1}}(h_i) = \frac{1}{y_{j_2}} \text{ and } \partial_{y_{j_2}}(h_i) = -\frac{y_{j_1}}{y_{j_2}^2}.$$

The strategy is to divide $\overline{\partial_{y_i}(g_i)}$ by y_{j_2}, add that into the $\partial_{y_{j_1}}$ line, or multiply it with y_{j_1}/y_{j_2}, the value computed in the node for y_i, and then subtract that from the ∂_{y_2} line. In other words, if $\circ_i = \div$ one needs 4 additional operations to go to the next level. There is one more issue that needs to be settled in the division case. Later substitutions for y_{j_2} must not cause a division by a function that is identical zero. This is true because the circuit computing the derivatives will only divide by quantities that the original program P divides by.

We now discuss how to accomplish the given length and depth measures. For each v_i in P, $n+1 \leq i \leq n+l$, we will introduce at most 5 instructions in our new program Q, one from the original program and at most 4 more to eliminate y_i. This leads to an upper bound of $5l$ for the length of Q, but l of these instructions either add the initial $\partial_{y_j}(g_{n+l}) = 0$ into $\overline{\partial_{y_i}(g_i)}(\partial_{y_j} h_i)(y_{j_1}, y_{j_2})$ or multiply $\partial_{y_{n+l}}(g_{n+l}) = 1$ into $\partial_{y_j}(h_{n+l})(y_{j_1}, y_{j_2})$. Since we have any instruction v_i participate in the computation of v_{n+l}, there are at least l trivial instructions that can be eliminated from such a Q. Note that if we only have subtractions on a line for ∂_{y_j} we pass the minus sign along to the level for the derivatives of g_{j-1}. On each line for ∂_{y_j}, $1 \leq j \leq n$, we might then have to negate the final result, costing us an additional instruction. However, we did not account for the savings at the starting level of those lines, and therefore we do not need more than $4\,l$ instructions overall. Lastly, we discuss how to accomplish the stated depth. First we observe that if we were to treat the lines in figure 1 on which we accumulate the ∂_{y_j} as single nodes, and if we were to treat the connections from ∂_{y_i} to $\partial_{y_{j_1}}$ and to $\partial_{y_{j_2}}$ as single edges, then the circuit to compute the derivatives would be a mirror image of the original circuit for f. Therefore, the depth of this view of the part of Q that implements the chain rules and which has "superedges" and "supernodes" is d. Now, on each "superedge" we only have a constant delay. Let t_j be the fan-out for v_j in P. Then in each supernode corresponding to the line for ∂_{y_j} we have exactly $t_j - 1$ addition and subtraction nodes. Separating the lines that get added from the ones that get subtracted, we can build with $t_j - 1$ nodes a tree that performs the same computation but which has $O(\log(t))$ depth (see figure 2). Hence the entire depth of Q can be made $O(\log(t)d)$. □

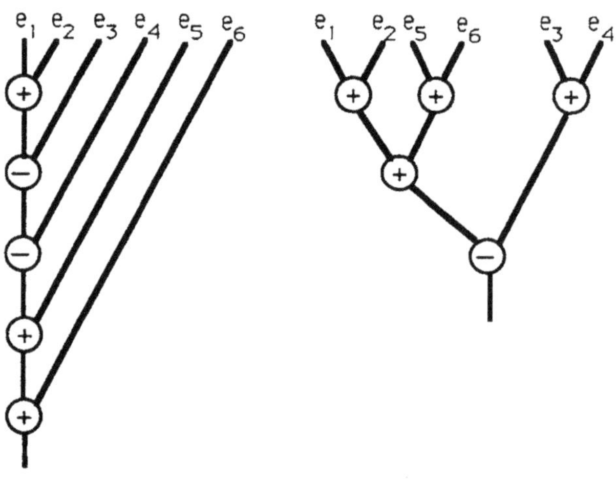

Figure 2: Balancing an Accumulation Tree

One can, at no expense in terms of length, reduce the depth of Q to $O(d)$. This follows from a construction by Hoover, Klawe, and Pippenger (1984) who show that for bounded fan-in circuits one can introduce duplication nodes behind the nodes of high fan out in such a way that both size and depth are preserved within a constant factor. In our case the situation is so special that we can accomplish this introduction of duplication nodes without increasing the size of the circuit.

Theorem 4.2. *Let f, P, l, and d be as in theorem 4.1. Then f and all derivatives $\partial_{x_1}(f)$, $\partial_{x_2}(f), \ldots, \partial_{x_n}(f)$ can be computed by a straight-line program Q of length no more than $4l$ and depth $O(d)$.*

Proof. We apply the transformation of Hoover et al. (loc. cit.) in an implicit way to the circuit constructed in the proof of theorem 4.1. Consider the lower part of Q that is a mirror image of P (see figure 1). Furthermore, assume that the subtrees in Q which perform additions on the ∂_{y_j} lines are again contracted to "supernodes". We suppose that subtractions are already separated out, and subtraction nodes on those lines remain untouched. Nonetheless, the depth of this abstraction of the circuit is still $O(d)$, the extra factor of $\log(t)$ in theorem 4.1 coming from the delay in the supernodes. Now we apply the construction by Hoover et al. to this view of the lower part of Q, reversing the flow of information. That construction will insert behind the nodes of high fan-out a binary tree of duplication nodes whose root is that node and whose leaves are the targets of the arcs leaving that node. Hoover et al. then show that if one optimizes the structure of that tree with respect to the distance of the target nodes to the output node in such a way that target nodes from which there are long paths to the output node are close to the root, one can overall retain depth $O(d)$. Once such duplication trees are in place behind the supernodes, all we have to do is reverse the flow of information and perform additions in both supernodes and duplication nodes. □

5. Conclusion

We have given size and depth efficient transformations from straight-line programs to such programs that compute either a higher order partial derivative in a single variable or all first order partial derivatives. These transformations are easily performed in polynomial time, but can also be executed in parallel themselves (for the tree optimization problem needed in the Hoover et al. fan-out reduction refer to (Teng 1987)). There are known computational limits to the derivation problem. The first discovered fact is that the formula size can grow exponentially if one allows exponentiation in the formula (Caviness and Epstein 1978). Even without this extra operator, the growth can be related to the $\#\mathcal{P}$-complete problem of evaluating the permanent. The example is the function given by the formula

$$f(y_1, \ldots, y_n, x_{1,1}, \ldots, x_{n,n}) = \prod_{i=1}^{n} \left(\sum_{j=1}^{n} y_j x_{i,j} \right)$$

of size $O(n^2)$, one of whose mixed iterated derivative is

$$\partial_{y_1} \cdots \partial_{y_n}(f) = \text{Permanent}(\left[x_{i,j}\right]_{1 \le i,j \le n}),$$

the generic permanent of the matrix in the $x_{i,j}$ (Valiant 1982). This means that if one had a solution with polynomial growth to transforming a straight-line program to one that computes iterated derivatives in different variables, then one would have a polynomial-time program for counting the number of satisfying assignments in a Boolean formula in random polynomial time, in particular one would have shown that $\mathcal{RP} = \#\mathcal{P}$. Applying Kronecker substitution to f one also can show that the k-th derivative in a single variable most likely cannot be computed in time polynomial in $\log(k)$ (Kaltofen 1987).

Following Baur and Strassen (1984) our result on simultaneously computing all first order partial derivatives was primarily motivated by the question on how to solve linear systems in parallel using an optimal number of processors. Given be a straight-line program† with $O(n^3)$ instructions of depth $O((\log n)^2)$ that computes the determinant

$$\Delta(x_{1,1}, \ldots, x_{n,n}) = \det(A), \quad A = \left[x_{i,j}\right]_{1 \le i,j \le n}.$$

Then the inverse of A can also be computed by a program of such asymptotic length and depth, namely

$$A^{-1} = \left[(-1)^{i+j} \frac{\partial_{x_{j,i}}(\Delta)}{\Delta}\right]_{1 \le i,j \le n}.$$

Morgenstern (1985) is one of the first to observe that the Baur and Strassen result generalizes to straight-line programs that also allow instructions like $v \leftarrow \exp(u)$, $v \leftarrow \log(u)$, $v \leftarrow \sqrt{u}$, etc. At least for fields K of characteristic zero, all our results can also be generalized to such an extended model of straight-line programs.

Acknowledgement: Part of the results were conceived in June 1988, while the authors were visiting the Research Institute for Symbolic Computation in Linz, Austria.

6. Literature Cited

AHO, A. V., STEIGLITZ, K., and ULLMAN, J. D., "Evaluating polynomials at fixed sets of points," *SIAM J. Comp.* **4**, pp. 533–539 (1975).

BAUR, W. and STRASSEN, V., "The complexity of partial derivatives," *Theoretical Comp. Sci.* **22**, pp. 317–330 (1983).

BRENT, R. P., "The parallel evaluation of general arithmetic expressions," *J. ACM* **21**, pp. 201–208 (1974).

CANTOR, D. G. and KALTOFEN, E., "Fast multiplication of polynomials over arbitrary rings," *Tech. Report* **87-35**, Dept. Comput. Sci., Rensselaer Polytechnic Institute, December 1987.

CAVINESS, B. F. and EPSTEIN, H. I., "A note on the complexity of algebraic differentiation," *Inf. Proc. Lett.* **7**, pp. 122–124 (1978).

CHISTOV, A. L., "Fast parallel calculation of the rank of matrices over a field of arbitrary characteristic," *Proc. FCT '85, Springer Lec. Notes Comp. Sci.* **199**, pp. 63–69 (1985).

COLE, R. and VISHKIN, U., "Optimal parallel algorithms for expression tree evaluation and list ranking," *Proc. AWOC 88, Springer Lec. Notes Comp. Sci.* **319**, pp. 91–100 (1988).

GALIL, Z. and PAN, V., "Parallel evaluation of the determinant and of the inverse of a matrix," *Inform. Process. Letters* **30**, pp. 41–45 (1989).

HOOVER, H. J., KLAWE, M. M., and PIPPENGER, N. J., "Bounding fan-out in logical networks," *J. ACM* **31**/1, pp. 13–18 (1984).

KALTOFEN, E., "Single-factor Hensel lifting and its application to the straight-line complexity of certain polynomials," *Proc. 19th Annual ACM Symp. Theory Comp.*, pp. 443–452 (1987).

KALTOFEN, E., "Greatest common divisors of polynomials given by straight-line programs," *J. ACM* **35**/1, pp. 231–264 (1988).

KAPLANSKY, I., *An Introduction to Differential Algebra*; Hermann, Paris, 1957.

KOSARAJU, S. R. and DELCHER, A. L., "Optimal parallel evaluation of tree-structured computations by raking," *Proc. AWOC 88, Springer Lec. Notes Comp. Sci.* **319**, pp. 101–110 (1988).

†To our knowledge, no such program is known at this time. The best field independent solution of depth $O((\log n)^2)$ appears to be Chistov's (1985) with length $O(n^4 \log(n))$, and the best solution for fields of characteristic zero appears to be Preparata's and Sarwate's (1978) with length $O(n^3\sqrt{n})$. In both cases fast matrix multiplication will improve the stated lengths (see also (Galil and Pan 1989)).

KUNG, H. T., "On computing reciprocals of power series," *Numer. Math.* **22**, pp. 341–348 (1974).

MILLER, G. L., RAMACHANDRAN, V., and KALTOFEN, E., "Efficient parallel evaluation of straight-line code and arithmetic circuits," *SIAM J. Comput.* **17/4**, pp. 687–695 (1988).

MILLER, G. L. and REIF, J. H., "Parallel tree contraction Part 1: Fundamentals," in *Randomness in Computation*, Advances in Computing Research **5**, edited by S. Micali; JAI Press Inc., Greenwich, CT., pp. 47–72, 1989.

MORGENSTERN, J., "How to compute fast a function and all its derivations," *SIGACT News* **16/4**, pp. 60–62 (1985).

PREPARATA, F. P. and SARWATE, D. V., "An improved parallel processor bound in fast matrix inversion," *Inform. Process. Letters* **7/3**, pp. 148–150 (1978).

SIEVEKING, M., "An algorithm for division of power series," *Computing* **10**, pp. 153–156 (1972).

STRASSEN, V., "Berechnung und Programm I," *Acta Inf.* **1**, pp. 320–335 (1972). (In German).

STRASSEN, V., "Vermeidung von Divisionen," *J. reine u. angew. Math.* **264**, pp. 182–202 (1973). (In German).

TENG, S.-H., "The construction of Huffman-equivalent prefix code in $\mathcal{NC}$," *SIGACT News* **18/4**, pp. 54–61 (1987).

VALIANT, L., "Reducibility by algebraic projections," *L'Enseignement mathématique* **28**, pp. 253–268 (1982).

VALIANT, L., SKYUM, S., BERKOWITZ, S., and RACKOFF, C., "Fast parallel computation of polynomials using few processors," *SIAM J. Comp.* **12**, pp. 641–644 (1983).

THE EVALUATION OF INTEGRALS OF THE HYPERGEOMETRIC FUNCTIONS

Adamchik V.S., Luchko Yu.F., Marichev O.I.
(Byelorussian University, Minsk, USSR)

The problem of analytical calculating of integrals is, one of the most complicated in computer algebra systems. Spacious bibliography of the analytical methods for calculating of integrals is represented in the article [9]. Probably, the most effective and the simpliest algorithm of analytical integration was made by O.I. Marichev [5]. Later it was realised in the reference-books [6,7,8]. It allows us to calculate definite and indefinite integrals of the products of elementary and special functions of hypergeometric type. It embraces about 80 per cent of integrals which are included in the world reference-literature.

The idea of our algorithm is in the following. The initial integral is transformed to contour integral from the ratio of products of gamma-functions by means of Mellin transform and Parseval equality. The residue theorem is used for the calculating of the received integral which due to the strict rules results in sums of hypergeometric series. The value of integral itself and the integrand functions are the special cases of the Meijer's G-function [4,5,8].

Programming packet is realized in programming languages PASCAL and REDUCE. It also offers the opportunity of finding the values for some classical integral transforms (Laplace, Hankel, Fourier, and etc.). The REDUCE's part of packet contains the main properties of the well-known special functions, such as the Bessel and gamma-functions and kindred functions, Anger function, Weber function, Whittaker functions, generalized hypergeometric functions. Special place in the packet is occupied by Meijer's G-function for which the main properties such as finding the particular cases and representation by means of hypergeometric series are realized.

1. Preliminary information. G-function is defined by the next formula:

$$G_{pq}^{mn}(z) = G_{pq}^{mn}\left[z \,\middle|\, \begin{matrix} (a_p) \\ (b_q) \end{matrix} \right] = \frac{1}{2\pi i} \int_{\mathcal{L}} g(s)\, z^{-s}\, ds, \qquad (1)$$

where

$$g(s) = \frac{\prod_{j=1}^{m}\Gamma(b_j+s)\prod_{j=1}^{n}\Gamma(1-\alpha_j-s)}{\prod_{j=1+n}^{p}\Gamma(a_j+s)\prod_{j=1+m}^{q}\Gamma(1-b_j-s)}, \quad 0\le m\le q,\ 0\le n\le p. \tag{2}$$

The function (1) was introduced and studied by C.S.Meijer in number of the articles of 1936-1956 years and later it was called Meijer's G-function. In its definition it is supposed that the contour $\mathcal{L}$ leaves all poles $s= -b_j- \ell$, $j=1, 2,\dots, m$, $\ell = 0, 1, \dots$ of the integrand function at the left and all poles $s= 1-a_j+ \ell$, $j= 1, 2,\dots, n$, $\ell= 0, 1, 2,\dots$, at the right.

Mellin-Barnes integrals, i.e. Meijer's G-functions (1),(2) produce the full analytical functions. Any finite linear combinations of these functions we shall call hypergeometric type functions. The fullest table of participal cases of G-functions containing about 1000 formulae are given in the reference-book[8]. In particular the following representations are true:

$$(1-x)_+^{\alpha-1} = \begin{cases}(1-x)^{\alpha-1}, & 0<x<1\\ 0, & x>1\end{cases} = \Gamma(\alpha)G_{11}^{10}\left[x\,\middle|\,\begin{matrix}\alpha\\ 0\end{matrix}\right], \tag{3}$$

$$e^{-x}= G_{01}^{10}\left[x\,\middle|\,\begin{matrix}\cdot\\ 0\end{matrix}\right], \qquad \ln(1+x)= G_{22}^{12}\left[x\,\middle|\,\begin{matrix}1, & 1\\ 1, & 0\end{matrix}\right], \tag{4}$$

$$\arctan(x)=\frac{1}{2}\, G_{22}^{12}\left[x^2\,\middle|\,\begin{matrix}1/2, & 1\\ 1/2, & 0\end{matrix}\right], \quad \operatorname{erf}(x)=\pi^{-1/2}G_{12}^{11}\left[x^2\,\middle|\,\begin{matrix}1\\ 1/2,0\end{matrix}\right], \tag{5}$$

$$\begin{Bmatrix}\sin(x)\\ \cos(x)\end{Bmatrix} =\sqrt{\pi}\, G_{02}^{10}\left[x^2/4\,\middle|\,\begin{matrix}\cdot\\ (1\pm1)/4,(1\mp1)/4\end{matrix}\right], \tag{6}$$

$$J_\nu(x)= G_{02}^{10}\left[x^2/4\,\middle|\,\begin{matrix}\cdot\\ \nu/2,-\nu/2\end{matrix}\right], \quad K_\nu(x)=\frac{1}{2}\, G_{02}^{20}\left[x^2/4\,\middle|\,\begin{matrix}\cdot\\ \nu/2, & -\nu/2\end{matrix}\right], \tag{7}$$

2. The computer's realization of the algorithm. The dialoguering regime which was realized in our INTEGRATOR system gives the possibility to make easier the first difficultest step: to elucidate the question about belonging the integrand function to participal cases of G-function (1), (2) or to product of two such cases. For this aim the special basis table of the participal cases of G-function which are wroten on the language PASCAL is used. Any man may choose necessary functions from this table and to form integrand expression of the following integral

$$\int_a^b x^{\alpha-1}G_1(\sigma x)G_2(\omega x^{l/k})\, dx, \quad 0\le a<b\le\infty. \tag{8}$$

The second step allows us to determine the convergence of studing integral. The convergence is understanding in the classical sense or in the sense of existence of the main part of the integral. Our programm realizes the check-up of the number conditions and put on the corresponding restrictions to the letteral parameters. This and next parts are wroten on RLISP language.

The third step realizes direct calculating of the integral by means of the substitution of the concrete values of the parameters for this case into the formula (8). Here the usual parameters counting off is realized. In the result of it the value of the integral (8) through Meijer's G-function with concrete parameters is written out.

The fourth step in this case is the inverse to the first step and so it is very difficult. Received G-function is transformed here into more famous and the simplest functions if such a change is possible. Here three possible variants of actions are realized. According to the first variant the participal cases of G-function may be transformed into the elementary and special functions with the help of special basic table the fragment of which is designated by the formulae (3)-(7). Received special functions may permit further simplifications. The given bellow chain of the equalities demonstrate these operations:

$$G^{20}_{04}\left[x \,\middle|\, \dot{1/4},\ \dot{3/4},\ \dot{0},\ \dot{1/2}\right] = x^{1/8}\, J_{1/2}(4x^{1/4}) =$$

$$= x^{1/8}(2/\pi)^{1/2}(4\ x^{1/4})^{-1/2}\ \sin(4x^{1/4}) = \frac{1}{\sqrt{2\pi}}\ \sin(4x^{1/4}).$$

It should noted that the attempt of the creation of computer's methods for the operations with the highest transcendental functions has been undertaken earlier in the connection with the working out of the system MACSYMA. For example the algorithms of the transformation of hypergeometric functions into elementary and special functions under some conditions on the parameters have been built /3/. These algorithms were based on the application of recurring formulae. Unfortunately we don't know the results of this great work. But we can emphasize that the approach proposed here differs from the previous one by the highest degree of generality because it is based on Meijer's G-function but not its participal cases, i.e. generalized hypergeometric series.

The second variant of actions is based on the representations of the integral (8) through the linear combinations of the generalized hypergeometric functions. Here the conditions on the paramters excluding the logarithmic cases must be fulfiled.

The third variant gives the possibility to write out the values of the integral (8) in the logarithmic cases when the right part of (8) is expressed through the series and sums containing the derivatives of the gamma-function and the degrees of logarithmic function. These formulae are repsesented in the work[1,2].

These given four steps have some else branches and additions. For example, if one of functions from the left part of (8) takes the fixed value of the type (3)-(7), then the programm will give the possibility to calculate the indefinite integrals, the integrals of fractional order and classical integral transforms of Stieltjes, Hilbert, Laplace, Fourier, Hankel and others.

3. Demonstrational examples.

$$\int_1^{\infty} x^{1-v} * (x**2-1)^{b-1} * Y_v(c*x)\, dx$$

```
1: in "start.red"$
   INTEGRAL IS CONVERGENT UNDER CONDITIONS :
C > 0 ;
Re(B) > 0 ;
Re(- 4*B + 2*V + 3) > 0 ;

   THE INTERMEDIATE RESULT :

GFM( (2 0 1 3) (( - V - 1)/2) (( - 2*B + V)/2,( - V)/2,( - V - 1)/2)

C**2/4 )

   THE FINISH RESULT :
  B                                   B
(2 *NEUM( - B + V,C)*GAMMA(B))/(2*C  )


2: sub(b=v+1/2,ws);

  B                                 B
(2 *GAMMA(( 2*V + 1)/2)*SIN(C))/(C SQRT(PI)*C  )
```

Here

NEUM(V,Z) = $Y_v(z)$, GAMMA(Z) = $\Gamma(Z)$

References.

1. Adamchik V.S., Kolbig K.S. A definite integral of a product of two polylogarifhms, SIAM J. Math. Anal. 1988. Vol. 19, No 4, p.926-938.
2. Adamchik V.S., Marichev O.I. Representations of functions of hypergeometric type in logarithmic cases. Vestsi Akad. Navuk BSSR. Ser. Fiz.-Mat. Navuk, 1983, No 5, p. 29-35 (in Russian).
3. Lafferty E.L. MACSYMA. 1979, p. 465-481.
4. Luke Y.L. The Special Functions and Their Approximations. In 2 vls. N.Y., Acad. Press, 1969. Vol. 1, 349 p.
5. Marichev O.I. Handbook of Integral Transforms of Higher Transcendental Functions, theory and algorithmic tables. Chichester, Ellis Horwood Ltd., 1983, 336 p.
6. Prudnikov A.P., Brychkov Yu.A., Marichev O.I. Integrals and Series. Vol. 1: Elementary functions, Gordon and Breach Sci. Publ., N. Y., London, Tokyo, 1986, 798 p. (first ed. in Moscow, Nauka, 1981).
7. Prudnikov A.P., Brychkov Yu.A., Marichev O.I. Integrals and Series. Vol. 2: Special functions, Gordon and Breach Sci. Publ., N. Y., London, Tokyo, 1986, 750 p. (first ed. in Moscow, Nauka, 1983).
8. Prudnikov A.P., Brychkov Yu.A., Marichev O.I. Integrals and Series. Vol. 3: More special functions, Gordon and Breach Sci. Publ., N. Y., London, Tokyo, 1989, 800 p. (first ed. in Moscow, Nauka, 1986).
9. Prudnikov A.P., Brychkov Yu.A., Marichev O.I. Calculus of integrals and Mellin transform. Itogi Nauki i Tekhniki, Math. analysis. Akad. Nauk SSSR. Inst. Nauchn. i Tekhn. Informatsii, Moscow, 1989, Vol. 27, p. 3-146 (in Russian).

ON A DEFINITE INTEGRAL OF THE PRODUCT OF TWO POLYLOGARITHMS

K.S. Kölbig

CERN, 1211 Geneva 23, Switzerland

1. Introduction

Indefinite integration of elementary functions is now, due to the powerful Risch algorithm, an established part of most symbolic algebra systems. For the integration of special functions, however, in both the indefinite and definite case, only limited progress towards symbolic evaluation has been achieved. For example, Piquette and Van Buren [7] and Piquette [6] have presented a technique for the evaluation of indefinite integrals containing products of Bessel, Legendre, and Hermite functions which can be used for symbolic computation. Geddes and Scott [2] have discussed the evaluation of definite integrals which can be expressed as repeated derivatives of products of quotients of gamma and zeta functions. Other results in this area have been presented by the present author [3], [4].

Taking into account the many disparate methods which are used for the evaluation of definite integrals, it is hardly surprising that their application in symbolic algebra systems is still only just beginning. Serious obstacles are encountered when analysing systematically the known results for certain integrals with a view to their suitability for symbolic algebra; in particular when these results contain "infinite" expressions- for example power series with parameters. The following integral illustrates some of these problems.

2. The Integral

We consider the integral

$$I_{nm}(\alpha, \sigma, \omega, r) = \int_0^\infty x^{\alpha - 1} Li_n(-\sigma x) Li_m(-\omega x^r) dx \tag{2.1}$$

where

$$Li_k(z) = \sum_{j=1}^{\infty} \frac{z^j}{j^k} \quad (k \geq 2, |z| \leq 1) \tag{2.2}$$

is the polylogarithm of order k, with

$$Li_1(z) = -\ln(1-z) \tag{2.3}$$

as a degenerate case. This integral can be transformed into a Mellin-Barnes integral which can be evaluated by residue techniques [1]. We give here the result for convenience:

Theorem. *Let $r \neq 0$ be real, α, $\sigma \neq 0$, $\omega \neq 0$ be complex with $|\arg \sigma| < \pi$, $|\arg \omega| < \pi$ and let $\zeta = \sigma\omega^{-1/r}$. Further, with $\mathbb{N} = \{1,2,3,\dots\}$, let $\mathbb{N}_0 = \{0, \mathbb{N}\}$.*

$$h(x) = \begin{cases} 1 & \text{if } x \in \mathbb{N}, \\ 0 & \text{otherwise}, \end{cases} \qquad \bar{h}(x) = 1 - h(x);$$

$$H(x) = \begin{cases} 1 & \text{if } x > 0, \\ 0 & \text{if } x < 0, \end{cases}$$

and let $k, l, K \in \mathbb{N}$.

Then, for $|\zeta| \neq 1$ and

$$\begin{array}{ll} -1 - r < \operatorname{Re} \alpha < 0 & \text{if } r > 0, \\ -1 < \operatorname{Re} \alpha < -r & \text{if } r < 0, \end{array}$$

the integral (2.1) can, for $\alpha \neq 0$, be expressed in the form

$$\begin{aligned}
&\int_0^\infty x^{\alpha-1} Li_n(-\sigma x) Li_m(-ax^r)\,dx \\
&= (-1)^n(\pm \operatorname{sgn} r)^{m+1}\pi\sigma^{-\alpha} \sum_{|r|k \pm \alpha \notin \mathbb{N}_0} (-1)^k \csc\,[\pi(\alpha \pm |r|k)](\alpha \pm |r|k)^{-n} k^{-m} \zeta^{\mp |r|k} \\
&+ (\pm 1)^{n+1}(-1)^{m+n+1}\frac{\pi}{|r|}\omega^{-\alpha/r} r^m \sum_{(l \mp \alpha)/|r| \notin \mathbb{N}_0} (-1)^l \csc\left(\pi\frac{\alpha \mp l}{r}\right)(\alpha \mp l)^{-m} l^{-n} \zeta^{\mp l} \\
&+ (-1)^{n+1}(\pm \operatorname{sgn} r)^{m+1}\sigma^{-\alpha} \sum_{|r|K \pm \alpha \in \mathbb{N}} (-1)^{K + |r|K \pm \alpha}(\alpha \pm |r|K)^{-n} K^{-m} \zeta^{\mp |r|K} \\
&\left(\pm \frac{m}{|r|K} + \frac{n}{\alpha \pm |r|K} + \ln \zeta\right) + H(\mp r)(-1)^{n+1}\sigma^{-\alpha}\pi^{m+1} r^m \alpha^{-n} \\
&\left\{ h(\pm\alpha)(-1)^\alpha \sum_{m_1=0}^{m+1} \frac{1}{m_1!}\left(-\frac{1}{\pi}\ln \zeta\right)^{m_1} \right. \\
&\sum_{m_2=0}^{m+1-m_1} \binom{m_2+n-1}{n-1}(-\pi\alpha)^{-m_2} \sum_{m_3=0}^{m+1-m_1-m_2} r^{-m_3} B^*_{m_3} B^*_{m+1-m_1-m_2-m_3} \\
&+ \bar{h}(\pm\alpha) \sum_{m_1=0}^{m} \frac{1}{m_1!}\left(-\frac{1}{\pi}\ln \zeta\right)^{m_1} \\
&\left. \sum_{m_2=0}^{m-m_1} \binom{m_2+n-1}{n-1}(-\pi\alpha)^{-m_2} \sum_{m_3=0}^{m-m_1-m_2} r^{-m_3} B^*_{m_3} C_{m-m_1-m_2-m_3}(\pi\alpha) \right\}
\end{aligned} \tag{2.4}$$

$$+ H(\pm 1)(-1)^m(\text{sgn } r)\omega^{-\alpha/r}\pi^{n+1}r^m\alpha^{-m}$$

$$\left\{h\left(-\frac{\alpha}{|r|}\right)(-1)^{n+\alpha/r}\sum_{n_1=0}^{n+1}\frac{1}{n_1!}\left(-\frac{1}{\pi}\ln\zeta\right)^{n_1}\right.$$

$$\sum_{n_2=0}^{n+1-n_1}\binom{n_2+m-1}{m-1}(\pi\alpha)^{-n_2}\sum_{n_3=0}^{n+1-n_1-n_2} r^{-n_3}B^{*}_{n_3}B^{*}_{n+1-n_1-n_2-n_3}$$

$$-\bar{h}\left(-\frac{\alpha}{|r|}\right)r^{-n-1}\sum_{n_1=0}^{n}\frac{1}{n_1!}\left(\frac{r}{\pi}\ln\zeta\right)^{n_1}$$

$$\left.\sum_{n_2=0}^{n-n_1}\binom{n_2+m-1}{m-1}\left(-\pi\frac{\alpha}{r}\right)^{-n_2}\sum_{n_3=0}^{n-n_1-n_2}(-r)^{n_3}B^{*}_{n_3}C_{n-n_1-n_2-n_3}\left(\pi\frac{\alpha}{r}\right)\right\},$$

and for $\alpha = 0$ *(which implies* $r < 0$*), in the form*

$$\int_0^\infty Li_n(-\sigma x)Li_m(-\omega x^r)\frac{dx}{x}$$

$$= -(\mp 1)^{m+n}\left\{|r|^{-n}\pi\sum_{|r|k\notin\mathbb{N}}(-1)^k\csc(\pi|r|k)k^{-(m+n)}\zeta^{\mp|r|k}\right.$$

$$+|r|^{m-1}\pi\sum_{l/|r|\notin\mathbb{N}}(-1)^l\csc\left(\pi\frac{l}{|r|}\right)l^{-(m+n)}\zeta^{\mp l} \tag{2.5}$$

$$\left.-|r|^{-n}\sum_{|r|K\in\mathbb{N}}(-1)^{K(1+r)}K^{-(m+n)}\zeta^{\mp|r|K}\left(\frac{m+n}{|r|K}\pm\ln\zeta\right)\right\}$$

$$+H(\pm 1)(-1)^{n+1}\pi^{m+n+1}r^m$$

$$\sum_{m_1=0}^{m+n+1}\frac{1}{m_1!}\left(-\frac{1}{\pi}\ln\zeta\right)^{m_1}\sum_{m_2=0}^{m+n+1-m_1} r^{-m_2}B^{*}_{m_2}B^{*}_{m+n+1-m_1-m_2},$$

where

$$B^{*}_j = \frac{|2^j-2|}{j!}|B_j|,$$

B_j *are Bernoulli numbers, where*

$$C_j(x) = \frac{1}{j!}\frac{d^j}{ds^j}\csc(s+x)\Big|_{s=0},$$

and where the logarithm is defined on its principal sheet.

In all the expressions above, the upper sign corresponds to $|\zeta| > 1$, the lower sign to $|\zeta| < 1$.

Although the Theorem is here given with the restriction $|\zeta| \neq 1$, it has been shown [1] that, at least for $\sigma = \omega = 1$ (and hence $\zeta = 1$), the integral is continuous. When looking at this formula, two facts become apparent: The formula is involved, but it is, at least in certain sections, suited for symbolic evaluation. More problematic, however, is the appearance of three rather complicated infinite series over k, l, and K which *a priori* seem to defeat all attempts at systematic treatment. In order to evaluate the integral completely by symbolic algebra using the above formula, a program would have to be written which could recognize whether or not, for a given set of parameters, the above series can be expressed in terms of a finite number of known functions. This would certainly be difficult, and might even be impossible.

In the absence of such a general approach, the only way to obtain specific results from (2.4) and (2.5) consists of trying to evaluate the series analytically, possibly using symbolic algebra for subsidiary calculations. Incidentally, proceeding in this way may yield mathematical insight or results which would be lost by a completely automatic approach. Thus, special cases discussed in [1] demonstrate that quite interesting results can be extracted from (2.4) and (2.5). For example,

$$\int_0^{\infty} x^{-3} Li_n(-x) Li_n(-x^2) dx = 2^{n-2}\pi \quad (n \geq 1)$$

and

$$\int_0^{\infty} Li_n(-x) Li_m(-x^{-2}) \frac{dx}{x} = 2^{-(n+2)} \frac{\pi^{n+m+1}}{(n+m-1)!} \left\{ |E_{n+m-1}| - \frac{2^{n+m+1}-2}{n+m+1} |B_{n+m+1}| \right\}$$

$$(n, m \in \mathbb{N},\ n+m \text{ odd}).$$

an integral which contains both the Bernoulli and Euler numbers.

3. Special cases

The special cases $r = -2,\ \alpha = 0;\ \ r = 2,\ \alpha = -2;\ \ r = 1,\ \alpha = -\frac{1}{2}$ have already been discussed in [1]. Here we consider some additional special cases. In order to simplify the formulae we shall set, at a certain stage, $\sigma = \omega = 1$, and thus $\zeta = 1$.

3.1 $r = -1,\ \alpha = \frac{1}{4}$**.** The summation conditions for the infinite series become

$$k \pm \frac{1}{4} \notin \mathbb{N}_0, \quad l \mp \frac{1}{4} \notin \mathbb{N}_0, \quad K \pm \frac{1}{4} \in \mathbb{N}$$

which permit all k and l, but exclude all K. Thus the series in (2.4) can be written as

$$\sum_{k=1}^{\infty} \zeta^{\mp k} k^{-m} (4k \pm 1)^{-n}\ , \quad \sum_{l=1}^{\infty} \zeta^{\mp l} l^{-n} (4l \pm 1)^{-m}. \tag{3.1}$$

Starting from the formulae [8] (No. 5.2.5.11, 13)

$$\sum_{k=0}^{\infty} \frac{x^{4k+\nu}}{4k+\nu} = \frac{1}{4} \ln \frac{1+x}{1-x} + \frac{1}{2} \varepsilon_\nu \arctan x$$

($\nu = 1,\ 3;\ \varepsilon_1 = 1,\ \varepsilon_3 = -1$) we obtain by repeated integration, using (2.2), for $j = 1, 2, 3, \ldots$,

$$\sum_{k=1}^{\infty} \frac{x^{4k}}{(4k \pm 1)^j} = \frac{1}{4} x^{\mp 1}\{Li_j(x) - Li_j(-x) \pm 2Ti_j(x)\} - \begin{cases} 1 \\ 0 \end{cases}$$

where $Ti_j(x)$ is the generalized arctangent integral (Lewin [5], p. 299):

$$Ti_2(x) = \int_0^x t^{-1} \arctan t\,dt \ , \ Ti_j(x) = \int_0^x t^{-1} Ti_{j-1}(t)dt.$$

By decomposing the series (3.1) into partial fractions, and using (2.2), we can express these series in closed form. This procedure is tedious, but is well-suited to a system like REDUCE. For reasons of simplicity, we have constructed such a program for $j \leq 2$ only. Setting $\sigma = \omega = 1$, and noting that $Li_2(1) = \pi^2/6$, $Li_2(-1) = -\pi^2/12$, we find

$$\int_0^\infty x^{-3/4} \ln(1+x) \ln(1+\frac{1}{x})dx = 8\pi\sqrt{2}\,(3 \ln 2 - 2)$$

$$-\int_0^\infty x^{-3/4} \ln(1+\frac{1}{x})Li_2(-x)dx = 2\pi\sqrt{2}\left(\frac{5}{3}\pi^2 - 16(3 \ln 2 + G - 2)\right)$$

$$-\int_0^\infty x^{-3/4} \ln(1+x)Li_2(-\frac{1}{x})dx = 2\pi\sqrt{2}\left(\frac{5}{3}\pi^2 + 16(3 \ln 2 + G - 4)\right)$$

$$\int_0^\infty x^{-3/4} Li_2(-x)Li_2(-\frac{1}{x})dx = 256\pi\sqrt{2}\,(3 - 3 \ln 2 - G)$$

where

$$G = Ti_2(1) = \sum_{k=0}^{\infty} \frac{(-1)^k}{(2k+1)^2} = 0.91596\ 55941\ \ldots$$

is Catalan's constant.

3.2 $r = -1,\ \alpha = \frac{1}{2}$. The summation conditions become

$$k \pm \frac{1}{2} \notin \mathbb{N}_0,\ l \mp \frac{1}{2} \notin \mathbb{N}_0,\ K \pm \frac{1}{2} \in \mathbb{N}$$

which permit all k and l, but exclude all K. The series can be written as

$$\sum_{k=1}^{\infty} \zeta^{\mp k} k^{-m} (2k \pm 1)^{-n} \ , \ \sum_{l=1}^{\infty} \zeta^{\mp l} l^{-n} (2l \mp 1)^{m}. \tag{3.2}$$

We find for $j = 1, 2, 3, \ldots$

$$\sum_{k=1}^{\infty} \frac{x^{2k}}{(2k \pm 1)^j} = \frac{1}{2} x^{\mp 1}(Li_j(x) - Li_j(-x)) - \begin{cases} 1 \\ 0 \end{cases}$$

and the series (3.2) can be obtained in closed form by decomposition into partial fractions. For $j \leq 2$, $\sigma = \omega = 1$, we find

$$\int_0^{\infty} x^{-1/2} \ln(1+x) \ln(1+\frac{1}{x}) dx = 4\pi(2 \ln 2 - 1)$$

$$-\int_0^{\infty} x^{-1/2} \ln(1+\frac{1}{x}) Li_2(-x) dx = 2\pi\left(\frac{1}{3}\pi^2 - 8 \ln 2 + 4\right)$$

$$-\int_0^{\infty} x^{-1/2} \ln(1+x) Li_2(-\frac{1}{x}) dx = 2\pi\left(\frac{1}{3}\pi^2 + 8 \ln 2 - 8\right)$$

$$\int_0^{\infty} x^{-1/2} Li_2(-x) Li_2(-\frac{1}{x}) dx = 16\pi(3 - 4 \ln 2).$$

3.3 $r = 1, \alpha = -1$. The summation conditions become

$$k \mp 1 \notin \mathbb{N}_0 \ , \ l \pm 1 \notin \mathbb{N}_0 \ , \ K \mp 1 \in \mathbb{N}$$

which exclude all k and l, but permit all $K \geq 2$ if $|\zeta| > 1$, and all K if $|\zeta| < 1$. Because of the continuity of the integral at $\zeta = 1$, we set $\sigma = \omega = 1$ for simplicity, and obtain (choosing arbitrarily the case $|\zeta| > 1$) the series

$$-m \sum_{K=1}^{\infty} K^{-n}(K+1)^{-(m+1)} - n \sum_{K=1}^{\infty} K^{-(n+1)}(K+1)^{-m}.$$

Using formula [8] (No. 5.1.24.8), which expresses these series in closed form, we obtain by using REDUCE the following examples:

$$\int_0^{\infty} [x^{-1} \ln(1+x)]^2 dx = \frac{1}{3}\pi^2$$

$$-\int_0^{\infty} x^{-2} \ln(1+x) Li_2(-x) dx = \frac{1}{3}\pi^2 + 2\zeta(3)$$

$$-\int_0^{\infty} x^{-2} \ln(1+x) Li_3(-x) dx = \frac{1}{30}\pi^4 + \frac{1}{3}\pi^2 + 2\zeta(3)$$

$$-\int_0^{\infty} x^{-2} \ln(1+x) Li_4(-x) dx = \frac{1}{30}\pi^4 + \frac{1}{3}\pi^2 + 4\zeta(5) + 2\zeta(3)$$

$$\int_0^{\infty} x^{-2} Li_2(-x) Li_3(-x) dx = \frac{1}{30}\pi^4 + \pi^2 + 6\zeta(3)$$

$$\int_0^{\infty} x^{-2} Li_3(-x) Li_4(-x) dx = \frac{2}{15}\pi^4 + \frac{10}{3}\pi^2 + 4\zeta(5) + 20\zeta(3).$$

Using well-known properties of $Li_k(z)$, one finds by partial integration that for $r = 1$, $\alpha = -1$, and all σ, ω

$$I_{n,m} = I_{n-1,m} + I_{n,m-1} \quad .$$

In particular,

$$I_{n,n} = 2I_{n-1,n} \quad .$$

3.4 $r = -p = -1, -2, -3, \ldots, \alpha = 0$. The summation conditions become

$$pk \notin \mathbb{N} \ , \ l/p \notin \mathbb{N} \ , \ pK \in \mathbb{N}$$

which exclude all k, but permit all l which are not multiples of p, and all K. The case $p = 1$ can be handled with polylogarithms; in particular for $\zeta = 1$ we obtain

$$\int_0^\infty Li_n(-x) Li_m\left(\frac{-1}{x}\right) \frac{dx}{x} = (m+n)\zeta(m+n+1).$$

The case $p = 2$ has been discussed in [1]. For all p, the series over K can be expressed in terms of polylogarithms. The series over l, however, leads for $p = 3$ and $\zeta = 1$ to

$$\sum_{j=0}^{\infty} (3j+1)^{-(m+n)} - \sum_{j=0}^{\infty} (3j+2)^{-(m+n)}. \tag{3.4}$$

Considering [8] (No. 5.2.4.9, 10)

$$\sum_{j=0}^{\infty} \frac{x^{3j+1}}{3j+1} = \frac{1}{6} \ln(1 + x + x^2) - \frac{1}{3} \ln(1 - x) + \frac{1}{\sqrt{3}} \arctan \frac{x\sqrt{3}}{x+2},$$

(and a similar formula for $3j + 2$), it does not seem likely that (3.4) can be expressed in terms of elementary functions for $m + n \geq 2$. The same is probably true for $p \geq 4$.

3.5 $r = -2, \alpha = -\frac{3}{2}$. The summation conditions become

$$2k \mp \frac{3}{2} \notin \mathbb{N}_0 \ , \ \frac{1}{2}\left(l \pm \frac{3}{2}\right) \notin \mathbb{N}_0 \ , \ 2K \mp \frac{3}{2} \in \mathbb{N}$$

which permit all k and l, but exclude all K. The series over k is

$$\sum_{k=0}^{\infty} (-1)^k \zeta^{\mp 2k} k^{-m} (4k \mp 3)^{-n},$$

which, because of [8] (No. 5.2.4.12, 14)

$$\sum_{k=0}^{\infty}(-1)^k \frac{x^{4k+\nu}}{4k+\nu} = \frac{1}{2\sqrt{2}} \arctan \frac{x\sqrt{2}}{1-x^2} + \frac{1}{4\sqrt{2}} \varepsilon_\nu \ln \frac{1+x\sqrt{2}+x^2}{1-x\sqrt{2}+x^2},$$

($\nu = 1, 3$; $\varepsilon_1 = 1$, $\varepsilon_3 = -1$) is unlikely to be expressible in closed form. The sine factor in the series over l gives rise to a sequence $+ + - - \ldots$ of signs for consecutive terms which probably excludes closed summation for this series too.

The last two examples (and others) show that the infinite series in (2.4) and (2.5) are likely to be representable in terms of known functions only for a very restricted set of parameters r and α. Further investigation will be necessary in order to get a more general picture.

References

1. V. S. Adamchik and K. S. Kölbig, A definite integral of a product of two polylogarithms. SIAM J. Math. Anal., 19 (1988) p.926-938.

2. K. O. Geddes and T. C. Scott, Recipes for classes of definite integrals involving exponentials and logarithms, *in* E. Kaltofen and S.M. Watts (Eds.), Computing and Mathematics. (Springer-Verlag, New York, 1989) p.192-201.

3. K. S. Kölbig, Explicit evaluation of certain definite integrals involving powers of logarithms. J. Symbolic Comp. 1 (1985) p.109-114.

4. K. S. Kölbig, Some types of definite integrals suitable for evaluation by symbolic algebra, *in* Proc. Int. Conf. on Computer Algebra and its Applications to Theoretical Physics. (JINR, Dubna, 1985) p.172-184.

5. L. Lewin, *Polylogarithms and associated functions*. (North-Holland, Amsterdam 1981).

6. J. C. Piquette, Applications of a technique for evaluating indefinte integrals containing products of special functions of physics. SIAM J. Math. Anal. 20 (1989) p.1260-1270.

7. J. C. Piquette and A. L. Van Buren, Technique for evaluating indefinite integrals involving products of certain special functions. SIAM J. Math. Anal. 15 (1984) p.845-855.

8. A. P. Prudnikov, Yu. A. Brychkov and O. I. Marichev, *Integrals and Series, vol. 1: Elementary Functions*. (Gordon and Breach, New York, 1986).

SYMBOLIC COMPUTATION OF THE GENERAL TERM OF TAYLOR SERIES

IVAN V. DIKOV

Institute of Mathematics with Computer Center
Bulgarian Academy of Sciences
"Akad. G. Bonchev" Str. bl. 8
1113 Sofia

ABSTRACT

AN APPROACH FOR SYMBOLIC COMPUTATION OF THE GENERAL TERM OF THE TAYLOR'S EXPANSION OF FUNCTIONS WITH USAGE OF KNOWLEDGE BASE IS PRESENTED. AN EXPERIMENTAL IMPLEMENTATION OF THIS APP ROACH USING THE SYSTEM FOR ALGEBRAIC COMPUTATION REDUCE 3.3 IS CONSIDERED.

1. INTRODUCTION.

Let's assume that the function $f(x) \in \mathbb{C}^{\infty}$ is smooth about the point $x_0 \in \mathbb{R}^1$. We'll consider functions of a single variable, because the case when $x_0 \in \mathbb{R}^n$ $(n>1)$ can be treated in the same way. It's well known that the coefficient a_n of the Taylor's expansion of the function $f(x)$ at x_0 is:

$$a_n = \frac{f^{(n)}(x_0)}{n!} \qquad (1)$$

The problem which is to be solved is to find such a formal representation of the coefficient a_n which doesn't contain the derivatives of the function $f(x)$. The formula (1) shows that it is sufficient to do this only with the nth derivative of the function $f(x)$ at the point x_0. In other words the problem is whether it's possible or not to find such a formal representation of the nth derivative of the function $f(x)$ at point x_0 so that it could be calculated without finding the derivatives of order lower than n that means to find such a function $F(n,x_0)$ for which the following is true:

$$F(n,x_0)=f^{(n)}(x_0) \qquad n \in \mathbb{N}$$

Of course while finding the function $F(n,x_0)$ we'll have to apply mathematical rules for obtaining the nth derivative of a function. It's evident that for a fixed value of n for the function $F(n,x_0)$ the corresponding derivative at x_0 will be found. What is more, if the function $f(x)$ is defined and smooth about the point x_0 then we can vary the value of the variable x about the point x_0. It is a good opportunity for studding the n^{th}derivative of the function f(x) about the point x .

2. Mathematical rules for finding the n^{th}derivative of the function f(x).

From now on we'll take for granted that the function f(x) has a nth derivative.

a. If $f(x)=\alpha g(x)+\beta h(x)$ then

$$f^{(n)}(x)=\alpha g^{(n)}(x)+\beta h^{(n)}(x) \qquad \alpha,\beta \in \mathbb{R},\ n \in \mathbb{N}$$

b. If $f(x)=g(x)h(x)$ then

$$f^{(n)}(x) = \sum_{i=0}^{n} \binom{n}{i} g^{(i)}(x)\, h^{(n-i)}(x) \qquad \text{(Leibniz)}$$

c. If $f(x)=g(h(x))$ then

$$f^{(n)}(x)=\sum_{i=1}^{n} g^{(i)}(h(x)) \frac{n!}{i_1!\ i_2!\dots i_p!} \prod_{t=1}^{p} \left(\frac{h^{(t)}(x)}{i!}\right)^{i_t},$$

where the summation is expanded over the all ordered p-tuple $(i_1,i_2,\dots,i_p)$, $p \in \mathbb{N}$ of not negative integers for which

$$\sum_{t=1}^{p} t\, i_t = n \quad , \quad i = \sum_{t=1}^{p} i_t$$

The rule a. is obvious because it is obtained from the basic property of the first derivative namely that is a linear operator. This property is valid for the n derivative.

The rule b. is well known as Leibniz's formula [6] for finding the nth derivative of two functions multiplied.

Finally, the rule c. is interesting task which can be solved by using some facts of the theory of the power series.

3. Rules for finding the nth derivatives of the basic mathematical function (BMF).

Here the problem is a little bit more complex, because it is difficult to define which functions are BMF. For example the functions cos(x) and sin(x) are described as BMF in a number of reference books [1,2] ,but the fact that the first one can be expressed by the other and vice versa suggests that such a division is quite relative. Some of them are:

$$(A^x)^{(n)}=\mathrm{Ln}^n(A)\ A^x \qquad \cos^{(n)}(x)=\cos(n\pi/2+x)$$

$$\mathrm{Ln}^{(n)}(x)=-(n-1)!/(-x)^n \qquad \sin^{(n)}(x)=\sin(n\pi/2+x)$$

A concrete realization of the approach offered above, can be seen in the program TAYLOR for Taylor's expansion of a certain function f(x). The program is written in SYMBOLIC MODE of the system for algebraic manipulation REDUCE3.3 on IBM-4341.

An important part of the program TAYLOR is the subroutine SIMPDF1 for finding the nth derivative of a function f(x). In it rules from parts 2 and 3 are used for formula manipulation.

4. A general functional scheme of the subroutine SIMPDF1.

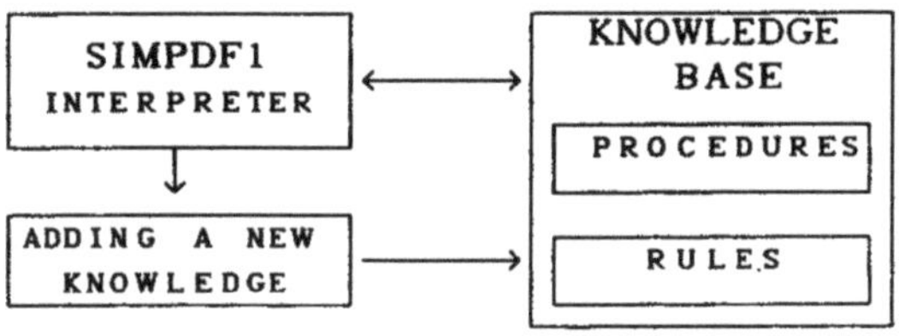

- SIMPDF1 - interpreter.

SIMPDF1 is a program which analyses the type of the function whose n^{th} derivative is to be found. She applies successively the procedures and rules for formula manipulation of the knowledge base. A basic working principle of SIMPDF1 is to derive the prefix operator from the prefix representation of the function f(x) in SYMBOLIC MODE. If it is an arithmetical operator - to call the corresponding handling procedure from the knowledge base; if it's a name of a BMF - from the rules. If there can't be found a simplifying rule in both bases one should call the program for adding new knowledge. This happens when the system either can't express a function by its familiar BMF or doesn't know its nth derivative.

- Knowledge base.

It includes two part - procedures and rules. The first part includes six procedures, for finding n^{th} derivative of two or more functions connected by the arithmetical operators +,-,*,/,** and composition of functions. Every one of these procedures works according to a rule or rules described in part 2. The result of the execution of a procedure returns in SIMPDF1 from where a formula manipulation in depth is done recursively.

The second part include production rules. It is an association list by name FN-DF, every element of which is:

((name of the function) (n^{th} derivative))

We'll suppose that the searching FN-DF has finished successfully if such an element is found that its left part coincide with the name of the function.

- A program for adding new knowledge.

It works in an interactive mode. It is advisable for the user to be acquainted with the structure and the condition of the base in advance.

5. Advantages and other problems.

Here we'll take into consideration only those advantages and problems which are directly connected with the Taylor's expansion of a function and the arithmetical operations with power series.

a. The first advantage over the conventional method of finding a concrete derivative is that in this case it's enough to substitute the variables n and x for fixed values and to compute the function F(n ,xo). For example, let's assume the function $f(x)=\cos(x)+e^{x}$, then we'll obtain :

$$F(n,xo)=\cos(n\pi/2+xo)+e^{xo}$$

Hence it will easy come out that the general term of the Taylor's expansion of the function is :

$$a_n=\cos(n\pi/2+xo)/n!+e^{xo}/n! \qquad (2)$$

b. From item a. it can be seen that since the explicit setting of the coefficient a_n is available can be solved the following problem to find that value N of the variable n for which the term $a_n(x-x_0)^n$ will not exceed a preliminary set number E for close values of x to x_0 and for any n>N . What is most interesting in this case is that this problem can be solved automatically for a certain function f(x).

c. It's easy and convenient to do arithmetical operations with power series. It's important because it allows one to find the general term of power series connected in one expression by arithmetical operations. This is directly connected with studding of the reminder term of the result within the boundaries of the given precision E.

d. The radius of convergence of the function f(x) expanded under Taylor in a power series can be defined automatically. It follows the rules of the mathematical analysis for determining the radius of convergence of a power series. For example Cauchy's rule:

$$1/R = \overline{\lim_{n \to \infty}} \sqrt[n]{|a_n|} \quad , \tag{3}$$

where $\overline{\text{Lim}}$ is limes superior.

Apart from these advantages there are some problems. All known systems which operate with power series work slowly [3,4,5] . With the offered approach finding the nth derivative and the general term of Taylor's expansion in a power series respectively is comparatively slow. Of course it depends on the type of the function but in general the ratio between the time of this approach and the standard one is 3 to 1. There is an explanation for this because with the offered approach a general formula is found and then there is a substitution for concrete values. It's also established that most of the time is taken up in substituting for the concrete values in general formula. This reminds that a further improvement of the computing procedures themselves is necessary.

The second problem is that there are some BMF whose n^{th} derivatives are to be expressed in a very complicated way. As for example the function arcsin(x). While for other BMF it's more expedient not to search their n^{th} derivative but to express them by such functions already familiar to the system. The choosing of an appropriate minimum class of BMF is an interesting mathematical problem which hasn't been solved so far.

REFERENCE.

1. BERMANT B.F., *SHORT COURSE ON MATHEMATICAL ANALYSIS,* NAUKA, (MOSCOW), 1964 p 57, (in Russian).

2. GELERT V., NESTNER H., *LEXIKON DER MATHEMATIKq* BIBLIOGRAPHISCHES INSTITUT, p 222, (LEIPZIG 1979).

3. KRIUKOV A., SHABLIGIN.G., *PROGRAM PACKAGE FOR POWER SERIES MANIPULATION* ,PREPRINT 02-112/88, (MOSCOW 1988).

4. BARNS A., PADGET J, *TRANCATED POWER SERIES,*(RECEIVED BY PROF. A. HEARN)

5. FELDMAR E., KÖLBIG K.S., REDUCE *PROCEDURES FOR THE MANIPULATION OF /GENERALIZED POWER SERIES*, CERN-DATA HANDLING DIVISION,DD/84/22, DECEMBER 1984.

6. JAMES AND JAMES, *MATHEMATICS DICTIONARY*, **THIRD EDITION**, D. VAN NOSTRAND COMPANY, p 321, (INC. 1968).

Translation Invariance of Nonlinear Equations and Computer Algebra

S.G.SHEKHOVTSOV
and
V.E.ZHURAVLEV
Institute of Radioengineering and Electronics
JE "Interface", ul. Klimashkina,9, 123557 Moscow, USSR

ABSTRACT

Many nonlinear equations of mathematical physics are invariant with respect to reference system translations. For such equations some procedure of qualitative analysis is developed. It is an infinite dimensional analog of the well known procedure of reducing to a normal form which goes back to Poincare. Namely, we consider a linearization by means of substitution of new variables. For the search of such variables a package was worked out in the frame of the mMATH system.

We shall consider nonlinear mappings connected with the equations which are translational invariant. It will be shown that this property allows to obtain explicit results by the traditional method of a normal form. For example, the mappings from Thomas and KdV equations

$$F:\ u \to u_{xy} + \alpha u_x + \beta u_y + \gamma u_x u_y\ ,\ (\alpha,\beta,\gamma{=}const)$$
$$G:\ u \to u_{xxx} + 6uu_x + u_y\ , \tag{1}$$

belong to such class. It's easy to see that the origin is also *a fixed point* and F, G are *analytic* in a neighborhood of the origin.

1. Formalism

Let F be a realization of some analytic mapping on a manifold for some local chart in a neiborhood of the fixed point. Another chart leads to another realization

$$\tilde{F} = S^{-1}FS\ ,\quad S:\ W \leftrightarrow W$$

where W is a coordinate space of the manifold. It is naturally to look for such S that $\tilde{F}$ has the simplest form. We study a possibility to find a linear realization.

It is convenient to formulate this problem in the algebraic (or formal) way. Let $\mathcal{A}$ be a set of formal mappings

$$F = \sum_{k\geq 1} F_k(u) \ , \qquad F_k(u) = f_k(u,\ldots,u) \ ,$$

where $u\in W$, f_k: $W^k \to W$ is k-linear (continuous) mapping, and let $\mathcal{R}\subset\mathcal{A}$ be the group of invertible elements of $\mathcal{A}$. Then for given $F\in\mathcal{A}$, we want to find $S\in\mathcal{R}$ such that $\tilde{F}_k=0$, $k>1$. Here the mapping S is obviously supposed to preserve the origin. In other words one should solve the equation

$$S\tilde{F}_1 = FS$$

or the equivalent recurrence system of so called homological equations (h.e.):

$$S_k\tilde{F}_1 - F_1S_k = (F'S)_k \ , \quad k=1,2,\ldots,$$

where $F'=F-F_1$, $(\ ):F \to F_k$. It is difficult to solve this system in general case, but the problem is simplified for translational invariant mappings. Indeed, let $\mathcal{A}_t$ be a set of such mappings:

$$\mathcal{A}_t = \left\{F\in\mathcal{A}\text{: } FT(\mathbf{a})=T(\mathbf{a})F, \ \forall\mathbf{a}\in\mathcal{G}\right\} \ ,$$

where $\mathcal{G}$ is a group of reference system translations and T is its action in W: $[T(\mathbf{a})u](\mathbf{x})=[u](\mathbf{x}+\mathbf{a})$. Let W be also an algebra of (real or complex) functions of several real variables. Then for many kinds of W k-linear mapping f_k ($\forall F\in\mathcal{A}_t$) can be represented as

$$f_k = \int_{\mathcal{G}^k} (d\mathbf{a})^k\varphi_k(\mathbf{a}_1\ldots\mathbf{a}_k)T(\mathbf{a}_1)\otimes\ldots\otimes T(\mathbf{a}_k) \ ,$$

where according to the definition $(T(\mathbf{a}_1)\otimes\ldots\otimes T(\mathbf{a}_k))=(T(\mathbf{a}_1)u_1)\cdots(T(\mathbf{a}_k)u_k)$ and φ_k is some proper distribution on $\mathcal{G}^k$. Transformations $T_i=id\otimes\ldots\otimes T(\mathbf{a}_i)\otimes\ldots\otimes id$ form n-parametric abelian group, so f_k may be perceived as an element of the group algebra. For definitness the group is further considered to be continuous and generated by k generators $\mathbf{d}_i$, $i=1,\ldots,k$. In this case we have

$$f_k = \int_{\mathcal{G}^k} (d\mathbf{a})^k \varphi_k(\mathbf{a}_1\ldots\mathbf{a}_k) T_1(\mathbf{a}_1)\cdots T_k(\mathbf{a}_k) =$$

$$= \int_{\mathcal{G}^k} (d\mathbf{a})^k \varphi_k(\mathbf{a}_1\ldots\mathbf{a}_k) \exp(\mathbf{a}_1\mathbf{d}_1+\ldots+\mathbf{a}_k\mathbf{d}_k) = f_k(\mathbf{d}_1\ldots\mathbf{d}_k) .$$

The symbolic expression $f_k(\mathbf{d}_1\ldots\mathbf{d}_k)$ is usually called *a symbol* of the corresponding k-linear mapping f_k. Thus for any $F\in\mathcal{A}_t$ the correspondence π: $F_k \to f_k(\mathbf{d}_1\ldots\mathbf{d}_k)$ allows us to use this set of symbols instead of F. In terms of this symbolic language a composition $H=FG$ of $F,G\in\mathcal{A}_t$ is represented as follows

$$h_n(\mathbf{d}_1\ldots\mathbf{d}_n) = \sum_{m=1}^{n} \sum_{k_1+\ldots+k_m=n} f_m(\mathbf{d}_1+\ldots+\mathbf{d}_{k_1},\ldots,\mathbf{d}_{n-k_m+1}+\ldots+\mathbf{d}_n)\cdot$$
$$g_{k_1}(\mathbf{d}_1\ldots\mathbf{d}_{k_1})\cdots g_{k_m}(\mathbf{d}_{n-k_m+1}\ldots\mathbf{d}_n) \qquad (2)$$

Since linear elements of $\mathcal{A}_t$ are commutative the first (linear) h.e. gives $F_1=\tilde{F}_1$ for abitrary invertible linear part of S; the other h.e. then become

$$[S_1,F_1] = (F'S)_k \ , \quad k=2,3,\ldots$$

This system is easily rewritten in a symbolic form

$$\Big[f_1(\mathbf{d}_1)+\ldots+f_1(\mathbf{d}_k)-f_1(\mathbf{d}_1+\ldots+\mathbf{d}_k)\Big] s_k(\mathbf{d}_1\ldots\mathbf{d}_k) \simeq \pi(F'S)_k(\mathbf{d}_1\ldots\mathbf{d}_k),$$

where $\simeq$ introduced because of ambiguity of π. In other words it is the system of equivalence conditions. It can be proved that all symbol operations are correct within the group algebra. The only exception is a "division" defined in a usual way; indeed, one may formally devide by symmetric expression only. Taking all these into account under symbol manipulations we can solve in turn h.e. begining from the bilinear one.

2. Elementary example

Consider for example F from Eq.1. In this case $F_1=f_1(\mathbf{d})=\xi\eta+\alpha\xi+\beta\eta$, $F_2=f_2(\mathbf{d}_1,\mathbf{d}_2)=\gamma\xi_1\eta_2$, where $\mathbf{d}=(\xi,\eta)$ is the generator of translations in the plane (x,y). Hence, bilinear h.e. is

$$(\xi_1\eta_2 + \xi_2\eta_1)s_2(\mathbf{d}_1,\mathbf{d}_2) \simeq -\gamma\xi_1\eta_2 s_1(\mathbf{d}_1)s_1(\mathbf{d}_2) \simeq$$

$$\simeq -\gamma/2(\xi_1\eta_2 + \xi_2\eta_1)s_1(\mathbf{d}_1)s_1(\mathbf{d}_2) .$$

According to the freedom in choosing S_1 we may take its symbol in the most convenient form. Now, let $s_1=1$ then $s_2=-\gamma/2$. It is easy to see that all symbols s_k do not depend on $\mathbf{d}$, so n-linear h.e.

$$\sum_{i\neq k} \xi_i\eta_k \, s_n(\mathbf{d}_1\ldots\mathbf{d}_n) \simeq$$

$$\simeq -\gamma \sum_{k=1}^{n-1} (\xi_1+\ldots+\xi_k)(\eta_{k+1}+\ldots+\eta_n)s_k(\mathbf{d}_1\ldots\mathbf{d}_k)s_{n-k}(\mathbf{d}_{k+1}\ldots\mathbf{d}_n) ,$$

ca be written as

$$n(n-1)\xi_1\eta_2 s_n = -\gamma\xi_1\eta_2 \sum_{k=1}^{n-1} k(n-k)s_k s_{n-k} .$$

The only solution of this equation is $s_n=(-\gamma)^{n-1}/n$. It immediately gives well known Thomas transformation $S: u \to u=\frac{1}{\gamma}ln(1+\gamma u)$. Similar situation takes place in the case of Burgers equation. In general, almost for all mappings connected with well known integrable nonlinear equations the following situation is observed: (1) a formal solution of the h.e. can be explicitly obtained; (2) this solution is such that the series for S which usualy has a very complicated structure is reduced to a summable power series; (3)the summ be valid also beyond a convergence region.

Similar results were first obtained by R.Rosales [1] by means of formal perturbation theory and independently by A.Bobylev [2].

All these results are valid for such classes of functions that linear mapping $f_1(\mathbf{d}_1)+\ldots+f_1(\mathbf{d}_k)-f_1(\mathbf{d}_1+\ldots+\mathbf{d}_k)$ is invertible on the corresponding tensor product. In the oposite case the normal form consist of a number of *reasonance* terms which cannot be eliminated by any described transformation.

3. About computer realization

For arbitrary equation it is very dificult problem to solve h.e. Moreover to calculate S_n for the given $s_n(d_1 \ldots d_n)$ one should represent the denominator of s_n as a product of linear forms. These problems can be solved by means of computer algebra system.

Under realization of the package we paid much attention to the effectiveness of algorithms. Thus, the key formula Eq.2 of the package is represented as

$$h_n(d_1 \ldots d_n) = \sum_{i=1}^{2^{n-1}} a_i \quad ,$$

where the structure of the terms

$$a_i = f_m(d_1 + \ldots + d_{k_1}, \ldots, d_{n-k_m+1} + \ldots + d_n)\; g_{k_1} \cdots g_{k_m}$$

is calculated directly from the binary representation of i. In this case $m = 1$ + a number of units in this representation of i and $k_i = 1$ + a number of zeros between i-th and $(i+1)$-th units in it. It is convenient to generate h_n from h_{n+1} by means of Young diagrams (Fig.1). Indeed, each term of h_n gives two terms for h_{n+1}:one by a new cell in the last column and the other by a new cell in the bottom row. At the base of this formula we constructed a computer version of the whole described algebra. The utilities of the package allow to perform any symbolic operations and to obtain different representations of s_n.

$f_3(d_1+d_2+d_3, d_4, d_5+d_6)\cdot$
$\cdot g_3(d_1, d_2, d_3) g_1(d_4) g_2(d_5, d_6)$

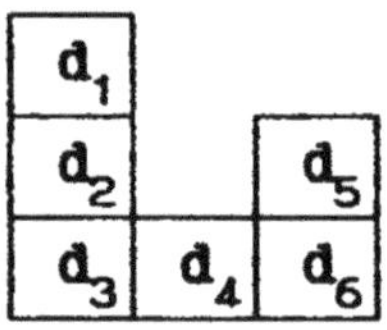

Fig.1

1. R.Rosales, *Stud.Appl.Math*, 59, p.117-151 (1978)
2.А.В.Бобылев, *Доклады АН СССР*, 256, N 6, p.1341-1346

ANALYTICAL APPROXIMATE SOLUTION OF SINGULAR ORDINARY DIFFERENTIAL EQUATIONS

K.Hantzschmann, N.X.Thinh

University of Rostock, Technical University of Dresden GDR

1. Introduction

Whereas in the Computer Algebra algorithms have precedence founded on algebraic concepts and aspiring to closed solutions, the Computer Analysis gives priority to determining of approximate solutions. In analogy to the Numerical Mathematics it holds to find approximate solutions for non closed solvable or only with non acceptable expense solvable problems and to analyse the quality of approximate solutions by error estimation. This purpose will be aspired in the Computer Analysis on the base of computer-aided formula work. By N.J.Lehmann/3/, who has supported very hard this work field internationally, the goal of the Computer Analysis was formulated as follows:

- finding of analytical approximate solutions which should reflect inherent properties of the given problem and the possible influence of parameters
- formula expressions should be transparent and simple provided with adequate precision
- securing of the accuracy of the approximate solutions by error boundaries evaluated completely by the computer.

Efficiency and quality of the algorithms developed in the Computer Analysis depend in a high degree on the possibility to the combination of the approved numerical with the symbolic computation. Mathematical approximation methods adapting to the given problem and producing transparent simple formula expressions with an acceptable expense have in the Computer Analysis great importance /1/.

Good results for the approximate solution of ordinary differential equations (DE) according the mentioned premises could be obtained in the last years by a two-step concept/4/. In the first step the given problem will be adapted by a suitable chosen "neighbour" problem whose closed determined solution forms the base for an adequate approximate ansatz in the second step. In this paper we shall inform about the use of this concept of the adaptive approximation for determining approximate solution in the neighbourhood of a singular point of the

initial value problem
$$L(y) \equiv \sum_{i=0}^{k} x^i R_i(x) y^{(i)}(x) = 0 \qquad 0 < x \leq b$$
$$y^{(i)}(b) = \beta_i \qquad (i = 0(1)k-1)$$

(R_i (i=o(1)k) in the neighbourhood of x=o analytical functions with $R_k(o) \neq o$). All the algorithms developed with this basis concept and the appertaining error estimations have been implemented, a program system on the basis of PL/1-FORMAC is available for practical applications /5/.

2. Algorithms for approximate solutions

Determining an analytical transparent approximate solution takes place in two steps. In the first step (denoted as adaption-step) a to the given DE adapting DE will be determined which can be solved automatically by the computer. In order to reflect the behaviour of the solution in the neighbourhood of the singular point adequately, an EULER's DE of sufficiently high order
$$L_n(y) \equiv \sum_{i=0}^{n} a_i x^i y^{(i)}(x) = 0 \qquad (n > k)$$
will be chosen. n>k creates a necessary free space for the desired accuracy. For determining an adapting EULER's DE auxiliary quantities
$$x^i y^{(i)} = g_i(x, y, y', \ldots, y^{(k-1)}) \qquad i = k(1)n$$
have been used. They can be determined without difficulty by differentiation the given DE and substitution of the $y^{(i)}$ for i>k from DE. With parameters $a_i \in R$ (i=o(1)n-1) we can give now
$$x^n y^{(n)} + \sum_{i=0}^{n-1} a_i x^i y^{(i)} = g_n(x, y, \ldots, y^{(k-1)}) + \sum_{i=0}^{n-1} a_i g_i(x, y, \ldots, y^{(k-1)}) \qquad *)$$
which is equivalent to
$$x^n y^{(n)} + \sum_{i=0}^{n-1} a_i x^i y^{(i)} = \sum_{i=0}^{k-1} \psi_i(x, a_0, a_1, \ldots, a_{n-1}) y^{(i)},$$
because the linearity of the g_i with respect to $y^{(i)}$ (i=o(1)k-1). The parameters a_i (i=o(1)n-1) can be computed by several "adaption criterions", the adapting DE can obtain in that way desired properties. In the approximation-step the fundamental system $\varphi_1, \varphi_2, \ldots, \varphi_n$ of the EULER's DE founded in the adaption-step will be determined and made an ansatz
$$y_n = \sum_{i=1}^{n} c_i \varphi_i$$
for the wanted approximate solution. The parameters c_i (i=1(1)n) can be evaluated by several "approximation criterions" according to the given problem under consideration of the initial conditions. In dependence of the choise the used criterions for determining the a_i (i=o(1)n-1) and c_i (i=1(1)n) in both steps we can deduce various algorithms for the determination of approximate solutions.

For securing a problem inherent analytical approximation we make use of informations about the singularity of the given DE. By substituting the ansatz

$$y(x) = x^r(d_0 + d_1x + d_2x^2 + \dots) \qquad (d_0 \neq 0,\ r \in \mathbb{C})$$

in the given DE we obtain by the FROBENIUS'method the determination equation for r

$$\sum_{i=0}^{k} R_i(0)\, r(r-1)\dots(r-i+1) = 0.$$

Let $R_2 = \min\{\mathrm{Re}\ r_i\ (i=1(1)k)\}$. The given DE has in the neighbourhood of x=o k formal solutions of either the form

$$y(x) = x^r \sum_{i=0}^{\infty} d_i x^i \qquad (d_0 \neq 0)$$

$$y(x) = x^r(\phi_0 + \phi_1 \ln x + \dots + \phi_j \ln^j x) \qquad (j < k)$$

(r root of the above equation, ϕ_i (i=o(1)j) power series).
As above mentioned many efficient algorithms can be developed by the basis concept of the adaptive approximation. Some of these algorithms will be presented.

Algorithm 1:
As adaption criterion we choose the vanshing of the defect to *) and its derivatives at the point x=b up to the order n-1. After solving the resulting linear equation system for a_i (i=o(1)n-1) we obtain the adapting DE in the determinant form

$$L_n(y) \equiv \begin{vmatrix} x^n y^{(n)}(x) & x^{n-1}y^{(n-1)}(x) & \dots & y(x) \\ x^n y^{(n)}/_b & x^{n-1}y^{(n-1)}/_b & \dots & y/_b \\ \vdots & & & \\ (x^n y^{(n)})^{(n-1)}/_b & (x^{n-1}y^{(n-1)})^{(n-1)}/_b & \dots & y^{(n-1)}/_b \end{vmatrix} = 0.$$

With the fundamental system $\varphi_1, \varphi_2, \dots, \varphi_n$ the ansatz $y_n = \sum_{i=1}^{n} c_i \varphi_i$ is made. The parameters c_i (i=1(1)n) will be determined so that holds

$$y_n^{(i)}(b) = y^{(i)}(b) \qquad (i = 0(1)n-1).$$

We can prove even $$y_n^{(i)}(b) = y^{(i)}(b) \qquad (i = 0(1)2n-1)$$

(complete TAYLOR-approximation).

Algorithm 2:
Let y_0 a starting solution (for instance determined with algorithm 1 or as solution for $\sum_{i=0}^{k} x^i R_i(0) y^{(i)} = 0$ with $y^{(i)}(b) = \beta_i$ i=o(1)k-1). The parameters a_i (i=o(1)n-1) will be determined so that holds

$$\| g_n(x,y,\dots,y^{(k-1)}) + \sum_{i=0}^{n-1} a_i g_i(x,y,\dots,y^{(k-1)})/_{y=y_0} \| \xrightarrow[a_i \in \mathbb{R}]{} \min$$

with

$$\|g\|^2 = \int_{\varepsilon}^{b} x^{\gamma} g^2(x)\,dx \qquad (\varepsilon, \gamma > 0).$$

It follows

$$L_n(y) \equiv \begin{vmatrix} x^n y^{(n)}(x) & x^{n-1}y^{(n-1)}(x) & \dots & y(x) \\ (g_n, g_{n-1}) & (g_{n-1}, g_{n-1}) & \dots & (g_0, g_{n-1}) \\ \vdots & & & \\ (g_n, g_0) & (g_{n-1}, g_0) & \dots & (g_0, g_0) \end{vmatrix} = 0.$$

The following approximation-step can be performed in analogy to algorithm 1.

Algorithm 3:
In the algorithms 1 and 2 it is not secured that only the in the solution of the given problem appearing singularities of the form $1/x^\lambda$ $(\lambda>0)$ can appear in the approximate solution. In the algorithm 3 is demanded that the real part of roots of the characteristic equation of $L_n(y)=0$ shall be greater than or equal to R_2. For realizing this demand is to transform in a condition on the parameters a_i $(i=0(1)n-1)$. The optimization problem with additional conditions for the determining the adapting DE is not trivial for general n. Therefore the order of the adapting DE will be increased step by step by 1 or by 2. Starting point is a solution from
$$L_0(y) \equiv \sum_{i=0}^{k} x^i R_i(0)\, y^{(i)} = 0 .$$

a) We now explain how $L_n(y)$ is to determine on the basis of $L_{n-1}(y)$. Let y_{n-1} the approximate solution determined from
$$L_{n-1}(y) \equiv x^{n-1} y^{(n-1)} + \sum_{i=0}^{n-2} a_i^{n-1} x^i y^{(i)} = 0 .$$
By multiplying the characteristic equation of L_{n-1} $\sum_{i=0}^{n-1} c_i^{n-1}\lambda^i=0$ with $\lambda-r$ we obtain the coefficients c_i^n $(i=0(1)n-1)$ in dependence of r. After computing the a_i^n $(i=0(1)n-1)$ from c_i^n $(i=0(1)n-1)$ for L_n it follows
$$x^n y^{(n)} + \sum_{i=0}^{n-1} a_i^n x^i y^{(i)} = \sum_{i=0}^{k-1} \psi_i(x,r)\, y^{(i)} .$$
The unknown r will be evaluated from
$$Z(r) = \Big\| \sum_{i=0}^{k-1} \psi_i(x,r)\, y_{n-1}^{(i)} \Big\|^2 \xrightarrow[r\in R,\; r\geq R_2]{} \min .$$
With the analytically evaluable solution r_{min} we obtain
$$L_n(y) \equiv x^n y^{(n)} + \sum_{i=0}^{n-1} a_i^n(r_{min})\, x^i y^{(i)} = 0 .$$
b) In the same way $L_n(y)$ is determined on the basis of $L_{n-2}(y)$. The characteristic equation of $L_{n-2}(y)$ is to multiply with $\lambda^2+r\lambda+s$. The optimization problem has then the following form:
$$Z(r,s) = \Big\| \sum_{i=0}^{k-1} \psi_i(x,r,s)\, y_{n-2}^{(i)} \Big\|^2 \xrightarrow[r+2R_2\leq 0,\; -s-rR_2-R_2^2\leq 0]{r,s\in R} \min .$$

Algorithm 4:
Good results could be achieved by other DE types in the adaption-step too. These lead contrary to the approximation in the space of fundamental solutions of EULER's DE to extended rational approximations. Choosing the ansatz
$$y_{mn}(x) = \frac{P_m(x)}{P_n(x)\, B(x)}$$
for the approximate solution where $B(x)=x^{-R_2}$ or $B(x)=x^{-R_2}/\ln x$ depending of the singularity structure (P_n, P_m polynomials of the degree

n,m), we have a DE $L_{mn}(y) \equiv (yP_nB)^{(m+1)}=o$ as reference problem. After determining the k-th order derivation (k=o(1)n) of this equation and substituting b for x we obtain a linear equation system for evaluating the coefficients of P_n. The coefficients of P_m can be determined from $y_{mn}^{(i)}(b)=y^{(i)}(b)$ (i=o(1)m). It is possible also to use the least-squares-approximation instead of the TAYLOR-conditions.
The practical tests of the developed algorithms has demonstrated that each algorithm has advantages and disadvantages. However all these algorithms are able to compute good approximate solutions with respect to the given premises of the Computer Analysis.

3. Error estimation
For error representation and error estimation we use the formula determinable GREEN's function of a neighbour EULER's DE k-th order. An algorithm for the formula computation of this GREEN's function has been developed and implemented. A known principle for error analysis for approximate solutions of DE, proposed by N.J.Lehmann/2/, could be modified so that approximate solutions of singular DE can be treated too.
Let g(x,s) the GREEN's function for given differential operator L(y) with the initial conditions at x=b. Then the error function $f=y_n-y$ can be described in the form

$$f(x) = \int_x^b g(x,s)\, d(y_n(s))\, ds$$

where $d(y_n(s))=L(y_n)$. Let G(x,s) the formula determinable GREEN's function to differential operator

$$L_k(y) = \sum_{i=0}^{k} a_i x^i y^{(i)} .$$

(Without restriction of the generality we assume the leading coefficients of L and L_k are equal to 1). Using the fact that $\| g(x,.)\|_x = (\int_x^b g^2(x,s)ds)^{1/2}$ can be estimated by the according integral norm of G(x,s) we can prove

$$|f(x)| \le \frac{\| G(x,\cdot)\|_x}{1 - \| L_\Delta (G(s,t))\|_x} \| d(y_n)\|_x ,$$

if

$$\| L_\Delta (G(s,t))\|_x = \left(\int_x^b \int_s^b L_\Delta^2 (G(s,t))\, dt\, ds \right)^{1/2} < 1$$

with $L_\Delta = L_k - L$.
The difficulties of numerical and analytical evaluating of error boundaries could be mastered with the aid of suitable measures. Details can not be presented in this paper unfortunately. Finally should be remarked that it is possible to modify a part of these algorithms for the application for boundary value problems.

References

/1/ Hantzschmann,K.: Computeranalytik - Konzepte und Algorithmen. Rostocker Inf.-Berichte H.10, Rostock (1989)

/2/ Lehmann,N.J.: Fehlerschranken für Näherungslösungen bei Differentialgleichungen. Num.Math. 10, S.261-288 (1967)

/3/ Lehmann,N.J.: Die Analytische Maschine - Grundlagen einer Computeranalytik. Sitzungsber. Sächs.Akad.d.Wiss. Bd.118,H.4, Berlin (1985)

/4/ Lehmann,N.J.: Computer Algebra and Practical Analysis. EUROCAL'85, Lect.Notes in Comp.Science 203, p.102-113, Berlin-Heidelberg-New York (1985)

/5/ Thinh,N.X.: Analytische Approximation für singuläre Anfangswertaufgaben gewöhnlicher Differentialgleichungen. Dissertation A, TU Dresden (1989)

Algorithms for Operations with Difference Operators and Grid Functions in Symbolic Form

V.G.Ganzha
Institute of Theoretical and Applied Mechanics,
Siberian Department of Academy of Science, USSR, Novosibirsk

A.V.Solovjov, M.Yu.Shashkov
Institute of Applied Mathematics,
Academy of Science, USSR, Moscow

The difference scheme construction for multidimensional equation of mathematical physics on nonorthogonal curvilinear meshes is connected with manipulation with large amount of symbolic information. The manual construction of calculation formulas and formula coding generate errors, and so it is desirable that the computer should be used for this action automatization.

The necessary condition of computer application to difference scheme construction in symbolic form is the possibility of this process formalization. One of universal difference scheme construction method assuming the formalization is basic operator method [1], [2]. The main idea of basic operator method is to construct the concordant system of difference analogues of fundamental invariant differential first order operators such as div, grad and rot. The operators concordance means that integral equations connected with the conservation laws are valid.

After the difference operators are built the difference scheme can be constructed by substitution of the difference operators in equations instead of differential ones.

Let us explain the fundamental ideas of basic operator method by example of difference scheme construction for thermal conduction. At first let us write the equation in invariant form:

$$\frac{du}{dt} = - \operatorname{div} \vec{W}, \quad \vec{W} = - \operatorname{grad} u, \tag{1}$$

where u is the temperature, $\vec{W}$ - thermal flow vector.

At second, it is necessary to find the properties of differential operators which should be kept for difference ones. For our equations there is the only conservation law:

$$\frac{dQ}{dt} = - \oint_s (\vec{W}, \vec{n}) \, ds, \quad \text{where } Q = \int_V u \, dv. \tag{2}$$

From formal view point the statement (2) follow from the equations (1) and the following property of operator div:

$$\int_V \operatorname{div} \vec{W} \, dv = \oint_S (\vec{W}, \vec{n}) \, ds.$$

The difference scheme will be conservative one if we'll construct difference analogue DIV of differential operator div on the basis of invariant definition

$$\operatorname{div} \vec{W} = \frac{\oint (\vec{W}, \vec{n}) \, ds}{V}$$

where V is mesh volume.

Excluded the temperature from equations (1) we get:

$$\frac{du}{dt} = \operatorname{div} \operatorname{grad} u = \Delta u,$$

where Δ is Laplacian.

It is known that $\Delta = \Delta^* > 0$. From formal view point this property is connected with that operators div and grad satisfy the equation

$$(3) \qquad \int_V u \operatorname{div} \vec{W} \, dv + \int_V (\vec{W} , \operatorname{grad} u) \, dv = \oint_S u (\vec{W} , \vec{n}) \, ds.$$

For the difference analogue of Laplacian to be selfadjoined and signfixed it is necessary that difference operators DIV and GRAD satisfy difference analogue of equation (3). It means that $GRAD = - DIV^*$ for zero temperature on the boundaries. So equation (3) does use for GRAD construction if operator DIV already is defined.

Thus the difference scheme for system of equations (1) looks as

$$\frac{du_h}{dt} = - DIV \vec{W}_h, \quad \vec{W}_h = - GRAD \, u_h$$

or

$$\frac{du_h}{dt} = DIV \, GRAD \, u_h.$$

Thus the difference scheme construction by the basic operator method is shown to reduce to production of conjugate operators, superposition of a few operators and arithmetical manipulations with them and with grid functions [3].

In this paper the grid functions and difference operators are analyzed as the object of symbolic transformations and fundamental algorithms for manipulations with them are presented.

From the view point of symbolic transformations the grid function is represented by variable with indexes that are changed within ranges defined by numbers or letters, for example $1 \le i \le N$. In what follows, for simplicity, we'll deal with grid function with the one index.

Grid function can have differ symbolic and numerical representation in differ intervals of index variation. For example

$$f_i = \begin{cases} a_i+1 & i=1,2 \\ x_i^2 & 3\le i\le N-2 \\ 5 & i=N-1,N \end{cases} \qquad g_i = \begin{cases} k & i=1 \\ y_{i-1} & 2\le i\le N-2 \\ 3 & i=N-1 \\ x_i & i=N \end{cases}$$

In what follows we shall consider the interval bound to be linear function of parameter N.

Thus we come to the new object - region of uniform symbolic representation of function or briefly *region of uniformity*. It is the region of index variation in which the function have one and the same symbolic representation. Function f have three regions of uniformity and function g - four. Denote regions of uniformity of function f as follows: $VF_1=\{1,2\}$, $VF_2=\{3,N-2\}$, $VF_3=\{N-1,N\}$. Analogously for function g: $VG_1=\{1\}$, $VG_2=\{2,N-2\}$, $VG_3=\{N-1\}$, $VG_4=\{N\}$.

In realizing arithmetic operations with grid functions the main problem is to define regions of uniformity for result grid functions. Let $r_i=f_i+g_i$, then it is easy to show that regions of uniformity for function r are represented non-empty intersections of regions VF_i and VG_j: $VR_k=VF_i\cap VG_j$, $i=1,..3$, $j=1,..,4$. As a result we have the following regions: $VR_1=VF_1\cap VG_1=\{1\}$, $VR_2=VF_1\cap VG_2=\{2\}$, $VR_3=VF_2\cap VG_2=\{3,N-2\}$, $VR_4=VF_3\cap VG_3=\{N-1\}$, $VR_5=VF_3\cap VG_4$. The intersections of pair of such regions can be found by using the REDUCE system.

Let us consider some actions with difference operators. Let we have operator $\mathbf{A}$: $H_1 \to H_2$, where spaces of grid functions H_1 and H_2 are connected with index sets IH_1 and IH_2 correspondingly. Operator $\mathbf{A}$ can be defined as follows:

$$(\mathbf{A}u)_i = \begin{cases} (\mathbf{A}^1u)_i & i \in VA^1 \\ (\mathbf{A}^2u)_i & i \in VA^2 \\ - - - - - - - - - - & \end{cases}$$

where expressions $(\mathbf{A}^ku)_i$ for $i\in VA^k$ have the same symbolic form. Regions VA^k are the regions of uniformity of operator $\mathbf{A}$. Connected to the each region of uniformity is stencil on which corresponding operators are defined. The stencil SA^k is the set of indexes which take part in expression $(\mathbf{A}^ku)_i$. For example, for $(\mathbf{A}^ku)_i=(u_{i+3/2}-u_{i-1/2})/2h$ we have $SA^k=\{i+3/2,i-1/2\}$. The stencil SA^k is described by two numbers:

$$LSA^k = \min_{s\in SA^k} s - i, \quad RSA^k = \max_{s\in SA^k} s - i,$$

which give us the distance with sign from grid index i to minimal and maximal stencil indexes. In our case we have $LSA^k=-1/2$, $RSA^k=3/2$.

Further let we have operator $\mathbb{B}$: $H_3 \to H_1$:

$$(\mathbb{B}u)_i = \begin{cases} (\mathbb{B}^1 u)_{i+1/2} & i+1/2 \in VB^1 \\ (\mathbb{B}^2 u)_{i+1/2} & i+1/2 \in VB^2 \\ - - - - - - - - - - - - - - - & \end{cases}$$

Our task is to find the regions of uniformity of operator $\mathbb{C}=\mathbb{A}*\mathbb{B}$: $H_3 \to H_2$. At first we construct the regions of uniformity in which the operator $\mathbb{C}$ can be represented in form $\mathbb{C}^k=\mathbb{A}^s*\mathbb{B}^l$. For this purposes for all regions VA^m we must define regions DA^m which involves indexes belonging to IH_1 and taking part in forming of expression $(\mathbb{A}^m u)_i$ for $i \in VA^m$:

$$DA^m = \left\{ \min_{s \in VA^m} s + LSA^m \ , \ \max_{s \in VA^m} s + RSA^m \right\}.$$

For construction of superposition $\mathbb{A}^s*\mathbb{B}^l$ to exist the set $PDA^{sl}=DA^s \cap VB^l$ must be non-empty. Let the set PDA^{sl} is non-empty. It is obviously that $PDA^{sl} \subset DA^s$. Therefore it is necessary to find corresponding to PDA^{sl} regions of uniformity of $VC^{sl} \subset VA^s$, i.e. it is necessary to find the ranges of function VC^{sl} according to the ranges of definition PDA^{sl}. It may be done as follows:

$$VC^{sl} = \left\{ \min_{r \in PDA^{sl}} r + LSA^m \ , \ \max_{r \in PDA^{sl}} r + RSA^m \right\}.$$

Non-empty sets of such type are the regions of uniformity of operator $\mathbb{C}$ in which $\mathbb{C}=\mathbb{A}^s*\mathbb{B}^l$. Besides, different stencil points being a member of different regions of uniformity of operator $\mathbb{B}$, operator $\mathbb{C}$ have special symbolic representation in each point from $IH_2 / \bigcup VC^{sl}$.

When we construct the sum of operators or the multiplication of an operator by grid function the same rules as one for arithmetic manipulations with grid functions are valid. When we construct the grid function being the result from effect of operator to other function the same rules as one for operator superposition construction are valid.

Let us consider briefly the process of conjugate operator constructing. Let we have operator $\mathbb{A}$: $H_1 \to H_2$. Then $\mathbb{A}^*$: $H_2 \to H_1$ and $(\mathbb{A}u,v)_{H_2} = (u,\mathbb{A}^*v)_{H_1}$ for all $u \subset H_1$ and $v \subset H_2$. Let us assume that inner product $(\cdot,\cdot)_{H_1}$ and the inner product in the space H_2 approximate the corresponding inner products in the continues function spaces, i.e. they approximate an integrals such as $\int u \ v \ dv$. These inner products we'll call *natural* ones. In addition we'll conconsider *formal* inner product defining as $[u,v]_{H_1} = \sum u_i v_i$, where summation is made by the index set corresponding to space H_1. Formal and natural inner products are connected by the equation $(u,v)_{H_1} = [M_1 u,v]_{H_1}$, where $M_1 = M_1^* > 0$. It may be shown that following equation is valid

$$A^* = M_1^{-1} \cdot A^{\bullet} \cdot M_2, \text{ where } [Au,v]_{H_2} = [u,A^{\bullet}v]_{H_1}$$

i.e. the operator $A^{\bullet}$ is the conjugate one to A for the formal inner product.

When we construct the operator $A^{\bullet}$ according to formula above we must build the inverse operator M_1^{-1}. We assume it possible only if the operator M_1 have diagonal form. The main problem in the operator A^* construction is to construct the region of uniformity of the operator $A^{\bullet}$. From formal view point the operator $A^{\bullet}$ construction is reduced to the summand regrouping in the sum $\sum (Au)_i \cdot v_i$ so as to pick the coefficient of u_i. It may be shown that the process of uniformity region construction is reduced to the manipulations with sets too.

Thus we showed that the manipulations with grid functions and difference operators are reduced mainly to manipulation with the sets which boundaries are the numbers or symbols. All these manipulations can being done by REDUCE system.

Described algorithms where realized as DISLAN system (DIfference Scheme LANguage) [4]. Let us give the DISLAN constructions examples. To the right we'll give the corresponding mathematical objects.

Grid function:

DEFINE SINGLE VARIABLE NAME: F INDEX DIAPASON I=1,N-1 VALUE X(I) INDEX DIAPASON I=N VALUE 1 END	$f_i = \begin{cases} x_i & i=1,..,N-1 \\ 1 & i=N \end{cases}$

Grid function spaces:

```
DEFINE SINGLE SPACE
NAME: HC
GLOBAL INDEX DIAPASON
I=1,N-1
SCALAR PRODUCT: (U,V)
INDEX DIAPASON
I=1,N-1
COMMON TERM
U(I)*V(I)*VC(I)
END
```

Grid function spaces with inner product

$$(u,v) = \sum_{i=1}^{N-1} u_i v_i VC_i$$

```
DEFINE SINGLE SPACE
NAME: HK
GLOBAL INDEX DIAPASON
I=2,N-1
SCALAR PRODUCT: (U,V)
INDEX DIAPASON
I=2,N-1
COMMON TERM
U(I)*V(I)*VK(I)
END
```

Grid function spaces with inner product

$$(u,v) = \sum_{i=2}^{N-1} u_i v_i VK_i$$

Difference operator:

```
DEFINE SINGLE OPERATOR
NAME: DIV
SPACE DEFINITION: HK
SPACE VALUE: HC
INDEX DIAPASON
I=2,N-2
COEFFICIENT IN KNOTS
KNOT: I
-1/VC(I)
KNOT: I+1
1/VC(I)
INDEX DIAPASON
I=1
COEFFICIENT IN KNOTS
KNOT: I+1
1/VC(I)
```

$$(DIV\ u)_i = \begin{cases} \dfrac{u_2}{VC_1} & i=1 \\ \dfrac{u_{i+1}-u_i}{VC_i} & i=2,\ldots,N-2 \\ -\dfrac{u_{N-1}}{VC_{N-1}} & i=N-1 \end{cases}$$

```
INDEX DIAPASON                          |
I=N-1                                   |
COEFFICIENT IN KNOTS                    |
KNOT: I                                 |
-1/VC(I)                                |
END                                     |
               Conjugate operator construction:
GRAD=DIV'                               |
                   Laplacian construction:
LAP=DIV*GRAD                            |
    Substitution of the grid function to the difference operator:
G=LAP(F)                                |
           Grid function and difference operator print:
PRINT G                                 |
PRINT LAP                               |
```

References

[1] Samarsky A.A., Tishkin V.F., Favorsky A.P., Shashkov M.Yu. The formulation of the differerence scheme of mathematical physics in operator form. Doklady Akademii Nauk, 1981, vol. 258, N 5.

[2] Samarsky A.A., Tishkin V.F., Favorsky A.P., Shashkov M.Yu. Operator difference scheme, Differential equations, 1981, vol. 17, N 7.

[3] Shashkov M.Yu., Shchenkov I.B. The symbolic transformation application for the construction and investigation of difference schemes. Journal of the calculation mathematics and mathematical physics. 1986, vol. 26, N 5.

[4] Shashkov M.Yu., Shchenkov I.B. DISLAN system, preprint N 23, Institute of applied mathematics, Moscow, 1985

THE USE OF META-LEVEL INFERENCE FOR SYMBOLIC SOLVING OF EQUATIONS.

P.Zielczyński, E.Pierzchała
Institute of Atomic Energy , 05-400 Otwock-Swierk, Poland.

1. Introduction

We have implemented in the Institute of Atomic Energy a system for trigonometric equation solving. The system is based on meta-level inference method proposed by the team of computer scientists from Edinbourgh who made the PRESS system for solving a large subset of algebraic equations /1//2/.

2. Meta-level inference

Usually a very large solving state space is required during symbolic solving of equations. In order to decrease this state space the use of two inference levels is proposed. In every step of equation transformation, first, with the aid of few conditions, the "method" (meta-level problem-solving operator) is chosen. Next, a low-level operator is chosen from a relatively small set of operators of this method (and not from a large set of all possible operators). After applying chosen operator the procedure is repeated for the next processing step. Conditions of particular methods are checked in the fixed order. If the condition is fulfilled, one of operators of this method is chosen for execution. After transformation of equation with chosen operator, new equation is treated in the same way (choose of method, choose of operator, execution of operator). If it is impossible to choose an operator, backtracking is performed.

Here are some examples of methods:

2.1. Isolation

a. Condition of the method:
Single appearance of unknown.

b. Goal of the method:
Elimination of the most external function from the side where is the unknown. For example in the equation:
1+sin(5*x+2)=0
the addition is the most external function, so isolation will transform this equation into:
sin(5*x+2)=-1
The next use of isolation will perform in turn:
- elimination of the sinus function,
- substraction of 2 from the both sides,
- both sides division by 5.

c. Examples of operators:
```
sin U = W     -->    U = arcsin W
U + A = W     -->    U = W - A
U * A = W     -->    U = W / A
- U = A       -->    U = - A
```

2.2. Factorisation

a. Condition of the method:
The equation is of the form:
U1 * U2 * U3 * U4 * ... = 0

b. Goal of the method:
Breaking the equation into components:
U1=0 U2=0 U3=0 U4=0 ...

2.3. Polynomial method

a. Condition of the method:
The equation has a form:
$A * x^2 + B * x + C = 0$.

b. Goal of the method:
Finding the two possible values of unknown.

2.4. Change of unknown.

a. Condition of the method:
Multiple appearance of unknown, all appearances in the same expressions.

b. Goal of the method:
Substitution of the expression with an old unknown by the new unknown, creation of two equations.

2.5. Collection

a. Condition of the method:
Multiple appearance of unknown.

b. Goal of the method:
Reduction of the number of appearances of unknown.
For example the expression cos(x)*sin(x)
can be replaced by 0.5*sin(2*x).

c. Examples of operators:
$(\sin U)^2 + (\cos U)^2$ --> 1
$(\cos U)^2 - (\sin U)^2$ --> cos(2*U)
sin U / cos U --> tan U

2.6. Attraction

a. Condition of the method:
Multiple appearance of unknown.

b. Goal of the method:
Attraction of one unknown to another.
For example $\sin A * (\cos x)^2 - \sin A * (\sin x)^2$
can be replaced by $\sin A * ((\cos x)^2 - (\sin x)^2)$.

c. Example of an operator:
A * U + A * W --> A * (U + W),
(unknown appears in U and in W).

2.7. Homogenisation

a. Condition of the method:
   Multiple appearance of unknown.

b. Goal of the method:
   Unification of expressions with unknown.
   For example $3\cos^2 x + \sin^2 x = 1$.
   can be replaced by $3\cos^2 x + 1 - \cos^2 x = 1$.

c. Example of an operator:
   $\sin^2 U \rightarrow 1 - \cos^2 U$

3. Example of program execution

Five methods were implemented in the prototype version of the system (isolation, factorisation, polynomial method, collection and attraction). Here is the example of solving an easy equation by this system:

```
Goal : solve("(cos(4 + x)) ^ 2 - (sin(4 + x)) ^ 2 = 0.33","x").
cos(2*(4+x))=0.33
(2*(4+x))=n*360+arccos(0.33)
2*(4+x)=n*360+arccos(0.33)
(4+x)=(n*360+arccos(0.33))/(2)
4+x=(n*360+arccos(0.33))/2
x=(n*360+arccos(0.33))/2-(4)
Result :     x=(n*360+arccos(0.33))/2-4
```

4. Conclusions

The system was implemented in two versions - in Lisp and in Prolog in order to compare usefullness of these languages to symbolic processing on IBM PC XT and AT /3/. As we expected, Prolog (Turbo Prolog version) was more compact for algorithm and knowledge base description, but the speed of compiled program was lower than the speed of analogical program written in Lisp (Golden Common Lisp version).

The system is based on expert system techniques what made a knowledge base with method and operator description very easy to mantain. It allows one to add new methods and operators in order to increase the power of the system.

The use of meta-level inference significantly decreases solving space: the time of solving average equation on IBM PC usually is not longer than one second.

References.

/1/ A. Bundy, R. Welham, "Using Meta-Level Inference for Selective Application of Multiple Rewrite Rule Sets in Algebraic Manipulation", Artificial Intelligence 16(1981) pp 189-212.

/2/ B. Silver, "Meta-Level Inference", North Holland, 1986.

/3/ E.Pierzchała, P.Zielczyński, "Zastosowanie wnioskowania dwupoziomowego do symbolicznego rozwiązywania wybranej klasy równań trygonometrycznych", IEA 0-58/CYF/87, Swierk 1987.

REPRESENTATION OF ALGEBRAIC CURVES *)

F. Winkler
RISC-LINZ, J. Kepler Universität, Linz, Austria

Abstract

If algebraic varieties like curves or surfaces are to be manipulated by computers, it is essential to be able to represent these geometric objects in an appropriate way. For some applications an implicit representation by algebraic equations is desirable, whereas for others an explicit or parametric representation is more suitable. Therefore, transformation algorithms from one representation to the other are of utmost importance.

We investigate the transformation of an implicit representation of a plane algebraic curve into a parametric representation. In the course of the transformation algorithm the coefficient field has to be extended algebraically. If the known parametrization algorithms are used uncritically, the algebraic extensions get so high that any computation becomes practically impossible. Our goal is to keep the degree of the necessary algebraic extension as small as possible.

Introduction

An algebraic variety V, the main object of study in algebraic geometry, can be represented in various different ways, for instance as the set of zeros of polynomial equations

$$V = \{(x,y) \mid 2x^4 - 3x^2y + y^2 - 2y^3 + y^4 = 0,\ x,y \in \mathbb{C}\},$$

or as the set of values of rational functions

$$V = \{(\phi(t),\chi(t)) \mid \phi(t) = -\frac{18t^4+21t^3-7t-2}{18t^4+48t^3+64t^2+40t+9},$$
$$\chi(t) = \frac{36t^4+84t^3+73t^2+28t+4}{18t^4+48t^3+64t^2+40t+9},\quad t \in \mathbb{C}\}.$$

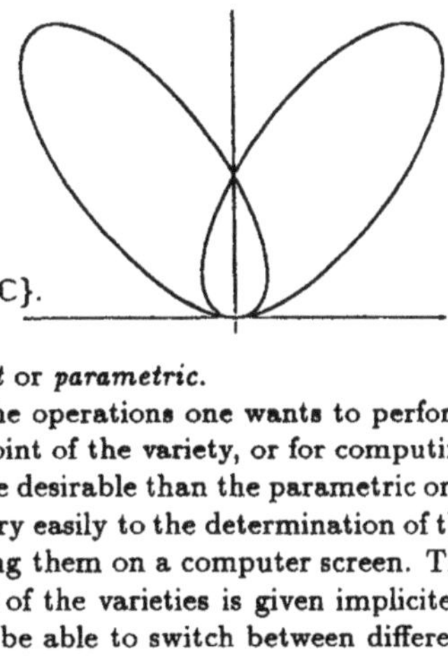

We call the first representation *implicit* and the second *explicit* or *parametric*.

The representation of choice is of course determined by the operations one wants to perform with the variety. For determining whether a given point is a point of the variety, or for computing singular points of the variety, the implicit representation is more desirable than the parametric one. On the other hand, the parametric representation lends itself very easily to the determination of the curvature, to tracing of varieties, and in particular to visualizing them on a computer screen. The intersection of varieties can be determined rather easily if one of the varieties is given implicitely and the other one explicitely. For this reason it is essential to be able to switch between different representations.

In /3/ the problem of computing the implicit equations from a given parametric representation is treated. Recently the application of the Gröbner basis method to the implicitization problem has been further investigated in /5/. The reverse problem, namely computing a rational parametrization from the given implicit equations, is a classical problem in algebraic geometry, see for instance /8/. In /2/,/4/ the problem of parametrization for space curves is reduced to the problem of parametrization of plane curves. In this paper we will deal only with the problem of parametrizing plane curves.

*) This research was supported by the Austrian *Fonds zur Förderung der wissenschaftlichen Forschung* under Projekt Nr. P6763.

Theoretically the problem of parametrization of plane curves is solved, and it is known that the parametrizable curves are exactly the curves of genus 0. In [8] also an algorithm is suggested for computing a rational parametrization. Instead of a pencil of degree $d-2$, as in [1], one could also use pencils of degree $d-1$ and d in the parametrization algorithm. In fact, these pencils are more attractive from a computational point of view. The determination of simple points on the curve introduces a lot of algebraic numbers. If they are not controlled, the computation soon becomes too inefficient. We show that a pencil can be passed through a set of points on the given curve without having to compute these points explicitly. It turns out that if F is a field in which the coordinates of the singular points of the curve C can be represented, then a rational parametrization of C can be found in an extension field of degree $\deg(C)$ over F. Our conjecture, however, is that for a curve of odd degree we need no algebraic extension of F. In this paper we can only state the results of our work. For details on the methods and proofs we refer to [7].

Let K be an algebraically closed field of characteristic 0. We will consider curves in the affine and projective planes over K. If the curve C is defined by $f(x,y)=0$ then the associated projective curve C^* is defined by $F(x,y,z)=0$, where F is the homogenization of f.

The irreducible affine curve C defined by the irreducible polynomial $f(x,y)\in\mathsf{K}[x,y]$ is *rational* iff there exist rational functions $\phi(t),\chi(t)\in\mathsf{K}(t)$ such that

(1) for almost all (i.e. for all but a finite number of exceptions) $t_0\in\mathsf{K}$, $(\phi(t_0),\chi(t_0))$ is a point on C, and

(2) for almost every point (x_0,y_0) on C there is a $t_0\in\mathsf{K}$ such that $(x_0,y_0)=(\phi(t_0),\chi(t_0))$.

If ϕ,χ satisfy the conditions (1) and (2), (ϕ,χ) is a *rational parametrization* of C.

The notion of rationality for affine curves can be extended in a natural way to a notion of rationality for projective curves. This is achieved by introducing a third rational function $\psi(t)$ and postulating the conditions (1),(2). In fact, a parametrization of C can be immediately obtained from a parametrization of the associated curve C^* in the projective plane and vice versa.

With this terminology we can state the problem of parametrization.

<u>Parametrization problem</u>:

given: an irreducible polynomial $f(x,y)\in\mathsf{K}[x,y]$ defining an irreducible affine algebraic plane curve C

decide: the rationality of C

find: (if C is rational) rational functions $\phi(t),\chi(t)\in\mathsf{K}(t)$ such that (ϕ,χ) is a rational parametrization of C. □

In the sequel we exclude the case where the degree of the polynomial f defining the curve C is 1, i.e. where C is a line. Obviously the parametrization of lines does not present a problem.

A singular point P of multiplicity r on the affine curve C defined by $f(x,y)$ is an *ordinary* singular point iff the r tangents to C at P are distinct. Otherwise P is called *non-ordinary*. The property of a singular point P of being ordinary or non-ordinary is called the *character* of P. An important result about singularities is the fact that if C is an irreducible projective or affine curve of degree d having multiplicities r_P at points P, then $(d-1)(d-2)\geq\sum_{P\in C}r_P(r_P-1)$. In the special case of an irreducible projective curve C^*, defined by $F(x,y,z)=0$, having only ordinary singularities one can characterize the rationality as follows. If $r_1,\ldots,r_n$ are the multiplicities of the singular points of C^*, C^* is rational if and only if $(d-1)(d-2)=\sum_{i=1}^{n}r_i(r_i-1)$. In the general case, for characterizing the rationality of a plane curve one usually introduces the concept of neighboring points. If we include also the neighboring singular points, then the above equation is a criterion for the rationality of a curve C.

A parametrization algorithm

Let us assume that the irreducible projective curve C^* of degree d defined by $F(x,y,z)=0$ is rational. If C^* has a $(d-1)$–fold point, then it is rational and a parametrization can be determined by cutting C^* with lines passing through this $(d-1)$-fold point. By Bezout's theorem there will be exactly one additional intersection point depending on the slope of the line, yielding the desired parametrization. This idea may be generalized. In the general situation one can also construct a pencil of curves such that for almost every curve in the pencil all its intersection points with C^*, except one, are predetermined. Moreover, all the predetermined intersection points are the same for every curve in the pencil. Thus, if one computes the intersection points of a generic element of the pencil with C^*, the expression of the unknown intersection point gives the parametrization of the curve by means of the parameter defining the pencil. Here we discuss only pencils of degree d.

Let us assume that D^* is a generic representative of a pencil of curves of degree d. Then in general D^* has d^2 intersections with C^*. We postulate that D^* satisfies the properties:

(1) every r–fold singular point (including the neighboring ones) on C^* is an $(r-1)$–fold point on D^*,

(2) there exist $3d-3$ simple points on C^* that are also simple points on D^*,

(3) C^* and D^* do not have a common component.

In this way, we force D^* to have some specific common points with C^*. In the sequel, we will refer to these points as the *fixed common points* of C^* and the pencil. The intersection multiplicity of C^* and D^* at the singular points P of C^* (including the neighboring ones) is at least $\sum r_P(r_P-1) = (d-1)(d-2)$, where r_P is the multiplicity of P on C^*. So by condition (2) we fix just so many simple intersection points of C^* and D^* as to leave at most one intersection point undetermined.

LEMMA 1: *The pencil of curves D^* of degree d satisfying (1) - (3) can be effectively computed and the coefficients of the pencil are polynomials in one free parameter. Almost every curve in the pencil intersects C^* in one additional point and for almost every simple point Q on C^* which is not one of the fixed common points there exists a curve in the pencil intersecting C^* at Q.* □

Having determined the pencil D^*, we want to compute a formula for the unknown intersection point of an arbitrary curve in the pencil with the given rational curve C^*. By resultant computations we will derive the rational parametrization from this formula. In the sequel we denote by $\mathrm{Res}_v(p,q)$ the resultant of the polynomials p and q w.r.t. the variable v.

Let t be the independent parameter of D^*. Let us also suppose that $P_i = (\lambda_i, \mu_i, \rho_i)$, $1 \le i \le n$, are the singular points of C^*, where P_i is a point of multiplicity r_i on C^*, and that $Q_i = (\bar\lambda_i, \bar\mu_i, \bar\rho_i)$, $1 \le i \le 3d-3$, are the fixed common simple points of C^* and D^*. Let $\bar r_i := r_i(r_i-1) + \sum_{P\in N(P_i)} r_P(r_P-1)$, $1 \le i \le n$, where $N(P_i)$ is the set of neighboring points of P_i w.r.t. C^*. Let C be the affine curve associated with C^*, i.e. C is defined by $f(x,y) = F(x,y,1)$. Let $h(x,y) = H(x,y,1)$ be the dehomogenization of the defining polynomial of the pencil D^*.

THEOREM 2: *With the notation introduced above there exist nonzero polynomials $m_1(t), m_2(t), n_1(t), n_2(t) \in \mathbb{K}[t]$ such that*

$$\mathrm{Res}_x(f,h) = \prod_{i=1}^{n}(\rho_i y - \mu_i)^{\bar r_i} \cdot \prod_{i=1}^{3d-3}(\bar\rho_i y - \bar\mu_i) \cdot (m_1(t)y - n_1(t)),$$

$$\mathrm{Res}_y(f,h) = \prod_{i=1}^{n}(\rho_i x - \lambda_i)^{\bar r_i} \cdot \prod_{i=1}^{3d-3}(\bar\rho_i x - \bar\lambda_i) \cdot (m_2(t)x - n_2(t)).$$ □

THEOREM 3: *Let C, C^*, D^*, F, H, f, h be as above. If $(u(t)x - v(t))$ and $(\bar u(t)y - \bar v(t))$ are the factors of $\mathrm{Res}_y(f,h)$ and $\mathrm{Res}_x(f,h)$ depending on t, respectively, then $\big(x(t) = v(t)/u(t),\ y(t) = \bar v(t)/\bar u(t),\ z(t) = 1\big)$ is a parametrization of C^* and $\big(x(t) = v(t)/u(t),\ y(t) = \bar v(t)/\bar u(t)\big)$ is a parametrization of C.* □

So we arrive at the following algorithm for computing a rational parametrization of an irreducible affine rational curve.

Algorithm PARAMETRIZE

The input is an irreducible affine rational curve C of degree d, defined by the irreducible polynomial $f(x,y)$. The output is a rational parametrization of C.

(1) Determine the singularities of the projective curve C^* associated with C, including the neighboring ones, and their multiplicities. Determine $3d-3$ simple points on C^*. Determine the

pencil D^* as it has been described above. Let $H(x,y,z)$ be the polynomial defining the pencil, $h(x,y) = H(x,y,1)$.

(2) Compute $S_1(y) = \mathrm{Res}_x(f,h)$ and $S_2(x) = \mathrm{Res}_y(f,h)$.

(3) Remove the factors corresponding to the common points of C^* and D^* from $S_1(y)$ and $S_2(x)$, as described in Theorem 2.

(4) Solve the linear system of equations $S_1(y) = 0$, $S_2(x) = 0$, where x, y are the unknowns. Let $(R_1(t), R_2(t))$ be the solution.

(5) Return the parametrization $(R_1(t), R_2(t))$. □

Of course we are interested in parametrizations which need only few, if any, algebraic numbers. Let F be a field of characteristic zero in which all the field operations can be carried out effectively and in which the singularities of the curve C can be represented. The remaining problem then is the selection of the simple common points of the curve C^* and the pencil D^*. We propose to use whole conjugacy classes of simple points in order to keep the necessary algebraic extension as low as possible.

LEMMA 4: *Let $q \in \mathsf{F}[s]$ with $\deg(q) = n \le a$, $\{\beta_1, \dots, \beta_n\}$ the roots of q, and $a_1, a_2, a_3, b_1, b_2, b_3 \in \mathsf{F}$.*

(a) The points $\{(a_1 + b_1\beta_i : a_2 + b_2\beta_i : a_3 + b_3\beta_i)\}_{i=1,\dots,n}$ are on D^* if and only if $q(s)$ divides $H(a_1 + b_1 s, a_2 + b_2 s, a_3 + b_3 s)$.

(b) If $\{(a_1 + b_1\beta_i : a_2 + b_2\beta_i : a_3 + b_3\beta_i)\}_{i=1,\dots,n}$ are common points of C^* and D^*, then $q(\frac{a_2 - a_3 y}{b_3 y - b_2}) \cdot (b_3 y - b_2)^n$ divides $\mathrm{Res}_x(f,h)$ and $q(\frac{a_1 - a_3 x}{b_3 x - b_1}) \cdot (b_3 x - b_1)^n$ divides $\mathrm{Res}_y(f,h)$. □

The algorithm constructing the classes of conjugate common simple points works as follows.

Algorithm SIMPLE

The input to SIMPLE is an irreducible rational curve C^* defined by the polynomial $F(x,y,z)$. The output consists of three distinct whole classes of conjugate simple points on C^*, each class containing $d-1$ points. One algebraic number β of degree at most d has to be introduced.

(1) Choose $b_1, b_2 \in \mathsf{F}$ such that for every singular point $(\rho_1 : \rho_2 : \rho_3)$ of C^* we have $b_2\rho_1 - b_1\rho_2 \neq 0$ (i.e. b_1, b_2 cannot be the first two coordinates of a singular point of C^*). Compute an irreducible monic factor of the univariate polynomial $F(b_1, b_2, s)$, say $q(s)$. Now, since b_1, b_2 are not the first two coordinates of a singular point of C^*, $P = (b_1 : b_2 : \beta)$, with $q(\beta) = 0$, is a simple point on C^*.

(2) Choose $\lambda_1, \lambda_2, \lambda_3, \mu_1, \mu_2, \mu_3 \in \mathsf{F}$, such that
   (a) $\lambda_i\mu_j - \lambda_j\mu_i \neq 0$ for $i \neq j$,
   (b) $\mathrm{Res}_s(\bar{q}_i(s), \bar{q}_i'(s)) \neq 0$ for $i = 1,2,3$, where $\bar{q}_i(s) = F(\lambda_i s + b_1, \mu_i s + b_2, s + \beta)$ and $\bar{q}_i'$ denotes the derivative of q_i w.r.t. s.

(3) For $i = 1,2,3$ set $q_i(s) := \bar{q}_i(s)/s \in \mathsf{F}(\beta)[s]$.

(4) Now $\{(\lambda_i\beta_i + b_1 : \mu_i\beta_i + b_2 : \beta_i + \beta)\}_{q_i(\beta_i)=0}$, $i = 1,2,3$, are three distinct whole classes of $(d-1)$ simple points each on C^*. □

THEOREM 5: *A parametrization of a curve of degree d can be found in an extension of degree d over* F.

Proof: Use algorithms PARAMETRIZE and SIMPLE. □

EXAMPLE: Applying the algorithms PARAMETRIZE and SIMPLE to the implicitely given curve at the beginning of this paper and choosing $(2/9, 4/9)$ as the simple point in step (1) of SIMPLE, we get the associated parametrization. □

A number theoretical problem

These considerations allow to keep the algebraic extension of the field small. But even the bound given in Theorem 5 might be far to pessimistic. J. Schico, a Ph.D. student at RISC-LINZ, has given the following reduction of the problem of finding a parametrization with few algebraic numbers to a number theoretical problem.

Let C be an affine rational curve of degree d. Let it have k singularities $P_1, \ldots, P_k$, including the neighboring ones, with multiplicities $s_1, \ldots, s_k$, respectively. Since the genus of C is 0, we have

$$\sum_{i=1}^{k} s_i(s_i - 1) = (d-1)(d-2). \quad (1)$$

Let S be the linear system of curves of degree a ($\leq d$) that have P_i, $1 \leq i \leq k$, as an r_i–fold point. The dimension of the system S is

$$\dim(S) = t = \frac{a(a+3)}{2} - \sum_{i=1}^{k} \frac{r_i(r_i+1)}{2}. \quad (2)$$

S has $\sum_{i=1}^{k} r_i s_i$ intersections with C. Thus,

$$t \leq ad - \sum_{i=1}^{k} r_i s_i, \quad (3)$$

because there is a curve in S with at least $t + \sum_{i=1}^{k} r_i s_i$ intersections with C, and the number of intersections cannot be greater than ad by Bezout's theorem. We are particularly interested in the case where equality holds in (3). Suppose

$$t = ad - \sum_{i=1}^{k} r_i s_i \geq 1. \quad (4)$$

Then we define a subsystem R of S intersecting C at $t-1$ simple points. R has dimension 1. All the intersection points of R and C, except one, are fixed. The pencil R can now be used to parametrize the curve C along the lines described in the previous chapter. In order to avoid algebraic field extensions, t should be as small as possible. So we arrive at the following number theoretical problem:

Pencil selection problem:
given: nonnegative integers $d, k, s_1, \ldots, s_k$ satisfying the condition (1),
find: $t, a(1 \leq a \leq d), r_1, \ldots, r_k$ that solve (2) and (4), where t is as small as possible. □

If d is odd and the curve C has at least $l = (d-3)/2$ double points, say $P_1, \ldots, P_l$, then there exists a solution to the pencil selection problem with $t = 1$, namely $a = d-2, t = 1, r_1 = \ldots = r_l = 2, r_i = s_i - 1$ for $l+1 \leq i \leq k$. So for curves of this type no algebraic extension is needed. If d is even and the curve C has at least $l = (d/2)-2$ double points, say $P_1, \ldots, P_l$, then there exists a solution to the pencil selection problem with $t = 2$, namely $a = d-2, t = 1, r_1 = \ldots = r_l = 2, r_i = s_i - 1$ for $l+1 \leq i \leq k$. So for curves of this type only an algebraic extension of degree d is needed. This bound on the field extension coincides with Theorem 5.

The conjecture, based on computer aided searches for solutions to the pencil selection problem, is that for every curve of odd degree a solution with $t = 1$ can be found, and for every curve of even degree a solution with $t = 2$ can be found.

References

1. Abhyankar, S.S., Bajaj, C.L. (1988). Automatic parameterization of rational curves and surfaces III: Algebraic plane curves. *Computer Aided Geometric Design* **5**, 309-321.
2. Abhyankar, S.S., Bajaj, C.L. (1990). Automatic rational parametrization of curves and surfaces IV: Algebraic space curves. To appear in *ACM Trans. on Graphics.*
3. Arnon, D.S., Sederberg, T.W. (1984). Implicit equation for a parametric surface by Groebner basis. *Proc. 1984 MACSYMA User's Conference*, V.E. Golden (ed.).
4. Garrity T., Warren J. (1989). On computing the intersection of a pair of algebraic surfaces. *Computer Aided Geometric Design* **6**, 137-153.
5. Kalkbrener, M. (1990). Implicitization of rational parametric curves and surfaces. Submitted to *AAECC-8.*
6. Sederberg, T.W. (1986). Improperly parametrized rational curves. *Computer Aided Geometric Design* **3**, 67-75.
7. Sendra, J.R., Winkler, F. (1989) *A Symbolic Algorithm for the Rational Parametrization of Algebraic Plane Curves.* Techn. Rep. RISC 89-41.0, RISC-LINZ, J.Kepler Univ., Linz, Austria.
8. Walker, R.J. (1978). *Algebraic Curves.* Springer-Verlag, New York-Heidelberg-Berlin.

GEOMETRICAL APPROACH TO THE ALGORITHMIC PROBLEMS OF TENSOR ALGEBRA MANIPULATIONS

V.Ilyin, A.Kryukov, A.Rodionov, A.Taranov
INP Moscow State Univ., 119899 Moscow, USSR

ABSTRACT

The unified approach based on simple geometrical ideas is suggested for symbolic handling symmetries, dummy indices and linear identities of tensors.

1. Introduction

There are two basic approaches to tensor manipulation in the computer algebra framework. In the first approach tensors are treated as multi-component objects i.e. arrays possibly having symbolic rather then numeric components. This approach though very efficient in some applications (see, for example, a paper by M.McCallum in this proceedings) has an intrinsic limitation, namely the dimension of space underlying tensor algebra must be fixed number. The second approach treats tensors as formally indexed values (refs.[1,2]). This allows one to work with unspecified space dimension, but also poses some specific problems. Considering a tensor as a formally indexed object makes very important to take into account all properties of tensors under consideration allowing to reduce tensor expressions. By these we mean here the following

S - symmetry properties of a tensor with respect to some permutations of its indices;
D - invariance of tensor monomials under renaming of dummy summation indices;
I - linear identities for tensors.

The most familiar (and probably most important) example of such properties is supplied by Riemann tensor:

S' - $\mathbf{R}_{abcd} = \mathbf{R}_{cdab}$, $\mathbf{R}_{abcd} = - \mathbf{R}_{bacd}$;

D' - $\mathbf{R}_{abcd} \cdot \mathbf{R}_{ckmn} \cdot \mathbf{R}_{dlps} = \mathbf{R}_{abdc} \cdot \mathbf{R}_{dkmn} \cdot \mathbf{R}_{clps}$;

I' - $\mathbf{R}_{abcd} + \mathbf{R}_{acdb} + \mathbf{R}_{adbc} = 0$.

Let's consider the following problem. Suppose we have two expressions. Whether these are equal or not? For example,

$$F_{abmnmncd} \equiv R_{abmn} \cdot R_{mncd} \quad \text{and} \quad F_{bakl cdlk} \equiv R_{bakl} \cdot R_{cdlk}$$

are equal since we can perform

$$R_{bakl} \overset{S}{\Longrightarrow} -R_{abkl}, \qquad R_{cdlk} \overset{S}{\Longrightarrow} -R_{cdkl} \overset{S}{\Longrightarrow} -R_{klcd},$$

$$R_{abkl} \cdot R_{klcd} \overset{D}{\Longrightarrow} R_{abmn} \cdot R_{mncd}.$$

Here letters S and D indicate use of symmetry properties and renaming dummy indices respectively.

Note that an application of linear identities can be considered as a multiterm substitutions and hence one such identity can produce several rewriting rules. For example the identity (I) could be used in forms

$$R_{abcd} + R_{acdb} + R_{adbc} \overset{I}{\Longrightarrow} 0, \quad \text{or} \quad R_{abcd} + R_{acdb} \overset{I}{\Longrightarrow} -R_{adbc},$$

$$\text{or} \quad R_{abcd} \overset{I}{\Longrightarrow} -R_{acdb} - R_{adbc}.$$

A naive approach to the problem of comparing of two tensors is to try to find some sequence of transformations, chosen among (S), (D) and (I), that reduce one tensor to the other. This approach can be illustrated by following diagram

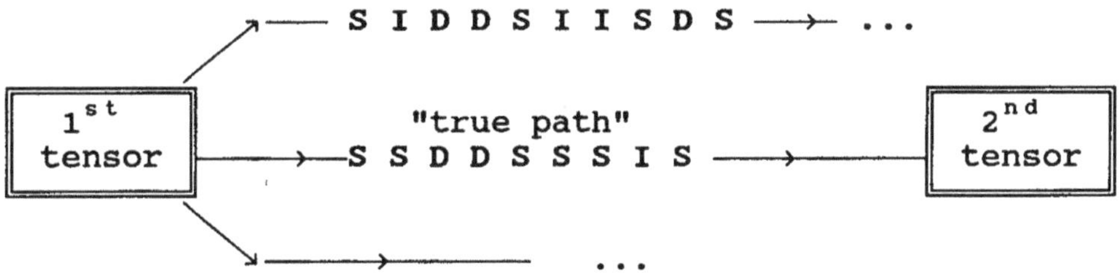

It is important to realize that due to linear identity application one can get intermediate expressions which look much more complicated than initial or final tensor. This explains why simple strategies of reduction may fail.

Note that if we deal with dummy indices and symmetry properties only the problem can be solved in a group-theoretic framework (in terms of double cosets, see ref.[2]). But if linear identities are to be taken into account this approach fails.

We are suggesting unified approach for handling symmetries, dummy indices and linear identities.

2. Definitions and notations

Let a tensor **F** has valence **n**. Denote its indices by $\mu_1, \ldots, \mu_n$. With this set of indices we are able to construct **n**! formally different objects

$$F_{\pi(\mu_1\ldots\mu_n)} \qquad \pi = \begin{Bmatrix} 1 & \ldots & n \\ \pi(1) & \ldots & \pi(n) \end{Bmatrix} \in S_n ,$$

S_n is the group of permutations of the set $(1,\ldots,n)$ and $\pi(\mu_1,\ldots,\mu_n)$ means the set $\left(\pi(\mu_1),\ldots,\pi(\mu_n)\right)$.

If **F** has a symmetry with respect to index permutations it means that a subgroup **H** of S_n and a representation d: $H \longrightarrow \mathbb{R}^1$ exists such that

$$F_{h\circ\pi(\mu_1\ldots\mu_n)} = d(h)\cdot F_{\pi(\mu_1\ldots\mu_n)},$$

$\forall$ $h\in H$ and $\pi\in S_n$, $h\circ\pi(\mu_1,\ldots,\mu_n)\equiv\left(\pi(h(\mu_1)),\ldots,\pi(h(\mu_n))\right)$.

Suppose now that among $(\mu_1,\ldots,\mu_n)$ there are some pairs of dummy indices. Without any loss of generality we may suppose that their names are already normalized in some way, i.e. they have fixed names. So the only remaining freedom is to interchange these names. Under this restriction the freedom to interchange dummy indices means that a subgroup $M\in S_n$ exists such that:

$$F_{\pi\circ m(\mu_1\ldots\mu_n)} = F_{\pi(\mu_1\ldots\mu_n)}, \qquad \forall\ m\in M \text{ and } \pi\in S_n.$$

Consider now linear identities which can be written as

$$\sum_{\pi\in S_n} \alpha^i_\pi \cdot F_{\pi(\mu_1\ldots\mu_n)} = 0 , \qquad i=1,\ldots,I \text{ and } \alpha^i_\pi\in\mathbb{R}^1.$$

3. Linear space

Consider linear space $\mathbb{R}^{n!}$ spanned by permutations $\pi\in S_n$, i.e. each $F_{\pi(\mu_1\ldots\mu_n)}$ is corresponded to unit vector $e_\pi\in\mathbb{R}^{n!}$. Now we have an explicit isomorphism between tensor expressions and points of $\mathbb{R}^{n!}$:

$$\sum_\pi \beta_\pi\cdot e_\pi \qquad \longleftrightarrow \qquad \sum_\pi \beta_\pi\cdot F_{\pi(\mu_1\ldots\mu_n)}, \qquad \beta_\pi\in\mathbb{R}^1 .$$

Rewrite S-D-I relations in the terms of $\mathbb{R}^{n!}$:

S - $\quad e_{h\circ\pi} - d(h)\cdot e_\pi = 0, \qquad h\in H, \quad \pi\in S_n;$

D - $\quad e_{\pi\circ m} - e_\pi = 0, \qquad m\in M, \quad \pi\in S_n;$

I - $\quad \sum_{k=}^{n!} \alpha^i_k\cdot e_k = 0, \qquad i=1,\ldots,I.$

4. S-D-I subspace

Now we can consider left hand sides of **S**, **D** and I relations as vectors defining subspaces in $\mathbb{R}^{n!}$. In terms of Euclidean metric we can split $\mathbb{R}^{n!}$ in orthogonal complements

$$\mathbb{R}^{n!}=\mathbf{I}^{SDI}\oplus\mathbb{C}.$$

Here $\mathbf{I}^{SDI}$ is linear subspace spanned by vectors

$$\mathbf{e}_{h\circ\pi}-\mathbf{d}(h)\cdot\mathbf{e}_{\pi}, \qquad \mathbf{e}_{\pi\circ m}-\mathbf{e}_{\pi}, \qquad \sum_{k=1}^{n!}\alpha_k^i\cdot\mathbf{e}_k;$$

$$h\in\mathbf{H}\ ,\quad m\in\mathbf{M}\ ,\quad \pi\in S_n \quad \text{and} \quad i=1,\ldots,I\ .$$

Let's remind that every point of $\mathbb{R}^{n!}$ corresponds to tensor expression. It is obvious that all points of $\mathbb{R}^{n!}$ lying on $\mathbf{I}^{SDI}$ correspond to tensor expressions which are equal to zero due to **S-D-I** properties. So any two points lying in the plane parallel to $\mathbf{I}^{SDI}$ correspond to equal tensor expressions.

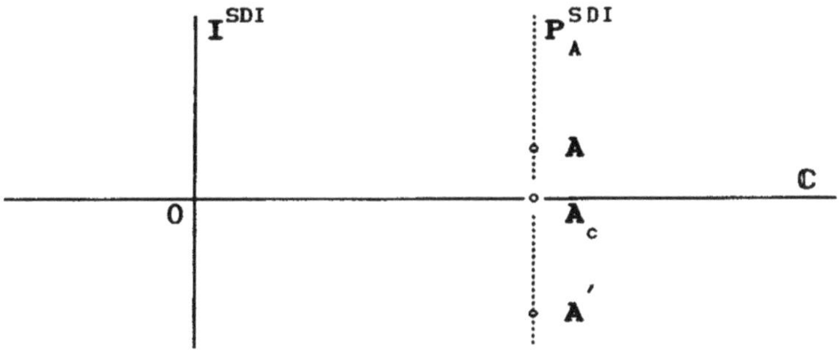

$$(\mathbf{A}-\mathbf{A}') \quad \longleftrightarrow \quad \sum_{\pi}\alpha_{\pi}^{SDI}\cdot\mathbf{F}_{\pi(\mu_1\ldots\mu_n)}=0\ .$$

Let's define $\mathbf{A}_c\in\mathbb{C}$ as a canonical representative of points of a plane $\mathbf{P}_{\mathbf{A}}^{SDI}$ (parallel to $\mathbf{I}^{SDI}$ and containing vector **A**). Having a canonical vector of a class of equivalent tensor expressions we solve the problem of comparing of tensors by comparing their canonical vectors.

5. Example

Consider simple example illustrating our approach in case when only **S**-relations are present. Let a tensor **F** with two indices has the symmetry

$$\mathbf{F}_{\mu\nu}=\frac{1}{2}\cdot\mathbf{F}_{\nu\mu}.$$

The corresponding space $\mathbb{R}^2$ is spanned on unit vectors

$$\mathbf{e}_1 \longleftrightarrow \mathbf{F}_{\mu\nu} \qquad \text{and} \qquad \mathbf{e}_2 \longleftrightarrow \mathbf{F}_{\nu\mu} \ .$$

S- relations have the form

$$\mathbf{e}_1 - \frac{1}{2}\cdot\mathbf{e}_2 = 0 \qquad \text{and} \qquad \mathbf{e}_2 - \frac{1}{2}\cdot\mathbf{e}_1 = 0 \ .$$

Now we immediately get $\mathbb{R}^2=\mathbf{I}^S$ and $\mathbb{C}=\varnothing$. This means that a canonical representative is zero for any point of $\mathbb{R}^2$ - our tensor equal to zero : $\mathbf{F}_{\mu\nu}\equiv 0$.

6. Sketch of an algorithm

Step 1:
calculate vectors of $\mathbb{R}^{n!}$ *spanning a subspace* $\mathbf{I}^{SDI}$.

Step 2:
construct basis $(\mathbf{c}_1,\ldots,\mathbf{c}_J)$ *of an orthogonal complement to* $\mathbf{I}^{SDI}$ *(J is dimension of* $\mathbb{C}$*).*

After these preliminary steps any pair of tensor expressions (points $\mathbf{A}_1$ and $\mathbf{A}_2$) is compared simply by

Step 3:
evaluate $(\mathbf{A}_1-\mathbf{A}_2,\mathbf{c}_j)$ *for* $j=1,\ldots,J$ *where (,) is a scalar product in* $\mathbb{R}^{n!}$.

Resume:
$\mathbf{A}_1= \mathbf{A}_2$ *(tensor expressions are equal) iff*

$$(\mathbf{A}_1-\mathbf{A}_2,\mathbf{c}_j) = 0 \qquad \textit{for each } j = 1,\ldots,J \ .$$

7. Conclusions

1. We have described an algorithm to solve the S-D-I (Symmetry-Dummy indices-linear Identity) problem in tensor algebra manipulations.

2. Symmetry coefficients $\mathbf{d}(h)$ and α_π^i in identities are usually rational numbers. It is important to understand that in this case all steps of the algorithm can be implemented using only rational arithmetic. One can use for this the obvious modification of familiar Gramm-Schmidt orthogonalization procedure skipping the normalization of vectors.

3. The algorithm may be easily parallelized or vectorised.

8. References

1. R.A.Bogen, *Journ. Symb. Comp.* **1** (1985) 1.
2. A.Rodionov and A.Taranov, *Lecture Notes in Comp. Sci.* **378** (Proceedings of EUROCAL'87, 1989) p.192.

ENLARGEMENT OF CAS REDUCE TO TENSORS OPERATIONS

V.V. Malanin, V.M. Mikryukov

The Perm State University, Perm, USSR

Anybody can't imagine scientific investigation in natural sciense's without analitical evalution. At last years methods analitical evalusion inculcate in differents parts of mechenic. One of then most development and wide-spread computer algebra system (CAS) is system REDUCE /1/.

Tensors operation ase necessary for decision practical task of mechanic's of systems of sollid bodys and of mechanic's of elastic body . For the easest work with tensors in REDUCE was create complex of program enlargement of CAS REDUCE. This complex of program was written on Standard LISP /2/. If we write it in format RLISP /1/ its length 907 lines. The complex was created for conduct analitical evalution with tensor final dimentionality. All operation with tensors are ussed in Algebraic mode REDUCE. Tensor is defined as new type of variable and expression, which has its own, change from other, inner from and its own quality.

In current time REDUCE, with help of the complex, permit do such command and operetions of tensor's enalysis, as :

- describe and/or conduct first initial of tensor object;
- define element's tensors one by one and/or by blocks: vector or matrix;
- execute all operations of tensor's analysis: letferine, summing up and subtract, external product, differentiate.tensor by tensor or by scalar and scalar by tensor;
- describe element of tensor as tensor;
- execute one-dementionality tensors or its elements operetions of vector's, diad's or scalar product;
- presentation of the results of tensors and/or matrix-vectors evalution, which were written in natural mathematic forms by scalar forms;
- for operation, which needs large resources of operation storage, change non-numerical elements by operators with the same name;
- presentation of the results as of FORTRAN's program.

For example, in this artical write text program and result it work, which receive equations of motion elliptical pendulum (two degrees of free).

```
% R - radius-vector length two, describe as of one :
   TENSOR R(2)$

% F - vector external farces length two :
   TENSOR F(2)$

% Q - vector generalized coordinetes length two :
   TENSOR Q(2)$

% U - vector equations of motion :
   TENSOR U$

% l,m - length and massa elliptical pendulum;
% t - time;
% g - accelerate free fall;
% h - displecement pole of elliptical pendulum elong X -oxis;
% ht,ht2  - first and second time-dirivatives of h;
% fi - angle by turn of elliptical pendulum about Z -oxis;
% ft,ft2 - first and second time-dirivatives of fi.

% Describe  vector  R  - radius-vector elliptical pendulum pole in
% absolute coordinates system :
   R(i) _= TENV( h + l*sin(fi), l*cos(fi) )$

% Describe  vector  F  -  vector  external  forces  in  absolute
% coordinates system :
   F(i) _= TENV( 0, m*g )$

% Describe  vector  Q - vector generalized coordinates in absolute
% coordinates system :
   Q(i) _= TENV( h, fi)$
% Describe depend h,ht,fi,ft  from t :
```

```
   DEPENDT( *(h,ht,fi,ft), *(t) )$

% Describe definition ht,ht2,ft,ft2 :
   let df(h,t)=ht, df(ht,t)=ht2, df(fi,t)=ft, df(ft,t)=ft2$

   let sin(fi)**2+cos(fi)**2=1$

% Write total equation dynamic's of point
%        d(R)/d(Q)*(m*W-F)=0,
% where W - accelerate :
   U(i) _= DFT(R(i),Q(j)) _: (m_*DFT(R(i),t,2) _- F(i))$

% Write equations, which were obtained :
   factor m$

   for i:=1:2 do write  U(i)," = 0"$

                                          2
m*(cos(fi)*l*ft2 - sin(fi)*l*ft  + ht2) = 0

m*l*(cos(fi)*ht2 + sin(fi)*g + l*ft2) = 0
```

References

/1/ Hearn A.C. REDUCE 3 User's Manual. RAND publication CP78 (4/83).

/2/ Marti J.B., Hearn A.C., Griss M.L., Griss C. Standard LISP Report. Utah Symbolic Computation Group, Report N UCP-60, 1978.

DECOMPOSITION THEOREMS FOR THE ZERO-SET OF AN ORDINARY OR DIFFERENTIAL POLYNOMIAL SET AND THEIR APPLICATIONS*

WU WEN-TSUN
(Institute of Systems Science, Academia Sinica)

Let us consider the following problems:

Prob 1. A tetrahedron with 6 edges given will have its volume well-determined. Find the explicit expression of its volume in terms of its 6 edges.

Prob 2. Given two curves C, C' in (1-1)-correspondence in the ordinary space. Find the possible relations existant between the torsions of the two curves at corresponding points.

Prob 3. The Kepler's equation in astromechanics is given by

$$E - e * SinE = M.$$

Solve the equation for the Kepler's angle E in terms of the mean anomaly M.

Prob 4. Suppose known the Kepler's observational laws of planet movements arround the sun. How to derive from them the underlying general laws of motion?

Despite the diversity of characters of the problems listed above they can all be solved by one and the same method. In fact the conditions of the problems can be turned, with suitable choices of coordinate systems for example, into the form of a system of equations $P_k = 0$. Here the P_k are either ordinary or differential polynomials (abbr. POLs) in certain indeterminates $X_i, i = 1, ..., n$, which are either variables in the ordinary case or are functions in a single independent variable in the differential case. The coefficients of the pols, ordinary or differential, are in a field F, resp. an ordinary or differential one. Let us denote the polynomial set (abbr. POLSET) consisting of the pols P_k by PS. Denote also the totality of zeros of PS in an arbitrary extension field, ordinary or differential, of F by $Zero(PS)$. Each zero in the set $Zero(PS)$ is then some geometrical configuration meeting the conditions of the problem, eventually in some extended imaginative space. It follows that once the structure of $Zero(PS)$ is well-determined then the problem in question will either be already settled or be relatively easy to be settled.

To illustrate this let us consider e.g. the Prob 1.

Let the vertices of the tetrahedron be $A_i, i = 0, 1, 2, 3$ with edge length $|A_i A_j| = a_{ij}$ and volume V. Take now a coordinate system so that A_i will be

*The present paper is partially supported by NSFC Grant JI85312

$$A_0 = (0,0,0), \quad A_1 = (a,0,0), \quad A_2 = (b,c,0), \quad A_3 = (d,e,f).$$

Then we have (taking $a > 0$)

$$a_{01} = a,\ a_{02}^2 = b^2 + c^2,\ a_{03}^2 = d^2 + e^2 + f^2,$$
$$a_{12}^2 = (b-a)^2 + c^2, a_{13}^2 = (d-a)^2 + e^2 + f^2, a_{23}^2 = (d-b)^2 + (e-c)^2 + f^2,$$
$$V^2 = [(f * c * a)/6]^2.$$

Replacing V, a_{ij}, etc. by $X_i, i = 1, ..., 13$, these equations will become a set $P_k = 0, k = 1, ..., 7$. Denote the set of P_k by PS. Then the problem is reduced to seek a relation $R = 0$ between V and a_{ij} which is true for any zero of $Zero(PS)$ or a pol R for which

$$Zero(PS) \subset Zero(R).$$

We are thus led to the problem of determining the structure of $Zero(PS)$ for a polset PS. To fix the ideas let us consider only the differential case so that by pols or polsets we mean all differntial ones. Consider thus a set of indeterminate functions $X_1, \cdots, X_n$ in the independent variable t over a (differential) field F all fixed in what follows. The i-th derivative of X_j will be denoted by D_iX_j with $D_0X_j = X_j$. For any pol P we can then form successive DERIVATIVES D_iP as well as various PARTIAL DERIVATIVES $dP/d(D_iX_j)$ in the usual formal manner.

For any non-constant pol P let c be the largest subscript c for which X_c occurs actually in P. Let r be then the highest order r for which D_rX_c of the above X_c occurs actually in P. Finally let d be the highest degree of the above D_rX_c which occurs actually in P. Such a pol can then be written in the form

$$P = I * (D_rX_c)^d \ + \ lower\ degree\ terms\ in\ D_rX_c,$$

for which the coefficients of powers in D_rX_c are all themselves pols involving only derivatives D_iX_j with either $j < c$ or $j = c$ and $i < r$. We call c, r, d and D_rX_c to be resp. the CLASS, the ORDER, the DEGREE and the LEAD of P, to be denoted resp. as $cls(P), ord(P), deg(P)$ and $lead(P)$. We call also I the INITIAL and the formal derivative $dP/d(D_rX_c)$ the SEPARANT of P, to be denoted resp. by $init(P)$ and $sep(P)$. In case P is a non-zero constant, then we define $c = 0, r = 0$, and $d = 0$, while init(P) and sep(P) will be left undefined.

A pol Q will be said to be REDUCED w.r.t. a pol P of class $c > 0$, if either the highest order m of derivative D_mX_c appearing in Q is $< ord(P) = r$, or, if D_rX_c appears in Q, then the highest degree of D_rX_c in Q is $< deg(P)$. On the other hand Q is not reduced w.r.t. any pol P of class 0.

Let us introduce now a partial ordering $\ll$ among the totality of such non-zero pols first according to the class, then to the order, and finally to the degree. If for two non-zero pols P_1, P_2 neither $P_1 \ll P_2$ nor $P_2 \ll P_1$ so that P_1 and P_2 are incomparable in this ordering, then we write $P_1 <> P_2$.

Consider now a finite sequence of non-zero pols

$$(AS) \quad P_1, P_2, \cdots, P_r.$$

(AS) will be called an ASCENDING SET (abbr. ASC-SET) if either

(a) $r = 1$ and $cls(P_1) = 0$, or

(b) $cls(P_1) > 0$, and for any $j > i, cls(P_j) > cls(P_i)$ and the initial of P_j is REDUCED w.r.t. P_i.

In case of (a) the asc-set is then said to be TRIVIAL.

For a non-trivial asc-set (AS) as above let $S_i = sep(P_i)$ and $I_i = init(P_i), i = 1, \cdots, r$. A pol G will be said to be REDUCED w.r.t. (AS) if it is reduced w.r.t each P_i in (AS). In particular all separants S_i and initials I_i are reduced w.r.t. (AS). For any pol G we have then the following REMAINDER FORMULA:

$$(PROD_i\ (I_i^{l_i} * S_i^{m_i})) * G = SUM_{jk}\ (Q_{jk} * D_j P_k)\ +\ R, \tag{I}$$

in which l_i and m_i are certain non-negative integers which will be taken to be as small as possible, and Q_{jk}, R pols with R reduced w.r.t. (AS). The product PROD is to be extended over $1 \le i \le r$, and the summation SUM is extended over only a finite number of terms. The pol R in (I) is accordingly called the REMAINDER (abbr. REMDR) of G w.r.t. (AS), to be denoted as $Remdr(G/ASC)$.

The asc-sets will also be arranged in a partial ordering $\ll$ according to the following

DEFINITION. Let a second asc-set

$$(AS)' \quad Q_1, Q_2, \cdots, Q_s$$

be given. Then $(AS) \ll (AS)'$ if one of the following cases occurs:

(a) There is some $k \le r$ and $\le s$ such that for $j < k, P_j <> Q_j$, while $P_k \ll Q_k$.

(b) $r > s$ and $P_j <> Q_j$ for $j \le s$.

If neither $(AS) \ll (AS)'$ nor $(AS)' \ll (AS)$ so that (AS) and $(AS)'$ are incomparable in this ordering, then we write $(AS) <> (AS)'$.

DEFINITION. For any polset PS of pols any asc-set AS of lowest order for which each pol belongs to PS is called a BASIC-SET (abbr. BAS-SET) of PS.

For any polset PS and any pol G let us set ($-$ in set sense)

$$Zero(PS/G) = Zero(PS) - Zero(G).$$

Given a polset PS we can now deduce a certain asc-set of particular interest in a mechanical way according to the following scheme:

$$\begin{array}{cccc} PS = PS_0 & PS_1 & \cdots & PS_m \\ BS_0 & BS_1 & \cdots & BS_m \\ RS_0 & RS_1 & \cdots & RS_m = empty. \end{array} \tag{II}$$

In (II) each BS_i is a bas-set of PS_i, each RS_i is the set of non-empty remainders, if there are any, of pols in PS_i w.r.t. BS_i. Finally, each PS_i is the preceding PS_{i-1}, enlarged by adjoining to it all the pols in RS_{i-1}, i.e.

$$PS_i = PS_{i-1} + RS_{i-1}, \tag{III}$$

in which + means set union. It is easy to prove that the procedure will ultimately terminate with some $RS_m = empty$.

DEFINITION. The final bas-set BS_m in the scheme (II) is called a CHARACTERISTIC -SET (abbr. CHAR-SET) of the polset PS.

The importance of the above char-set, say CS, lies in the following

WELL-ORDERING PRINCIPLE. For the zeros of PS we have

$$Zero(PS) = Zero(CS/K) + SUM_k\ Zero(QS_k), \tag{IV}$$

in which K is the product of all initials and separants of pols in the char-set CS, and QS_k are the enlarged PS with one of the initials or the separants adjoined to it. We have besides

$$Zero(CS/K) = Zero(PS/K) \subset Zero(PS) \subset Zero(CS), \tag{V}$$

More general than (V), we have also

$$Zero(PS) = Zero(PS_k) \subset Zero(BS_k), \tag{V'}$$

for any polset PS_k and bas-set BS_k appearing in scheme (II).

REMARK. During the procedure it is convenient to remove certain factors to make pols occuring in the reduction not too high in term numbers. In such case the SUM_k in (IV) should run over also all removed factors besides initials and separants and the product K should include all such factors too.

There are various refined forms of the above principle which may be formulated in the form of ZERO-DECOMPOSITION THEOREMS as given below:

$$Zero(PS) = SUM_j\ Zero(AS_j/Kj) = SUM_k\ Zero(IRR_k/K_k) = Zero[IRR_k], \qquad \text{(VI)}$$

in which AS_j and IRR_k are certain asc-sets resp. irreducible asc-sets determined in a mechanical way from PS. Though these formulas (VI) comprise the most general form for the determination of the zero-set of a polset the formulas (IV), (V), (V') are already sufficient for diverse applications, in particular for the settling of the Problems 1-4 above. We shall thus leave away for the explanations of (VI) and show below in turn how (I)-(V') can be applied to these problems.

Let us remark first that the above differential case will be reduced to the ordinary case if all the pols involved are ordinany ones and the coefficient field F is also an ordinary one, or a differential field with trivial differentiations.

Return now to the Prob 1. Let the notations be so deliberately chosen such that $a_{ij} = x_1, \cdots, x_6$ and $V = x_7$, are the first 7 $x's$ with smallest subscripts. The char-set CS of the corresponding PS is consisting of 7 pols for which the first one C_1 is easily found, after returning back to the original notations, to be

$$C_1 = -144*V^2 - SUM_1\ [a_{01}^2*a_{23}^4 + a_{23}^2*a_{01}^4] + SUM_2\ [a_{02}^2*a_{01}^2*a_{13}^2] - SUM_3\ [a_{01}^2*a_{02}^2*a_{12}^2].$$

The SUM's are summations over respective ranges below:

SUM_1 over 3 pairs of opposite edges like (a_{01}, a_{23}),

SUM_2 over 12 non-closed triples of edges like (a_{02}, a_{01}, a_{13}),

SUM_3 over 4 triples of edges forming a triangle like (a_{01}, a_{02}, a_{12}).

From (V) it follws that $Zero(PS) \subset Zero(C_1)$ and $C_1 = 0$ is thus the relation expressing V in terms of the 6 edges a_{ij} as required which has been found in an automatic manner.

Consider now the Prob 2. Let us refer the curves C and C' to their arc-lengths t and t' as parameters so that under the correspondence $X \leftrightarrow X'$ t' will be a function of t. Associate to the corresponding points X, X' the Frenet trihedrals (X, E_1, E_2, E_3) and (X', E_1', E_2', E_3') then we have for some G,

$$X' = X + G.E_2.$$

Denote the angle between the tangents at corresponding points by A. Denote also the curvatures and torsions of C and C' by K, K', T, and T' resp. Then a comparison of Frenet formulas at corresponding points will give rise to a set of equations $P_k = 0, k = 1, \cdots, 12$, in the functions $dt'/dt, G, CosA, SinA, K, K', T$ and T' for which

$$P_1 = dt'/dt * CosA - 1 + G * K,$$

$$P_2 = -D_1 G,$$
$$P_3 = dt'/dt * SinA - G * T, etc.$$

the last one is

$$P_{12} = Cos^2 A + Sin^2 A - 1.$$

Let us use x_k to represent the various entities $dt'/dt, CosA, SinA, etc.$ as above. In order to find the possible yet unknown relation between the torsions T and T' of C and C' the x_k will be so chosen so that T and T' are the x of smallest subscripts but otherwise arbitrary. It turns out that the first pol C_1 of the char-set CS of PS is, returning back to the original notations, given by

$$C_1 = T * D_1 T' + T' * D_1 T.$$

Again by (V) $Zero(PS) \subset Zero(C_1)$. Hence we have $C_1 = 0$ or $T * T' = const$. We have therefore discovered in an automatic manner the following

Theorem of Schell. For a pair of curves in (1-1)-correspondence with coincident principal normals at corresponding points the product of torsions at corresponding points is a constant.

Consider next the Prob 3. The Kepler's equation, important in astronomical computations, is usually solved by means of numerical or graphical methods, and also by Lagrange series of E in e with coefficients derivatives of powers of $SinM$ w.r.t. M. We furnish now a solution by our general method entirely different from the preceding ones as follows. Let $x_1 = E$ and $x_2 = SinE$ so that, with $M = t$ as the independent variable we find easily, besides the Kepler's equation $P_1 = 0$, also the equation $P_2 = 0$ with

$$P_1 = x_1 - e * x_2 - M,$$

$$P_2 = D_1 x_1 * D_2 x_2 - D_2 x_1 * D_1 x_2 + (D_1 x_1)^3 * x_2.$$

The first pol C_1 of the char-set of $PS = \{P_1, P_2\}$ is now found to be, again returning back to the original notations,

$$C_1 = E'' + (E - M) * E'^3$$

with prime denoting derivation w.r.t. M. This gives us thus an ordinary differential equation $C_1 = 0$ of E with initial values easily seen to be $E(0) = 0, E'(0) = 1/(1-e)$ and may thus be solved as a series in terms of M:

$$E = (1/(1-e)) * M - (e/(1-e)^4) * M^3/3! + \cdots$$

Consider finally the Prob 4. Let us take rectangular coordinates with sun at the origin so that the elliptic orbit equation of the planet may be written in the form

$$r = p + e * x, \qquad r^2 = x^2 + y^2. \tag{1}$$

With prime denoting derivation w.r.t. time t we have then

$$e' = 0, \qquad p' = 0. \tag{2}$$

The Kepler's law of area swept by the planet may be put in the form

$$x * y' - y * x' = h, \qquad h' = 0. \tag{3}$$

Denote the acceleration of the planet by A so that

$$A^2 = x''^2 + y''^2. \tag{4}$$

That the acceleration is always along the line sun-planet follows directly from (3). To see what will be the general law governing A we may set the various functions A, r, x, y, p, e, h as x_k with $r = x_1, A = x_2$ as the $x's$ with smallest subscripts. The equations (1)-(4) will then be turned into a set $P_k = 0$, $k = 1, \cdots, 7$. During the formation of the char-set of $PS = \{P_k\}$ we see that the first pol B_{41} of the 4-th bas-set BS_4 is given by, again returning to original notations,

$$B_{41} = 2 * r' * A + r * A'.$$

By (V') we see that $B_{41} = 0$ or $r^2 * A = const.$ for such planet movement. We have thus discovered in an automatic manner the Newton's inverse square law of planet motions.

Our method is of a general character and has been applied to diverse problems besides the above ones. The application which has been done in a gigantic experimental scale is for the mechanical theorem proving and discovering in geometries, elementary or differential. For other applications we may cite: inequalities proving, non-linear programming, inverse kinematic equations-solving of robot motions, linkage synthesis, pole-assignment of control theory, Liapounov constants of polynomial ODE in the plane, etc. Much more are expected to be done in recent times. For more details of our method we refer to the relavent papers of the author, and the MM-Preprints (Mathematics-Mechanization Preprints) of the Institute of Systems Science, Academia Sinica. Below we list only the references to the 4 problems treated above.

REFERENCES.

1. Wu Wen-tsun, A mechanization method of geometry and its applications, 1. Distances, areas, and volumes in euclidean and non-euclidean geometries, Kexue Tongbao, 32 (1987) 436-440.

2. Wu Wen-tsun, A mechanization method of geometry and its applications, 2. Curve pairs of Bertrand type, Kexue Tongbao, 32 (1987) 585-588.

3. Wu Wen-tsun et al, A mechanization method of geometry and its applications, 5. Solving transcendental equations by algebraic methods, MM-Preprints, No.3 (1989) 30-32.

4. Wu Wen-tsun, Mechanical derivation of Newton's Gravitational Laws from Kepler's Laws, MM-Preprints, No.3 (1989) 30-32.

THE DIFFERENCES OF ROOTS OF POLYNOMIALS AND SOLUTIONS OF ODE

V.I.Galiev, A.F.Polupanov, I.E.Shparlinsky
Institute of Radioengineering and Electronics
& J/E INTERFACE, Academy of Sciences of the USSR,
Marx av.18, Moscow, 103907, USSR

ABSTRACT

We present a polynomial algorithm for the determination of the existence of nonzero integers among differences of roots of the polynomial $f(x) \in \mathbb{Z}[x]$. This problem arises when constructing algorithms for solving some systems of ODE.

The problem of the determination of the existence of nonzero integers among differences of roots of the polynomial $f(x) \in \mathbb{Z}[x]$ often arises when constructing formal solutions of some systems of ordinary differential equations. For example, the properties of solutions in a neighbourhood of a regularly singular point (see [1-4]) and the algorithm of their construction (see [5,6]) depend drastically on the answer to this question.

Really, let us consider the following equation

$$w \cdot r^2 \frac{d^2R}{dr^2} + a_0 \cdot r \frac{dR}{dr} + (b_0 + \sum_{m=1}^{M} b_m \cdot r^m) R = 0 \tag{1}$$

where $R = (R_\alpha)$ is an d-component column; w, a_0 and b_n are constant (d×d) matrices. They look for the solutions of this equation in the neighbourhood of the regularly singular point r = 0 in the form $R \sim r^\rho \cdot V$, $r \to 0$ where V is a d-component column, $V \to$ const, $r \to 0$. Define the following matrix of the dimension (d×d)

$$\Gamma(\rho) = \rho \cdot (\rho-1) w + a_0 \cdot \rho + b_0 .$$

Let $\rho_1, \ldots, \rho_{2d}$, $\mathrm{Re}\rho_1 \geq \mathrm{Re}\rho_2 \geq \ldots \mathrm{Re}\rho_{2d}$ are the roots of the following characteristic equation

$$\det\Gamma(\rho) = 0. \tag{2}$$

Note that the character of its roots determines

completely the character of solutions of Eq. 1. If all differences $\rho_k-\rho_m$ between the roots of Eq. 2 are not equal to an integer then all solutions of Eq. 1 has the form of power series with the first term being proportional to r^{ρ}, where ρ is the corresponding root of Eq. 2. However another, the most complicated situation often takes place in the applications, i.e. when all differences $\rho_k-\rho_m$ are equal to integers and some of the roots may be multiple. The solutions of the Eq. 1 are in this case of absolutely different form, viz. the combinations of power series and positive powers of logarithmic function [5,6]. This takes place e.g. for radial Schroedinger equation of the form of Eq. 1 in the case of total Hamiltonians being quadratic form of the momentum operator $\vec{p}$ [6]. However though in this last case one can just derive all roots ot the Eq. 2 exactly, in the general case the problem of the determination of the existence of nonzero integers among differences of roots of given polynomial is not solved. Besides, for a numerical stability of this algorithm an exact lower bound for the deviation of the named differences from nonzero integers must be known. Note that since the exact analytical solutions of the equations like Eq. 1 are found in [5,6] in the most complicated case when differences $\rho_k-\rho_m$ are equal to integers, then the solution of the problem of roots completes the construction of the solutions of Eq. 1.

In the present paper we show that for any polynomial $f(x) \in \mathbb{Z}[x]$ this problem can be solved in time bounded by a polynomial in degree $n = \mathbf{deg}\ f$ and the logarithm of its height

$$H(f) = \mathbf{max}\{|a_n|, \ldots, |a_0|\},$$

where $f(x) = a_n x^n + \ldots + a_0$, i.e. a polynomial in the size of its writing in bits.

Let us denote by $\|\gamma\|$ the distance from a complex γ to the nearest integer (usually this function is defined only for real γ, in our case may be $\|\gamma\| > 1/2$).

For the polynomial $f(x) \in \mathbb{Z}[x]$ we put

$$\delta(f) = \min_{1\le i<j\le m} \|\lambda_j - \lambda_i\|,$$

where $\lambda_1, \ldots, \lambda_m$ are all pair wise different roots of $f(x)$.

<u>Theorem</u>. **For any polynomial $f(x) \in \mathbb{Z}[x]$, degf $= n$, one can find in time $(n\log[H(f)+1])^{O(1)}$ such δ that $\delta \le \delta(f) \le 2\delta$**

<u>Proof.</u> Let us denote by $L(g)$ the length of a polynomial $g(x) = b_m x^m + \ldots + b_0$, i.e. $L(g) = |b_m| + \ldots + |b_0|$

Let α and β are two different roots of $f(x)$, $\varphi(x)$ and $\psi(x)$ being their minimal polynomials over $\mathbb{Z}$.

For an algebraic γ we denote by $\mathsf{L}(\gamma)$ its length, i.e. the length of its minimal polynomial over $\mathbb{Z}$.

We have (see [7], Theorem 4.3)

$$\mathsf{L}(\alpha - \beta) \le 3^{n^2}[\mathsf{L}(\alpha)\mathsf{L}(\beta)]^n = 3^{n^2}[L(\varphi)L(\psi)]^n.$$

Since $\varphi | f$ then from [7], Theorem 3.3, one has $L(\varphi) \le 2^n L(f)$. The analogous inequality for ψ also holds. Therefore

$$\mathsf{L}(\alpha - \beta) \le 3^{n^2}[2^{2n}L^2(f)]^n \le (n+1)^{2n} 12^{n^2} H^{2n}(f),$$

(if $\varphi \ne \psi$ then this bound can be improved).

Let $\gamma = \alpha - \beta$ and $F(x)$ be its minimal polynomial over $\mathbb{Z}$. One has

$$H(F) \le L(F) = \mathsf{L}(\alpha - \beta) \le (n+1)^{2n} 12^{n^2} H^{2n}(f). \quad (3)$$

If $\gamma \in \mathbb{Z}$ then it is clear that $\delta(f) = 0$.

Let us show that in opposite case the following bound holds

$$\|\gamma\| \ge \left[(n+1)^{2n+1} 2^n 12^{n^2} \left(H(f) + 2\right)\right)^{3n-1}\Big]^{-1} \quad (4)$$

Because the polynomial $F(x)$ is irreducible over $\mathbb{Z}$ and $\gamma \notin \mathbb{Z}$ we have $F(k) \ne 0$ for any $k \in \mathbb{Z}$. Let k be the nearest integer to γ. We consider the circle $C(\gamma)$ with the radius 1/2 and the center in the point γ on the complex plane. If $k \notin C(\gamma)$ (this may be for a complex γ) then $\|\gamma\| \ge 1/2$ and Eq. 4 holds. Since $F(k)$ is a nonzero integer in the opposite case we have

$$1 \le |F(k)| = |F(k) - F(\gamma)| \le$$

$$\le |k - \gamma| \max_{z \in C(\gamma)} |F'(z)| = \|\gamma\| \max_{z \in C(\gamma)} |F'(z)|$$

Taking into account that absolute values of roots of $f(x)$ do not exceed $H(f) + 1$ (see [8]) we obtain $|\gamma| \le 2[H(f) + 1]$. Hence

$$\max_{z\in C(\gamma)} |F'(z)| \le nH(F) \max_{z\in C(\gamma)} (|z| + 1)^{n-1} \le$$

$$\le nH(F)(|\gamma| + 3/2)^{n-1} \le n2^{n-1}H(F)[H(f) + 2]^{n-1}$$

This inequality and Eq. 3 yield the bound (4).

Let us define

$$\vartheta(f) = \left[(n + 1)^{2n+1}2^{n}12^{n^2}\left(H(f) + 2\right)^{3n-1}\right]^{-1},$$

then either $\delta(f) = 0$ or $\delta(f) \ge \vartheta(f)$.

Let us calculate the approximations $\mu_1, \ldots, \mu_m$ to all roots $\lambda_1, \ldots, \lambda_m$ of $f(x)$ with the error less than $\varepsilon = \vartheta(f)/8$, i.e. $|\lambda_i - \mu_i| < \varepsilon$, $i = 1, \ldots, m$, and denote $\tau = \min_{1\le i<j\le m} \|\mu_j - \mu_i\|$.

Then we can define

$$\delta = \begin{cases} \tau-2\varepsilon, & \text{if } \tau \ge 6\varepsilon, \\ 0, & \text{if } \tau < 6\varepsilon. \end{cases}$$

It is clear that $\delta \le \delta(f) \le 2\delta$. It has been proven in [9,10] that for any $\varepsilon > 0$ all roots of the polynomial $f(x)$ can be calculated in time $\left(n\log[H(f) + 1]\log[\varepsilon^{-1} + 1]\right)^{O(1)}$ with error less than ε. Since $\log\left[\vartheta(f)\right]^{-1} = \left(n\log[H(f) + 1]\right)^{O(1)}$ the proof is complete.

Let us note that the proof of the Theorem is analogous to the proof of the Liouvill's theorem (see [7,11]). More strong variants of this theorem, theorem of Thue-Sigel-Roth or theorem of Feldman based on Baker's results(see [7,11]), can not be applied to our problem (the first one is noneffective, the second one has very large depending on n constants).

We suppose that the following lower bound

$$\delta(f) \ge \exp\left(-Cn\log[H(f) + 1]\right)$$

holds, where C is an absolute constant. Let us also point out that it will be important to find the algorithm which determines whether $\delta(f) = 0$ or not in time linear on $n\log[H(f) + 1]$.

Note that our bound is a generalization of the

well-known bound of [12] for

$$\Delta(f) = \min_{1 \le i < j \le n} |\lambda_j - \lambda_i|.$$

(it is clear that $\Delta(f) > \delta(f)$).

Probably, more complex algorithm of [13] also can be applied to this problem.

Acknowledgement

We are grateful to A.Ya. Shulman for stimulating discussions.

References

1. W. Wasow, *Asymptotic expansions for ordinary differential equations*, (Interscience publishers, New York, 1965).
2. M.V. Fedorjuk, *Asymptotic methods for the solution of ordinary differential equations*, (Nauka, Moscow, 1983).
3. F.R. Gantmaher, *Theory of matrices*, (Nauka, Moscow, 1988).
4. D.V.Chudnovsky and G.V.Chudnovsky, *On expansion of algebraic function in power and Puiseux series*, 2, J. of Complexity, 3 (1987), 1
5. V.I.Galiev, A.F.Polupanov and I.E.Shparlinskiy, *On the construction of solutions of systems of linear ordinary differential equations in the neighbourhood of a regular singularity*, Preprint IRE AS USSR, 17(518) (1989), 1-21.
6. V.I.Galiev and A.F.Polupanov, *The exact expressions for the wave functions of matrix Hamiltonians with Coulomb term*, Preprint IRE AS USSR, 18(519) (1989), 1-38.
7. N.I.Feldman, *Approximations of algebraic numbers*, (Moscow Univ. Pres., Moscow, 1981).
8. H.G.Zimmer, *Computational Problems, Methods, and Results in Algebraic Number Theory, Lecture Notes in Mathematics*, v. **262**, (Springer-Verlag, Berlin, 1972).
9. V.Pan, *Algebraic complexity of computing polynomial zeros*, Comput. Math. Applic., **14**(**4**) (1987), 255
10. A.Shonhage, *The fundamental theorem of algebra in terms of computational complexity*, Preprint of Depart. of Math., Univ. of Tubingen, Tubingen, West Germany, 1982.
11. W.Schmidt, *Diophantine Approximations*, (Springer-Verlag, Berlin, 1980).
12. M.Mignotte, *Some useful bounds*, *Computer Algebra*, (Springer-Verlag, Berlin, 1983).
13. J.Hastad, B.Just, C.J.Lagarias and C.P.Schnorr, *Polynomial-time algorithm for finding integer relations among real numbers*, SIAM J. Comp., **18**(5) (1989), 859

COMPUTING IN POLYNOMIAL RINGS OF SOLVABLE TYPE

HEINZ KREDEL
UNIVERSITÄT PASSAU, D-8390 PASSAU, GERMANY

Abstract: Polynomial rings of solvable type provide a framework for various important data structures in physics, e. g. Clifford algebras, iterated Ore extensions of the ground field and universal enveloping algebras of finite dimensional Lie algebras. We present implemented effective algorithms for computation in such rings: non–commutative products with respect to commutator relations, (left) reduction, left and two–sided Gröbner bases according to [KRW88]. Examples computed in the MAS system are included.

1. Introduction. One of the most important tools in the algorithmic theory of commutative polynomial rings is the calculation of Gröbner bases by Buchbergers algorithm [B65]. Gröbner bases for non–commutative polynomial rings have first been considered by [M85]. Kandri–Rody and Weispfenning introduced one and two–sided Gröbner bases for non–commutative polynomial rings of solvable type [KRW88]. A solvable polynomial ring is an ordinary commutative polynomial ring $\mathbf{R} = \mathbf{K}[X_1, \ldots, X_n]$ equipped with a new non–commutative multiplication $*$. The field $\mathbf{K}$ is assumed to be commutative and to commute with the indeterminates $X_1, \ldots, X_n$. The $*$–product is defined via commutator relations such that $f * g - g * f$ is smaller than the commutative product $f \cdot g$ in the sense of an arbitrary fixed admissible term order on $\mathbf{R}$.

The class of solvable polynomial rings includes e. g. enveloping algebras of finite dimensional Lie algebras. Gröbner Bases for enveloping algebras have been studied by Apel and Lassner [AL88]. An implementation of their algorithms was given by Petermann and Apel [AP88] in the LOGLAN system [B83].

In this article we describe an implementation of the theory of Kandri–Rody and Weispfenning in the MAS system [K90]. The implementation is based on an earlier implementation in the ALDES / SAC–2 system [CL80]. The correctness proofs for the algorithms are included in a full version of the paper. To improve efficiency of the non–commutative product algorithm, we introduce the method of incrementaly computed relation tables. We show examples computed with the MAS system.

The correctess of the algorithms can be proved using the definitions and propositions of [KRW88]. In order to make the paper selfcontained we summarize all required items. The reader interested in the proofs should refer to the article of [KRW88] and the full version of present paper.

The correctness proof of the $*$–product algorithm is different to the existence proof in the original paper, since the induction hypothesis does not coincide with the recursive application of the $*$–algorithm. Also the original $*$–algorithm can not be modified to include the relation table handling with out making the proof invalid.

The computation of a reduced Gröbner base can be made more efficient when the Gröbner base properties are exploited. Finaly a more efficient version of the algorithm for the computation of two–sided GB's is presented.

In more detail the main implementation design considerations can be summarized as follows:

1. The algebras of solvable type are commutative polynomial rings equipped with a non–commutative product. In the implementation we use an ordinary commuta-

tive (distributive) polynomial representation. Actually the Distributive Polynomial System of [GK83], implemented in the SAC–2 / ALDES system is used.

2. The non–commutative product $*$ is defined via relations, which are elements of a free associative algebra. These relations are implemented as triples (u, v, p) of (commutative) terms u, v and a (commutative) polynomial p, such that $u * v = p$.
3. Besides the defining relations between variables of the non–commutative product, many relations between powers of variables and terms are derived during computation. These relations are incrementaly stored in a so called relation table. Each time a product of terms is to be computed the relation table is scanned for an applicable relation. Missing relations are treated as if the two variables commute.
4. For the rest of the Gröbner base algorithms the existing ones from the Buchberger algorithm system of [GK84] could be used as a starting point. However great care was inorder to assure that for no algorithm the input parameters where commuted. Further the non–commutative product modifies the leading coefficients of the product polynomials, so the order of computation steps had also to be checked.
5. The computation of two–sided Gröbner bases uses an improved way of including right variable multiples during the main Buchberger loop instead of iterating the left Buchberger algorithm on bases given by right variable multiples of polynomials.

The plan of the paper is as follows: In section 2 we will summarize some of the theory of [KRW88] and discuss some fundamental representation issues. In section 3 we will discuss the non–commutative product algorithm, the relation table handling and some examples for the complexity of the $*$–product. Section 4 will contain the reduction algorithm for polynomals and sets of polynomials, section 5 the S–polynom computation and the Buchberger algorithm for left–sided Gröbner bases and section 6 will consist of a description of the two–sided Gröbner base construction. In section 7 a small example showing the usage of the algorithms in the MAS system is discussed. In the final section 8 we will summarize some computing times and draw some conclusions.

The author would like to thank V. Weispfenning, T. Becker, J. Apel and T. Mora for valuable discussions on commutative and non–commutative Gröbner base theory.

2. Polynomial Rings of Solvable Type. In this section we first summarize some of the theory of Kandri–Rody and Weispfenning. Since we will have to refer to their items, we use the notation KW n.m. for item 'm' in section 'n' of [KRW88].

As stated already, a solvable polynomial ring is an ordinary commutative polynomial ring $\mathbf{R} = \mathbf{K}[X_1, \ldots, X_n]$ equipped with a new non–commutative multiplication $*$. The field $\mathbf{K}$ is assumed to be commutative and to commute with the indeterminates $X_1, \ldots, X_n$. The set T of terms (power–products of indeterminates) is supposed to be linearly ordered by an admissible order $<_T$.

For a fixed term order $<_T$, $(\mathbf{R}, *)$ is a solvable polynomial ring if the following axioms are satisfied:

AXIOMS 2.1 (KW 1.2.).

1. *$(\mathbf{R}, 0, 1, +, -, *)$ is an associative ring with* 1.
2. *For all $a, b \in \mathbf{K}$, $1 \leq h \leq i \leq j \leq k \leq n$, $t \in T(X_i, \ldots, X_j)$,*
   (a) $a * bt = bt * a = abt$,
   (b) $X_h * bt = bX_h t$,
   (c) $bt * X_k = btX_k$.
3. *For all $1 \leq i \leq j \leq n$ there exist $0 \neq c_{ij} \in \mathbf{K}$ and $p_{ij} \in \mathbf{R}$ such that $X_j * X_i = c_{ij} X_i X_j + p_{ij}$ and $p_{ij} <_T X_i X_j$.*

Any admissible order satisfying condition (3) will be called *$*$-compatible*. Solvable polynomial rings will be denoted by $\mathbf{R} = \mathbf{K}\{X_1, \ldots, X_n\}$ or if Q denotes the set of commutator relations of axiom 2.1(3), by $\mathbf{K}\{X_1, \ldots, X_n; Q\}$. The commutative multiplication on $\mathbf{R}$ will be denoted by '$\cdot$'.

By the following lemma, the definition of the $*$-multiplication is extended to arbitrary polynomials in $\mathbf{R}$.

LEMMA 2.2 (KW 1.4.). *Let $\mathbf{R} = \mathbf{K}\{X_1, \ldots, X_n\}$ be a solvable polynomial ring, let $<_T$ be a $*$-compatible admissible term order, and let $f, g \in \mathbf{R}$. Then there exists an $h \in \mathbf{R}$ such that*

$$f * g = c \cdot f \cdot g + h$$

and $h <_T f \cdot g$. Moreover, c and h are uniquely determined by f and g. Notation $c = fctr(f, g)$.

The lemma is proved by noetherian induction on $f \cdot g$ with respect to $<_T$. The $*$-multiplication is distributed over the terms of the polynomials and $*$-products of terms are handled by case distinction according to the axioms 2.1.

For the complete proof and further details we must refer the reader to KW 1.4.

2.1. Implementation. The non-commutative polynomials are represented as ordinary commutative polynomials. The implementation of the algorithms uses the Distributive Polynomial System [GK83]. A polynomial in distributive representation is a list of so called exponent vectors and so called base coefficients. The coefficient field $\mathbf{K}$ is the field of rational numbers in the current implementation.

The commutator relations from 2.1(3) are implemented as triples of commutative polynomials. More precisely a relation $X_j * X_i = c_{ij} X_i X_j + p_{ij}$ for some $1 \leq i \leq j \leq n$ is represented as triple $(X_j, X_i, c_{ij} X_i X_j + p_{ij})$ of distributive polynomials. Missing relations are treated as if the relation $X_j * X_i = X_i X_j$ was specified, i. e. as if a relation with $c_{ij} = 1$ and $p_{ij} = 0$ was defined.

The set of all commutator relations is stored in a list called *relation table*. So in order to compute the product $X_j * X_i$ one has to look for a triple starting with (X_j, X_i, r) and the take the third polynomial in this triple.

3. The Non-commutative Product. In this section we will sketch the product algorithm. In a seperate subsection the relation table handling is discussed.

The product of two non-commutative polynomials is formed as the sum over the products of the terms (power-products). The product of terms are formed as commutative product or by looking for appropriate commutator relations and recursive application of the (polynomial) product algorithm.

During the algorithm many products of powers of variables are formed. In order to save computation time, these products are remembered in the relation table. The next time such a product is needed it can be taken from the table without being recomputed. Only commutator relations of the form $X_j^{e_j} * X_i^{e_i} = q$ are stored since the storage of arbitrary product relations would waste memory to much.

More precisely let u, v be two terms, and assume we want to compute $u * v$:

1. If by axioms 2.1 $u * v = u \cdot v$ then we are done, otherwise proceed as follows:
2. Split each term into three parts $u * v = u_1 \cdot u_2 \cdot u_3 * v_3 \cdot v_2 \cdot v_1$. Where $u_2 \cdot u_3 = X_j^{e_j}, v_3 \cdot v_2 = X_i^{e_i}, i < j$. u_3, v_3 are determined by the largest existing relation table entry for the product of powers of variables. Let this product be p.

3. Compute the product $u_2 * (p * v_3)$ by two fold recursive application of the product algorithm. Let the results be q and q'.
4. Add the triples $(u_3, X_i^{e_i}, q)$ and $(X_j^{e_j}, X_i^{e_i}, q')$ to the relation table. Note that during the recursion all triples $(X_j^{l_j}, X_i^{l_i}, q')$ with $l_j < e_j$ and $l_i < e_i$ are already added to the relation table.
5. Finally form the product $u_1 * (q * v_1)$ by recursion.

Although this algorithm seems to be quite effective, it is difficult to prove its correctness. The proof of lemma 2.2 can not be applied directly to the algorithm since the usage of the induction hypothesis in the proof does not coincide with the recursive application of the product algorithm. On the other hand, in the product algorithm in KW 2, which is designed after the proof, the products $X_j^{e_j} * X_i^{e_i}$ are not computed, and therefore they can not be put into a relation table.

The algorithm is called DINPPR, its listing and the correctness proof can be found in the full version of the article. We turn now to the relation table algorithms. An example for the complexity of the $*$–product, is included in section 3.3.

3.1. Relation Tables. As noted in the section on the product algorithm the relation table must be maintained through recursive applications of the $*$–product algorithm. Naturally we want to use all computed relations to be accessible at any time during further recursive calls.

The relation table is implemented as a list of distributive polynomials:

$$T = (u_1, v_1, p_1, \ldots, u_t, v_t, p_t)$$

where the $u_i = X_{j_i}^{e_i}$, $v_i = X_{k_i}^{l_i}$ and $p_i = c_i \cdot X_{k_i}^{l_i} \cdot X_{j_i}^{e_i} + p_i'$. The table entries are partially ordered with respect to divisibility of the relation heads (u_i, v_i).

DEFINITION 3.1. *Let T be a relation table, then the following condition must hold:*

$$\forall_{1 \leq i \leq t} i : \neg \exists_{i < j \leq t} j : u_i \mid u_j \textit{ and } v_i \mid v_j$$

I. e. relation heads which are 'later' in the table may divide relation heads, which come 'earlier' in the table.

If T is empty at the beginning, i.e. in this case **all** variables commute, then no non–commuting relation will ever be computed during DINPPR and T will remain empty.

The search for a product relation goes from left to right in the list so one finds a relation with maximal exponents. The search is successful, if both exponents of u_i and v_i divide the exponents of the relation we look for. If no relation matches, we assume the variables to commute, i.e. we assume $c = 1$ and $p = 0$.

The relation table update algorithm determines the correct position if the new relation according to condition 3.1. Then it inserts the relation without modifying the list pointer to the table. This is important to make the information which is stored in recursive calls of DINPPR available to all further (top–level) calls of the product algorithm.

One *optimization* of the table implementation would be to built separate table entries for each pair of non–commuting *variables*. Then the search for a relation could be restricted to a certain sub table. However for the scope of examples which can be computed the current table implementation suffices.

3.2. Input and Output. Since polynomials are internally represented by lists over atoms (integers with absolute value less than 2^{29}) we need some facilities to display or input polynomials in more natural form.

For the display routines the standard display routines for the (commutative) distributive rational polynomials can be used. For the input routines the non-commutative product algorithm can be incorporated to respect the order of the variables in the polynomials.

The syntax (in EBNF) of a polynomial accepted by the input routines is:

```
poly   = ( rat | var | "(" sum ")" )
sum    = term { ( "+" | "-" ) term }
term   = factor { [ "*" ] factor }
factor = pol [ "**" nat ]
rat    = int [ "/" int ]
```

nat denotes a positive atom, `int` denotes an integer, `rat` denotes a rational number and `var` denotes a variable name defined in the polynomial variable list.

The specification of the multiplication operator * is optional. In any case the product is the non-commutative $*$-product. During input the polynomials are multiplied out to obtain the internal canonical form of distributive rational polynomials. In the output the polynomials are therefore represented as sums over monomials. See section 7 for an example.

3.3. Complexity. During the computation of a $*$-product of two terms many new multiplications of terms may occur. E. g. if the commutator relations contain high degree polynomials or if the commutator relations are dense polynomials even of low degree. So we discuss two examples to give some impression of the complexity of the algorithm.

1. Let $\mathbf{R} = \mathbf{Q}[x, y, z]$ be a solvable polynomial ring with the commutator relations $y*x = xy+x^d, z*x = xz+y^d, z*y = yz+y^d$, where $<_T$ is the inverse lexicographical term order. In this example the commutator relations are 'sparse' with 'high' degree.
2. Let $\mathbf{R} = \mathbf{Q}[x, y, z]$ be a solvable polynomial ring (a Lie algebra) with the commutator relations $y*x = xy+x+y+z, z*x = xz+x+y+z, z*y = yz+x+y+z$, where $<_T$ is the total degree (inverse graduated) term order. The commutator relations are 'dense' of 'low' degree.

Suppose we want to compute the product $z^c * y^c * x^c$ for varying input paramaters d and c. We will observe the following output paramaters:

r the number of commutator relations after the computation of the product,
p the number of terms in the product, and
t the computing time on an Atari ST.

The results are summarized in table 1. They show, that the complexity depends both on the degree and density of the commutator relations. In example 1 more commutator relations are computed compared to example 2. In example 2 at a certain stage no new commutator relations are computed and only the relations from the relation table are used.

4. Left Normal Form and Left Irreducible Sets. Let $\mathbf{R}$ be a solvable polynomial ring, $f, f', p \in \mathbf{R}$. The reduction $f \longrightarrow_p f'(t)$ is defined as follows: Let $t \in T(f)$ such that $HT(p)$ divides t (in the commutative sense). Now let a be the coefficient of t in f and b be the leading base coefficient of p. Further let $s = \frac{HT(p)}{t}$ and $c = factr(s * HT(p))$ then $f' = f - \frac{a}{bc} s * p$. Since s is multiplied from the left to p $f \longrightarrow_p f'(t)$ is a *left* reduction. For a *right* reduction s is multiplied from the right to p. If $d = factr(s * p)$ then f' can also be written as $f - \frac{a}{d} s * p$.

Example	d	c	r	p	t
1	2	1	5	6	0
	2	2	21	15	8
	2	3	45	28	26
	2	4	73	45	68
	3	1	9	7	2
	3	2	37	38	18
	3	3	92	90	96
	4	1	15	8	6
	4	2	61	46	36
	4	3	134	110	204
	6	1	33	10	18
	6	2	104	62	96
	6	3	314	150	958
2	1	1	3	10	0
	1	2	10	41	14
	1	3	23	105	158
	1	4	42	214	1412

TABLE 1
Complexity of the $*$ product

The reductions with respect to a polynomial $f \longrightarrow_p f'$ and with respect to a set P of polynomials $f \longrightarrow_P f'$ are defined as in the commutative case. If f can not be reduced with respect to P we say that f is irreducible with respect to P or that f is in normalform with respect to P. P is irreducible or in normalform, if all $f \in P$ are irreducible with respect to $P \setminus \{f\}$.

We will now turn to the algorithms which implement these reductions. Once the algorithm for the non-commutative product is available, the implementation of the reduction algorithm for polynomials and sets of polynomials are straight forward modifications of the respective algorithms in the commutative case [GK84].

The algorithm `DINLNF` computes a normalform R of a polynomial S with respect to a set (list) of polynomials P. Every term t in S (or some reduct of S) is checked if there exists a polynomial q in P such that $HT(q)$ divides t. If such a q exists, the a one step reduction is carried out, otherwise the term t is irreducible and it is placed in the output polynomial R.

The algorithm `DINLIS` computes an irreducible set (or a normalform of a set) P of polynomials. The number of irreducible polynomial is initially set to zero. Then every p in P is checked if it is in normalform with respect to $P \setminus \{p\}$. If this is the case, then the polynomial is counted as irreducible, otherwise the polynomial was reducible and might now be able to reduce further polynomials in $P \setminus \{p\}$. Therefore the number of irreducible polynomials must be reset to zero. The algorithm terminates, if the number of irreducible polynomials is equal to the total number of polynomials.

This algorithm makes maximal use of the fact, that for reducibility of a polynomial only the head terms of the reducing set of polynomials have any influence.

In this view the loop consists of two stages: a *first* stage which reduces all head terms (and probably many terms in the rests) and a *second* stage (in the same loop) where the (remaining) rests are reduced in $length(P')$ steps.

5. S–Polynomials and Left Gröbner Bases. Gröbner bases, introduced by [B65], are defined as sets of polynomials P such that the reduction $\longrightarrow_P^*$ is confluent.

It is shown, that it is sufficient for P to be a Gröbner base, that all S–polynomials of polynomials $f, g \in P$ reduce to zero with respect to P. S–polynomials are defined before KW 3.9 and the definition is similar to the commutative case.

The implementation of the S–polynomial algorithm is straight forward. Only some care is needed to get the coefficients right.

5.1. Buchberger algorithm. Buchberger's algorithm for constructing Gröbner bases is based on the following theorem. In the commutative case it was proved in [B65,B85]. In the case of enveloping algebras of Lie algebras and a total degree admissible term order is was proved by [AL88].

THEOREM 5.1 (KW 3.11.). *Let G be a finite set of polynomials in $\mathbf{R}$. Then G is a left Gröbner base iff for all $f, g \in G$, $LSP(f,g) \longrightarrow_G^* 0$.*

The algorithm of Buchberger takes as input a finite set of polynomials and delivers a Gröbner base as output. In a main loop in the algorithm for all pairs of polynomials the S–polynomials are constructed and reduced to their normal form modulo the set of polynomials. If the resulting polynomial is non zero, the polynomial is added to the set of polynomials. The algorithm terminates when all S–polynomials of all pairs can be reduced to zero.

Termination is assured by Dickson's lemma and upon termination the condition of theorem 5.1 shows that the output polynomials set is a Gröbner base.

Buchberger showed also, that the construction of certain S–polynomials and their reduction could be avoided if some conditions on the head terms of the polynomials and the sequence in which the S–polynomials are constructed are fulfilled. But we cannot go into detail here.

The implementation of the algorithm is a straight forward modification of the commutative Buchberger algorithm in [GK84] and is omitted here.

5.2. Left reduced Gröbner bases. Besides the application of the `LRED` in KW 2. or our `DINLIS` algorithm to compute a left reduced Gröbner base there is some cheaper way to obtain the same result. As in the commutative case by [B85], one can exploit the fact that we already have a Gröbner base.

LEMMA 5.2. *Let G be a left GB in $\mathbf{R}$ and let $g, h \in G$ such that $HT(h)$ divides $HT(g)$. Then $G \setminus \{g\}$ is still a Gröbner base.*

With this lemma it is sufficient to remove polynomials with head terms that are divided by some other polynomials in the base. This avoids the reduction of these polynomials. For the remaining polynomials only the reductas must be reduced to normal form to obtain a reduced basis. This can be done by `LRED` or preferably `DINLIS`, since `DINLIS` needs only one loop through the base.

6. Two–sided Gröbner Bases. Besides the left ideals and left Gröbner bases in solvable polynomials rings, Kandri–Rody and Weispfenning characterize also two–sided ideals and two–sided Gröbner bases constructively. Their main observation is that two–sided GB's can be obtained by left GB construction combined with taking right variable multiples of all polynomials in the base.

More precisely let $I(G)$ denote the two–sided ideal generated by $G \subset \mathbf{R}$. Then the finite set G of polynomials in $\mathbf{R}$ is a two–sided Gröbner base if satisfies the equivalent conditions of the theorem KW 5.4:

THEOREM 6.1 (KW 5.4.). *Let G be a finite set of polynomials in* **R**. *Then the following assertions are equivalent:*

(4) *For all $f \in I(G)$, $f \longrightarrow^*_G 0$ (left reduction).*
(6) *G is a left GB and for all $1 \le i \le r$, $p \in G$, $p * X_i \longrightarrow^*_G 0$.*

Based on this theorem it is easy to give an algorithm for the construction of two-sided GB's. Iterate the computation of left GB's G and the combination with polynomials $p * X_i$: $G_{k+1} := G_k \cup \{p * X_i \longrightarrow^*_G \mid p \in G,\ 1 \le i \le r\}$ until $G_{k+1} = G_k$. This gives the algorithm `GROEBNER` in KW 2.

The following algorithm is slightly different: Take the algorithm for the computation of left GB's and start the algorithm with the combined set $G \cup \{p * X_i \longrightarrow^*_G \mid p \in G,\ 1 \le i \le r\}$. At any time a S-polynomial is reduced to a polynomial $h \ne 0$, combine G not only with $\{h\}$ but additionally with the set $\{h * X_i \longrightarrow^*_G \mid 1 \le i \le r\}$. Further modify the lists `B` and `D` as appropriate.

The algorithm is called `DINCGB`. Note, that singleton sets of polynomials may not be two-sided GB's any more. So the respective case detections in algorithm `DINLGB` can not be carried over to algorithm `DINCGB`.

This completes the overview of the algorithms. We turn now to a small example.

7. Example. So far we have discussed the non-commutative product and Buchberger algorithm. We will now discuss an example computed with the MAS system and give some computing times on several machines.

The listing of an example is given below. The input of non-commutative polynomials consists of two steps:

1. the input of the commutator relations together with the list of variables of the polynomial ring and the desired term order,
2. the input of the non-commutative polynomials itself.

The commutator relations are a list of commuting polynomials which are read by the MAS function `PREAD`. `PREAD` reads from the current input stream and returns a list of distributive rational polynomials in internal representation. `PREAD` expects the following items:

1. The variable list: `(a, x, y)`. A variable name may consist of a alpha-numerical character sequence starting with a letter. All variables occurring in the polynomials must be specified.
2. The desired term order: `L`. The term order may be one of the following:
   L for the inverse lexicographical term order
   G for the inverse graduated term order
   polynomial list a list of univariate (integral) polynomials in the variable `T` (this name is fixed). The list length must be the same as the number of variables. The list is interpreted as the coefficients of a linear form which defines a term order. See [W87] for details.
   There is no check if the term order is compatible with the commutator relations or if the linear form defines an admissible term order.
3. The commutator relations themselves: `( y ), ( x ), ( x y + a )`. This relation is interpreted as $y * x = xy + a$. Not specified relations are interpreted as if the variables commute. In this example a commutes both with x and y. The relations must be given as a list (!) of polynom triples.

The second input consists of the non-commutative polynomials. NPREAD takes as input a relation table and reads a list (!) of polynomials from the current input stream. The output is a list of distributive rational polynomials. ** denotes exponentiation, the multiplication operator * may be omitted. I.e. x y denotes x * y. All multiplications of variables mean the non-commutative *-product. It is possible to specify also more complex polynomial expressions. See 3.2 for the accepted syntax. Be sure to include enough parenthesis to avoid ambiguities.

```
(* Commutator relations: *)
t:=PREAD().
(a,x,y) L
(
( y ), ( x ), ( x y + a ),
)
PWRITE(t).

(* Non-commutative polynomials: *)
p:=NPREAD(t).
(
( y**3 + x**2 y + x y ),
( x**2 + x )
)
PWRITE(p).

(* Left Normalform: *)
c:=LIRRSET(t,p).
PWRITE(c).

(* Left G-base: *)
c:=LGBASE(t,p,1).
PWRITE(c).

(* Two sided G-base: *)
c:=TSGBASE(t,p,1).
PWRITE(c)..
```

Next three of the above discussed algorithms are called:

1. LIRRSET Left IRReducible SET, algorithm DINLIS. The input parameters are t the relation table and p the polynomial list. The output c is the left irreducible set of the input.
2. LGBASE Left Gröbner BASE, algorithm DINLGB, The input parameters are t the relation table, p the polynomial list and 1 a trace flag (0 = on trace, 1 = trace reduced S-polynomials, >1 = trace as much as possible). The output c is the left Gröbner base of the input.
3. TSGBASE Two-Sided Gröbner BASE, algorithm DINCGB. The input parameters are t the relation table and p the polynomial list. The output c is the two-sided Gröbner base of the input.

Note that during the computations the number of known commutator relations is increased. At the end 5 new commutator relations have been computed.

In any case the output polynomials are printed to the current output stream with the procedure PWRITE. PWRITE prints the actual variable list, the actual term order and the list of polynomials, each polynomials starting on a new line.

The following output is produced:

```
Polynomial in the variables: (a,x,y)
Term ordering:  inverse lexicographical.
Polynomial list:
```

Left irreducible set:

```
( y**3 -2 a x - a )
( x**2 + x )
```

Left Gröbner base:

```
  a**2
( x**2 + x )
  a y**2
( y**3 -2 a x - a )
```

Two-sided Gröbner base:

Algorithm	IBM 9370/VM ALDES / SAC-2	Atari ST MAS	IBM AT/386 MAS	Used Cells
DINLIS	0.03	< 1.0	< 1.0	0.3
DINLGB not irred.	1.47			22.4
DINLGB + irred.	1.47	18.0	13.0	22.8
DIN1GB	1.93			29.7
DINCGB	0.58	8.0	5.0	8.6

Computing time in seconds. Cells in thousand.

TABLE 2

Statistics for the algorithms

```
      a
(  x**2 + x )
   y**3
```

8. Summary. A summary of computing times for ALDES on IBM 9370/VM and for MAS on an Atari 1040 ST and an IBM AT/386SX are given in table 2. The timings are obtained with a fresh list of commutator relations in each case, i. e. not in the sequence suggested by the above input listing.

'DINLGB not irred.' means that the algorithm did not compute a reduced (irreducible) GB. 'DIN1GB' denotes the MAS algorithm corresponding to algorithm 'GROEBNER'. 'DINCGB' is superior to 'DIN1GB' due to the fact, that the polynomials $p * X_i$ are added to the base at the beginning of the computation, so much more polynomials will be reducible later on.

The two examples from [AL88] need 2 respectively 6 seconds for a reduced left Gröbner base on an Atari ST.

A more complicated example from [S89] needs 54 seconds on an Atari ST. The problem is to compute a left Gröbner base of the polynomials

```
( v5 v6 - v2 v3 )
( v4 v5 - v1 v3 )
```

in an exterior algebra over a vector space generated by `(v1,v2,v3,v4,v5,v6)`. The computation can be done by specification of the commutator relations as $c_{ij} = -1$ and $p_{ij} = 0$, $(1 \leq i < j \leq n)$ and adding the polynomials v_i^2 $(1 \leq i \leq n)$ to the ideal base.

8.1. Conclusion. We have presented efficient algorithms for the computation in solvable polynomial rings. The algorithms may be used for computations of ideal bases in enveloping algebras of finite dimensional Lie algebras, in iterated differential operator rings or in Clifford algebras.

Further the algorithms may serve as a base for the implementation of further constructive methods such as computation of syzygies, quotient fields, residue class rings, etc. Due to limitations in space we must refer the interested reader to the full version of the paper.

REFERENCES

[AL88] J. Apel, W. Lassner, *An extension of Buchberger's Algorithm and calculations in enveloping fields of Lie algebras*. J. Symb. Comp., 6, pp 361–370, 1988.

[AP88] J. Apel, U. Petermann, *A program for algebraic computations in quotient skew fields of enveloping algebras of Lie algebras*. Karl-Marx-University Leipzig, preprint 1988.

[B83] W. M. Bartol, *et. al.*, ***Report on the Programming Language LOGLAN 82.*** Polish Scientific Publications, Warsaw, 1984.

[B86] W. Böge, R. Gebauer, H. Kredel. ***Some Examples for Solving Systems of Algebraic Equations by Calculating Gröbner Bases.*** J. Symbolic Computation 1, pp 83-98, 1986.

[B65] B. Buchberger. ***Ein Algorithmus zum Auffinden der Basiselemente des Restklassenringes nach einem nulldimensionalen Polynomideal,*** Dissertation, University of Insbruck 1965.

[B85] B. Buchberger, ***Gröbner Bases: an algorithmic method in polynomial ideal theory.*** In: Progress, directions and open problems in multidimensional systems theory. (N.K.Bose ed.) Reidel Publ. Comp., 1985, pp 184–232.

[CL80] G. E. Collins, R. Loos, ***ALDES and SAC-2 now available.*** ACM SIGSAM Bulletin Vol. 12, No. 2, p 19, 1980.

[KRW88] A. Kandri–Rody, V. Weispfenning, ***Non-Commutative Gröbner Bases in Algebras of Solvable Type.*** J. Symb. Comp., 9, pp 1–26, 1990. also available as: Technical Report University of Passau, MIP–8807. March 1988.

[GK83] R. Gebauer, H. Kredel, ***Distributive Polynomial System.*** Several Technical Reports University of Heidelberg, 1983.

[GK84] R. Gebauer, H. Kredel, ***Buchberger Algorithm System.*** Technical Report University of Heidelberg, 1984. (See also ACM SIGSAM Bulletin Vol. 18, No. 1, p 19.)

[K88] H. Kredel, ***Distributive Polynomial Interpreter System.*** Technical Report University of Passau, 1988.

[K90] H. Kredel, ***MAS Modula-2 Algebra System.*** Proc. DISCO '90, Capri 1990, LNCS 429, pp 270–271, 1990. Manuals and Technical Reports University of Passau, 1990.

[M85] T. Mora, ***Gröbner Bases for non-commutative polynomial rings.*** Proc. AAECE–3, Grenoble 1985, LNCS 229, pp 353–362.

[S89] T. Stokes, ***Gröbner Bases in exterior algebra.*** preprint University of Tasmania, 1989.

[W87] V. Weispfenning, ***Admissible orders and linear forms.*** ACM SIGSAM Bull. 21, pp 16–18, 1987.

ONE PARTICULAR CODE FOR POLYNOMIAL MANIPULATIONS

J. Obdržálek

Faculty of Mathematics and Physics – Charles University

Ke Karlovu 3, 121 16 Prague, Czechoslovakia

1. Problematics.

Among different applications of the symbolic manipulations, polynomials manipulation plays a particular role. Polynomials with complex, real or integer coefficients generate a ring, i.e. both summing and multiplying polynomials yields back polynomials, too. Similarly, the substitution of a polynom for a variable or a term in a polynom leads to a polynom and there are many other operations transforming polynomials back to the polynomials.

The polynomial manipulation has among other particular meaning in the theoretical physics. The polynomials function can be either a fair approximation of a more complex function or even an exact description, eventually after some "helping" by introducing new variables. In the latter case, some additional relations may be done among variables; denoting $s = \sin t$, $c = \cos t$, the identity $c * c + s * s - 1 = 0$ could simplify some expressions.

All those manipulations may be done by general languages like REDUCE, mu-Math etc. In this case, the list form of coding is usually used; namely, the expression $a^2b - 6abd$ is coded in a form like

```
a%  ^%  2%  *%  b%  -%  6%  *%  a%  *%  b%  *%  d%,
```

resp. using RPN,

```
a%  2%  ^*  b%  *%  6%  a%  *%  b%  *%  d%  *%  -%,
```

where the symbol % represents a pointer to the next atom.

Handling polynomials only, more compact manners may be used. An example is in

POLYMAN /1/, using for the above mentioned expression code

1.000	2	1	0	0
-6.000	1	1	0	1
0.000	0	0	0	0

i.e. each term coded by one real coefficient and set of integers giving the exponents by corresponding variables. Terms are ordered lexicographically, a zero term denoting the end of the polynomial. This manner is speed increasing and space saving, both factors being crucial for small computers. It is supposed that the number of variables is limited and kept fixed in frame of one calculation. Also the maximal degree of variables is limited to the maximal value of the integer type; usually, this is not a serious problem. Sophisticated methods can be used to overgo "compatibility problems" for gluing more calculations together.

2. Another code for polynomials.

Particular physical problem presented by Cordts, Deus and Frei [2] led to an evaluation of rather hugh expressions, namely the determinants of matrices M, M^2, M^3,, where 24 * 24 matrix M is constructed from round 50 different elements A_1.. H_6. Variables A..H may be further expressed as simple polynomials in 10 variables a, b... This problem has been solved by a program written in PASCAL, handling *homogeneous* polynomials of limited degree (here 3) in many (50) variables coded as

1.000	1	1	2
2.000	1	2	4
0.000	0	0	0

i.e. each term of the homogeneous polynomial (of the degree of D = 3, in this case) being coded by one real coefficient and three integers denoting the ordinal number of the variables (resp. an index of the variable). Terms are ordered lexicographically, zero term denoting the end of the polynomial. Non-homogeneous polynomials of degree less than D may be handled as well using a symbolic index equal to zero.

3. Internal organization.

Both summation and multiplication needs finding terms having identical exponents; any form of term ordering by exponents will speed up those operations. Whereas the result is kept always as a well ordered polynomial, speed improving has been reached using not fully ordered polynomials during all the work. Namely, the multiplications use the semi-ordered structure of the heap; the complexity falls down from n^2m to nm. lg_2n. The heap of terms is organized as follows:

```
   T_k                      T_k+1
  /   \                    /     \
T_2k  T_2k+1           T_2k+2  T_2k+3
```

where relations between terms $T_k < T_{2k}$, $T_{2k} < T_{2k+1}$ are kept during all manipulations rather than the "full order scheme" like $T_k < T_{k+1}$ for any k.

Summary.

Above described form of coding polynomials is very economic concerning both memory needs and working speed. It enabled even a 8-bit microcomputer BBC Acorn with only 64k core memory to perform rather large symbolical calculations connected to the topics of /2/.

References:

1. Obdrzalek J., Luzny J.: POLYMAN - a system for polynomials.
Computer Physics Communications 50 (1988), 255.
2. Cordts W., Deus P., Frei V.: An extended dynamic Keating matrix of the chalcopyrite lattice.
Czech J. Phys. B35 (1985), 1346-54.

THE TWO CLASSICAL SUBRESULTANT PRS METHODS
(Extended Abstract)

Alkiviadis G. Akritas*
Moscow State University
Department of Mathematics and Mechanics
Moscow 119899, USSR

---

*On a Fulbright grant for the Spring Semester 1990. Permanent address: University of Kansas, Department of Computer Science, Lawrence, KS 66045-2192.

Abstract. We present in some detail the two classical subresultant prs methods that exist in the literature for computing polynomial remainder sequences (prs) and greatest common divisors (gcd) of polynomials over the integers. Both methods are based on Sylvester's paper of 1853 [12]; the first method makes use of pseudodivisions whereas the second one triangularizes a matrix corresponding to the resultant of the two polynomials under consideration. The following figure demonstrates the relation of these two methods [3]

```
<------- Sylvester's paper of 1853 ------->
↓                                         ↓
pseudodivisions              matrix-triangularization
```


The two methods. We restrict our discussion to univariate polynomials with integer coefficients and to computations in $\mathbb{Z}[x]$, which is *not* a Euclidean domain. Given the polynomial $p(x) = c_n x^n + c_{n-1}x^{n-1} + \ldots + c_0$, its degree is denoted by $\deg(p(x))$ and c_n, its leading coefficient, by $lc(p(x))$; moreover, $p(x)$ is called *primitive* if its coefficients are relatively prime. Consider now $p_1(x)$ and $p_2(x)$, two primitive, nonzero polynomials in $\mathbb{Z}[x]$, $\deg(p_1(x)) = n$ and $\deg(p_2(x)) = m$, $n \geq m$.

The Pseudodivisions Subresultant PRS Method. We know that the polynomial division algorithm (with remainder), call it **PD**, that works over a field, cannot be used in $\mathbb{Z}[x]$ since it requires exact divisibility by $lc(p_2(x))$. So we use *pseudodivision*, which always yields a pseudoquotient and pseudoremainder; in this process we have to premultiply $p_1(x)$ by $lc(p_2(x))^{n-m+1}$ and then apply algorithm **PD**. That is, we have:

$$lc(p_2(x))^{n-m+1}p_1(x)=q(x)p_2(x)+p_3(x), \qquad deg(p_3(x)) < deg(p_2(x)) \quad (1)$$

Applying the same process to $p_2(x)$ and $p_3(x)$, and then to $p_3(x)$ and $p_4(x)$, etc (Euclid's algorithm), we obtain a *polynomial remainder sequence* (prs)

$$p_1(x),\ p_2(x),\ p_3(x),\ \dots\ ,p_h(x),\ p_{h+1}(x) = 0$$

where $p_h(x)$ nonzero is a greatest common divisor of $p_1(x)$ and $p_2(x)$. If $n_i = deg(p_i(x))$ and we have $n_i - n_{i+1} = 1$, for all i, the prs is called *complete*, otherwise, it is called *incomplete*. The problem with the above approach is that the coefficients of the polynomials grow exponentially and hence slow down the computations. We wish to control this coefficient growth. Observe that equation (1) can be also written in a more general form as

$$lc(p_{i+1}(x))^{n_i-n_{i+1}+1}\ p_i(x) = q_i(x)p_{i+1}(x) + \beta_i p_{i+2}(x),$$
$$deg(p_{i+2}(x)) < deg(p_{i+1}(x)), \quad (2)$$

$i = 1, 2, \dots , h-1$. That is, if a method for choosing β_i is given, the above equation provides an algorithm for constructing a prs.

In Sylvester's approach of 1853 we have [12]

$$\beta_1 = 1 \text{ and } \beta_i = lc(p_i(x))^2,\ i = 2, 3, \dots , h-1, \quad (3)$$

which is ideally suited for complete prs's; for incomplete prs's we can easily modify (3) to obtain

$$\beta_1 = 1 \text{ and } \beta_i = lc(p_i(x))^{n_{i-1}-n_i+1},\quad i = 2, 3, \dots , h-1 \quad (3')$$

It should be noted that using (3′) we obtain smaller coefficients than those obtained by (3), but still, we do not obtain the smallest possible coefficients. This was achieved by Habicht in 1948 [8].

In Habicht's approach we have

$$\beta_1 = (-1)^{n_1 - n_2 + 1} \quad \text{and} \quad \beta_i = (-1)^{n_i - n_{i+1} + 1} lc(p_i(x)) H_i^{n_i - n_{i+1}},$$

$$i = 2, 3, \ldots, h-1, \qquad (4)$$

where

$$H_2 = lc(p_2(x))^{n_1 - n_2} \quad \text{and} \quad H_i = lc(p_i(x))^{n_{i-1} - n_i} H_{i-1}^{1-(n_{i-1} - n_i)}$$

$$i = 3, 4, \ldots, h-1.$$

In the case of incomplete prs's using (4) we obtain the smallest possible coefficients without coefficient gcd calculations. (Also note that in the case of complete prs's using (3), (3′) and (4) we obtain the same coefficients.)

In both cases above what we did was to divide $p_{i+2}(x)$ by the corresponding β_i before we use it again. The proofs that the β_i's shown in (3) and (4) exactly divide $p_{i+2}(x)$ have been presented in a very complicated fashion in [5], [6], and [7] and have up to now obscured simple divisibility properties [10]; see also [1].

The Matrix Triangularization Subresultant PRS Method. In this case we make use of Sylvester's form of the resultant (for the two polynomials $p_1(x)$ and $p_2(x)$ mentioned above)which can be expressed in the following form [2]:

$$res_S(p_1, p_2) = \begin{vmatrix} c_n & c_{n-1} & \cdots & c_0 & 0 & 0 & \cdots & 0 \\ d_n & d_{n-1} & \cdots & d_0 & 0 & 0 & \cdots & 0 \\ 0 & c_n & \cdots & & c_0 & 0 & \cdots & 0 \\ 0 & d_n & \cdots & & d_0 & 0 & \cdots & 0 \\ & & & \vdots & & & & \\ 0 & \cdots & 0 & c_n & c_{n-1} & & \cdots & c_0 \\ 0 & \cdots & 0 & d_n & d_{n-1} & & \cdots & d_0 \end{vmatrix} \qquad (S)$$

Note that $p_2(x)$ has been transformed into a polynomial of degree n by introducing zero coefficients and that this is a matrix of order 2n (as opposed to n+m for the forms of the resultant encountered in the literature). This form of the resultant appears in Sylvester's paper of 1853 [12] and is only once mentioned in the literature by Van Vleck [13].

Van Vleck showed that, if we have the polynomial remainder sequence $p_1(x), p_2(x), \ldots, p_h(x)$ we can obtain the (negated) coefficients of the (i+1)th member of the prs, $i = 0, 1, 2, \ldots, h-1$, as minors formed from the first 2i rows of (S) by successively associating with the first 2i-1 columns (of the 2i by 2n matrix) each successive column in turn. Moreover, it has been proved by Habicht [8] that the coefficients obtained in this way are the *smallest* possible without coefficient gcd computations; see also [9].

Using Bareiss's integer-preserving transformation algorithm [4] (see also Malashonok's preprint [11]) and *bubble pivot* we have shown the following:

Theorem. Let $p_1(x)$ and $p_2(x)$ be two polynomials of degree n and m respectively, $n \geq m$. Using Bareiss's algorithm transform the matrix $\mathbf{M}_S$ corresponding to $res_S(p_1(x), p_2(x))$ into its upper triangular form $T(\mathbf{M}_S)$; let n_i be the degree of the polynomial corresponding to the ith row of $T(\mathbf{M}_S)$, $i = 1, 2, \ldots, 2n$, and let $p_k(x)$, $k \geq 2$, be the kth member of the (*complete or incomplete*) polynomial remainder sequence of $p_1(x)$ and $p_2(x)$. Then if $p_k(x)$ is in row i of $T(\mathbf{M}_S)$, the coefficients of $p_{k+1}(x)$ (within sign) are obtained from row i+j of $T(\mathbf{M}_S)$, where j is the smallest integer such that $n_{i+j} < n_i$. (If $n = m$ associate both $p_1(x)$ and $p_2(x)$ with the first row of $T(\mathbf{M}_S)$.)

For a proof and examples see [3].

References

[1] Akritas, A.G.: A simple proof of the reduced prs algorithm. Computing 38, 369–372, 1987.

[2] Akritas, A.G.: A new method for computing greatest common divisors and polynomial remainder sequences. Numerische Mathematik 52, 119–127, 1988.

[3] Akritas, A.G.: *Elements of Computer Algebra with Applications.* John Wiley, New York, 1989.

[4] Bareiss, E.H.: Sylvester's identity and multistep integer-preserving Gaussian elimination. Mathematics of Computation 22, 565–578, 1968.

[5] Brown, W.S.: On Euclid's algorithm and the computation of polynomial greatest common divisors. Journal of ACM **18**, 476–504, 1971.

[6] Brown, W.S.: The subresultant prs algorithm. ACM Transactions On Mathematical Software **4**, 237–249, 1978.

[7] Collins, G.E.: Subresultants and reduced polynomial remainder sequences. Journal of ACM **14**, 128–142, 1967.

[8] Habicht, W.: Eine Verallgemeinung des Sturmschen Wurzelzälverfahrens. Commentarii Mathematici Helvetici **21**, 99–116, 1948.

[9] Laidacker, M.A.: Another theorem relating Sylvester's matrix and the greatest common divisor. Mathematics Magazine **42**, 126–128, 1969.

[10] Loos, R.: Generalized polynomial remainder sequences. In: Computer Algebra Symbolic and Algebraic Computations. Ed. by B. Buchberger, G.E. Collins and R. Loos, Springer Verlag, Wien, New York, 1982, Computing Supplement **4**, 115–137.

[11] Malashonok, G.I.: Solution of a system of linear equations in a commutative ring. (In Russian.) Preprint FMI AS UkrSSR, Lvov 1986.

[12] Sylvester, J.J.: On a theory of the syzygetic relations of two rational integral functions, comprising an application to the theory of Sturm's functions, and that of the greatest algebraical common measure. Philoshophical Transactions **143**, 407–548, 1853.

[13] Van Vleck, E.B.: On the determination of a series of Sturm's functions by the calculation of a single determinant. Annals of Mathematics, Second Series, Vol **1**, 1–13, 1899–1900.

PRACTICAL APPLICATION OF GROEBNER BASES FOR THE SOLUTION OF POLYNOMIAL EQUATION SYSTEMS

Herbert Melenk

Konrad-Zuse-Zentrum für Informationstechnik Berlin
Heilbronner Str. 10, D-1000 Berlin 31, Federal Republic of Germany
E-mail: melenk@sc.zib-berlin.dbp.de

1. Introduction

Multivariate polynomial equation systems of the type

$$0 = f_j(x_1,\ldots,x_k) = \sum_j c_{jl} x_i^{i_{1jl}} \cdots x_k^{i_{kjl}}, \quad l = 1,\ldots,n, \; i_{mjl} \in \mathbb{N}, \; c_{jl} \in \mathbb{Q} \text{ or } \mathbb{Z} \qquad (1)$$

arise in many areas of computational sciences. They are derived, for example, from the following problem classes: Geometrical descriptions, steady state analysis of differential equation systems, truncated power series.

2. Algebraic Background

As we are primarily interested in "real life" problems, we concentrate on the case of multivariate polynomials with rational/integer coefficients. For theses domains the ideal theory and the Groebner bases are well intoduced. One important property of the Buchberger algorithm is that at each step of the computation the system of polynomials is a complete basis of the ideal.

Properties of Groebner Bases. An overview of Groebner basis properties is given in /4/ and /5/. Some properties which are most relevant in the context of equation solving are:

- Two ideals are identical if their *GB*'s are equal.
- The *GB* defines an algorithm for computation modulo ideal.
- The *GB* defines an algorithm for the decision of ideal membership.
- The dimension of the ideal (~cardinality of solutions) can be read off from the Groebner basis immediately.
- The polynomial p in ideal (f_i) which is minimal *wrt* the order is guaranteed to be explicit member of the *GB*.
- When the ordering defines a lexical relation between two variables, e.g.

  $$x^n \succ_{\text{ord}} y^m \quad \text{for all} \quad n, m > 0,$$

  and there are polynomials in the ideal which do not depend on x, then in the *GB* all multiples of x are eliminated from the biggest possible part of the polynomials.

- If the complete lexical ordering is used, the dependency pattern is of triangular shape

$$
\begin{array}{llll}
g_1(x_1, & x_2, \ldots, & x_n, x_{n+1}, \ldots, x_k) \\
 & \vdots & & \\
g_l(& x_2, \ldots, & x_n, x_{n+1}, \ldots, x_k) \\
 & & \vdots & \\
g_m(& & x_n, x_{n+1}, \ldots, x_k)
\end{array}
$$

- The complete triangular pattern ($n = k$) is guaranteed in the case of zero dimensional ideals (solution is finite set of isolated points).

3. Equation solving with Groebner Bases

Based on Lexicographical Ordering. The simplest approach to problem (1) is to compute

$$G = GB_{\mathrm{lex}(x_1,\ldots,x_k)}\big(f_1(x_1, \ldots, x_k), \ldots, f_n(x_1, \ldots, x_k)\big)$$

or the lexicographical GB for some permutation of the $(x_1, \ldots, x_k)$. If the solution space is zero dimensional, the last polynomial is a univariate polynomial in x_k. If of degree up to 4, it can be solved algebraically with the Cardan formulae, even if the coefficient domain contains formal parameters. If the degree is higher, the roots can be approximated to arbitrary accuracy using one of the classical methods, e.g. the Sturm or Uspensky techniques. These roots then can be substituted into the remaining elements of the GB, giving now a univariate polynomial in x_{k-1}. So step by step all points of the zero dimensional solution set are determined either as closed forms or as approximations with arbitrary precision.

Example:

Modified Edelstein (chemical reaction network) with symbolic coefficients.

$$\alpha c_1 - \beta c_1^2 - \gamma c_1 c_2 + \epsilon c_3, \; -\gamma c_1 c_2 + \epsilon c_3 + 2\theta c_3 - 2\eta c_2^2, \; \gamma c_1 c_2 - \epsilon c_3 - \theta c_3 + \eta c_2^2$$

The Groebner Basis *wrt* (c_3, c_2, c_1) here is

$$\{c_3\epsilon - c_2 c_1 \gamma, \; c_2^2 \epsilon\eta - c_2 c_1 \gamma\theta, \; c_1^2\beta - c_1\alpha\}$$

Here the third polynomial can be decomposed formally and with backsubstituting we immediately get the solutions

$$
\begin{array}{l}
\{ \{c_3 = 0, c_2 = 0, c_1 = \alpha/\beta\} \;\; \{c_3 = 0, c_2 = 0, c_1 = 0\}, \\
\;\; \{c_3 = (\alpha^2\gamma^2\theta)/(\beta^2\epsilon^2\eta), \; c_2 = (\alpha\gamma\theta)/(\beta\epsilon\eta), \; c_1 = \alpha/\beta\} \}
\end{array}
$$

In the following example (Arnborg, dim=4)

$$
\begin{array}{c}
X + Y + Z + T, \; XY + XT + YZ + ZT, \\
XYZ + XYT + XZT + YZT, \; XYZT + 1
\end{array}
$$

the ideal is not zero dimensional: its *lex* basis is

$$\{X+Y+Z+T,\quad Y^2+2YTT+T^2,$$
$$YZ-YTT-Z^2T^4+ZT-2T^2,\quad YT^4+Y+T^5+T,$$
$$Z^3T^2+Z^2T^3+Z+T,\quad Z^2T^6+Z^2T^2+T^4+1\}$$

where the last polynomial is bivariate in (z,t). Here the structure of the solution set, which is a manifold parameterized by either z or t, becomes obvious. If we fix either z or t, the corresponding isolated roots then can be determined by backsubstitution.

Alternatives to the Direct Lexicographical Approach. Generally, the lexicographical Groebner basis gives the best insight into the structure of the ideal, as far as the solution of (1) is concerned. On the other hand, lexicographical bases are the hardest to compute: their calculation often explodes in space and time because of an enormous growth of intermediate expressions (number of polynomials, coefficient size). Here different methods have to be used instead. The most important technique is decomposition of the ideal by factorization, which will be discussed in the following section in more detail. But other orderings may help here (sometimes in connection with decomposition techniques) too.

A graduated *GB* has no guaranteed elimination property, but instead it finds the polynomial(s) with minimal total degree in the ideal, which often is a very useful information too.

- **Zero dimensional ideal.** The graduated *GB* can be converted to a lexical one by the *FGLM* algorithm of FAUGERE/GIANNI/MORA/LAZARD /9/, by solving linear systems derived from the congruence.

- **Zero dimensional ideal with numerical coefficients.** From the congruence established by the graduated *GB*, the isolated solutions can directly be approximated by the second part of the algorithm by Auzinger/Stetter /1/ (eigenvalues of a linear transformation modulo the ideal).

For special questions (e.g. specific eliminations) special orderings can be applied.

4. Decomposition of Ideals

Factorization. If we are able to find in the ideal I with an intermediate basis $(h_1(x),\ldots,h_n(x))$ a polynomial $g(x)=g_1(x)*g_2(x)$ with non constant g_1 and g_2, we know that g can only vanish for some x if either $g_1(x)$ or $g_2(x)$ vanishes. On the other hand, we can add $g(x)$ to any basis of I and so we get a basis of the form $(g_1(x)*g_2(x),h_1(x),\ldots,f_n(x))$. So we can see immediately that a solution in I is a solution either of

$$I_1=\big(g_1(x),h_1(x),\ldots,h_n(x)\big)\quad\text{or of}\quad I_2=\big(g_2(x),h_1(x),\ldots,h_n(x)\big).$$

The Groebner bases for I_1 and I_2 now can be computed in separate cycles. The decomposition effect is twofold: The solution space is decomposed, giving a better insight in its structure, and the order of the algorithm is lowered. Each branch can fork in a recursive style, if more factorable polynomials are found.

Not in any case all branches lead to significant parts of the solution. If one of the factors does not have a zero in common with the rest of the basis, its branch will end with the ideal (1); this in general will be determined soon, and the effect of the lower order for the other branch remains in effect. And the Ideal I_2 may be identical with I_1, although g_2 is not identical g_1; in this case the branch is superfluous and this property can be definitely stated, when g_1 is found as member of I_2.

Passive Factorization. The simplest approach for finding factorizations is to inspect all intermediate polynomials during the run of Buchberger algorithm. In the success case forking takes place. This technique was used in several contexts /7/, /8/ /13/, and especially /8/ describes an important approach for lowering the costs of that analysis.

Example (chemical reaction network):

$$-c_1^2\,\beta - c_1\,c_2\,\gamma + c_1\,\alpha + c_3\,\epsilon - c_1\,c_2\,\gamma - 2\,c_2^2\,\eta + c_3\,\epsilon + 2\,c_3\,\theta c_1\,c_2\,\gamma + c_2^2\,\eta - c_3\,\epsilon - c_3\,\theta \ .$$

Here the algorithm soon detects a H-polynomial $-c_1^2\,\beta + c_1\,\alpha$, it can be factorized into a product *variable* $*$ *polynomial*, here $c_1\,(-c_1\,\beta + \alpha)$. The algorithm detects more factorizations of this type and generates separate calculation branches, which result in three Groebner bases:

$$\{c_3, c_2, c_1\},\ \{c_3, c_2, c_1\beta - \alpha\},\ \{-c_3\beta^2\epsilon^2\eta + \alpha^2\gamma^2\theta,\ c_2\beta\epsilon\eta - \alpha\gamma\theta, c_1\beta - \alpha\}\ .$$

The third basis offers the unique positive solution

$$c_3 = \frac{\alpha^2\gamma^2\theta}{\beta^2\epsilon^2\eta}\ ,\quad c_2 = \frac{\alpha\gamma\theta}{\beta\epsilon\eta}\ ,\quad c_1 = \frac{\alpha}{\beta}\ .$$

Sometimes, especially with high variable numbers, so many factorizations are found, that the problem tree explodes too. This effect can be controlled by a careful combination algorithm /13/ which at the branching point looks forward into the further computation and tries to detect common parts, especially common future factorizations which then can be included in the fork design.

Active Factorization. With the above approach, the forking is more or less casual; it depends highly from the strategy decisions taken. There are many cases where the ideal contains factorable polynomials but they will not be found by any strategy. For example, the system $\{x^2 + 2xy + 101, y^2 - 101\}$ is a Groebner basis with *lex* ordering over $\mathbb{Z}$ already. It is obvious that the sum of the two polynomials nicely factors, but this sum will never occur in the Buchberger algorithm.

There are two novel approaches for a more systematical factorization:

• **Testfactors.** If there is the assumption, that a factor $q(x)$ might lead to a significand decomposition of the ideal, the expressions $r(x) =$ remainder $(h(x), q(x))$ can be investigated; if there is a syzygie among these remainders (that is a linear dependency in the ideal of remainders), the corresponding combination of the h's is (if not zero) a polynomial in I that has q as a factor. So we have a way to look actively for multiples of $q(x)$ in I. This technique corresponds algebraically to an ideal quotient computation, which here is performed quasi on line. And it can be executed for several testfactors simultaneously.

Testfactors often arise from obvious properties of the polynomials: for example homogeneous systems tend to monomial factors. Or they are derived from the insight in the (e.g. physical) background of the underlying application.

• **Symmetries.** In many applications, e.g. from physics, the systems are symmetric *wrt* some or all variables: the ideal is invariant under some permutations of its variables (rotation, pairwise, exchange, ...). In /10/ a new approach for some classes of this type is given, where polynomials in the ideal are detected which in advance are known to factor. This leads to a decomposition of the ideal before the Buchberger algorithm starts and thus in many cases to a dramatical decreasing of the complexity/computing requirements, such that much larger systems get solvable.

Arithmetic restrictions. A second source for decomposition is the formal restriction that only nonnegative real values are allowed as components of solutions. This restriction often arises if the variables represent nonnegative physical quantities like masses, distances, ... So we can exploit a set of inequalities $x_i \geq 0, \ \ i = 1, \dots, N$. In generalizing Descartes rule of signs, we can state for a multivariate polynomial

$$p = a_0 + a_1 x_1^{i_{11}} \cdots x_k^{i_{1k}} + \cdots + a_m x_1^{i_{m1}} \cdots x_k^{i_{mk}}$$

where all a_i are real and have the same sign:

a) if $a_0 \neq 0$ then $p(x_1, \dots, x_k) = 0$ implies that at least one $x_i < 0$;

b) for $a_0 = 0$ then $p(x_1, \dots, x_k) = 0$ and $x_i \geq 0$ implies, that each monomial in p is zero already.

For example, the polynomial $17xy^2r \ + 4x^2z$ leads to the decomposition

$$x = 0 \ \lor \ (y = 0 \ \land \ z = 0) \ \lor \ (r = 0 \ \land \ z = 0)$$

which describes the nonnegative real solutions.

Arithmetic restrictions together with testfactors and symmetries are means to make present knowledge about the problem to be solved available and exploitable for the solver, enabling computations which are impossible without. These techniques were successfully applied for complicated problems given by H. CAPRASSE /6/ and A. NOONBURG /16/, SPEER /17/.

5. Conclusion

The recent progress of the Groebner technology has made it a useful tool of computer algebra. Especially the decomposition techniques have moved the horizon of computability dramatically, such that real life problems of moderate size can be handled effectively.

Because of the limited space only a few small examples for Groebner applications could be mentioned here. More examples are found in /2/, /5/, /15/ and in the test series for the REDUCE Groebner package available via electronic mail from the REDUCE netlib.

References

1. Auzinger/Stetter: *An Elimination Algorithm for the Computation of all Zeros of a System of Multivariate Polynomial Equations.* In: Numerical Mathematics Singapore 1988, R. Agarwal, Y. M. Chow, S. J. Wilson (eds.), ISNM, vol. 86, Birkhäuser Verlag Basel (1988).

2. W. Boege, R. Gebauer, R., H. Kredel.: *Some Examples for Solving Systems of Algebraic Equations by Calculating Groebner Bases.* J. Symbolic Computation **1**, pp. 83-98, (1986).

3. B. Buchberger: *An Algorithm for Finding a Basis for a Residue Class Ring of a Zero-Dimensional Polynomial Ideal (German).* Ph.D.Thesis, Univ. of Innsbruck, Austria, Dept. of Mathematics (1965).

4. B. Buchberger: *Groebner Bases: An Algorithmic Method in Polynomial Ideal Theory.* In: (Bose, N. K. ed.) Progress, directions and open problems in multidimensional systems theory. pp. 184–232. Dordrecht: Reidel, (1985).

5. B. Buchberger: *Applications of Groebner Bases in Non-Linear Computational Geometry.* In: (Janssen, R. ed.) Trends in Computer Algebra, pp. 52–80. Berlin, Heidelberg (1988).

6. H. Caprasse, J. Demaret: *Private communication* (1989).

7. S.R. Czapor: *Solving Algebraic Equations via Buchberger's Algorithm.* In: (Davenport, J.H. ed.) Eurocal'87, pp. 260–269, Berlin, Heidelberg (1989).

8. J. H. Davenport: *Looking at a Set of equations.* Bath Computer Science Technical Report 87–06 (1987).

9. J. C. Faugère, P. Gianni, D. Lazard, T. Mora: *Efficient Computation of Zero-Dimensional Groebner Bases by Change of Ordering.* Preprint (1989).

10. K. Gatermann: *Symbolic solution of polynomial equation systems with symmetry.* Konrad–Zuse–Zentrum für Informationstechnik Berlin. Preprint SC 90–3 (1990).

11. R. Gebauer, H. M. Möller: *A New Implementation of Buchberger's Algorithm.* J. Symbolic Computation **6**, 275-286 (1988) .

12. Anthony C. Hearn: *REDUCE User's Manual, Version 3.3.* July 1987. The RAND Corporation CP78, Santa Monica.

13. H. Melenk, H. M. Möller, W. Neun : *On Groebner Bases Computation on a Supercomputer Using REDUCE.* Konrad–Zuse–Zentrum für Informationstechnik Berlin, SC 88-2 (January 1988).

14. H. Melenk, H. M. Möller, W. Neun : *Symbolic Solution of Large Stationary Chemical Kinetics Problems.* Impact of Computing in Science and Engineering **1**, 136–167 (1989).

15. H. Melenk, H. M. Möller, W. Neun : *GROEBNER. A Package for Calculating Groebner Bases.* Version January 1990. Available from data base REDUCE–netlib at rand.org.

16. V.W. Noonburg: *A Neural Network Modeled by an Adaptive Lotka–Volterra System.* To appear in Siam J. Appl. Math.

17. E. R. Speer: Private communication.

PARAMETRIC GRÖBNER BASES FOR NON-COMMUTATIVE POLYNOMIALS

H. Kredel, V. Weispfenning
Universität Passau
D - 8390 Passau, FR Germany

Introduction. It was shown in [KRW] that the Gröbner basis technique with all its immediate applications (see [B]) can be extended from commutative polynomials to a large class of non-commutative polynomial rings over fields. More precisely, the method works for one- **and** two-sided ideals in **solvable algebras**, a class of rings that comprises e.g. enveloping algebras of finitely dimensional Lie algebras, Weyl algebras and iterated Ore-extensions of fields (compare also [K]). So the method is a strong generalization of the results in [AL].

On the other hand, it is well-known that for commutative polynomials with parametric coefficients, Gröbner bases are extremely unstable under specialization of the parameters. This problem was overcome by the second author by the construction of **comprehensive Gröbner bases**, i.e. bases for commutative polynomial ideals that are Gröbner bases in every specialization of the parametric coefficients (presented at the CoCoa-II conference, Genova 1989, see [W2]).

In this paper, we outline

1. the construction of parametric solvable algebras that may serve as "universal objects" for large classes of solvable algebras.

2. the construction of comprehensive Gröbner bases in parametric solvable algebras, where both the coefficients of the polynomials in question and the commutator relations defining the algebra are parametric.

This has the following far-reaching consequence: Let $d, k \in \mathbf{N}$ be given bounds; then for all solvable algebras S' over a suitable class of fields and with commutator relations below the degree bound d, and for all finite polynomial systems F in S' with $|F| \leq k$ and $deg(F) \leq d$, the computation of Gröbner bases for the left, right, or two-sided ideal generated by F can be reduced to a single computation of a comprehensive Gröbner basis G in a single parametric solvable algebra S, from which all the desired Gröbner bases in the algebras S' can be obtained simply by specialization. The same applies to all the applications of one- and two-sided Gröbner bases in solvable algebras indicated in [KRW] and [AL] and [W2]. Moreover, the construction of G can be performed for a fixed term-order or even an arbitrary class of term-orders; in the latter case one arrives at a **universal comprehensive Gröbner basis** (compare [W1]).

1 Parametric Solvable Algebras

Solvable algebras were introduced in [KRW] as a wide class of non-commutative algebras, in which most of the Gröbner basis theory for commutative polynomial rings (compare [B]) is valid for one- and two-sided ideals.

Parametric solvable algebras arise from solvable algebras when the ground field is replaced by a quotient ring of a multivariate polynomial ring over an integral domain, so that the commutator relations of the algebra may involve parameters. More precisely, let K be an integral domain, let $K[U_1, ..., U_m]$ be the (commutative) polynomial ring in $U_1, ..., U_m$, let I be an ideal in $K[U_1, ..., U_m]$ with $I \cap K = \{0\}$, and let $R = K[U_1, ..., U_m]/I$, so that $R = K[u_1, ..., u_m]$ with $u_i = U_i + I$. Next let $S = R[X_1, ..., X_n]$ be the polynomial ring in $X_1, ..., X_n$ over R, and let $<$ be a term-order on the set T of all terms (power-products) in $X_1, ..., X_n$. Let $*$ be a new, possibly non-commutative, multiplication on S such that the following conditions are satisfied:

Definition 1.1 *(i)* $(S, 0, 1, +, -, *)$ *is an associative ring extending* R.

(ii) For all $a, b \in R$, $1 \leq h \leq i \leq j \leq k \leq n$, $t \in T(X_i, ..., X_j)$, $a * bt = bt * a = abt$, $X_h * bt = bX_ht$, $bt * X_k = btX_k$.

(iii) For all $1 \leq i < j \leq n$ *there exist* $0 \neq c_{ij} \in R$, $p_{ij} \in S$ *such that* $X_j * X_i = c_{ij}X_iX_j + p_{ij}$ *and* $p_{ij} < X_iX_j$ *in the quasi-order on* S *induced by the term-order on* T. *Moreover, if* $C = \{c_{ij} : 1 \leq i < j \leq n\}$ *and* C' *is the multiplicative subset generated by* C *in* R, *then* $0 \notin C'$.

Then we call $(S, *)$ *a* **parametric solvable algebra over** R *and denote it by* $R\{X_1, ..., X_n\}$. *The special case of a* **solvable algebra over** K *is obtained by specializing* K *to a field and taking* $m = 0$. *We put* $P = \{p_{ij} : 1 \leq i < j \leq n\}$.

The following lemma shows that the multiplication $*$ of $S = R\{X_1, ..., X_n\}$ is uniform in the parameters $u_1, ..., u_m$.

Lemma 1.2 *Let* K *be an integral domain* $m \geq 1$, $R = K[u_n, ..., u_m]$; *let* $S = R\{X_1, ..., X_n\}$ *be a parametric solvable algebra satisfying 1.1; let* $0 \neq f, g \in S$. *Then one can compute* $c \in C'$ *and* $p \in S$ *such that* $f * g = c \cdot f \cdot g + p$ *and* $p < f \cdot g$. c *and* p *are uniquely determined by these properties, and the coefficients of* p *in* R *are* *polynomials in all* c_{ij}*, the coefficients in* R *of all* p_{ij} *and of* f, g*. Furthermore these polynomials are formed uniformly, independently of the ring* R.

Moreover, if for some $1 \leq i \leq n$, $f \in R\{X_1, ..., X_i\}$ *and* $g \in R\{X_i, ..., X_n\}$, *then* $f * g = f \cdot g$.

Proof. By analyzing the proof of lemmas 1.3 and 1.4 in [KRW]. □

Left reduction $f \underset{H}{\longrightarrow} f'$ for $f, f' \in S$, $H \subseteq S$ is defined similar as in the commutative case except that we avoid division in R; so $d \cdot f - f'$ is in the left ideal $I_L(H)$ generated by H for some $0 \neq d \in R$. A similar convention applies to the **left S-polynomial** $SPol(f,g)$ of $0 \neq f, g \in S$ (see [KRW] and [W2] for details).

Corollary 1.3 *Let* $S = R\{X_1, ..., X_n\}$ *be a parametric solvable algebra, let* $0 \neq f, g \in S$, $H \subseteq S$*. Then the following hold:*

(i) *If* $f \underset{H}{\longrightarrow} f'$ *in* S*, then coefficients of* f' *in* R *are polynomials in the coefficients in* R *of the polynomials in* $\{f\} \cup H \cup C \cup P$*. Moreover,* $f' < f$.

(ii) *The coefficients in* R *of* $h = SPol(f,g)$ *are polynomials in the coefficients in* R *of the polynomials in* $\{f,g\} \cup C \cup P$.

A **specialization** of $R = K[u_1, ..., u_m]$ is a ring homomorphism $\sigma : R \longrightarrow K'$, where K' is a field. Let $S = R\{X_1, ..., X_n\}$ be a parametric solvable algebra over R given as above, and let $S' = K'\{X_1, ..., X_n\}$ be a solvable algebra over K' such that for $1 \leq i < j \leq n$, $X_j * X_i = c'_{ij} X_i X_j + p'_{ij}$ with $0 \neq c'_{ij} \in K'$, $p'_{ij} \in S'$, $p'_{ij} < X_i X_j$. Let $\overline{\sigma} : S \longrightarrow K'[X_1, ..., X_n]$ be the natural extension of σ obtained by applying σ coefficientwise. Then we call σ **admissible for** S and S', if for $1 \leq i < j \leq n$, $\sigma(c_{ij}) = c'_{ij}$ and $\overline{\sigma}(p_{ij}) = p'_{ij}$.

Lemma 1.4 *Let* $\sigma : R \longrightarrow K'$ *be a specialization of* R *that is admissible for* $S = R\{X_1, ..., X_n\}$ *and* $S' = K'\{X_1, ..., X_n\}$*. Then* $\overline{\sigma} : S \longrightarrow S'$ *is a ring homomorphism (with respect to the new multiplication* $*$*). So whenever* $0 \neq f, g \in S$, *and* $f * g = c \cdot f \cdot g + p$ *and* $\overline{\sigma}(f) * \overline{\sigma}(g) = c' \cdot \overline{\sigma}(f) \cdot \overline{\sigma}(g) + p'$ *are as in 1.2, then* $c' = \sigma(c)$ *and* $p' = \overline{\sigma}(p)$.

Proof. Immediate from 1.2. □

Let $K'\{X_1, ..., X_n\}$ be a solvable algebra over an arbitrary field K', let $T'_{ij} (1 \leq i < j \leq n)$ be finite subsets of T and put $T' = \{T'_{ij}\}_{1 \leq i < j \leq n}$. Then we say, $K'\{X_1, ..., X_n\}$ **is of type** T', if all terms occuring in some polynomial p_{ij} $(1 \leq i < j \leq n)$ of definition 1.1 (iii) are elements of T'.

As an application of the preceeding lemmas, we can now show that for fixed T', every solvable algebra of type T' over an arbitrary field K' can be obtained from a single parametric solvable algebra by specializations:

Theorem 1.5 *Let $T' = \{T'_{ij}\}_{1\le i<j\le n}$ be a family of finite subsets of T and let $<$ be a term-order on T such that $t < X_iX_j$ for every $t \in T_{ij}$ $(1 \le i < j \le n)$. Let K be an integral domain. Then there exists a ring $R = K[u_1, ..., u_m] = K[U_1, ..., U_m]/I$ (where $K[\mathbf{U}]$ is a polynomial ring over K) and a parametric solvable algebra $S = R\{X_1, ..., X_n\}$ such that for all fields K', all solvable algebras $S' = K'\{X_1, ..., X_n\}$ of type T' and all ring homomorphisms $\varphi : K \longrightarrow K'$, φ has an extension to a specialization $\sigma : R \longrightarrow K'$ that is admissible for S and S'. If σ is surjective, then we have in particular $S' = \overline{\sigma}(S)$.*

Proof. For all $1 \le i < j \le n$, we let c_{ij} be an indeterminate and p_{ij} the most general polynomial with indeterminate coefficients containing all terms in T'_{ij}. Let $(U_1, ..., U_m)$ be a list of these indeterminates. For every pair of terms $s, t \in T$ let c_{st} and p_{st} be determined by 1.2 in such a way that

$$(1) \quad s * t = c_{s,t}st + p_{s,t} \text{ and } c_{st} \in C', \quad p_{s,t} < st$$

in every parametric solvable algebra $S = R\{X_1, ..., X_n\}$, satisfying 1.1, where $R = K[u_1, ..., u_n]$.

Then (1) together with 1.1 (ii) determines the structure constants in R of S with respect to the basis T of S. By [J], proposition 1, page 4, there exists a set Q of quadratic polynomials in these structure constants with the following property: $q(u_1, ..., u_m) = 0$ for all $q \in Q$ iff the algebra on the free R-module with basis T and multiplication $*$ defined by (1) is associative. Let now I be the ideal generated by Q in the polynomial ring $K[U_1, ..., U_m]$. Then $I \cap K = \{0\}$, (since there exists the solvable algebra $K'[X_1, ..., X_n]$ over $K' = Quot(K)$). Put $R = K[U_1, ..., U_m]/I = K[u_1, ..., u_m]$ and define the R-algebra S with basis T over R by (1). Then by construction, S is associative and hence a parametric solvable algebra over R satisfying 1.1.

Let now $S' = K'\{X_1, ..., X_n\}$ be a solvable algebra of type T' over a field K' satisfying for $1 \le i < j \le n$

$$X_j * X_i = c'_{ij}X_iX_j + p'_{ij} \text{ with } 0 \ne c'_{ij} \in K', \quad p'_{ij} \in S', p'_{ij} < X_iX_j.$$

Let $v_1, ..., v_m$ be the m-tuple consisting of c'_{ij} and the coefficients of the p'_{ij} listed in the same order as the m-tuple $(U_1, ..., U_m)$ for the c_{ij} and p_{ij}. Let $\varphi : K \longrightarrow K'$ be a ring-homomorphism and let $\overline{\varphi} : K[U_1, ..., U_m] \longrightarrow K'$ be its canonical extension with $\overline{\varphi}(U_i) = v_i$. Then by definition of Q, $\overline{\varphi}(q(U_1, ..., U_m)) = 0$ for all $q \in Q$. Consequently $ker(\overline{\varphi}) \supseteq I$, and so $\overline{\varphi}$ induces a specialization $\sigma : R \longrightarrow K'$ with $\sigma(u_i) = v_i$. By construction σ is admissible for S and S', and so by 1.4 $\overline{\sigma} : S \longrightarrow S'$ is a ring-homomorphism. Moreover $\overline{\sigma}$ is surjective, if σ is surjective. □

To indicate the scope of this theorem we mention the following special case:

Let $1 \le n \in \mathbf{N}$ be fixed and let K be a field. Let $\mathcal{U}_n$ be the class of all universal enveloping algebras of some n-dimensional Lie-Algebra $\mathcal{A}$ over K with basis $X_1, ...X_n$

(see [J]). Every $S \in \mathcal{U}_n$ is a solvable algebra $S = K\{X_1, ..., X_n\}$ over K such that for $1 \leq i < j \leq n$, $X_j * X_i = X_i X_j + p_{ij}$ where p_{ij} is a linear form in $X_1, ..., X_n$ and the term-order ist e.g. total-degree-lexicographical. Then the theorem yields:

Corollary 1.6 *There exists a parametric solvable algebra $S = R\{X_1, ..., X_n\}$, where $R = K[u_1, ..., u_m]$ and $m = \frac{1}{2}n^2(n-1)$ such that for every $S' \in \mathcal{U}_n$ there exists a specialization $\sigma : R \longrightarrow K$ that is admissible for S and S' such that $\sigma \mid K = id$ and $S' = \overline{\sigma}(S)$.*

2 Comprehensive Gröbner Bases

Let $S = R\{X_1, ..., X_n\}$ be a parametric solvable algebra over $R = K[u_1, ..., u_m]$ with respect to a fixed term-order $<$ on T as in 1.1.. For $F \subseteq S$ we let $I_L(F)$, $I_R(F)$, $I(F)$ denote the left, right, and two-sided ideal generated by F in S, respectively. For the definition and characterizations of left, right, and two-sided Gröbner bases for an ideal, we refer the reader to [KRW].

Definition 2.1 *Let I be a left (right, two-sided) ideal in S and let G be a finite subset of I. Then G is a* **comprehensive left (right, two-sided) Gröbner basis of** *I, if for all solvable algebras $S' = K'\{X_1, ..., X_n\}$ over a field K' and all specializations $\sigma : R \longrightarrow K'$ that are admissible for S and S', $\overline{\sigma}(G)$ is a left (right, two-sided) Gröbner basis of the left (right, two-sided) ideal $I_L(\overline{\sigma}(I))$, $(I_R(\overline{\sigma}(I))$, $I(\overline{\sigma}(I)))$ in S'. (Here and in the following, $\overline{\sigma}$ is the canonical extension of σ to a ring-homomorphism $\overline{\sigma} : S \longrightarrow S'$.)*

Our main result is the following:

Theorem 2.2 *Let $S = R\{X_1, ..., X_n\}$ be a parametric solvable algebra as above and let F be a finite subset of S. Then one can construct a comprehensive left (right, two-sided) Gröbner basis G of $I_L(F)$, $(I_R(F)$, $I(F))$ in S. The construction is algorithmic relative to the term-order $<$ on T and computations in the ground ring R.*

The **proof** combines the constructions in [KRW] and [W2] with the uniformity results on parametric solvable algebras. All these constructions taken together form a large and fairly complicated system of interrelated algorithms. Moreover, in the present situation, any specialization σ of the parameters $(u_1, ..., u_n)$ of R changes not only the construction of Gröbner bases, but even the computation of products in $\sigma[R]$. So, in addition, these algorithms have to be interlocked. Hence we can present here only an outline of the ideas and steps involved.

Very roughly speaking, the argument follows a pattern of folding and unfolding: In the folding phase one constructs a Gröbner system and in the unfolding phase one extracts a comprehensive Gröbner basis from the Gröbner system by deleating all information on case distinctions.

A left (right, two-sided) Gröbner system GS **for** F is a finite tree, whose vertices are pairs (γ, G) with the following properties:

(i) γ is a condition, i.e. finite set of polynomial equations and inequations in the parameters $(u_1, ..., u_m)$ of the ring R.

(ii) G is a finite subset of $I_L(F)$ (of $I_R(F)$, of $I(F)$) extending F.

(iii) The root of GS is $(\emptyset, F)$.

(iv) Any leaf (γ, G) of GS has the following property: Whenever $\sigma : R \longrightarrow K'$ is a specialization that is admissible for S and S' (where $S = K'\{X_1, ..., X_n\}$ is a solvable algebra) and is compatible with the condition γ, then G is a left (right, two-sided) Gröbner basis of $I_L(F)$ (of $I_R(F)$, of $I(F)$) in S'.

From any such Gröbner system, a left (right, two-sided) comprehensive Gröbner basis G of $I_L(F)$ ($I_R(F)$, $I(F)$) can be extracted by "forgetting" all the conditions γ in the leaves of GS and uniting all the corresponding polynomial sets G; in other words

$$G = \bigcup \{G : (\gamma, G) \text{ is a leaf of } GS\}.$$

Next we discuss the construction of a left and a two-sided Gröbner system for F.

First, we recall from [KRW] the steps involved in the construction of a left (two-sided) Gröbner basis from a finite polynomial system F in a solvable algebra $S' = K'\{X_1, ..., X_n\}$:

(i) Formation of left S-polynomials $h = SPol(f, g)$ for $f \neq g$.

(ii) Successive left reduction of h to a normal form.

(iii) (in the case of two-sided Gröbner bases). Formation of products $g * X_i$ $(1 \leq i \leq n)$ and successive left reduction of these products to a normal form.

We want to perform the analogous steps in the parametric solvable algebra S in such a way that they remain valid under every admissible specialization of R. This forces us to adjoin to each step a case distinction on the vanishing or non-vanishing of certain coefficients that are relevant to this step. In this way a single step is transformed into a finite fan of single steps with associated conditions. Iterating this transformation, one may expect to arrive at a tree that forms Gröbner system for F.

This is, however, not the case for the following reason: If one follows a path trough this tree, the computation steps are performed according to the associated conditions. This means e.g. that by certain polynomial equations in a condition, certain monomials of a polynomial in S have to be cancelled. This in turn means that the corresponding computation step - regarded as a computation in R - will lead outside of the ideal $I_L(F)$ (or $I(F)$, respectively). So property (ii) of a Gröbner system will be violated.

The way out of this problem is as follows:
Each computation step in the tree to be constructed is performed in a "schizophrenic" manner. Instead of cancelling a monomial m according to a condition γ, we "colour" the monomial m according to γ: m is coloured green (red, white) if γ "says" that m should be zero (γ "says" that m should be non-zero, γ leaves m undetermined). In no case, however, a monomial is cancelled. Instead the computation step is guided by the colours, where care has to be taken that all monomials relevant to the step have been coloured green or red before.

In other words, the **definition** of the step is guided by the specializations compatible with the condition in question; its **formal computation** is, however, done in S in such a way that it never leads outside the ideal $I_L(F)$ (or $I(F)$, respectively).

Finally, one has to guarantee that the modified tree constructed in this way is indeed finite. To show this, let B be a branch of this tree T. Modify the branch by cancalling all monomials that have been coloured green by their corresponding condition. Then the modified branch B' constitutes a correct left (two-sided) Gröbner basis computation with respect to any admissible specialization $\overline{\sigma} : S \longrightarrow S'$ that is compatible with all conditions in the branch B'. So by the results of [KRW], B' is finite, and hence B is finite. Thus we know that every branch of T is finite. Moreover, by construction, T is finitely branching in every vertex; so by König's tree lemma, the whole tree T is finite.

This completes our proof sketch. □

Next we indicate an extension of theorem 2.2 that concerns comprehensive Gröbner bases in S that have this property for **all** term-orders, as far as they are compatible with the solvability of the algebra S.

Recall from [W1] that a finite subset G of a solvable algebra $S' = K'\{X_1, ..., X_n\}$ is a **universal left (right, two-sided) Gröbner basis**, if G is a left (right, two-sided) Gröbner basis for all term-orders $<$ on T that are compatible with 1.1 (iii). In an analogous way, we define a finite subset G of S to be a **universal comprehensive left (right, two-sides) Gröbner basis**, if G is a comprehensive left (right, two-sided) Gröbner basis for all term-orders $<$ on T that are compatible with 1.1 (iii).

We denote this class of term-orders by $Ord(S)$.

Let us now revisit the proof sketch above with the goal of constructing a **universal** comprehensive left (or two-sided) Gröbner basis of $I_L(F)$ (of $I(F)$). We modify the construction of a Gröbner system to that of a **universal Gröbner system:**
We extend conditions to include finitely many inequalities $s < t$ for $s, t \in T$. At each vertex of the tree, we distinguish cases (by constructing finitely many new conditions in the extended sense) in such a way that, besides the vanishing or non-vanishing of coefficients, we also take into account the possible order relations between the terms under consideration for term-orders compatible with the extended condition at the given vertex. Then the remaining arguments are still valid.
So we obtain:

Theorem 2.3 *Let $S = R\{X_1, ..., X_n\}$ be a parametric solvable algebra as above and let F be a finite subset of S. Then one can construct a universal comprehensive left (right, two-sided) Gröbner basis G of $I_L(F)$, $(I_R(F)$, $I(F))$ in S. The construction is algorithmic relative to computations in the ground ring R.*

3 Applications and Conclusions

We have shown in the preceedings sections that the Gröbner basis theory in solvable algebras can be performed in a very uniform, parametric way. This suggests a number of applications both of theoretical and practical interest.

1. Gröbner basis calculations in non-commutative solvable algebras tend to be costly due to the great number of $*$-products of terms that has to be computed. In part, this problem can be overcome by storing all the values of products that occur during a specific computation and updating the list of products as necessary (see [K]). By theorem 1.5, these computations have to be performed only once for every solvable algebra S' of given type T'. The same applies to Gröbner bases computations in S' for polynomial systems of bounded degree by 2.1.

2. The following applications of comprehensive Gröbner bases in commutative polynomial rings in [W2] can be carried over to parametric solvable algebras with only minor changes:

(i) Parametric ideal membership
(see [W2], theorem 6.2).

(ii) Parametric modules of syzygies
(see [W2], theorem 6.4).

(iii) Deformation of residue spaces and residue algebras
(see [W2], theorem 6.8).

Example: Let $n, d \in \mathbf{N}$ be fixed. Consider the problem of transforming a right fraction ab^{-1} into a left fraction $c^{-1}d$ in the enveloping field of an arbitrary n-dimensional Lie algebra $\mathbf{A}$ and arbitrary polynomials a, b of degree $\leq d$ (see [AL]). Then the problem can be solved by the computation of a **single** comprehensive Gröbner basis in a suitable parametric solvable algebra and a **single** computation of a finite generating system for the pointwise modules of syzygies (see [W2], section 6).

3. In commutative polynomial rings, comprehensive Gröbner bases provide also a "fast" method for quantifier elimination in algebraically closed fields (see [W2], theorem 6.1). In a modified and somewhat restricted sense, such a method can also be obtained for existentially closed skewfields, using comprehensive Gröbner bases in parametric solvable algebras. Research on this topic is not yet completed and will be presented in a subsequent report.

References

[AL] J. Apel, W. Lassner, An extension of Buchberger's algorithm and calculations in enveloping fields of Lie algebras, J. Symb. Comp. 6 (1988), 361-370.

[B] B. Buchberger, Gröbner bases: An algorithmic method in polynomial ideal theory, chap. 6 in *Recent Trends in Multidimensional System Theory*, N. K. Bose Ed., Reidel Publ. Comp., 1985.

[J] N. Jacobson, *Lie algebras*, Interscience Publishers, 1962.

[KRW] A. Kandri-Rody, V.Weispfenning, Non-commutative Gröbner bases in algebras of solvable type, 1986, J. Symb. Comput. 9 (1990), pp. 1 - 26.

[K] H. Kredel, Computing in polynomial rings of solvable type, this volume.

[W1] V. Weispfenning, Constructing Universal Gröbner Bases, in Proc. AAECC-5, Springer LNCS col 356, pp. 195 - 201.

[W2] V.Weispfenning, Comprehensive Gröbner bases, preprint March 1990, University of Passau, MIP-9003.

Computer Algebra Applications

AUTOMATIC CALCULATION OF FEYNMAN AMPLITUDE

Toshiaki Kaneko

Meiji-Gakuin University, Totsuka, Yokohama, Japan

Abstract

A program package has been developed which generates Feynman diagrams and source code for the corresponding amplitudes in the Weinberg-Salam model and QCD. We have applied this package to several scattering processes in which the most complex one is $e^+e^- \to W^+W^-\gamma$ and obtained a satisfactory reliability in numerical calculation.

1. Introduction

New accelerators for high energy physics have been developed and have increased available energies for scattering processes. In the analysis of data obtained by experiments, the more theoretical calculations are required for the more complicated processes. The number of Feynman diagrams grows quickly with the complexity of the processes. Sometimes even counting the number of them is not easy task and a checking procedure of it is required. When one looked what kinds of works were needed in the course of calculation, he would recognize that many parts of them can be done mechanically. Therefore it would be more clever to leave them to a computer than to do by hands.

Some parts of the work have been executed on a computer so far; integration of differential cross-sections is performed by a numerical method and Dirac-matrices are processed on symbolic manipulating systems. However, these have not been connected with in an organic way.

This paper reports a program package being developed by the theory group at TRISTAN. It automates the perturbative calculation of cross-sections of various scattering processes in the Weinberg-Salam model and QCD. We have applied it to several practical scattering processes and checked its validity.

We consider the required structure of such program package in the section 2. In the section 3, the current status of our program package is described. Test calculations for practical scattering cross-sections are shown in the section 4. We summarize our results and discuss the future development in the last section.

2. Structure of the program package

The procedure of calculating differential cross-sections is divided into the following four steps:

1. First of all one has to draw all the Feynman diagrams necessary for each reaction.
2. A source program is constructed to calculate the Feynman amplitudes in accordance with these diagrams.
3. In order to integrate differential cross-sections, one has to choose the integration variables, which are usually assumed to vary between 0 and 1, and convert them into kinematical variables.
4. The total cross-section is obtained by the integration over the phase space. An event generator should be also constructed in the manner consistent with the differential cross-sections.

We are considering to achieve all of these steps on a computer. Some parts of them have already been processed by a computer as mentioned.

For the first step of the calculation, we have programs which generate Feynman diagrams automatically within Quantum Electro-Dynamics(QED)[/1,2/]. They produce necessary and sufficient Feynman diagrams for any scattering processes of arbitrary higher order of coupling constants. In order to apply them to the standard model of the particle physics, these programs should be extended to models including Non-Abelian gauge theory. The kinds of particles in the model should not be fixed, as some of them may be ignored or some extra particles may be added in practical calculations. Graphical expression of generated Feynman diagrams is also necessary as an output in order to check the results of calculation by eyes.

In the usual manner of calculation, the second step is realized by the following procedure: First one converts Feynman diagrams into mathematical expressions. These expressions contain Dirac-matrices and Lorentz indices. They are processed by symbolic manipulation systems or by hands. The obtained expressions are coded as a computer program.

Once Feynman diagrams are obtained by a computer program, it is not hard to convert them into mathematical expressions. This process is realized for QED by a program package described in the Ref.2. Its output is a source code for REDUCE[/3/]. However, symbolic manipulation systems do not necessarily fit to our aim. In actual calculation, they often requires huge main memory exceeding capacity of available computers. In such case, one has to divide input code into small pieces or to rewrite it in some ways to save the memory. Though this problem is not an intrinsic characteristic of symbolic manipulating systems, an additional work is needed to modify the program into a suitable form so as to get rid of the memory trouble. A program package is desired which does not need any complicated modification for the program to be executed, once it is written in a simple-minded way. This property is especially important when the program becomes very large for a complicated process. Unfortunately, however, we do not yet know a system which satisfies our requirements. In order to circumvent this complexity, one possible way which we can do at present is to adopt a calculating method of Feynman amplitudes which enables direct generation of source code for numerical calculation.

For the third step of the calculation, we have to standardize interfaces of the modules of kinematics and construct a library of them. These modules give a great influence on the efficiency of the integration and are also dependent on the integration package used.

The fourth step is realized in Ref.4. It is made of two components, BASES and SPRING. BASES is a program package for multi-dimensional integration which utilizes the adoptive Monte-Carlo method. SPRING is an event generator; it generates events by calculating the same FUNCTION as used in BASES. The generation of events is accelerated by looking up the table of the differential cross-section made by BASES. Thus once a source code for the differential cross-section is constructed, the event generator is immediately obtained.

3. Development of the program package

General and efficient algorithm is not known to generate Feynman diagrams containing self-interactions which appear in the Non-Abelian gauge theory or even in the ϕ^3-theory. It would be not so hard to generate a set of diagrams, if one allowed the appearance of the same diagrams. However, it is hard to find and delete duplicated diagrams from the given set of diagrams in the efficient way enough for practical use. The number of comparisons would grow as the factorial of the number of diagrams in the set and the number of elements within a diagram.

By this reason we wanted to construct a special but practical program rather than a general but unpractical one and we have limited ourselves to deal with only tree and one-looped diagrams. Within this restriction we have developed an automatic generator of Feynman diagrams[6] for Weinberg-Salam model and QCD. This program reads the order of coupling constants and the external particles appeared in a physical reaction. The table of particles can be altered if new particles are to be added.

```
* 5150      G  G   => U   UB  G      TREE
CORDER       3
INITIAL GL   2
FINAL   UQ   1
FINAL   UQB  1
FINAL   GL   1
END
```

Figure 1. The input commands for the process $gg \to Q\bar{Q}g$.

We show an example of the input data for the process of $gg \to Q\bar{Q}g$ in Figure 1.

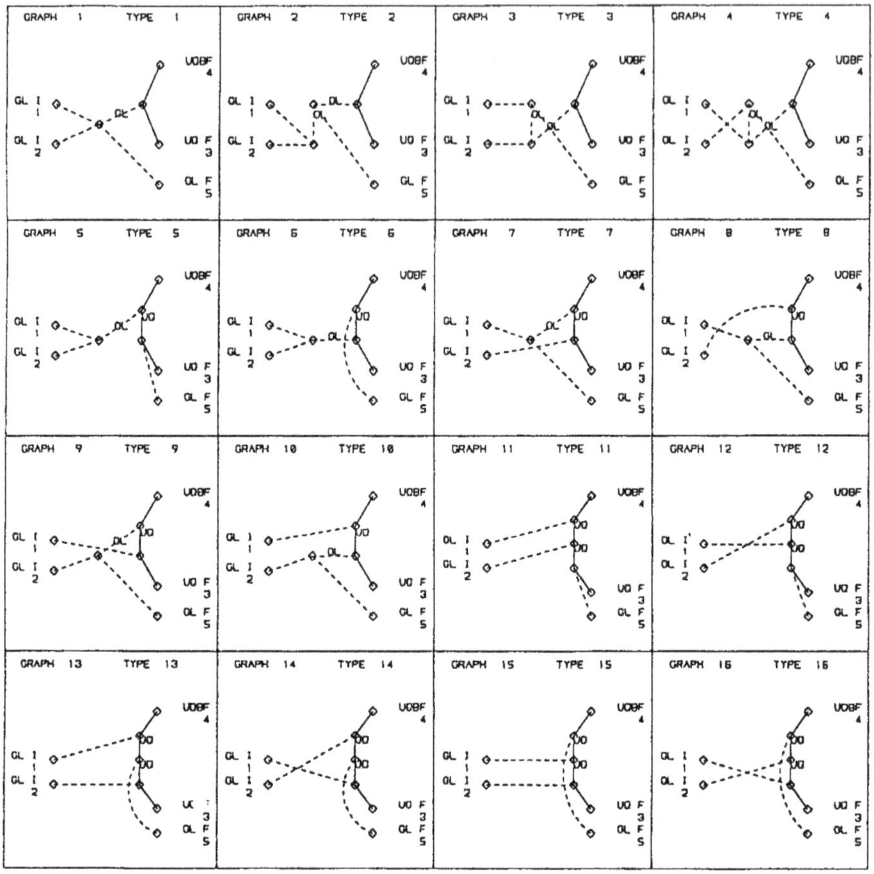

Figure 2. The input commands for the process $gg \to Q\bar{Q}g$.

In the following, we take this process as an example. The program produces a data file, which contains information about the set of generated Feynman diagrams. An another program draws these diagrams on a graphic device as shown in Figure 2.

We have adopted a program package CHANEL described in the Ref. 7 and 8 in calculating Feynman amplitudes. This method of calculation has advantages over the usual one with symbolic manipulating systems in two features. One of them is that we can calculate the helicity amplitudes directly. All particles can be either massive or massless. The total amount of calculation and the total length of the source code is proportional to N for N Feynman diagrams. On the other hand it is proportional to N^2 when amplitudes are squared before numerical cal-

```
************************************************************************
      SUBROUTINE AMPTBL
** 5150      G  G   => U   UB  G       TREE
      IMPLICIT REAL*8(A-H,O-Z)
      INCLUDE  (INCL1)
      INCLUDE  (INCL2)
*-----------------------------------------------------------------------
      NG = 0
*   EXTERNAL LINES
      CALL SMVECT(LEPEXG,AMGL,PE0001,EP0001,EW0001,IGAUGL)
      CALL SMVECT(LEPEXG,AMGL,PE0002,EP0002,EW0002,IGAUGL)
      CALL SMEXTF(LOUTGO,AMUQ,PE0003,PS0003,CE0003)
      EW0003(1) = LPRTCL
      CALL SMEXTF(LINCOM,AMUQ,PE0004,PS0004,CE0004)
      EW0004(1) = LANTIP
      CALL SMVECT(LEPEXG,AMGL,PE0005,EP0005,EW0005,IGAUGL)
* GRAPH NO.    1 -    1 (    1)
      IF (JSELG(    1).NE.0) THEN
      NG = NG + 1
*   INTERNAL MOMENTA
      DO     1 I = 1, 4
      PE0011(I) =  +PE0004(I) +PE0003(I)
    1 CONTINUE
      APROP(NG) = 1.0
      CALL SMPROP(PE0011,AMGL,AGGL,IGAU00,APROP(NG))
*   INTERNAL LINES
      CALL SMVECT(LEPING,AMGL,PE0011,EP0011,EW0011,IGAUGL)
*   VERTICES
      CALL SMFFV(LEXTRN,LEXTRN,LEPING,EW0004,EW0003,AMUQ,AMUQ,
     &           CQQG  ,CE0004,CE0003,PS0004,PS0003,EP0011
     &           ,LT0001,AV0001)
      CALL SMGGGG(LEPING,LEPEXG,LEPEXG,LEPEXG,CQCDSQ
     &            ,EP0011,EP0001,EP0002,EP0005,LT0002,AV0002)
*   CONNECT VERTICES.
      CALL SMCONV(LT0002,LT0001,   1,    3,EW0011,AV0002,AV0001,
     &            LG(0, NG),AG(0, NG))
      APROP(NG) = -APROP(NG)
      INDEXG(   1, NG) =    1
      INDEXG(   2, NG) =    2
      INDEXG(   3, NG) =    5
      INDEXG(   4, NG) =    4
      INDEXG(   5, NG) =    3
      ENDIF
* GRAPH NO.    1 -    2 (    2)
      IF (JSELG(    2).NE.0) THEN
      NG = NG + 1
*   INTERNAL MOMENTA
      DO     2 I = 1, 4
      PE0017(I) =  +PE0004(I) +PE0003(I)
    2 CONTINUE
```

Figure 3 A part of the generated FORTRAN source code.

culation. Their difference is critical when the physical process is complicated and the number of diagrams becomes large.

The other advantage of CHANEL is its simplicity in the construction of source code. At first one calls subroutines corresponding to the vertices, internal lines and external lines of the Feynman diagram. They calculate wave function of the external line, divide internal line into a set of components and calculate matrix element of the vertices. Matrix elements of neighboring vertices are summed after being multiplied together with components of internal lines. This is corresponding to connecting two vertices. When all vertices are connected with each other by internal lines, the amplitude of the diagram is obtained. Thus what we should do is to arrange appropriate subroutine calls in accordance with the structure of a given diagram. There is no necessity of modification of the algebraic expressions by hand or by symbolic manipulating systems.

We have developed a program which generates source code by this method for the lowest order diagrams in the Weinberg-Salam model and QCD. For QCD, we have to calculate color factors in addition to the momentum part of the amplitudes. A Feynman diagram which includes k four-point vertices of gluons is divided into 3^k diagrams in corresponding to the three terms of the four-point vertex. It makes the color factor to global one of each diagram. Color factors are calculated automatically and are multiplied when numerical value of the amplitudes are squared. A part of an example of the generated source code is shown in Figure 3.

4. Testing the program package

As shown in the previous section, we have developed the required program package except for the library of kinematics. At present the latter is not generated automatically but is written by hand. We can now test them for a practical physical process with subroutines of kinematics.

First we have checked the self-consistency of the generated source code. We can change gauge-parameters of the covariant gauge and can select the axial gauge in addition for QCD. The generated source code are kept unchanged with the selection of the gauge, which is controlled by parameters for CHANEL. As we have to take a specific Lorentz frame in the momentum space to calculate amplitudes numerically, we have checked the Lorentz invariance of the cross section by changing the frame in various ways.

Next we have checked the generated source code by calculating the matrix element squared for the following four processes (only tree diagrams are considered):

1. $e^+e^- \to W^+W^-$.

   The programs to be compared with the generated code are as follows;

   a) a program based on the analytic expression obtained by REDUCE for unpolarized process.

b) a program based on CHANEL but written by hand in order to achieve maximal optimization (for any combination of particle helicity states).

2. $\gamma\gamma \to W^+W^-$.

We repeated a similar check for this process.

a) the same kind of program as the previous item (a)/9/.

b) the same kind of program as the previous item (*b*)

3. $e^+e^- \to \mu^+\mu^-\gamma$.

In this case we prepared three programs for the comparison:

a) a program produced by REDUCE for the unpolarized cross-section.

b) a program written in the similar way as the previous item (*b*).

c) a program produced by REDUCE which calculates *amplitudes* directly.

4. $e^+e^- \to W^+W^-\gamma$./8/

The calculation of the generated code is compared with the following programs:

a) a program obtained by REDUCE for unpolarized case.

b) a program written in the similar way as the previous item (*b*).

5. $gg \to Q\bar{Q}g$.

We have prepared two programs in different gauges obtained by REDUCE for unpolarized process:

a) a program in the 'tHooft-Feynman gauge.

b) a program in the axial gauge.

The first two processes are rather simple but the rest are complicated and check of them is not trivial. The total cross-section as well as various kinds of distributions obtained by phase space integration are compared. For all processes, we got good agreement among them in the significant six digits in the integrated total cross-section with the calculation in double precision (64bits real data).

In making the point by point check of the fourth process in the phase space, all three programs agreed with each other in the accuracy of 28 digits in quadruple precision (the same check for the third process gave the agreement of 14-15 digits in double precision). Since analytic calculation of the soft photon emission cross-section is trivial, we can make another test for the program. We have used the generated program for large photon energy ($k > k_c$) and the analytic form for small photon energy ($k < k_c$). We have checked the consistency of the result by varying k_c from 0.001GeV to 0.1GeV and observed that the sum of soft and hard parts of the total cross-section is not changed when the photon mass is kept fixed.

The generated code for the fourth process takes 30msec to calculate Feynman amplitudes for one point in the four dimensional phase space. While the programs 4-a and 4-b take 30msec and 10msec, respectively.

5. Summary and discussions

We have developed a program package which generates automatically FORTRAN source code of the Feynman amplitude of scattering processes. The input for the program package is the order of coupling constants and the external particles of the scattering process. We have checked the reliability of numerical results of our program package for several of actual processes and confirmed the validity of it.

More tests are left to a variety of physical processes, especially taking care in the accuracy of the numerical result.

The generated source code consumes rather longer CPU-time than those of hand writing amplitude or produced by REDUCE. The algorithm of code generation is very simple one in the current stage. It is not so hard task to improve it and get a more optimized source code (we are expecting that it becomes about two times faster than the current version). It is obvious that more drastic improvement of the efficiency will be achieved by using vector processors[10,11].

Constructing a general library of kinematics, which is common and applicable for any mass combination of external particles, contains the problem of mass singularity. Some cares must be taken in performing the numerical integration, when its integrand has steep peaks which give rise to large contribution to the integration. These peaks appear not only as points in the phase space (*e.g.* soft photon singularity), but also as higher dimensional objects (collinear singularity). Many of existing integration packages need to locate these peaks at the edge of the integration volume. Even much more intelligent packages have to assume that singularities are placed parallel to the co-ordinate axes in the integration volume. These restrictions should become hard obstacles when there take place one or more dimensional peaks which are crossing each other in the phase space. In order to construct the general library of kinematics, it is unavoidable to develop an integration package which is able to manage any location of steep peaks. One possible algorithm for such integration has been recently developed in ref. 12.

In order to check a numerical calculation of cross-section, it is desirable to have two or more programs which are constructed by independent methods. Therefore symbolic manipulating systems do never loose its importance in our field, in spite of their limitation in the current status. However the spirit of their design seems so far assuming the interactive usage rather than the treatment of very long mathematical expressions. We hope a new system will be developed which bears very long inputs and creates optimized (much shorter) source code.

We are dealing with only tree diagrams in this paper. The next development would be to include looped diagrams. To calculate them, one has to face to loop integration and renormalization. One possible way is to process them algebraically in a symbolic manipulating system. Another possibility is again to rely on numerical method[13].

Acknowledgement

This work was done by computer physics subgroup of theoretical group for TRISTAN centered at KEK. The author expresses his thanks to the members of this group for useful discussions and continuous encouragement. Especially he is indebted to S. Kawabata, Y. Shimizu and H. Tanaka for valuable comments.

References

1 T. Sasaki, *J.Comput.Phys.* **22** (1976) 189.

2 T. Kaneko, S. Kawabata and Y. Shimizu, *Comput.Phys.Commun.* **43** (1987) 279.

3 A.C. Hern, REDUCE User's Manual, version 3.2 (The Rand Corporation, 1985).

4 S. Kawabata, *Comput.Phys.Commun.* **41** (1986) 127.

6 T. Kaneko, in preparation.

7 H. Tanaka, *Comput.Phys.Commun* **56** (1990) 337.

8 H. Tanaka, T. Kaneko and Y. Shimizu, KEK preprint 90-29H/D, May 1990.

9 R. Najima, Yokohama College of Commerce Preprint YCCP-9001, 1990; I.F. Ginzburg *et al.*, Nucl.Phys.B228(1983),285.

10 S. Kawabata and T. Kaneko, *Comput.Phys.Commun.* **48** (1988) 353.

11 T. Kaneko and S. Kawabata, *Comput.Phys.Commun.* **55** 1989 141.

12 S. Kawabata, in preparation.

13 Y. Oyanagi, T. Sasaki, S. Kawabata, Y. Shimizu and T. Kaneko, in *"Perspectives of Particle Physics"*, edited by S. Matsuda, T. Muta and R. Sasaki *et al.*, (World Scientific, 1988) 369; K. Kato, private communication.

THE USE OF COMPUTER ALGEBRA FOR CALCULATING THE QUANTUM CHEMISTRY INTEGRALS

L.V. Artamonov, E.V. Mozdor, A.B. Roitsin, Yu.S. Fishman

Section No.8 of the Institute of Semiconductors,
the Ukr.SSR Academy of Sciences,
Prospect Nauki 45, Kiev-28, 252650, USSR

A brief summary on the method of exact analytical calculation of the multicentered integrals and its realization by means of computer algebra is given.

1. One of the basic problems arising in the process of determining the energy structure and wave functions of perfect crystal, imperfect crystals, molecules and various complex combinations is calculation of the so-called quantum chemistry integrals (QCI). The latter represent definite matrix elements (ME) of operators of energy interaction between electrons, and between electrons and atomic nuclei (cores of nuclei). They arise during the calculation of energy spectrum of multiatomic multielectron systems as the compound elements of more complex ME of the form:

$$M_{ij} = \int \Psi_i^* \hat{H} \Psi_j \, dv$$

where Ψ_k and $\hat{H}$ are multielectron functions and Hamiltonian, V is the region of integration variables changing. If Ψ_k would be represented as a linear combination of product of one-particle (atomic) functions φ, and $\hat{H}$ - as a sum of one- and two-particle interactions $\hat{h}$, then M_{ij} would be represented as linear combination of one- and two-electron "elementary" integrals which are the QCI. The general equation for one-electron integral has the form:

$$[ab/c] = \int \varphi_{a(\nu)}^{(1)*} \hat{h}_c(\nu) \varphi_{b(\nu)}^{(2)} dV_\nu$$

where ν are coordinates of the ν-th electron; a, b, c are coordinates of nuclei at which functions φ and operator $\hat{h}$ are centered; the superscript over φ indicates the possible different form of the functions. The example of operator $\hat{h}_c(\nu)$ is the Coulomb interaction between electron ν and nucleus c. In the general case, one-electron integrals are triple ones, and in the case of different location of centers ($a \neq b \neq c \neq a$), they can also be three-centered ones. In particular cases, the multiplicity and centering can be smaller. The general equation for two-electron integral has the form:

$$[ab/cd] = \int \varphi_{a(\nu)}^{(1)} \varphi_{b(\nu)}^{(2)} \hat{h}(\nu,\mu) \varphi_{c(\mu)}^{(3)} \varphi_{d(\mu)}^{(4)} dV_\nu \, dV_\mu \quad \text{I}$$

The example of operator $\hat{h}(\nu,\mu)$ can be Coulomb interaction between electrons μ and ν. In the common case, two-electron integrals are sextuple ones, and in the case of different location of centers, they can also be four-centered ones. Thus, the maximum multiplicity and centering depend on the kind of electromagnetic interaction within multiatomic system and on the method of representation of multielectron wave function through the one-electron ones. Three- and four-centered integrals are also assumed to be called multicentered integrals (MCI).

2. The calculations of one- and two-centered QCI present no difficulties. Many of them can be represented by the so-called closed form, i.e. as the finite linear combination of elementary and special functions. For a long time, there was no success in integrating MCI in the same way. This fact was associated both with the lack of general approach to the calculation of integrals of this kind and with awkwardness of analytical calculations, the computer-unsupported accomplishment of which appeared to be impossible in practice. For this reason, they are

calculated: by means of approximation formulae; by means of formulae representing expansion into the infinite series; on the basis of semiempirical equations; by the method of integral transformations and with the help of Gaussian functions; and also by direct numerical integration of input equation[1,2]. Alongside with it, the application of the closed equations simplifies the subsequent numerical calculations, multiplies their accuracy, makes it possible to obtain the desired values analytical dependences on the different parameters, and, in particular, it makes it possible to avoid difficulties associated with substraction of closely spaced large numbers.

Recently[1,3,4], it was demonstrated that different MCI can be calculated in the same way, and difficulties associated with the awkwardness of calculations can be overcome with the help of analytical computer methods. This made it possible to develop the method of calculating the most complex QCI - i.e. four-centered one (at Slater orbitals) - and to calculate accurately three types of similar integrals[2,5,6] for the first time.Since the used equations are unwieldy, we shall present the principal scheme of calculation only, then we shall remark upon some peculiarities inherent to each MCI type, and shall discuss the result.

3. The task of accurate analytical calculation of MCI comes to the representation of input sextuple integral (or triple one, in the case of one-electron integral) as the finite linear combination of elementary and special functions, i.e. to the equation in which the further integration is principally no longer possible. The calculation of each MCI has its own peculiarities, but the common concept of calculation is one and the same in all cases - it means consecutive reduction of complex equations into simpler ones. The corresponding calculation

stages, the illustrations of which can be found in [1-6], are: a. Preliminary integration of equation Eq.1 for coordinates of one electron (the application of equation for one-electron three-centered integral); b. Introduction of substitution resulting in linear dependence on integration variables in index of exponential function degree; c. Representation of subintegral equation in the form of the sum of summands containing functions of various complexity degrees (elliptic integrals; exponential, logarithmic and exponential integral functions; polynomials), and reduction, in parts, the summands having complex subintegral equations to the summands having simpler ones in the process of integration; d. Obtaining and application of recurrent relationships with the aim of reduction of the used integrals to their minimum quantity; e. Utilization of computers orientated at automatization of alphabetic computations in accomplishing time-consuming analytical transformations.

In the process of work, the procedures were singled out for the problems class under consideration, and the system of programs was composed in "Analytic-79" and "Analytic-75" languages. The realization of this system with the help of "CM-1410" and "Mir-2" computers made it possible to automize the accomplishment of clumsy calculations: a. Removing the brackets, reduction of simplitude; b. Formation of random algebraic equation in accordance with wanted variable; c. Representation of fractional rational function as the sum of partial fractions; d. Obtaining the reduced algebraic equation in the case of the defined value of one of its arguments; e. Multiplication polynomials by the sum of polynomial and fractional rational function, division of polynomials; f. Integration of product of polynomial by exponential; g. Integration of exponential divided by binomial; h. Integration by the method of consequent application of recurrent relationships. The technical difficulties, that have appeared, are associated mainly with a great number of different and frequently repeated procedures; with high powers of polynomials (binomials) and polynomial dependence of their coefficients on the parameters of the problem; with quite awkward (up to eleven significant digits) numerical coefficients, fractions as a rule. Partially because of the mentioned difficulties, and also because of the novelty of the task, the computer calculations were carried out in stages in accordance with the definite procedure. However, they showed that the creation of the single program of analytical computer calculations of MCI is possible.

The calculations of the elementary four-centered integrals[2,5,6] have shown that input sextuple integral reduces to the different-kind linear combination of products of elementary and special functions (polynomials, exponentials, logarithms, exponential integral functions, elliptical integrals of the 1st, 2nd and 3rd type). The arguments of these equations are parameters of wave functions and distances between centers. In the case of three-centered integrals, the equations become much simpler, in particular the elliptical integrals disappear. The results of test calculation of input and final equations were the same for some parameter values of the problem. These calculations have also shown that integrals with smaller multiplicity of integration are more suitable for programming, and demend less machine time (despite the fact that their number grows and corresponding equations become more awkward).

References:

1. L.V. Artamonov, A.M. Grekhov, A.B. Roitsin, Ukr.Fiz.Zh. 23(1978)215
2. L.V. Artamonov, A.B. Roitsin, Ukr.Fiz.Zh. 28(1983)1738
3. L.V. Artamonov, A.M. Grekhov, A.B. Roitsin, Teor.Eksp. Khim. 14(1977)396
4. L.V. Artamonov, Ukr.Fiz.Zh. 27(1982)1741
5. L.V. Artamonov, E.V. Meinarovich, E.V. Mozdor, A.B. Roitsin, Teor.Eksp.Khim. 22(1986)491
6. L.V. Artamonov, V.D. Gorovoi, E.V. Meinarovich, E.V. Mozdor, A.B. Roitsin, Yu.S. Fishman, Zh.Struk.Khim.30(1989)3

TWO-LOOP CALCULATIONS IN THE LARGE MASS EXPANSION IN THE FRAMEWORK OF THE QCD SUM RULES METHOD

P.Baikov, V.Ilyin, V.Smirnov, A.Taranov
INP, Moscow St. Univ., 119899 Moscow, USSR

and

K.Chetyrkin
INR Acad. Sci. USSR

ABSTRACT

Non-leading two-loop contributions to the coefficient function of the operator G^2 in the heavy quark mass expansion of the correlators of two currents (vector, scalar, and pseudo scalar) are symbolically calculated. Our corrections in the vector case imply that value of gluon condensate must be increased by 2÷3 times.

Consider the product of two vector (V) currents

$$T_{\mu\nu}(q,m) \equiv i\cdot\int dx\cdot e^{iqx}\cdot T(J_\mu(x)\cdot J_\nu(0)) \ , \quad J_\mu=\bar{\psi}\gamma_\mu\psi \qquad (1)$$

For other currents we have

$J_\mu \longrightarrow \bar{\psi}\psi$ (S- scalar),

$J_\mu \longrightarrow i\bar{\psi}\gamma^5\psi$ (PS- pseudo scalar),

$J_\mu \longrightarrow \bar{\psi}\gamma_\mu\gamma^5\psi$ (PV- pseudo vector).

Calculations in PV case are now in progress.

To be specific let us imply the vector case. In (1) m is the heavy quark mass. The expression (1) may be represented with the help of the operator expansion in inverse powers of the large mass $M=2m$:

$$T_{\mu\nu}(q,m) \underset{m\to\infty}{\longrightarrow} (q_\mu\cdot q_\nu - g_{\mu\nu}\cdot q^2)\cdot\Big(C_0(t,\frac{M}{\mu},\alpha_s) + \sum_i C_i(t,\frac{M}{\mu},\alpha_s)\cdot\frac{O_i}{M^4}\Big) + o(1/M^4) \ , \qquad t\equiv -q^2/M^2.$$

Here μ is a normalization point, α_s is the strong coupling constant,

$$O_1=G^a_{\mu\nu}\cdot G^{a\mu\nu}, \quad O_2=A^a_\nu\cdot\nabla^{ab}_\mu G^b_{\mu\nu}-\partial_\mu\bar{c}^a\cdot\partial^\mu c^a, \quad O_3=(\nabla^{ab}_\mu\partial_\mu\bar{c}^b)\cdot c^a,$$

and $G^a_{\mu\nu}$ is the gluon curvature tensor, A^a_ν is the gluon field, $\bar{c}^a$ and c^a are the ghost fields.

To pick out the C_1 we have used the formula

$$C_1 = \frac{1}{(1-D)\cdot q^2}\cdot\Pi_1(T)\cdot Z^{-1}_{G^2}\ , \qquad (2)$$

where

$$\Pi_1(T) \equiv \frac{1}{4D(1-D)}\cdot\frac{\partial}{\partial p^\nu}\,\frac{\partial}{\partial p'_\nu}\langle\tilde{A}^a_\alpha(p)\cdot T^\mu{}_\mu\cdot\tilde{A}^{a\alpha}(p')\rangle_0^{amp}\Bigg|_{p=p'=0},$$

and D=4-2ε is the space-time dimension, <...> means the vacuumexpectation value, "amp" implies that external gluon lines are amputated; Z is the renormalization constant of G^2 (this UV-divergent factor cancels infrared divergences that appear after the action of the projector Π_1). For the Wilson expansion the formula of type (2) was proposed in ref.[1] (see also ref.[2]).

We have calculated the first seven coefficients of the expansion of C_1 in $t\equiv -q^2/M^2$ at the 2-loop level:

$$C_1(t,2,\alpha_s) = \frac{\alpha_s}{\pi}\sum_{n=0}^{\infty} a_n\cdot\left(1+\frac{\alpha_s}{\pi}\cdot b_n\right)\cdot t^n.$$

For simplicity we take μ=m.

The 1-loop coefficients are equal to (see ref.[3]):

$$a_n = \frac{2^{-n}\cdot(n+3)!\cdot(n+1)}{48\cdot(2n+5)!!} \qquad \text{for V-currents,}$$

$$= -\frac{3\cdot 2^{-n}\cdot(n+2)!\cdot(3n+7)}{32\cdot(2n+5)!!} \qquad \text{for S-currents,}$$

$$= -\frac{2^{-n}\cdot(n+2)!\cdot(n-3)}{96\cdot(2n+3)!!} \qquad \text{for PS-currents.}$$

Our results are symbolic calculations for coefficients b_n $(n=0,1,\ldots,6)$ in the $\overline{\text{MS}}$-scheme. It was necessary to calculate 26 Feynman diagrams of five types (see ref.[4]). The symbolic (exact) $\overline{\text{MS}}$ results for V-currents were calculated and published in ref.[4]. However we had made in this publications a mistake in handling the "on-shell mass" values.

There are the following steps in our calculations.

A. Typing Feynman amplitudes - handily, REDUCE.
B. Trace calculation - standard REDUCE HEP-package.
C. Π_1 projection - original REDUCE program LDFC in algebraic mode (≃100 lines).
D. Large mass expansion for each diagram - original program LMEXP in symbolic mode (≃1000 lines on RLISP).
E. Collecting results for coefficients b_n - original REDUCE program TOTRES in algebraic mode.

Results are storing automatically after each step.

We have used REDUCE 3.3 system[5] and a computer with 5 Mflops productivity (IBM compatible 370 clone).

Run time: V-currents≃30h, S- and PS-currents≃10h.

Run time distribution: B≃5%, C≃5%, D≃90%.

The package RATAR, written by A.Ya.Rodionov, was employed for rational arithmetics.

All programs were written and tested in 1988. They haven't been changed up to present time. All calculations were carried out later. The following criteria were applied to verify the obtained results:

1) the same programs (LDFC, LMEXP) were used for 1-loop calculations (a_n ,n = 0,1,..6) and were compared with well-known analytical formulae mentioned above;

2) since the diagrams are 2-loop they could involve $1/\varepsilon^2$ terms but due to the action of Π_1 they should be canceled and this was verified for any diagram;

3) all $1/\varepsilon$ terms for QED like diagrams are removed by corresponding 1-loop counter terms which have been calculated handily and by our programs LDFC and LMEXP;

4) the $1/\varepsilon$ terms in the total 2-loop result are removed by the corresponding 1-loop mass and current's renormalizations (the latter for S- and PS-currents);

5) there were carried out in V-case the calculations (by using LMEXP) for "incorrect" projector

$$\Pi_1'(T) \equiv \frac{1}{4D(1-D)}\cdot\frac{\partial}{\partial p_\mu}\,\frac{\partial}{\partial p'_\nu}\langle\tilde{A}^a_\alpha(p)\cdot T_{\mu\nu}\cdot\tilde{A}^{a\alpha}(p')\rangle_0^{amp}\Big|_{p=p'=0}.$$

The zero results must be for each a_n and b_n in this case, and this criterion has been indeed satisfied;

6) in the V-case, there were carried out the calculations for $(q_\mu\cdot q_\nu\cdot T^{\mu\nu})$ and the projector Π_1 and there were obtained zero results for a_n and b_n, which is connected with transversality of the correlator;

7) in the V-case, there were carried out the calculations for 2-loop corrections to the coefficient C_0 of the unit operator and the well known analytical formula (see ref.[3]) was checked for n=0,1,..,6.

We show our results in tables. The $\overline{\rm MS}$ results are shown in the first column. These are exact results obtained by symbolic computations. In the next two columns there are the values of b_n for two other frequently used normalization of the heavy quark mass: "the on-shell mass":

$$m(p^2=m^2) = m_{\overline{MS}}\cdot\left\{1 + \frac{\alpha_s}{\pi}\cdot\left[\frac{4}{3} + \ln\left(\mu^2/m^2_{\overline{MS}}\right)\right]\right\} ;$$

and "the off-shell mass":

$$m(p^2=-m^2) = m_{\overline{MS}}\cdot\left\{1 + \frac{\alpha_s}{\pi}\cdot\left[\frac{4}{3} - 2\cdot\ln 2 + \ln\left(\mu^2/m^2_{\overline{MS}}\right)\right]\right\} .$$

For the V-case the last column contains the values corresponding to the asymptotical (in n) formula[6]:

$$b_n = \pi^{3/2}\cdot\frac{197}{192}\cdot\sqrt{n} .$$

Results:

Vector case

n	$\overline{MS}$		$m(p^2=m^2)$	$m(p^2=-m^2)$	asymp.
0	$\frac{281}{162}$	≈ 1.73	7.07	1.52	—
1	$\frac{6179}{12960}$	≈ 0.48	8.48	0.160	5.7
2	$-\frac{53}{56}$	≈ -0.95	9.72	- 1.37	8.1
3	$-\frac{105629}{42525}$	≈ -2.48	10.85	- 3.01	9.9
4	$-\frac{195616567}{47628000}$	≈ -4.11	11.89	- 4.74	11.4
5	$-\frac{702046171}{121080960}$	≈ -5.80	12.87	- 6.54	12.8
6	$-\frac{86325933143}{11442150720}$	≈ -7.54	13.80	- 8.39	14.0

It should be noted that our values of b_n approach rather quickly the corresponding asymptotical ones.

Scalar case

n	$\overline{MS}$		$m(p^2=m^2)$	$m(p^2=-m^2)$
0	$\frac{95}{504}$	≈ 0.188	5.52	- 0.023
1	$-\frac{487}{450}$	≈ -1.08	6.92	- 0.140
2	$-\frac{57809}{21840}$	≈ - 2.65	8.02	- 3.07
3	$-\frac{211969}{48384}$	≈ - 4.38	8.95	- 4.91
4	$-\frac{2989}{480}$	≈ - 6.23	9.77	- 6.86
5	$-\frac{13851863}{1698840}$	≈ - 8.15	10.5	- 8.90
6	$-\frac{383732201}{37837800}$	≈ -10.14	11.2	- 10.0

Pseudoscalar case

n	$\overline{MS}$		$m(p^2=m^2)$	$m(p^2=-m^2)$
0	$\frac{19}{24}$	$\approx$ 0.792	6.125	0.580
1	$-\frac{23}{45}$	$\approx$ - 0.511	7.49	- 0.829
2	$-\frac{54127}{15120}$	$\approx$ - 3.58	7.09	- 4.00
3	—		—	—
4	$-\frac{3175927}{1663200}$	$\approx$ 1.91	17.9	1.27
5	$-\frac{188412479}{113513400}$	$\approx$ - 1.66	17.0	- 2.40
6	$-\frac{17101943}{4324320}$	$\approx$ - 3.95	17.4	- 4.80

Let us briefly discuss our results. If, in accordance with ref.[7] the value of gluon condensate is determined from the moments at n=5÷6 with the "off-shell mass normalization" then our corrections in the vector case imply that this value is increased by 2÷3 times.

References

1. S.Gorishny et al., *Phys.Lett.* **124B** (1983) 217; S.Gorishny, S.Larin, *Nucl.Phys.* **B283** (1987) 452.
2. K.Chetyrkin et al., *Phys.Lett.* **110B** (1982) 407; F.Tkachov, *Phys.Lett.* **124B** (1983) 212; K.Chetyrkin, *Teor.M.Fiz.* **75**(1988)26; **76**(1988)207; V.Smirnov, *Preprint* **89-48/125** (NPI MSU, Moscow, 1989); *Comm.Math.Phys.* (1990) (to be published).
3. L.Reinders et al., *Phys. Lett.* **94B** (1980) 203.
4. K.Chetyrkin et al., *Yad. Fiz.* **49** (1989) 1515; K.Chetyrkin et al., *Phys. Lett.* **225B** (1989) 411.
5. A.Hearn, *The REDUCE user's manual, version 3.3*, (The Rand Publication CP78, Rev. 7/87, USA).
6. M.Voloshin, *Yad.Fiz.* **36** (1982) 247.
7. M.Shifman et al., *Nucl.Phys.* **B147** (1979) 385.

SOFTWARE EXTENSIONS OF REDUCE FOR OPERATOR CALCULUS IN QUANTUM THEORY

M. Warns
Physikalisches Institut der Universität Bonn
Nussallee 12, D-5300 BONN 1 (F.R.G.)

1 Introduction

The complexity of the problems occurring in modern theoretical physics has led to an increasing interest of the research community in computer tools for performing standard but tedious and time-consuming algebraic manipulations of large formulae. To cover these requirements a wide range of computer algebra systems (CAS) are nowadays available which can roughly be classified in two categories:

- "Special-purpose" CAS which have been designed to solve a specific class of problems (e.g. SCHOONSHIP to simplify high energy physics expressions). They are usually very efficient because the code is optimized for the task to handle but it is very difficult to extend their abilities to applications which are beneath their original scope.

- On the other hand there exists a variety of "general-purpose" CAS (e.g. REDUCE, MACSYMA, ...) which provide the user with a new (more or less complete) programming language to formulate his tasks. It is nowadays common standard for all these systems to perform complex arithmetic operations (including factorization of polynomials), integration and differentiation of algebraic expressions and to solve equation systems. However these systems due to their general scope are obviously not specially adapted to a particular class of problems. It is therefore left to the user to reformulate his particular task in order to use the features of the CAS in the most efficient way.

Among the last category, the CAS REDUCE developed by A.C. Hearn [1] is one of the most popular systems used in physical research groups all over the world. The possibility given to the user to program his own functions and procedures in the REDUCE language allows to adapt, at least partially, the environment to the particular problem one wishes to solve. However, when dealing with large problems involving specific methods beyond the framework of standard algebra like e.g. differential geometry calculus or noncommutative algebra, this way of proceeding becomes very complicate and unefficient because a lot of new features have to be added to the standard environment. A way to get out of this trouble in REDUCE is to take advantage of its software architecture. The programming language of REDUCE is implemented in the computer in a Meta-LISP language called RLISP. This provides the possibility to extend the standard instruction set of REDUCE very efficiently by adding a package of new RLISP functions defining e.g. new data types or new algebraic operations. In this way, one can in a certain sense combine the advantages of a "special-purpose" CAS designed for specific tasks with the power of a "general-purpose" CAS to perform the standard algebra calculations. Many of such "user contributed extension packages" (UCEP) to REDUCE have been developed in the past few years to handle various tasks e.g. differential geometry calculations (EXCALC [2]). In the following I will present such an UCEP called PHYSOP which has been developed to perform operator calculus in quantum theory.

The paper is organized as follows: In section 2 the general requirements for such a program from the physical point of view are discussed. Section 3 is a brief review of the methods of linking such an UCEP into the standard REDUCE system. Section 4 describes the various implemented features and shows some application example.

2 Using REDUCE to perform operator calculus in quantum theory

One problem occurring frequently in quantum theory is the need to perform algebraic manipulations on large expressions containing operators. The main difference to standard algebra calculations lies in the fact that most of these operators are noncommutative and obey definite commutation relations which have to be considered in the reordering and multiplication procedures. To be more precise we have to deal with a set of operators, e.g.:

$$M, \vec{X}, \vec{P}, \Gamma_\mu, T^{\mu\nu}, \ldots$$

which are contained in an expression, e.g. a Hamiltonian:

$$H := M + \frac{(\vec{P} - e\vec{A}(\vec{x},t))^2}{2M} - \frac{(\vec{P} - e\vec{A}(\vec{x},t))^4}{8M^3} \tag{2.1}$$

The aim is now first to manipulate and simplify this expression so that the operators appear finally in some canonical order (e.g. move all differentiation operators to the right) and then to apply it on a state $|\Psi\rangle$. The final step is then to calculate matrix elements:

$$\langle\Psi'|H|\Psi\rangle$$

which then can be further used e.g. for numerical calculations.

From this one can derive the general requirements for a computer algebra program designed for this class of problems. The user input to the system should consist in defining the elementary commutation relations between the operators, e.g.:

$$[X_i, P_k] = i\delta_{ik}$$

and their action on a state vector, e.g.:

$$P_k|\Psi(\vec{y})\rangle = -i\frac{\partial}{\partial y^k}|\Psi(\vec{y})\rangle.$$

The program should be able to simplify any complicated expression down to the level where these elementary relations could be inserted and then to rebuild the resulting expression in the correct way. From the algorithmic point of view the problem is rather simple: One has to redefine the multiplication operation in such a way that the program checks the ordering of operators introducing the related commutators if necessary. Furthermore simplification routines for complex commutators and for the application of operator expressions on states are needed.

In principle these tasks could be realized at least for simple expressions using the standard instruction set of REDUCE. However, the following trivial example shows already some basic problems inherent to REDUCE which make the handling of more complicated expressions very difficult.

Consider the following commutator occurring in the Hamiltonian (2.1):

$$C := \left[\frac{\vec{P}^2}{2M}, \frac{e}{4M^2}\vec{P}\cdot\vec{A}\right] \tag{2.2}$$

$\vec{P}$ represents the momentum operator and M the mass operator acting e.g. on some one–particle state. The electromagnetic field $\vec{A}$ appearing in the commutator depend on the spatial coordinate $\vec{x}$, so that the operators $\vec{P}$ (which is the canonical conjugate of the space operator $\vec{X}$) and $\vec{A}$ do not commute. Setting:

$$[P_i, A_j] = k_i A_j \tag{2.3}$$

where $\vec{k}$ is the momentum of the photon field, a short calculation leads to the result:

$$C := \frac{e}{8M^3}\left(\vec{k}^2\vec{k}\cdot\vec{A} + \vec{k}^2\vec{A}\cdot\vec{P} + 2(\vec{k}\cdot\vec{A})(\vec{k}\cdot\vec{P}) + 2A_j(\vec{k}\cdot\vec{P})P_j\right) \tag{2.4}$$

Fig. 1 shows a sample REDUCE program which gives the expected result following the steps outlined above. A careful look at this listing displays the main problems:

1. The most important problem is connected with the way REDUCE handles noncommuting objects. The NONCOM statement introduced in the program displayed in Fig. 1 should from the user's point of view be interpreted by the system in such a way that the two operators $\vec{A}$ and $\vec{P}$ do not commute which each other. Unless stated explicitly, it is assumed that both operators commute with themselves and with all other objects present in the expression. The NONCOM statement in REDUCE however puts a global NONCOM flag on each of its arguments regardless of the others. This is the reason why the last two terms in the expression C do not cancel. Introducing operator ordering through the KORDER statement does not help in this case. The problem can only be bypassed by an additional substitution rule for the operator $\vec{P}$.

2. Another important issue is the way how to introduce commutation relations into the REDUCE system. The standard way is to define substitution rules (with the LET operator) for product terms (see Fig. 1). At this point several difficulties arise. First, one cannot simultaneously define LET rules for the two terms of a commutator, i.e. coding:

```
FOR ALL J,L LET P(J)*A(L) = A(L)*P(J) + K(J)*A(L),
                A(L)*P(J) = P(J)*A(L) - K(J)*A(L);
```

   would obviously lead to an infinite loop. So the way in which the LET assignments are coded automatically fixes the ordering of operators in the expression. This in turn means that one has to know this ordering in advance to code the right substitution rules and to change all these rules when the operator ordering has to be changed. Other problems are connected with commutators between components of the same operator, e.g.:

$$[\sigma_j, \sigma_k] = 2i\epsilon_{jkl}\sigma_l$$

   which demand for a careful index checking to avoid infinite loops.

3. There is also a need for more specific data types adapted to the particular problem in the example displayed in Fig. 1. With the available type declarations in REDUCE, one cannot make any distinctions in the program between physical operators and other objects like e.g. indexed scalars. REDUCE in its current release does not provide the user with an easy way to define new data types. However it is clear that when one wants to handle algebraic problems involving abstract objects the introduction of new specific data types is necessary in order to allow clear distinction between the different terms involved in a calculation and to get the possibility to associate special functions and operators to each data type.

```
     OFF MCD; OFF ALLFAC;
% define the operators  and declare P and A noncommuting
     OPERATOR P,A,K,M;
     NONCOM P,A;

%define the elementary commutation relation
     FOR ALL J,L LET P(J)*A(L) = A(L)*P(J) + K(J)*A(L);

% fix the ordering of variables
     KORDER E,M,K(N),A(N),K(J),A(J),P(J);
     ORDER E,M,K(N),A(N),K(J),A(J),P(J);

% this is the commutator to be evaluated
% N and J are dummy summation indices to
% simulate scalar product of vectors

     C := P(N)*P(N)/(2*M)*E/(4*M**2)*(P(J)*A(J))
             -E/(4*M**2)*(P(J)*A(J))*P(N)*P(N)/(2*M);

                   -3      2                          -3      2
     C := 1/8*E*M   *K(N) *K(J)*A(J) + 1/8*E*M   *K(N) *A(J)*P

                          -3                                  -3
          (J) + 1/4*E*M   *K(N)*K(J)*A(J)*P(N) + 1/4*E*M   *K(

                                               -3              2
          N)*A(J)*P(N)*P(J) - 1/8*E*M   *A(J)*P(J)*P(N)   + 1/

                 -3           2
          8*E*M   *A(J)*P(N) *P(J)

% additional LET rule to get the right result
    LET P(N)*P(J) = P(J)*P(N);

C;
           -3      2                          -3      2
 1/8*E*M   *K(N) *K(J)*A(J) + 1/8*E*M   *K(N) *A(J)*P(J) +

           -3                                     -3
 1/4*E*M   *K(N)*K(J)*A(J)*P(N) + 1/4*E*M   *K(N)*A(J)*P(J)*P(N)
```

Figure 1: A REDUCE program to calculate the commutator (2.2)

Thus, a software package extending the abilities of REDUCE to perform operator calculations should include the following features: a better definition of the notion of noncommutativity than in standard REDUCE, a clear und unambigous procedure to define commutation relations between elementary objects, a type definition and checking mechanism for new operator types, and further simplification routines for e.g. complex commutators, scalar products of vector operators, handling of inverse operators (this problem has been "solved" in the example by setting the switch MCD to off), substitution rules, etc. depending on how general the field of application should be.

3 Software modifications and extensions of REDUCE

The software structure of the REDUCE CAS is displayed schematically in Fig. 2. All commands and evaluation procedures available to the user in the REDUCE environment (also called the algebraic mode of REDUCE) are implemented in form of procedures written in the Meta–LISP language RLISP which has an ALGOL–like syntax much easier to read than the usual LISP syntax. These RLISP statements are then translated by an interface to the LISP language available on the computer.

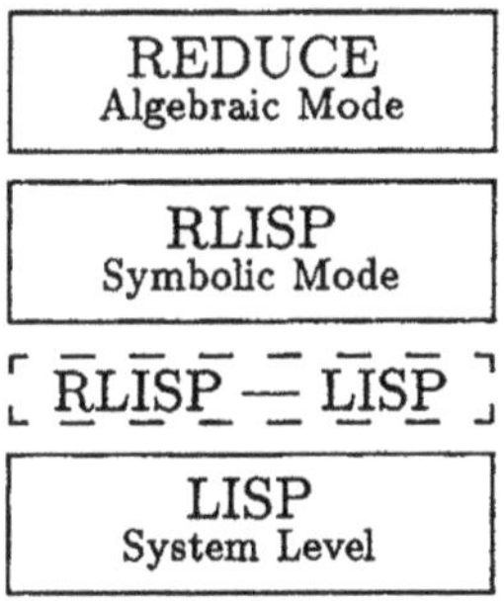

Figure 2: The structure of the REDUCE CAS

The algebraic mode of REDUCE can be altered and extended in three different ways:

First one could recode some of the standard RLISP procedures in order to change e.g. the definition of a REDUCE command, the evaluation procedures of some algebraic operator or the way substitution rules are applied. However this should be avoided as much as possible for the following obvious reasons: Due to the very high degree of interdependency between the various RLISP procedures, one very often generates in this way unwanted side effects in another place of the system which may be very difficult to discover and correct. Furthermore, if one modifies some of the standard REDUCE features the system would be much harder to use for noninitiated users. Finally the modifications become highly dependent on the current release of the REDUCE software. Therefore this way of proceeding should be limited mainly to the correction of (real or apparent) "bugs".

Another more practicable way is to add a set of additional commands and functions to the REDUCE environment to perform special tasks keeping available the standard instruction set of REDUCE. The new procedures coded in RLISP are then made available to the REDUCE environment by putting the appropriate indicators on the property list of the procedure name. For example, if one wants to add a new REDUCE declaration command MAKEANEWTHING to the system, the following sequence has to be coded in RLISP:

```
SYMBOLIC PROCEDURE MAKEANEWTHING U;

 % this procedure does the fancy new things
 % with the arguments contained in the list U

        (procedure body)

END;

% the following statement make the declarator MAKEANEWTHING
% available to REDUCE

PUT('MAKEANEWTHING,'STAT,'RLIS);
```

Several UCEPs of this kind have been designed, e.g. GENTRAN [3] generating FORTRAN code from REDUCE expressions or TRI [4], an interface between REDUCE and the TEX text processing system.

Finally, one can also introduce new data types into the programming environment. The standard data types defined in REDUCE are strings, domain elements, scalars, lists, operators and arrays as well as matrices and vectors for high energy physics calculations. For each data type one can define a complete set of evaluation procedures for the standard algebraic operators (i.e. addition, multiplication etc.) and add further special functions for some particular tasks (e.g. calculating the determinant of a matrix). When the master simplification routine called by the REDUCE supervisor detects that an expression contains objects of a certain data type, it passes control over that expression to another simplification procedure designed especially for that data type. This procedure can now in turn analyse the input expression and call the relevant evaluation procedures for the different algebraic operators and functions occurring in the expression. The final result after evaluation and simplification can then either be printed out by special printing routines designed for the data type or passed back to the standard REDUCE supervisor. In Fig. 3 the data flow through the different procedures is illustrated with a simple example taken from the PHYSOP package described in this paper.

In this way one can add a "special purpose" CAS designed for a specific data type to REDUCE without altering any of the standard system features. Moreover, it is possible to use part of the standard REDUCE system, e.g. the pattern matcher, in the simplification of expressions involving objects of the new data type. A complete software package for the handling of a new data type contain the following parts:

- Type declaration commands.
- A basic simplification procedure to which the input expression is passed by the REDUCE supervisor.
- Evaluation procedures for the algebraic operators which are called by the basic simplification procedure if the argument list of the operator contains an object of the new data type.
- An assignment routine which allows to assign the evaluated expression to a new identifier and gives it the correct data type.
- Additional functions for special operations (dot product, determinant calculation ...) and output prettyprinting.

```
% declare A and B to be of the new type
    SCALOP A,B;
% declare A and B noncommuting
     NONCOM A,B;
% define the commutator
     LET COMM(A,B) = C;
% evaluate the product;

     B*A;

% this is the master simplification routine of REDUCE
% the epression to be calculated is passed in form of an
% algebraic expression in LISP syntax
O*** ARGUMENTS OF AEVAL
     ((TIMES B A))

% the routine calls first the REDUCE type checking procedure
O*** ARGUMENTS OF GETRTYPE
     ((TIMES B A))
O*** VALUE OF GETRTYPE
     PHYSOP

% returned type is PHYSOP so the  argument list is passed
% to the basic simplification routine of the PHYSOP data type
O*** ARGUMENTS OF !*PHYSOPSM!*
     ((TIMES B A) NIL)

% this routine calls the special evaluation procedure
% of the algebraic operator TIMES (the multiplication)
O*** ARGUMENTS OF PHYSOPTIMES
     ((B A))
O*** VALUE OF PHYSOPTIMES
     (!*SQ ((((A . 1) ((B . 1) . 1)) ((C . 1) . -1)) . 1) T)

% value of the evaluation routine is passed back in prefixed
% standard quotient form indicating that no further evaluation
% is needed
O*** VALUE OF !*PHYSOPSM!*
     (!*SQ ((((A . 1) ((B . 1) . 1)) ((C . 1) . -1)) . 1) T)
O*** VALUE OF AEVAL
     (!*SQ ((((A . 1) ((B . 1) . 1)) ((C . 1) . -1)) . 1) T)

% and this is the final result as it appears in REDUCE

     A*B - C
```

Figure 3: A simplified traceback map to show the relations between the REDUCE system procedures and the procedures of the UCEP PHYSOP

The UCEP VECTOR [5] designed to perform three dimensional vector arithmetic is an example of such an implementation.

4 The UCEP PHYSOP for operator calculations in quantum theory

The UCEP PHYSOP has recently been designed to meet the requirements of various people working on constituent quark models, QCD sum rules and statistical field theory to have a powerful tool for the manipulation of very large expressions involving operators. Previous attempts to use the standard instruction set of REDUCE in such calculations have revealed the need for a more efficient implementation of the operator calculus along the guidance lines described in Section 2.
In its current release, the package is split into two parts:

The first part called NONCOM2 is an independent subpackage which redefines the handling of noncommutative objects by REDUCE in such a way that problem 1 described in section 2 is resolved. Noncommuting operators have now to be declared pairwise and each operator gets a new property list where the names of the other operators which do not commute with it are kept. Some multiplication and ordering routines have also been modified so that the check for noncommutativity is now performed on the basis of this new property lists. Finally, this new implementation also demands some modifications of the pattern matching routines. On the user side there are no new commands, the original NONCOM statement is kept as the declarator for noncommuting operators. This subpackage has the flaw to redefine standard system procedures so that the comments made in the last section apply here. However, these modifications are minor and seem not to produce any side effect as far as the tests have shown. This subpackage could also be used independently from the PHYSOP program.

The main part of the PHYSOP package implements the tools needed for the operator calculus. The link to the standard REDUCE system is made through a new data type called PHYSOP. The most important new features available to the user are:

- Type declaration commands for states (STATE), scalar operators (SCALOP), vector operators (VECOP), tensor operators (TENSOP) and special indices (PHYSINDEX). The indices are needed to address specific components of a vector or tensor operator. Tensor operators have to be declared together with their rank which fixes the number of special indices they are allowed to carry. All operators may carry other indices to allow further distinction.

- Elementary (anti-)commutation relations between operators are introduced to the system with a new operator COMM (ANTICOMM). Values can be assigned to these (anti-)commutators through ordinary LET statements. Commutators can only be declared between scalar objects (i.e. scalar operators or components of vector and tensor operators). Since both the commutator and the anticommutator for the same couple of operators could be defined, the system has a switch ANTICOM to control whether commutators or anticommutators should be used in simplification routines.

- Ordering of operators in an expression can be defined using a new OPORDER command. The syntax is very similar to the KORDER command available in standard REDUCE. The ordering routines have been changed so that every quantity which is not a PHYSOP object is ordered ahead of the operators contained in an expression.

- Special evaluation routines have been designed for the standard algebraic operations in order to deal with the algebra of operators. The multiplication procedures have been modified to include commutation relations. Quotients of scalar operators are transformed into products of their inverse. Integer powers of vector and tensor operators are expanded introducing dummy summation indices. A new operator DOT is available for scalar products of vector operators. Adjoint operators are defined together with every operator declaration.

- Two new algebraic operators have been defined for the simplification of complex operator expressions: COMMUTE and ANTICOMMUTE.

- A switch CONTRACT is provided to contract over dummy indices in the final output of an expression in those cases where this is not prevented by the defined operator ordering.

- Substitution rules for operator expressions can be introduced in the standard way through LET statements. Correct operator type assignment is automatically performed in substitution and assignment statements.

In Fig. 4 some of the features described above are demonstrated in a program which solves the example (2.2) using the PHYSOP UCEP package.
Furthermore the package also includes an operator OPAPPLY (infix version |) to perform matrix element calculations Its different applications are:

1. To define the action of elementary operators on states through LET statements.
2. To calculate the action of a complex operator expression on a given state
3. To calculate the matrix element of an operator expression sandwiched between states.

Example:

```
% declare some states
     STATE PSI,PHI;

% define the action of the operator P on PSI
     FOR ALL N,Q LET P(N) | PSI(Q) = DF(PSI(Q),Q(N));

% Calculate the action of an operator expression on a state
          C | PSI;

% Calculate the matrix element of an operator expression
          PHI | C | PSI;
```

The package is currently tested on several large problems and should be released in a complete version including documentation before the end of the year. It will then be placed into the REDUCE Network Library for free use by the user community.

Acknowledgements

Participation to the conference was supported by a grant from the Deutsche Forschungsgemeinschaft (DFG).

References

[1] A.C. Hearn, *REDUCE User's Manual Version 3.3* , Rand Publication CP78 (rev. 7/87) , The Rand Corporation, Santa Monica CA 90406-2138 1987.

[2] E. Schrüfer, F. W. Hehl and J. D. McCrea, General Relativity and Gravitation **19** (1987) 197.

[3] B. L. Gates, in: *Proc. SYMSAC '86* (1986), p. 94-99.

[4] W. Antweiler, A. Strotmann and V. Winkelmann, SIGSAM Bulletin **23** (1989) 26.

[5] D. Harper, Comp. Phys. Comm. **54** (1989) 295.

```
% define the operators
    VECOP P,A,K;
    SCALOP M;

% declare P and A to be pairwise noncommutative
    NONCOM P,A;

% define the commutation relation between P and A
    FOR ALL J,L LET COMM(P(J),A(L))=K(J)*A(L);

% declare the operator ordering for the final output
    OPORDER M,K,A,P;

% calculate the commutator
    C:= COMMUTE(P**2/(2*M) , E/(4*M**2)*(P DOT A));

    C := (E*(M-1*M-1*M-1*K(IDX2)*K(IDX2)*K(IDX1)*A(IDX1)

             + M-1*M-1*M-1*K(IDX2)*K(IDX2)*A(IDX1)*P(IDX1)

             + 2*M-1*M-1*M-1*K(IDX2)*K(IDX1)*A(IDX1)*P(IDX2)

             + 2*M-1*M-1*M-1*K(IDX2)*A(IDX1)*P(IDX2)*P(IDX1

             )))/8

% The program has introduced two dummy indices IDX1 and IDX2
% into the expression during the expansion of the vector
% squared P**2 and the dot product P DOT A.
% The object M-1 has been inserted by the program and represents
% the inverse of the scalar operator M.  The quotient 1/M**3
% has been expanded to a product of the inverse operator M-1.

% switch on the contract mode over dummy indices and powers
    ON CONTRACT;

    C;
           3  2               3  2                 3
    (E*(M-1 *K *K DOT A + M-1 *K *A DOT P + 2*M-1 *K(IDX2)*

                                      3
       K DOT A*P(IDX2) + 2*M-1 *K(IDX2)*A(IDX1)*P(IDX2)*P(

       IDX1)))/8
% contraction is not complete because of the chosen operator order
```

Figure 4: The example (2.2) calculated with the PHYSOP package

Some Computer Realizations of the REDUCE-3 Calculations for Exclusive Processes in SQCD

V.A.Rostovtsev
Joint Institute for Nuclear Research

Ya.Z. Darbaidze, Z.V. Merebashvili
Tbilisi State University

The REDUCE-3 program for the calculation of the squared gauge invariant set of tree diagrams is given in the α^3 order of the perturbation theory. The necessity of using suchprogram packages as "Factorizator", "COLOR-Factor" and so on is shown. The correctness of calculation for the infrared radiation corrections as compared with manual cal - culations is discussed. An example of apllying the programs is given for the matrix and noncommutative algebras when the well-known suppersymmetric commutative relations is proved.

The article is published at full length in *Fortschritte der Physik,* **38** (1990) 9, 717-731

COMPUTER-AIDED CLASSIFICATION OF GEOMETRIES IN GENERAL RELATIVITY

M.A.H. MacCallum

School of Mathematical Sciences
Queen Mary and Westfield College, University of London
Mile End Road, London E1 4NS, U.K.
E-mail: mm@maths.qmw.ac.uk

1 INTRODUCTION

There are three main ways of approaching a theory of mathematical physics (assuming we are not ambitious enough to be constructing one of our own) namely to prove general theorems about it, to apply (analytic or numerical) approximation schemes, and to find (necessarily very special) exact solutions. The first of these is meritorious but very difficult, the second is what most applied mathematicians and theoretical physicists do although it frequently suffers, in the case of non-linear theories like general relativity, from lack of a rigorous basis for the approximation scheme used, and the third, because it takes exact account of the non-linearities, may be useful in providing simple models, in generating comparisons to test against the approximation schemes (see e.g. [1]), and in other ways, despite the very special character of the solutions.

In texts on relativity the arbitrariness of coordinates is stressed from the start, and it naturally prompts the question of how to handle solutions in a way that is coordinate independent, but this question is not discussed in most texts, and books and journal articles still frequently characterize solutions in terms of hypotheses about coordinate forms of the metric. What we now describe is the way in which this obvious gap can be filled.

Some years ago I was an author of a book on the known exact solutions in general relativity [2]. The solutions are found by simplifying the equations in various ways, which is essential given the large number of terms for a completely general metric. This may be done (e.g.) by imposing symmetry or some special algebraic structure, and leads to the basic problem of recognising the same solution in two coordinate systems (the "equivalence problem"), or of characterizing a solution uniquely in a coordinate-free way.

The method described here does the second of these, i.e. performs the characterization and hence identification of an individual solution. For a sufficiently differentiable metric this is an algorithmic procedure; the subsequent comparison of two metrics (a task left to the humans at present) is a well-defined problem[1], but has no algorithmic solution. The reason for this is that it involves determining whether or not there are solutions of a set of simultaneous equations, which may be non-linear or even transcendental, but for certain classes of functions no algorithm for such a problem is possible. It should be noted that since we are dealing with solutions of the equations in the sense of expressions for metrics our results can only decide local equivalence, i.e. we will not be able to distinguish, by these methods, between a plane and a flat torus, which differ only in global topology.

The problem in fact has (at least) four aspects, namely

(a) the pure mathematical problem (more precisely, the differential geometric problem) of uniquely characterizing a spacetime,
(b) the algorithmic problem of finding effective means of performing the necessary calculations,
(c) the actual performance of the algorithms, which in practice needs algebraic computing packages and therefore poses programming and computing problems, and finally the most interesting and difficult of all,
(d) how does the mathematical characterization relate to the physics which it in some way encapsulates?

(To be honest little progress has been made on the last of these, using the classifications produced by our method: the work in [4], [5] and [6, 7] is really all that has been done so far.) Thus the work mixes pure mathematics, software engineering, and physics.

It may be worth noting that the inverse process, that is to say specifying a set of invariants and recovering the metric in some coordinate system given by the invariants, has been considered by Karlhede, Lindström,

[1] provided certain differentiability requirements [3], which I will not present in detail here, are satisfied. Instead I will assume these requirements hold for the solutions we deal with.

Bradley and others, though I so far find the practical examples unconvincing in that they concern only previously known solutions [8, 9, 10].

The work on this problem has constituted a very long-term project, and I have therefore given previous reviews of it on a number of occasions. Alert readers will detect the inevitable substantial overlap between this and previous articles.

The early work on the equivalence problem was due to Christoffel [3, 11] and was the context of the invention of the Christoffel symbols; it showed that, for metrics without symmetry, equivalence of two metrics could be decided by computing the Riemann tensor and a finite number of its derivatives in coordinate bases, and at most the 20th derivatives were required. That limit could involve calculating 29 320 310 074 020 terms, of which 79 310 are independent when the Bianchi and Ricci identities are taken into account, and it was clearly impractical.

An important reduction is made by using frames with constant metric components (e.g. orthonormal frames) in which case the required order of differentiation is not more than 10 and the resulting maximum number of possible independent terms 8690: this was first realized by Cartan [12], who also showed how the method applied to spacetimes with isometries. Its practical application was proposed by Brans [13] but he suggested first classifying the 10th derivatives, which is still a rather impractical method.

A renewed effort was made from about 1978 onwards at Stockholm prompted by Bertel Laurent. Anders Karlhede re-investigated the theory [14] while Jan Åman developed the required programs. By 1979 they had a working version for vacuum solutions [15, 16] and it is these programs, much extended, which form the package CLASSI; the extensions are the work of a number of people, including me, but by far the largest role was, and continues to be, played by Åman. CLASSI is SHEEP with the classification programs and other auxiliary programs loaded: the division between SHEEP and CLASSI is more a boundary between different authors' contributions than an useful division of capabilities. In fact CLASSI now does many things other than provide the tools for classification (see [17]), but this is still the main impetus behind its development.

The next section discusses the pure mathematical and algorithmic problems, and the final section gives some information about the practical application of this theory using CLASSI, and the form of the results obtained.

2 THE THEORY OF THE "EQUIVALENCE PROBLEM"

The pure mathematics of Cartan's method have been reviewed elsewhere [14, 18, 19]. I do not propose to give the proofs here, only to give a statement of the method motivated by heuristic arguments. The basic idea, which all relativists would immediately suggest, is to use invariants, e.g. the value of the Ricci scalar, but most people would first think of using the so-called scalar polynomial invariants (i.e. scalar invariants which are polynomials in the curvature and its derivatives) and unfortunately this idea fails. The example of a plane wave and flat space shows that scalar polynomial invariants are not sufficient to distinguish two spacetimes (because in both of these spaces all scalar polynomial invariants of the curvature vanish).

This shows us we need tensor components, and these of course depend on the choice of the basis in the tangent vector space (the "frame") as well as the space-time position. Thus we must work on a frame bundle, which, for a manifold M, is a manifold $F(M)$ of points each of which represents a pair given by a point of M and a frame for M at that point, and try to provide a unique characterization of location in this larger space. Space-time tensor components are invariant scalars on $F(M)$, i.e. given a point in M and a frame at that point, the components are completely determined. Now we need to provide a set of these scalars uniquely defined by the geometry of M. In fact what we will find are a set of one-forms uniquely defined by the connection, and from them we obtain a set of scalars (on the frame bundle $F(M)$, i.e. not necessarily scalars in the space-time M).

Those scalars turn out to be components of the Riemann tensor and its covariant derivatives, which is not surprising since these quantities give the only possible invariants of a Riemannian geometry, a result familiar from the construction of Riemannian normal coordinates (extension to other problems which can be thought of in terms of a connection or equivalence classes of connections is quite direct [19]).

A point in the frame bundle has coordinates which are the coordinates of the appropriate spacetime point $\{x^a\}$ together with coordinates describing the choice of frame, the frame itself being given as a set of one-forms $\{\omega^b\}$. For the frames, we use coordinates giving a parametrization of the allowed transformation group (e.g. the Lorentz group), since the members of the set of all frames at a point (the "fibre" of the bundle) can obviously be labelled in this way. The space-time connection can be used to define a "horizontal" in the frame bundle, which is obtained by considering the parallel transport of frames between neighbouring points of M, and horizontal one-forms can then be defined on $F(M)$ (and denoted by the same symbols $\{\omega^b\}$ as the frames on M) so that at each point P of $F(M)$ they project to the particular frame on M which is represented by P. The connection itself (described as a set of one-forms on M, $\{\omega^a{}_b\}$) can be "lifted" to the bundle by including terms dependent on the coordinates in a fibre: these terms represent the effect of transformations of frames on the vectors of those frames, without change of space-time position. The "lifted" connection one-forms and the one-forms $\{\omega^b\}$ on $F(M)$ representing the frames together provide a basis of the (co)tangent space of $F(M)$ *which is uniquely defined by the connection of the space-time.*

Now if two spacetimes are isometric, they will define identical frame bundles with identical sets of one-form fields. Thus two spacetimes M and M' will be equivalent only if their frame bundles can be identified, with (e.g.) $\omega^1 = \omega^{1'}$ and $\omega^1{}_2 = \omega^1{}_2{}'$, and so on (the prime denoting quantities defined from M'). If we denote the $(\omega^i, \omega^i{}_j)$ collectively by σ^J, the requirement that for each J, $\sigma^J = \sigma^{J'}$ implies that $d\sigma^J = d\sigma^{J'}$ and since $d\sigma^J = \gamma^J{}_{KL}\sigma^K \wedge \sigma^L$ we must also have

$$\gamma^J{}_{KL} = \gamma^J{}_{KL}{}' \ . \tag{1}$$

Now the $\gamma^J{}_{KL}$ will (from Cartan's structure equations) be either universal constants (since there are some coefficients which are always 0 or ± 1) or components on $F(M)$ of the curvature of M. If the equations (1) give insufficient information (which they do) then one has to take the exterior derivatives $d\gamma^J{}_{KL}$ on $F(M)$ and equate components of these quantities, and so on. It turns out that these are expressed just in terms of components on $F(M)$ of covariant derivatives on M of the curvature and already known quantities (see [14] for the details), as the heuristic arguments above suggested. Thus we can produce a set of scalar equations between the two sets of coordinates of two explicitly given frame bundles.

Also it is easy to see that if the components of the $(p+1)$-th derivative, considered as scalars on the frame bundle, depend on the derivatives up to the p-th, all subsequent derivatives also depend on the first p (by repeated differentiation of the relation for the $(p+1)$-th and back-substitution). Thus we will at worst need to differentiate up to $p = d$ where d is the dimension of the frame bundle considered, and we in any case stop as soon as we take a derivative that gives no new functions on $F(M)$. This implies that unless $p = d$, in which case we know no new functions can arise from a further differentiation, we will always have to calculate one derivative that does not give new independent functions but only gives functional dependencies of derivatives. If at this stage, we still have less than d functions the reason is that the space-time has isometries. (Part of Cartan's contribution was to show that this still uniquely characterized the manifold.)

An alternative, and more heuristic, way of describing what is happening is that we are finding a set of coordinates on the frame bundle in an invariantly-defined way, these coordinates being some chosen set of components of the $\gamma^J{}_{KL}$ and their derivatives which are functionally independent on $F(M)$. From this point of view, we could then in principle determine the coordinate transformation to the invariant coordinates and compute the metrics in these coordinates – though note that in order to achieve this we would need to find a way to cope with transformations of the ignorable coordinates in the case of manifolds with symmetry. The definition of the metric involves taking exterior derivatives of the coordinates, and this is equivalent to the need to take one more derivative than the last one in which new information arises, i.e. to calculate the $(p+1)$-th derivative. In practice, the need for explicit calculation of coordinate transformations makes this approach impractical.

For four dimensions, the coordinate frame bundle has dimension 20, while a bundle of frames with a constant metric (e.g. orthonormal frames or Newman-Penrose null tetrads) will be 10-dimensional. These numbers explain the dimensions in the results of Christoffel and Cartan quoted earlier. As already mentioned, the technique is not restricted to Riemannian manifolds, or to Lorentzian signature, or to four dimensions, but those are the conditions for which the specific classification programs described below are written, and in them advantage is taken of these extra conditions.

The essential improvement (also based on ideas of Cartan [12]) provided by Karlhede was the idea that the dimension of the frame bundle to be dealt with could be reduced by fixing the frame at each step of calculation so that the Riemann tensor and its derivative are at each step in a canonical form. (Determination of the possible forms is of course an algebraic problem of a classic type.) Working this way one can easily show (for spacetimes, though the same ideas can be adapted to more general manifolds) that at most the seventh derivatives are required (with a maximum of 3156 independent terms), and experience so far suggests that no more than the third derivatives (at worst 430 terms) are really needed. (I have an idea that may lead to a proof that this bound is rigorous.)

Thus the procedure is to start with an order of differentiation $q = 0$. Then:

1. Calculate the derivatives of the Riemann tensor up to the q-th.
2. Find the canonical form of the Riemann tensor and its derivatives.
3. Fix the frame as far as possible by this canonical form, and note the residual frame freedom (the group of allowed transformations is the isotropy group H_q). The dimension of H_q is the dimension of the remaining "vertical" part of the frame bundle.
4. Find the number t_q of independent functions of space-time position in the components of the Riemann tensor and its derivatives in canonical form, and record that. This tells us the remaining "horizontal" freedom.
5. If the isotropy group and number of independent functions are the same as at the previous step, let $p+1 = q$ and stop, (for relativity, the case $p = d$ never arises) or otherwise (or if $q = 0$) increment q by 1 and go to step 1.

The space-time is then characterized by the canonical form used, the successive isotropy groups and independent function numbers, and the values of the non-zero components of the Riemann tensor and derivatives which are found. (Note that the "horizontal" and "vertical" dimensions are not treated in the same way, since the independent vertical functions are removed by restricting the choices of frame.) Since there are t_p essential space-time coordinates, clearly the remaining $4-t_p$ are ignorable, so the isotropy group of the spacetime will have dimension $s = \dim H_p$ and the isometry group has dimension $r = s + 4 - t_p$. (Note that axisymmetric spaces do *not* have isotropy, the definition of isotropy being that the same sort of isotropy transformation can be applied at *every* point, not only at points on an axis or some other special subset.)

To compare two space-times one can first compare the discrete properties such as the sequence of isotropy groups, and only if those all match does one have to check whether the set of equations obtained by equating corresponding components of the Riemann tensor and its derivatives in canonical form have a solution. Note that *while the discrete properties can prove inequivalence of manifolds, they are insufficient to prove equivalence in general.*

It is important in dealing with examples that to prove two spaces are equivalent one need not *solve* the equations given by (1) and the higher derivatives. One has only to show a solution exists, i.e. that the equations are consistent. What consistency means here is perhaps most simply shown by an example: the equations $x = y$ and $x^2 = y^2$ are consistent because for every y there is a corresponding x, whereas $x = y$ and $x^2 = y$ are not because they are only both true if $x = 0 = y$ or $x = 1 = y$. The non-algorithmic step of deciding the consistency in fact has turned out to be solvable in practical examples. (We have no right to expect this in general. For algebraic (i.e. polynomial) equations in the variables one can use Gröbner basis techniques.)

3 CLASSIFYING SPACETIMES IN PRACTICE

3.1 THE ALGORITHMS

The implementation of the Cartan-Karlhede procedure in CLASSI uses the Newman-Penrose formalism. The reason for this choice is essentially that it exploits the simplifications, especially of the symmetries of tensors, arising from the use of the homomorphism from $SL(2, C)$ to the Lorentz group. The resulting formalism, which can be considered in terms of spinors [20, 21] or self-dual bivectors, also has the minor advantage that complex quantities replace real ones, so a smaller number are required. For instance, the first 7 derivatives of the Riemann tensor have at most 1668 complex (3156 real) terms, and the first 3 have at most 234 spinor (430 real) terms. (Note that the number of spinor terms is not simply half the number of real terms, because some of the spinor components are diagonal elements of Hermitian arrays and hence real.) These numbers were computed by Åman and myself [22].

Step 1 of the algorithm described at the end of the last section for $q = 0$ is straightforward, but steps 2 and 3 then need careful investigation of the Petrov and Segre classifications and efficient methods of locating the canonical frames.

The Petrov and Segre classifications both depend on the handling of quartics which may have multiple roots. The first algorithm for Petrov classification was that of d'Inverno and Russell-Clark [23] but it is far from optimal. The first, but incomplete, improvements were those of Fitch [24], which used ideas similar to those later (re-)introduced by others. The main sources of improvements (due to Hon [25], Åman et al. [26], and Letniowski and McLenaghan [27]) are as follows.

First, cases can be separated according to whether or not one or more of the Newman-Penrose components Ψ_A of the Weyl tensor vanish. In many cases this immediately gives the Petrov type. The basic idea was introduced by Hon [25], but his work unfortunately remained unavailable outside the University of Waterloo until recently, so that this idea had to be re-discovered [26].

Secondly, and relatedly, one arranges the tests so that any calculation of the full discriminant (a sixth degree polynomial, a sextic, in the components Ψ_A) is put off until it becomes unavoidable [25, 26].

Thirdly, one can build up more complicated expressions in the components Ψ_A by a series of binary operations followed by simplification. This will in general save both time and computer memory in the calculations of quantities of high degree [26].

Finally, one must look very closely at the tests required to determine the Petrov type in those remaining cases where a number of types are possible. The overall algorithms of [25, 26, 27] differ essentially only at this point (though the presentation arising from the formulation in Letniowski and McLenaghan's paper is clearly better). Some further small improvements in the version used in CLASSI have recently been made by Seixas (unpublished).

The Segre classification [28] makes use of the same methods for quartics as the Petrov classification. It has two parts. The first part classifies the Plebanski spinor

$$\Phi_{(AB}{}^{C'D'}\Phi_{CD)C'D'}$$

(where $\Phi_{ABA'B'}$ is the Newman-Penrose spinor equivalent of the tracefree Ricci tensor) into types algebraically equivalent to the Petrov types, called Plebanski-Petrov types; it uses programs identical to those for Petrov classification. Unfortunately different Ricci tensor types give the same Plebanski-Petrov type, so a second stage is usually required in which particular methods for separating the subcases are used. These are based on considering a second quartic, the eigenvalue equation for the tracefree Ricci tensor. The method aims only at separation into the different Segre types, as it was felt that the refinements discussed by (e.g.) Penrose and Rindler [21] are not necessary for our purposes. Seixas (unpublished) has made some useful improvements on the method as described in [28].

Having thus dealt with step 2 for $q = 0$ of the procedure given in the previous section, we now have to bring the Riemann tensor into canonical form by Lorentz transformations. Because there are many vacuum solutions and few conformally flat solutions, we start by putting the Weyl tensor into canonical form. Although this is in principle algorithmic (since quartics have an algorithmic solution, and the quartic for Petrov classification has as its roots the parameters of the required null rotations) it is in practice often the most difficult step, especially in the Petrov type I case. We are helped by the fact that the method used for Petrov classification is essentially based on the Euclidean algorithm for simultaneous roots of polynomials, so that when there are equal roots they are readily found.

The fourth and fifth steps are straightforward in principle, though the programmming was still a substantial effort. Functional dependence is tested by finding the rank of the Jacobians between the possibly independent functions, and the isotropy can be found from the canonical form.

Now we reach step 1 again for $q = 1$. For $q \geq 1$ it is very important to cut down the number of quantities to be calculated by taking only a minimal set of derivatives of the Riemann tensor, which can be done because the derivatives obey the Ricci and Bianchi identities and their differential concomitants. An explicit minimal set of derivatives was found by Åman and myself [22], using some ideas from Penrose [29]. In terms of the Newman-Penrose quantities $\Psi_{ABCD}, \Phi_{ABA'B'}$ and Λ it can be described as follows. One has to take, in the general case, the following components for $q \geq 0$ (where the term "totally symmetrized" means that a spinor is to be symmetrized over all its free primed indices and over all its free unprimed indices):

1. The totally symmetrized q-th derivatives of Λ
2. The totally symmetrized q-th derivatives of Ψ_{ABCD}
3. The totally symmetrized q-th derivatives of $\Phi_{ABA'B'}$
4. For $q \geq 1$, the totally symmetrized (q-1)-th derivatives of
   $\Xi_{ABCD'} = \nabla^D{}_{D'}\Psi_{DABC}$
   which is one side of one of the Bianchi identities.

5. For $q \geq 2$ the d'Alembertian $\nabla^{AA'}\nabla_{AA'}$ applied to all the quantities calculated for the derivatives of order $q-2$.

In the program the notation uses only a single primed and a single unprimed index on each quantity, defined in the same way as in the Newman-Penrose notation, so that, for example, $\Psi_2 = \Psi_{(0011)}$ and $\Phi_{11'} = \Phi_{(01)(0'1')}$.

Steps 2 and 3 of the procedure for $q \geq 1$ are at present handled in a manner which is capable of refinements that would probably be useful in creating more precise classifications. In fact, in most cases all that is tested is whether any invariance at the $q = 0$ stage persists, and if it does not, new canonical forms are not found. Steps 4 and 5 continue to be in principle straightforward, if more time-consuming (because of the increasing numbers of components to be tested).

There is a further extension of the results which has not yet been fully implemented (though it has been carried through for a few cases, with some hand calculation, and work on the full implementation is in progress). This is the method due to Karlhede and me [30] for extracting additional information, in particular the Lie algebra structure of the symmetry group, for spaces with isometries. The starting point is to consider a Lie group G acting on itself (by multiplication). As well as its generators, vector fields $\{\xi_\alpha\}$ say, on G one can find another set of vector fields $\{\eta_\beta\}$, which also form a basis of the tangent space at each point of G, such that

$$[\xi_\alpha,\ \eta_\beta] = 0$$

for all α and β: the $\{\eta_\beta\}$ are called the generators of the reciprocal group, and have the same Lie algebra as the group G. (A fuller introduction to these ideas is in [2], chapter 8.)

When a group G acts "simply-transitively" on a spacetime (i.e. so that for any pair of points p and q equivalent under the group, there is exactly one transformation T moving p to q) then each orbit of the group (the orbit of a point p being the set of all points $T(p)$ for transformations T in G) is (locally) isomorphic to the group itself (take a map such that the identity e in G is mapped to p and the point T in G to $T(p)$). Moreover, the frames $\{\omega^a\}$ constructed by the classification procedure outlined above will necessarily have the property that $[\xi_\alpha,\ \mathbf{e}_a] = 0$ where ξ_α now means the image on the orbit of the vector field of the same name on G, and the $\{\mathbf{e}_a\}$ are the vector fields dual to the $\{\omega^b\}$. (Here Latin indices are spacetime indices and Greek indices run over the dimension of the symmetry group G, which could be, but is not necessarily, the same as that of the spacetime.) Hence to find generators of the reciprocal group on the orbit we need only project the $\mathbf{e}_a$ vectors into the orbit and take a minimal basis of them, and then to get the Lie algebra we have only to take the commutators of these vectors (actually we can do all this using the dual one-forms without calculating the vectors explicitly).

This will not work directly in cases with isotropy: for example, if there is a rotational symmetry about every point in the $(x,\ y)$ plane it is impossible to find basis vectors in the plane invariant under this symmetry, because they all get rotated. Isometry groups which include rotations or other isotropies are called "multiply-transitive" because there will be more than one transformation T in G taking a point p to another point q in the orbit of p. In particular the basis one-forms $\{\omega^a\}$ on M will *not* be invariant under all the Killing vectors if the spacetime M has a multiply-transitive isometry group.

The way out of this dilemma is the same as that for classification: use of a frame bundle. Groups acting multiply-transitively on M act simply-transitively on $F(M)$. Thus one has only to apply the recipe for the simply-transitive case, but this time on $F(M)$. So one needs to take the basis one-forms on $F(M)$, including the vertical coordinates, and project them onto the orbits in $F(M)$ and form the commutators. (Note that since our classification procedure would eliminate all vertical coordinates if the group is simply-transitive on M, the same statement applies to both cases.)

As well as the Lie algebraic structure of the group, the method leads to information on the curvature of the orbits, and to some ways of dealing with the ignorable coordinates (see [30]). It is perhaps worth noting one way in which one can actually perform the projections, since this is not discussed in [30]. The orbits in $F(M)$ are known in terms of certain functions (those components of the Riemann tensor and its derivatives which are functionally independent on $F(M)$) which are constant on each orbit. Calling these functions I^α, we can find a new basis of $F(M)$ which contains their exterior derivatives dI^α. This is achieved by the following method.

1. Take a particular I^α and the original basis of one-forms of $F(M)$ (the σ^J in the notation used earlier).
2. Form dI^α
3. Replace the first remaining one of the original basis one-forms by this dI^α.

4. Check if the resulting set of one-forms is still a basis (i.e. linearly independent). If so, keep that replacement, and if not try replacing the next σ^J (it must be possible to replace one of them), until dI^α has been inserted into the basis.
5. If there are no more I^α, the procedure is finished. Otherwise, take a new α and go to step 2.

When all the dI^α have been put into the basis, take the dual basis. The vectors $\mathbf{e}_J$ in this dual basis corresponding to the surviving σ^J will necessarily be a basis of the orbits as required (since they satisfy $dI^\alpha(\mathbf{e}_J) = 0$ and inherit invariance under the group from the invariance of the σ^J and I^α).

The classification of a given metric will usually not succeed just from the first version of the input data. There are two main reasons. One is that while the input will usually be in a tetrad convenient for the form in which the metric was found, this will not usually be the one in which the Riemann tensor takes its canonical form. The other is that the substitutions necessary to enable zero expressions to be detected as zero by the system, and to cut the size of expressions down to a manageable length, will usually not have been completely foreseen. These problems have, of course, been overcome in the metric files distributed with SHEEP and CLASSI, so that just running one of them through the system may give a misleading impression of the amount of work needed to classify a metric.

CLASSI has mechanisms to help the user recognise these problems. It issues a warning if the Riemann tensor is not in a standard form, which is typically because a further position-dependent tetrad transformation is required. Similarly all functions performing classification can output a list of those quantities among the ones calculated which must be genuinely non-zero for the classification to be correct, and which therefore must be checked to see if simplifications are possible. Some considerable effort, especially in the case of Petrov classification, has been put into making these lists minimal, i.e. not every quantity the computer believed to be non-zero is output, since the conclusion may still be correct although some terms are actually zero.

Among the facilities of CLASSI are some to help the user overcome difficulties in classification, the most frequent of which is a need to change the tetrad. For this there is a package which provides automatic transformation of the Riemann tensor to canonical form, when that is possible; this uses the already mentioned fact that when there are repeated roots they are easily found.

Some details of sample calculations with CLASSI have been given in earlier reviews [31, 17], the latter of which provides very detailed information on the use of the CLASSI programs.

3.2 RESULTS OF CLASSIFICATIONS

While a complete comparison of two metrics which are equivalent will necessitate the comparison of the values of the various quantities calculated in the classification process, it is very helpful to have briefer summaries which may be, and often are, sufficient to prove inequivalence and also would be useful in answering requests for a list of, say. all known Petrov type D perfect fluid solutions. For this reason output in the form of a one-line summary has been developed, and the main purpose of this section is to explain what it contains. Sample classifications are given in Table 1.

The coding of the Petrov types is the usual one, except that types I, II and III are coded 1, 2 and 3. The numbers r, s and t_i are just integers (except that if $r = 10$ the Latin X is used). The coding of the isotropy groups and energy-momentum tensor types, which was the result of several discussions between Åman, myself, Joly, Skea and others, is as follows. (Here the isotropy groups are listed in order of decreasing dimension and the energy-momentum tensors in order of decreasing generality.)

The whole Lorentz group is coded as 6; the group $SO(3)$ of spatial rotations is p (because it is the rotational symmetry of a perfect fluid energy-momentum), the group $SO(2,\ 1)$ similarly is t for tachyons, and the group generated by two null rotations and a spin about one spatial axis is r (for radiation); the group of boost and spin is called e (by analogy with electromagnetism), and the group of two null rotations n (for null fields); the one-dimensional groups of spin, boost and null rotation are called respectively s, b and k. 0 has the obvious meaning.

The codings p, t, r, e, s, and b are also used for the Ricci tensors with those isotropy groups (and real eigenvectors), and the other cases are as follows: g is used for the general case with 4 real eigenvectors, z for the case with a complex pair, and 1 where these cannot be distinguished; f is used when p and t cannot be distinguished; c is used for the case with a spin symmetry and a pair of complex eigenvalues, i being used for the situation where s and b cannot be distinguished and d for the case where c and i cannot be distinguished; 2 is used for the Segre type [11,2], h for its special case with s symmetry and n for its special

Table 1: Sample output from the classification programs. The column headings, which we have added, label columns containing the following data: 'name' is a short name for the metric (with a limit of 6 characters), usually built from the name(s) of the author(s) who first found it; Φ represents the energy-momentum tensor type; R indicates whether the Ricci scalar is 0 or not; Ψ is the Petrov type; r the dimension of the isometry group; s the dimension of the isotropy group; H_i for $i = 0, 1, 2, 3$ are the isotropy groups for successive derivatives of the Riemann tensor; and t_i the number of independent functions of space-time position they contain. The coding is described in the main text. In the last 8 columns '-' denotes a value not calculated because the algorithm terminated. Some well-known metrics are in this sample; the line named 'wavezk' is Kundt's example [32], which needs three derivatives of the Riemann tensor for its classification.

name	$\Phi R \Psi$	r	s	H_i t_i
allnut	p13	1	0	00--33--
antdes	110	X	6	66--00--
bertot	e00	6	2	ee--00--
cylema	e01	3	0	00--11--
datray	r02	3	0	00--11--
defris	r1N	6	2	nn--00--
desitt	110	X	6	66--00--
einsta	f10	7	3	ff--00--
frwcl	f10	6	3	ff--11--
gottin	e1D	6	2	ee--00--
hartso	00N	6	2	nn--00--
kasner	001	3	0	00--11--
kerdeb	113	3	0	000-011-
kerr	00D	2	0	e00-222-
kottle	11D	4	1	ess-111-
leroyn	11N	3	0	n00-011-
melnic	f1D	4	1	ss--11--
minkow	000	X	6	66--00--
petvac	b1D	4	1	bb--11--
renord	e0D	4	1	ess-111-
schwar	00D	4	1	ess-111-
sik2	r12	3	0	00--11--
sphsta	s1D	4	1	ss--11--
szeksp	f1D	3	1	sss-122-
taub	001	3	0	00--11--
tol4	p1D	4	1	ss--11--
tol7	p1D	4	1	ss--11--
vaidya	r0D	3	1	sss-122-
wavez	00N	5	2	nnn-011-
wavezk	00N	1	0	n0000233

case with k symmetry; 3 is used for Segre type [1, 3] and 4 for its special case with all eigenvalues equal. The reasons for the name choices can be found in [28].

I should emphasize that each line in Table 1 is calculated by the computer from a input metric without extra information from hand calculations being typed in. At the time of writing there are well over 100 solutions in the full table and this number is still growing.

One might ask where one goes next with this work. There seem to me to be the following directions

1. Completion of the catalogue. This is in progress.
2. Studying interpretation of the invariants. This needs both hard work and inspiration. Preliminary examples were mentioned above.
3. Development of specialised techniques for handling particular classes of solutions within the general scheme, e.g. for the very complicated forms of the metric obtainable by application of the generating techniques.
4. Extensions of the method, (a) to other gravity theories or gauge theories or Euclidean spaces [33, 34], (b) to obtain extra information, and (c) to other areas such as differential equations [35].
5. Inversion of the procedures as a way of finding exact solutions. The first attempts have so far only re-discovered known solutions, as mentioned earlier.

References

[1] Centrella, J.M., Shapiro, S.L., Evans, C.R., Hawley, J.F. and Teukolsky, S.A., "Test-bed calculations in general relativity", in *Dynamical spacetimes and numerical relativity*, ed. J.M. Centrella, Cambridge University Press, London and New York, 326-344 (1986).

[2] Kramer, D., Stephani, H., MacCallum, M.A.H. and Herlt, E., "Exact solutions of Einstein's field equations", Deutscher Verlag der Wissenschaften, Berlin, and Cambridge University Press, London and New York (1980) (Russian edition: ed. V.S. Vladimirov, Energoisdat, Moscow 1982.)

[3] Ehlers, J., "Christoffel's work on the equivalence problem for Riemannian spaces and its importance for modern field theories of physics", in *E.B. Christoffel*, ed. P.L. Butzer and F. Fehér, 526-542, Birkhaüser Verlag, Basel (1981).

[4] Karlhede, A., Lindström, U. and Åman, J.E., "A note on a local effect at the Schwarzschild sphere", *Gen. Rel. Grav.*, **14**, 569-572 (1982).

[5] Skea, J.E.F., "Anisotropic cosmology and curvature invariants", Ph. D. thesis, University of Sussex (1987).

[6] Araujo, M.E. and Skea, J.E.F., "The automorphism groups for Bianchi universe models and computer aided invariant classification of metrics", *Classical and Quantum Gravity*, **5**, 537-550 (1988).

[7] Araujo, M.E. and Skea, J.E.F., "Automorphisms in action: spatially homogeneous Einstein-Maxwell plane waves", *Classical and Quantum Gravity*, **5**, 1073-1096 (1988).

[8] Karlhede, A. and Lindström, U., "Finding space-time geometries without using a metric", *Gen. Rel. Grav.*, **15**, 597-610 (1983).

[9] Bradley, M., "Construction and invariant classification of perfect fluids in general relativity", *Class. Quant. Grav.*, **3**, 317-334 (1986).

[10] Bradley, M. and Karlhede, A., "On the curvature description of gravitational fields", *Class. Quant. Grav.*, **7**, 449-463 (1990).

[11] Thomas,T.Y., "Differential invariants of generalised spaces", Cambridge University Press, Cambridge (1934).

[12] Cartan, É., "Leçons sur la geomètrie des éspaces de Riemann", Gauthier-Villars, Paris (1946).

[13] Brans,C.H., "Invariant approach to the geometry of spaces in general relativity", *J. Math. Phys.*, **6**, 95-102 (1965).

[14] Karlhede, A., "A review of the geometrical equivalence of metrics in general relativity", *Gen. Rel. and Grav.*, **12**, 693 (1980).

[15] Karlhede, A. and Åman, J.E., "Progress towards a solution of the equivalence problem in general relativity", Pp. 42-44 in *EUROSAM 79. Symbolic and algebraic computation* (Lecture Notes in Computer Science, **72**), ed. Ng, E. Springer-Verlag, Berlin and Heidelberg (1979).

[16] Åman, J.E. and Karlhede, A., "A computer-aided complete classification of geometries in general relativity. First results", *Phys. Lett A*, **80**, 229-231 (1980).

[17] MacCallum, M.A.H. and Skea, J.E.F., "SHEEP: a computer algebra system for general relativity", to appear in *Proceedings of the first Brazilian School on computer algebra*, ed. M.J. Rebouças, Oxford University Press (1990).

[18] MacCallum, M.A.H., "Computer-aided classification of exact solutions in general relativity", in *Gravitational collapse and relativity*, ed. T. Nakamura and H. Sato, World Scientific, Singapore (1986).

[19] Kamran, N., "Contributions to the study of the equivalence problem of Élie Cartan and its applications to partial and ordinary differential equations", *Princeton University preprint* (1988).

[20] Penrose, R. and Rindler, W., "Spinors and space-time I: Two-spinor calculus and relativistic fields", Cambridge University Press, London and New York (1984).

[21] Penrose, R. and Rindler, W., "Spinors and space-time II: Spinor and twistor methods in space-time geometry", Cambridge University Press, London and New York (1985).

[22] MacCallum, M.A.H. and Åman, J.E., "Algebraically independent n-th derivatives of the Riemann curvature spinor in a general space-time", *Class. Quant. Grav.*, **3**, 1133-1141 (1986).

[23] d'Inverno, R.A. and Russell-Clark, R.A., "Classification of the Harrison metrics", *J. Math. Phys.*, **12**, 1258 (1971).

[24] Fitch, J.P., "An algebraic manipulator", Ph. D. thesis, Cambridge University (1971).

[25] Hon, E., "Application of REDUCE system to some problems in general relativity", M. Math thesis, University of Waterloo, (1975).

[26] Åman, J.E., d'Inverno, R.A., Joly, G.C. and MacCallum, M.A.H., "Quartic equations and algorithms for Riemann tensor classification", in *EUROSAM 84: Proceedings of the 1984 European conference on symbolic and algebraic manipulation*, ed. J. Fitch, Lecture Notes in Computer Science, **174**, 47-58, Springer Verlag, Berlin and Heidelberg (1984).

[27] Letniowski, F.W. and McLenaghan, R.G., "An improved algorithm for quartic equation classification and Petrov classification", *Gen. Rel. Grav.*, **20**, 463-484 (1988).

[28] Joly, G.C. and MacCallum, M.A.H., "Computer-aided classification of the Ricci tensor in general relativity", *Class. Quant. Grav.*, **7**, 541-556 (1990).

[29] Penrose, R., "A spinor approach to general relativity", *Ann. Phys. (N.Y.)*, **10**, 171 (1960).

[30] Karlhede, A. and MacCallum, M.A.H., "On determining the isometry group of a Riemannian manifold", *Gen. Rel. Grav.*, **14**, 673-682 (1982).

[31] MacCallum, M.A.H., "Algebraic computing in general relativity", in *Classical general relativity*, ed. W.B. Bonnor, J.N. Islam and M.A.H. MacCallum, Cambridge University Press, London and New York, 145-171 (1984).

[32] Kundt, W., "Note on the equivalence problem", in *Recent developments in general relativity: essays in honour of L. Infeld*, 307-311, PWN, Warsaw and Pergamon, Oxford (1962).

[33] Karlhede, A., "Classification of Euclidean metrics", *Class. Quant. Grav.*, **3**, L1-4 (1986).

[34] Karlhede, A., "On the potential field strength relation", *Class. Quant. Grav.*, **3**, L27-30 (1986).

[35] Kamran, N. and Shadwick, W.F., "The solution of the Cartan equivalence problem for $dy^2/d^2x = F(x, y, dy/dx)$ under the pseudo-group $\bar{x} = \phi(x)$, $\bar{y} = \psi(x, y)$", in *Field theory, quantum gravity and strings*, Lecture Notes in Physics, **246**, ed. H.J. de Vega and N. Sanchez, Springer-Verlag, Berlin and Heidelberg (1986).

ANALYTICAL COMPUTATIONS FOR THE LOCAL METHODS OF NON-LINEAR ANALYSIS IN CELESTIAL MECHANICS

A.G.Sokolsky, N.N.Vasiliev, S.R.Karimov, I.I.Shevchenko
Institute of Theoretical Astronomy,
USSR Academy of Sciences, Leningrad

Abstract. The CELESTA system, now being developed on the multilingual basis (REDUCE, FORTRAN, C, PASCAL, etc.) in the Institute of Theoretical Astronomy, is described. It is intended for use on the IBM compatible personal computers. The first version of the system provides solution of systems of algebraic, transcendental and differential equations by means of the method of a small parameter, normalization of canonical systems in the neighbourhood of the point of equilibrium by means of Lie transforms, averaging in multifrequency canonical systems.

The computer algebra methods are already traditional in celestial mechanics. The reason of it is in the bulky volumes of the processed symbolic information as well as in the relative easiness of automatization of routine analytical computations with functions and mathematical expressions of homogeneous type. Universal computer algebra systems (such as REDUCE), as well as numerous specialized computer algebra systems worked out for application in specific classes of celestial mechanics problems [1], are widely applied to the solution of problems connected with development of analytical theories of motion of natural and artificial celestial bodies.

The modern software standards present strict requirements to the means of analytical programming. The user-friendly interface, transportability of algorithms, exchange of analytical processing results, informational support, etc., ought to be provided.

Therefore it is vital to create the integrated computer algebra systems intended for application in specific fields of problems. These integrated systems ought to combine the efficiency of specialized systems with universality of systems of REDUCE type, to provide informational support in the form of an expert system, to provide opportunities of work with widely spread universal CAS and libraries of analytical algorithms.

Local methods of non-linear analysis [2] are an example of a suitable field of application of such a system. They represent a totality of methods connected with solution of an equation of the kind

$$N(x,\varepsilon) = 0, \tag{1}$$

where ε is a small parameter, x is a vector of unknown variables, N is some non-linear operator (it may be functional, differential, integral, etc.). It is assumed that the solution of the unperturbed system with $\varepsilon = 0$ is known, i.e. the vector

$$y: N(y,0) = 0 \tag{2}$$

is known.

It is needed to find a solution $x(\varepsilon)$ of the equation (1) such as $x(0) = y$. Smoothness of the operator N being assumed, the solution of equation (1) is sought in the form of a series of powers of the small parameter ε. In the common case the series contain round powers only, in degenerate (resonant and other) cases the series may contain fractional powers as well.

The classical iteration methods, the modifications of the method of a small parameter, the methods of normalization of systems of ordinary differential (in particular, Hamiltonian) equations, the methods of averaging, the special methods of investigation of systems of ordinary differential equations in the neighbourhood of the point of equilibrium or in the neighbourhood of a periodic solution, the analysis of singularities of transforms and singularities of systems of ordinary differential equations, etc., enter in this general scheme. These methods differ one from another by the field of application (which is determined by the type of the operator N) as well as by the schemes of construction of approximations. What do they have in common, is that their application in specific problems implies bulky volumes of routine analytical computations.

The reasons of application of CAS for solution of problems of this type are as follows:

1) realization of the scheme itself of one or another local method may need a cumbersome iterational procedure of construction of the next approximation from the preceding one;

2) a dependence on some additional parameters may be present in the initial equation (1), and then it is used to be reasonable to find the dependence of the solution upon these parameters in the analytical form.

The local methods of non-linear analysis are particularly widely used in the qualitative and analytical celestial mechanics.

The CELESTA system for the analytical support of the local methods of non-linear analysis in celestial mechanics is now being developed in the Institute of Theoretical Astronomy of the USSR Academy of Sciences. It is intended for use on the IBM compatible personal computers. The following concepts are founded in the basis of the development of the system

- functioning: common environment of the system (the monitor), friendliness of the system, compatibility of modules and the modular structure of the library of programs;

- formation: hierarchy of the table of contents (the menu), bases of symbolic data, continuity of versions, and openness of the system;

- programming: multilinguity (REDUCE, FORTRAN, PASCAL,

C, etc.) supported by the symbolic-numerical interface.

The CELESTA system consists of a shell and a library of program modules. The shell consists of

- a supervisor program (which conducts the dialogue with the help of a multiwindow interface, carries out the docking of modules, forms the answer);

- an input/output subsystem (which accomplishes the contact with the user in the natural language, input of initial data and output of the solution);

- a library of editing procedures (which edit the intermediate and final results, transform the representation of expressions of various types);

- a "helper" in the form of the library of help-files, which prompt the user how to operate with the system on various occasions; it contains also a brief description of the local methods of non-linear analysis implemented in the system and their application in the problems of celestial mechanics.

The functional filling of the first version of the CELESTA system is compiled of the special libraries of REDUCE procedures, which accomplish

- solution of non-degenerate systems of algebraic, transcendental and differential equations of the standard type by means of the method of a small parameter;

- solution of systems of differential equations by means of the Picar iterational method;

- solution of integral equations by means of the Neuman iterational method;

- normalization of canonical systems of differential equations in the neighbourhood of the point of equilibrium by means of Lie transforms;

- auxiliary modules of general mathematical procedures and expressions.

The programs implementing the method of a small parameter [3] allow to find the solutions of the systems of algebraic and transcendental equations

$$F(x,\varepsilon) = 0, \tag{3}$$

as well as to find the solutions of the systems of differential equations

$$\dot{x} = \varepsilon\, F(x,\varepsilon,t), \qquad x\big|_{t=0} = x_0 \tag{4}$$

in the form of series of natural powers of the small parameter ε.

The programs implementing the Picar method [6] allow to solve the systems of differential equations of the first order

$$y' = F(x,y), \qquad y(x_0) = y_0 \tag{5}$$

by means of the recurrent formula

$$y_p = y_0 + \int_{x_0}^{x} F(x,y_{p=1})\, dx, \qquad p = 1,\ 2,\ \dots .$$

The programs implementing the Neuman method allow to solve the Fredholm non-homogeneous integral equations of the second kind

$$U(x) - \lambda\int_a^b K(x,\xi)U(\xi)\, d\xi = F(x), \tag{6}$$

where the core of the integral transform $K(x,\xi)$, the function $F(x)$, integration limits a and b, and the parameter λ are known. It is needed to find the function $U(x)$. The problem is solved by means of the method of successive approximations (in the form of the Neuman series) according to the recurrent formula

$$U_p(x) = F(x) + \lambda\int_a^b K(x,\xi)U_{p-1}(\xi)\, d\xi, \quad U_0(x) = F(x),$$

$$p = 1,\ 2,\ \dots .$$

The programs of non-linear normalization of autonomous Hamiltonian systems implement the Deprit-Hori method of construction of the normalizing transform with the help of Lie series [7,8]. It is assumed that the initial Hamiltonian represents the truncated expansion in powers of coordinate and momentum variables, and its quadratic component is the sum of Hamiltonians of non-coupled linear oscillators. As a result, the normal form of the Hamiltonian, the generator of the canonical transform, the normalizing and the inverse transforms of variables are derived in the analytical form. The algorithm of computations implemented in the programs gives the principal opportunity to normalize the system with any given number of degrees of freedom and up to any needed order of expansion. The resonances, if they exist, are treated automatically.

In order to support the enumerated methods of local analysis, the CELESTA system contains auxiliary modules of general mathematical procedures and modules dealing with special classes of expressions, such as elliptic integrals and functions, expansions in fractional powers of independent variables (generalized polynomials), special functions of celestial mechanics.

The programs intended for work with elliptic integrals and Jacobi elliptic functions compute their analytical approximations in the form of truncated series of various kinds [4].

The programs intended for work with series of fractional powers of independent variables are based on the algorithms described in the package [5] of a similar purpose. For operations using several variables simultaneously, such as inversion and raising to negative powers, iterational algorithms are used.

Though the CELESTA system is intended for solution of

problems of celestial mechanics, it may be applied as well to the other fields of scientific research, where the local methods of non-linear analysis are used.

The forthcoming versions of the CELESTA system will differ from the first ones by the more voluminous library of procedures of the local methods of non-linear analysis first of all.

REFERENCES

1. Brumberg V.A., Tarasevich S.V., Vasiliev N.N. Specialized celestial mechanics systems for symbolic manipulation. - Celest. Mech., 1989, 45, p. 149.
2. Bruno A.D. The local method of non-linear analysis of differential equations. - Moscow: Nauka, 1979.
3. Giacaglia G.E.O. Perturbation methods in non-linear systems. - Berlin: Springer-Verlag, 1972.
4. Gradshtein I.S., Ryzhik I.M. Tables of integrals, sums, series, and products. - Moscow: Nauka, 1971.
5. Feldmar E, Kolbig K.Z. REDUCE Procedures for the manipulations of generalized power series. - Comp. Phys. Comm., 1986, 39, p. 267.
6. Korn G.A., Korn T.M. Mathematical handbook for scientists and engineers. - New York: McGraw-Hill, 1968.
7. Markeev A.P., Medvedev S.V., Sokolsky A.G. Methods and algorithms of normalization of differential equations. - Moscow: Moscow Aircraft Institute Publishers, 1985.
8. Markeev A.P., Sokolsky A.G. Some computational algorithms of normalization of Hamiltonian systems. - Preprint of the Institute of Applied Mechanics of the USSR Academy of Sciences, 1976, No. 31.

COMPUTER ALGEBRA, INTEGRABLE SYSTEMS AND ALGEBRAIC CURVES

N.A. KOSTOV

Institute of electronics
Bulgarian Academy of Sciences
72 bul. Lenin , Sofia 1784, Bulgaria

Abstract

We study the family of commuting differential operators with elliptic coefficients of the Halphen type and also with spectral parameter. The integrability condition of the pair of equations of Halphen type gives the large class of nonlinear differential equations of Lax-Novikov type. The algorithmic solution of the problem of constructing Baker-Akhiezer function associated to N-sheeted coverings of the elliptic curve is given. This algorithm is implemented using computer algebra system REDUCE 3. Many examples, including the construction of new elliptic solutions of completely integrable dynamical systems related to one-dimensional Schrodinger equation with Lame potential are presented.

References

1. E.D. Belokolos, A.I. Bobenko, V.Z. Enol'skii, V.B. Matveev, *Algebro-geometrical principles of superposition of finite-gap solutions of integrable nonlinear equations,* Russian Math. Surveys, 41,1986, 1-49.

2. V.Z. Enol'skii, N.A. Kostov *Differential equations in a spectral parameter and the coverings of the torus,* Preprint of the Technical University of Denkmark, 1989, 53p.

3. N.A. Kostov, *Quasi-periodic solutions of the integrable dynamical systems related to Hill equation,* Lett.Math.Phys. 17, 95-104.

4. V.P. Gerdt, N.A. Kostov, *Computer algebra in the theory of ordinary differential equation of Halphen type,* in: Computer and Mathematics, ed. E.Kaltofen and S.M. Watt, Springer-Verlag, 1989, 277-283.

5. V.Z. Enol'skii, N.A. Kostov, *On the geometry of elliptic solitons,* ed. I.M. Krichever, Springer-Verlag, 1991, in press.

OPERATOR ORDERING AND SYMBOL REPRESENTATIONS OF ENVELOPING ALGEBRAS

Wolfgang Lassner

Sektion Informatik, Karl-Marx-Universität

Leipzig 7010

Abstract

Phase space methods in quantum theory make use of symbol representations of the Weyl algebra. The method can be generalized to enveloping algebras U(L) of Lie algebras L different from the Heisenberg Lie algebra. The method starts from a one-to-one correspondence between the linear space of the enveloping algebra U(L) and that of the symmetric algebra S(L) over L. The algebra S(L) can be equipped with a so-called twisted product so that it becomes isomorphic to U(L).

Different ordering rules correspond to different bases in U(L). The elements of U(L) are represented by different symbols in dependence on the ordering rule chosen. We generalize the notion of an ordering defining function ϕ. This helps to determine a transformation between different types of symbols and allows to use the same fast multiplication algorithm for different orderings.

1. Introduction

Operator symbol representations have their origin in mathematical literature in the theory of pseudo-differential operators and in physical literature in so-called phase space methods of quantum mechanics. Phase space methods in quantum theory make use of symbol representations of the Weyl algebra. A restriction to a pure algebraic view point, i.e. to symbols of algebra elements rather than of operators, allows to extent the class of non-commutative algebras for which symbol representations exist. Furthermore, the twisted product techniques can be extended to subrings of the quotient division ring of the Weyl algebra. The method has been elaborated /1/ for calculations in computer algebra systems and was applied to quantum mechanical problems (see /1/ and ref. therein), to Gröbner bases calculations in non-commutative algebra[/2/], and to Lie optics in order to calculate aberrations (see /3/ and ref. therein).

The method can be generalized to enveloping algebras U(L) of Lie algebras L different from the Heisenberg Lie algebra. The method starts from a one-to-one correspondence between the linear space of the enveloping algebra U(L) and that of the symmetric algebra S(L) over L. The algebra S(L) can be equipped with a so-called twisted product so that it becomes isomorphic to U(L).

Operator orderings have been described in mathematical and physical literature by various techniques[/4-6/]. Symbol representations are powerful tools in this context[/4/]. A generalization to enveloping algebras U(L) resp. to differential operators over Lie groups will be of theoretical and practical interest.

The ordering problem consists first of all in a description of different operator orderings. Different ordering rules correspond to different bases in U(L). The elements of U(L) are represented by different symbols in dependence of the ordering rule chosen. An appropriated twisted product depends on the ordering rule in a non-trivial way. We generalize the notion of an ordering defining function ϕ. It is possible to calculate ϕ for various ordering rules. This helps to determine a transformation between different types of symbols and allows to use the same fast multiplication algorithm for different orderings. The fast multiplication algorithm for the non-commutative algebra must not be related to twisted product techniques. A very fast algorithm[/7/] makes uses of the fast integer arithmetic of special processors. The algebra elements are represented by integers in that case. The representation is related to a given basis. The mentioned transformation allows to use the same fast algorithm also if the algebra elements are given in another basis related to a different ordering.

2. Basic notions and notations

Let L be a finite dimensional Lie algebra over the field K . The Lie algebra L is given by its structure constants c_{ij}^{k} in a basis $\{X_1,X_2,\ldots,X_n\}$, $[X_i,X_j]=\sum_{k=1}^{n} c_{ij}^{k}X_k$ for $i,j=1,2,\ldots,n$.

We denote by U(L) the enveloping algebra of L over K. One gets U(L)= T(L)/J from the tensor algebra T(L) of L by factorization according to the ideal J generated by the set $\{\ X\otimes Y - Y\otimes X - [X,Y] \mid X,Y \in L\ \}$. The enveloping algebra $U(L)$ has a graded structure $U(L) = \oplus_n U_n$, $U_n = U^n \backslash U^{n-1}$, where $U^n = \{X \in U(L) \,|\, \deg X \le n\}$. The set $\{X_1^{i_1} X_2^{i_2} \cdots X_n^{i_n}\}$,$(i_1, i_2, \ldots, i_n) \in N_0^n$ is a basis of $U(L)$ according to the Poincaré-Birkhoff-Witt theorem. We use the convention $X_1^0 X_2^0 \cdots X_n^0 = 1$, and we denote by capital letters X_k also the elements of L in their embedding $L \xrightarrow{i} U(L)$ in $U(L)$. So, there is no problem to differ non-commutative multiplication of elements $X_1^{i_1} X_2^{i_2} \cdots X_n^{i_n}$ in $U(L)$ from commutative multiplication in the symmetric algebra $S(L)$. The basis elements of $S(L)$ are denoted by small letters $x_1^{i_1} x_2^{i_2} \cdots x_n^{i_n}$. The symmetric algebra $S(L)$ over L can be obtained from the tensor algebra $T(L)$ by factorization $S(L) = T(L)/I$ according to the ideal I generated by the set $\{X \otimes Y - Y \otimes X \mid X, Y \in L\}$. $S(L)$ is isomorphic to the algebra of polynomials $K[x_1, x_2, \ldots, x_n]$.

We use multiindex notation:

$$x = (x_1, x_2, \ldots, x_n),\ \ \xi = (\xi_1, \xi_2, \ldots, \xi_n),\ \ \xi x = \sum_{k=1}^{n} \xi_k x_k\ ,\ \ \alpha = (\alpha_1, \alpha_2, \ldots, \alpha_n) \in N^n$$

$$|\alpha| = \sum_{k=1}^{n} \alpha_k\ ,\quad x^\alpha = x_1^{\alpha_1} x_2^{\alpha_2} \cdots x_n^{\alpha_n}\ ,\quad \frac{\partial^{|\alpha|}}{\partial \xi^\alpha} = \frac{\partial^{|\alpha|}}{\partial \xi_1^{\alpha_1}\, \partial \xi_2^{\alpha_2} \cdots \partial \xi_n^{\alpha_n}}\ .$$

3. Generalized Weyl ordering

A one-to-one correspondence between the linear spaces of $S(L)$ and $U(L)$ is defined by the Gelfand map[/8/]:

$$\omega :\ S(L) \longrightarrow U(L)$$

given by

$$\omega(x_{i_1} x_{i_2} \ldots x_{i_r}) = \frac{1}{r!} \sum_{\pi} X_{i_{\pi(1)}} X_{i_{\pi(2)}} \cdots X_{i_{\pi(r)}} \tag{1}$$

and by linearity of ω. The sum runs over all permutations of r elements. The r.h.s. of (1) is the complete symmetrization $\langle X_{i_1} X_{i_2} \ldots X_{i_r} \rangle$ of $X_{i_1} X_{i_2} \ldots X_{i_r}$.

This complete symmetrization will be synonymously called Weyl ordering. The map (1) will be called Weyl quantization due to applications in physics. The element $y \in S(L)$ is called Weyl symbol of $\omega(y) \in U(L)$. The notions Weyl ordering and Weyl quantization are now generalizations of the corresponding notions known for the special case if L is the Heisenberg Lie algebra H_n . H_n is the (n+1)-dimensional Lie algebra over R given by

$[P_i, Q_j] = \delta_{ij} Z$ with $\delta_{ij} = \begin{cases} 1 \text{ if } i=j \\ 0 \text{ if } i \ne j \end{cases}$ $(i,j = 1,2,..,n)$, where Z is a central element.

Example 1: Let L be the Lie algebra so(3) with the basis $\{X_1,X_2,X_3\}$, $[X_1,X_2]=X_3$, $[X_2,X_3]=X_1$, $[X_3,X_1]=X_3$.
Then

$$\omega(x_1x_2) = \langle X_1X_2\rangle = \frac{1}{2}(X_1X_2+X_2X_1) \quad .$$

The element $x_1x_2 \in S(L=so(3))$ is the Weyl symbol of

$$\frac{1}{2}(X_1X_2+X_2X_1) = X_1X_2 - \frac{1}{2}X_3 \qquad \in U(so(3)).$$

4. Twisted product for Weyl symbols

We introduce in S(L) a new non-commutative multiplication which makes S(L) isomorphic to U(L). Since ω is bijective we are allowed to define

$$y_1 * y_2 = \omega^{-1}(\ \omega\ (y_1)\cdot\omega\ (y_2)\ ) \tag{2}$$

A key idea in symbol representations is to calculate the non-commutative multiplication between symbols in a direct way without the non-commutative multiplication in U(L).

For the usage of computer algebra systems it is very convenient that the multiplication coincides with the following convolution product /9/. Note that only differentiations occur.

$$(y_1 * y_2)(x) = e^{x\,D(\ \partial_{x'}\ ,\ \partial_{x''})}\qquad y_1(x')\ y_2(x'')\ |_{x'=x''=x} \tag{3}$$

The differential operator D in the exponential series is formally given by

$$D(\partial_{x'},\partial_{x''})= \ \tau(\partial_{x'},\partial_{x''}), \tag{4}$$

where $\tau(\xi,\eta) = \lambda(\xi,\eta)-\xi-\eta$ is the non-linear part of the exponent in the Baker-Campbell-Hausdorff formula (see for instance /9/)

$$e^{\xi X} \circ e^{\eta X} = e^{\lambda(\xi,\eta)X} \tag{5}$$

Since we calculate twisted products of polynomials, the formal exponential series of Eq. 3 will be used only up to a finite order. A proof, that the twisted product of Eq. 3 coincides with that defined in Eq. 2 will be given in section 6 together with a justification of the notation used. The operator D will be given by a truncated series.

The twisted product in combination with the linear map ω from Eq. (1)

allows to execute tedious calculations in U(L) using now only symbols. This can be done easily in almost all computer algebra systems, since only differentiation and commutative polynomial algebra facilities are required. Of course, D has to be determined. However, if D is calculated once up to an order high enough in dependence of degree of the polynomials, then D can be used in all subsequent calculations. Moreover, useful formulae as applied in a LOGLAN package /10/ for calculations in enveloping algebras and Lie fields can be derived from the twisted product.

5. Orderings defined by a function

In order to apply the twisted product techniques also for orderings different from that defined by the map ω of Eq. 1, we introduce the notion of an ordering defining function ϕ . For this reasons we consider another way to define the map ω. The method is partially known in literature for the special case if L is the Heisenberg Lie algebra (see /1/ and references therein).

Let $\xi x=\sum_{k=1}^{n}\xi_k x_k$ be a general homogeneous element of degree 1 in S(L). Replacing the small x's by capital X's and not touching the parameters ξ we get a linear element in U(L). If this will be done in the exponential series $e^{\xi x}$, so this is just the way to describe Weyl quantization for exponential series

$$\Omega(e^{\xi x}) = e^{\xi X} \quad . \tag{6}$$

From Eq. 6 we derive the Gelfand map of Eq. 1, i.e. the Weyl quantization of polynomials. Formal differentiation $\frac{\partial^{|\alpha|}}{\partial\xi^{\alpha}}\ldots|_{\xi=0}$ applied on $e^{\xi x}$ gives the monomial $x^{\alpha}= x_1^{\alpha_1}x_2^{\alpha_2}\cdots x_n^{\alpha_n}$ whereas the same operator applied on $e^{\xi X}$ gives just the complete symmetrization $\langle X_1^{\alpha_1}X_2^{\alpha_2}\cdots X_n^{\alpha_n}\rangle = \omega(x_1^{\alpha_1}x_2^{\alpha_2}\cdots x_n^{\alpha_n}) \in U(L)$ of $X_1^{\alpha_1}X_2^{\alpha_2}\cdots X_n^{\alpha_n}$. That will not wonder, since the Weyl quantization $\omega(x_1^{\alpha_1}x_2^{\alpha_2}\cdots x_n^{\alpha_n})$ of $x_1^{\alpha_1}x_2^{\alpha_2}\cdots x_n^{\alpha_n}$ occurs as "coefficient" of $\xi^{\alpha} = \xi_1^{\alpha_1}\xi_2^{\alpha_2}\cdots\xi_n^{\alpha_n}$ in the exponential series $e^{\xi X}=\sum_i\frac{(\xi X)^i}{i!}$, $(|\alpha|=i)$. Differentiation with respect to ξ at $\xi=0$ picks out just that coefficient:

$$\omega(x^{\alpha}) = \frac{\partial^{|\alpha|}}{\partial\xi^{\alpha}}\,\Omega(e^{\xi x})|_{\xi=0} = \langle X_1^{\alpha_1}X_2^{\alpha_2}\cdots X_n^{\alpha_n}\rangle \quad . \tag{7}$$

So we receive the linear map ω by formal differentiation from Ω as shown in the following diagram.

$$\begin{array}{ccc} e^{\xi x} & \xrightarrow{\quad\Omega\quad} & e^{\xi X} \\ \frac{\partial^{|\alpha|}}{\partial\xi^{\alpha}}\cdots\Big|_{\xi=0}\Big\downarrow & & \Big\downarrow\frac{\partial^{|\alpha|}}{\partial\xi^{\alpha}}\cdots\Big|_{\xi=0} \\ x_1^{\alpha_1}x_2^{\alpha_2}\cdots x_n^{\alpha_n} & \xrightarrow{\quad\omega\quad} & \langle X_1^{\alpha_1}X_2^{\alpha_2}\cdots X_n^{\alpha_n}\rangle \\ & & \text{complete symmetrization} \end{array}$$

We refer to $x_1^{\alpha_1}x_2^{\alpha_2}\cdots x_n^{\alpha_n}$ as the Weyl symbol of the symmetrized basis element $\langle X_1^{\alpha_1}X_2^{\alpha_2}\cdots X_n^{\alpha_n}\rangle$. Note that the map ω defines the Weyl basis $\{\langle X_1^{\alpha_1}X_2^{\alpha_2}\cdots X_n^{\alpha_n}\rangle \mid \alpha \in N^n\}$ of U(L) according to Eq. (1). We introduce an ordering defining function ϕ and generalize the map Ω as follows

$$\Omega_{\phi}\,(\,e^{\xi x}\,) = e^{\xi X + \phi(\xi)X} \quad , \tag{8}$$

where $\phi=(\phi_1(\xi),\ldots,\phi_n(\xi))$ has to be appropriately chosen. We will not try to characterized the class of allowed functions ϕ. However, ϕ can be explicitly determined for various orderings. The particular case $\phi = 0$ describes Weyl ordering. Formal differentiation to ξ at $\xi=0$ can be used to generate a basis in U(L). The basis differs from the Weyl basis if ϕ is not zero or a constant.

We define the map ω_{ϕ} for polynomials by

$$\omega_{\phi}(x^{\alpha}) = \frac{\partial^{|\alpha|}}{\partial\xi^{\alpha}}\,\Omega_{\phi}(e^{\xi x})\,\Big|_{\xi=0} = \langle X_1^{\alpha_1}X_2^{\alpha_2}\cdots X_n^{\alpha_n}\rangle_{\phi} \tag{9}$$

and by linearity of ω.

The method is summarized in the following diagram.

$$\begin{array}{ccc} e^{\xi x} & \xrightarrow{\quad\Omega_{\phi}\quad} & e^{\xi X + \phi(\xi)X} \\ \frac{\partial^{|\alpha|}}{\partial\xi^{\alpha}}\cdots\Big|_{\xi=0}\Big\downarrow & & \Big\downarrow\frac{\partial^{|\alpha|}}{\partial\xi^{\alpha}}\cdots\Big|_{\xi=0} \\ x_1^{\alpha_1}x_2^{\alpha_2}\cdots x_n^{\alpha_n} & \xrightarrow{\quad\omega_{\phi}\quad} & \langle X_1^{\alpha_1}X_2^{\alpha_2}\cdots X_n^{\alpha_n}\rangle_{\phi} \end{array}$$

We now speak of $x_1^{\alpha_1}x_2^{\alpha_2}\cdot\cdot x_n^{\alpha_n}$ as the ϕ-symbol of the ϕ-ordered basis element $\langle X^{\alpha_1}X^{\alpha_2}\cdot\cdot X^{\alpha_n}\rangle_\phi$ of U(L) .

Example 2: Standard ordering, Poincaré-Birkhoff-Witt basis.

In many practical cases and with almost all computer algebra systems it seems much easier to work with the Poincaré-Birkhoff-Witt basis $X_1^{\alpha_1}X_2^{\alpha_2}\cdot\cdot X_n^{\alpha_n}$ of U(L) than with the symmetrized elements of the Weyl basis. The function ϕ for the Poincaré-Birkhoff-Witt basis (PBW-basis) or standard basis can be calculated iteratively using the Baker-Campbell-Hausdorff formula. We will do it step wise so that the origin and nature of ϕ becomes transparent.

Because of $X_i^{\alpha_i} = \frac{\partial^\alpha}{\partial\xi_i^{\alpha_i}} e^{\xi_i X_i}|_{\xi_i=0}$ for $i=1,2,\ldots,n$, we obtain

$$X_1^{\alpha_1}X_2^{\alpha_2}\cdot\cdot X_n^{\alpha_n} = \left(\frac{\partial^{\alpha_1}}{\partial\xi_1^{\alpha_1}} e^{\xi_1 X_1}|_{\xi_1=0} \right)\cdots\left(\frac{\partial^{\alpha_n}}{\partial\xi_n^{\alpha_n}} e^{\xi_n X_n}|_{\xi_n=0} \right) . \tag{10}$$

Since the differentiation concerns different parameters this equals to

$$X_1^{\alpha_1}X_2^{\alpha_2}\cdot\cdot X_n^{\alpha_n} = \frac{\partial^{|\alpha|}}{\partial\xi^\alpha} \left(e^{\xi_1 X_1}\cdots e^{\xi_n X_n} \right)|_{\xi=0} . \tag{11}$$

In order to find the function ϕ such that

$$X_1^{\alpha_1}X_2^{\alpha_2}\cdot\cdot X_n^{\alpha_n} = \frac{\partial^{|\alpha|}}{\partial\xi^\alpha} \left(e^{\xi X + \phi(\xi)X} \right)|_{\xi=0} , \tag{12}$$

we iteratively apply the BCH-formula

$$e^{\xi_1 X_1}\cdot e^{\xi_2 X_2}\cdots e^{\xi_n X_n} = e^{\xi_1 X_1 + \xi_2 X_2 + \tau(\underline{\xi}_1,\underline{\xi}_2)X}\, e^{\xi_3 X_3}\cdots e^{\xi_n X_n} , \tag{13}$$

where $\underline{\xi}_i=(0,..,0,\xi_i,0,\ldots,0)$ denotes the n-tuple with ξ_i in the i-th component. Then the "vector valued" function ϕ is given by

$$\phi(\xi) = \tau^{(1)} + \ldots + \tau^{(n-1)} , \tag{14}$$

where $\tau^{(k)}$ is recurrently defined by

$$\tau^{(k)}=\tau(\underline{\xi}_1+\ldots+\underline{\xi}_k+\tau^{(1)}+\ldots+\tau^{(k-1)},\underline{\xi}_{k+1}) \tag{15}$$

with

$$\tau^{(1)}=\tau(\underline{\xi}_1,\underline{\xi}_2) .$$

The function ϕ is the key to transformations between symbol for different orderings. Let $\phi(\xi)$ be the ordering defining function for the

Poincaré–Birkhoff–Witt basis, i.e. for the standard ordering of example 2. The transformation $e^{x'\phi(\partial_x)} \ldots|_{x'=x}$ applied on a standard symbol of an element Y of U(L) gives just the Weyl symbol of the same element Y.

$$e^{x'\phi(\partial_x)}\, y_s(x)|_{x'=x} = y_w(x) \,. \tag{16}$$

The proof starts again from exponentials. Note that $e^{\xi x}$ is the ϕ–symbol of $e^{\xi X+\phi(\xi)X}$ according to Eq. 8. The following identity

$$e^{x'\phi(\partial_x)}\, e^{\xi x}\,|_{x'=x} = e^{\xi x + \phi(\xi)x} \tag{17}$$

may be verified by comparison of coefficients.

The r.h.s. of Eq. 17 is just the Weyl symbol of $e^{\xi X + \phi(\xi)X}$.

Formal differentiation $\frac{\partial^{|\alpha|}}{\partial\xi^\alpha} \ldots |_{\xi=0}$ of Eq. 17 gives for the l.h.s. $e^{x'\phi(\partial_x)}\, x^\alpha\,|_{x'=x}$ where x^α is the ϕ–symbol of $\langle X_1^{\alpha_1} X_2^{\alpha_2} \cdot\cdot X_n^{\alpha_n}\rangle_\phi = X_1^{\alpha_1} X_2^{\alpha_2} \cdot\cdot X_n^{\alpha_n}$ for the standard ordering. The same formal differentiation gives for the r.h.s. of Eq. 17 a polynomial in $x_1,\ldots,x_n$ which is just the Weyl symbol of $\langle X_1^{\alpha_1} X_2^{\alpha_2} \cdot\cdot X_n^{\alpha_n}\rangle_\phi = X_1^{\alpha_1} X_2^{\alpha_2} \cdot\cdot X_n^{\alpha_n}$.

6. Twisted products

The idea to derive the twisted product for polynomials starting from exponential series is the key for very simple proofs. First we show that for Weyl ordering the twisted product of Eq. 2 is given by Eq. 3, where D has to be the nonlinear part of the BCH–formula. So we start with an Ansatz that the twisted product is given by an operator e^{xD} for an appropriate D and we have to demand

$$\Omega(e^{\xi x}) \circ \Omega(e^{\eta x}) = \Omega(e^{x\, D(\partial_{x'},\, \partial_{x''})}\, e^{\xi x'} e^{\eta x''}|_{x'=x''=x}) \,. \tag{18}$$

Note that the following identity is valid.

$$e^{x\, D(\partial_{x'},\partial_{x''})}\, e^{\xi x'} e^{\eta x''}|_{x'=x''=x} = e^{xD(\xi,\eta) + \xi x + \eta x} \tag{19}$$

This identity may be easily verified by comparison of coefficients in the exponential series. Substitution of Eq. 19 into the r.h.s of Eq. 18 and Weyl quantization (Ω) gives

$$e^{\xi X} e^{\eta X} = e^{(\xi + \eta + D(\xi,\eta))X} . \tag{20}$$

Hence, $D(\xi,\eta)$ has to coincide with the nonlinear part $\tau(\xi,\eta)$ in the BCH-formula.

In order to justify the notation used, we remember that according to the BCH-formula $e^{\xi X}e^{\eta X}=e^{(\xi+\eta+\tau(\xi,\eta))X}$ the nonlinear part $\tau(\xi,\eta)$ defines an analytic map from L x L into L. $\tau(\xi,\eta)$ may be obtained by an absolutely convergent series $\tau(\xi,\eta) = \sum_{n=2}^{\infty} c_n(\xi,\eta)$, where c_n are recurrently defined polynomial maps from L x L into L of degree n. We refer to /11/ for details. Formal differentiation $\frac{\partial^{|\alpha|}\partial^{|\beta|}}{\partial\xi^{\alpha}\,\partial\eta^{\beta}}\ldots|_{\xi=\eta=0}$ applied on the rewritten Eq. 18

$$\Omega(e^{\dot{x}\,D(\partial_{x'},\partial_{x''})}\, e^{\xi x'}e^{\eta x''}|_{x'=x''=x}) = e^{\xi X}\, e^{\eta X} \tag{21}$$

gives

$$\omega(x^{\alpha} \bullet x^{\beta}) = \omega(x^{\alpha})\, \omega(x^{\beta}) \tag{22}$$

according to Eq. 7. So the twisted product of Eq. 2 coincides with that of Eq. 3 due to linearity of ω. □

7. Conclusions

If the basis of U(L) is given not in the Weyl ordering but in the standard ordering (resp. ϕ-ordering) there exist in principle two procedures for the multiplication of two elements of U(L).

Algorithm I:

a) Replace the elements of U(L) by their symbols according to the ordering rule given. These ϕ-symbols are immediately given by replacing the capital X's by small ones.

b) Transform the ϕ-symbols by $e^{x'\phi(\partial_x)}\ldots|_{x'=x}$ into Weyl symbols.

c) Compute the twisted product. The result is given by its Weyl symbol.

d) Transform the Weyl symbol to a ϕ-symbol by $e^{-x'\phi(\partial_x)}\ldots|_{x'=x}$.

e) The (commutative) ϕ-symbol written in ϕ-ordering gives just the result by replacing small x's by capital ones.

Algorithm II:

This algorithm differs from Algorithm I by the fact that one defines a twisted product immediately for ϕ-symbols. This may be done by the same techniques and it will be presented in a forthcoming paper.

Note that the transformation between different orderings helps also if another fast multiplication for a given basis of U(L) is available in a computer algebra system.

The Goldberg-Eriksen algorithm for BCH-series is an effective one/12/ for the computation of D up to a certain order/13/. Several simplification for the twisted product realization in REDUCE/14/ and criteria for the truncation of the series are applicable/13/. Note that the twisted product operator acts on polynomials. So the BCH-series for D and the exponential series e^D are used only up to a certain order in dependence of the degree of the polynomials.

8. References

1. W. Lassner, Symbol representations of non-commutative algebras, Lect. Notes in Comp. Sci. **204** (1988), 99-116.
2. J. Apel, W. Lassner, An extension of Buchberger's algorithm and calculations in enveloping fields of Lie algebras, J. Symb. Comp. 6 (1988), 361-370.
3. K. B. Wolf, "Group-theoretical methods in aberrating systems" in Proc. of XIII Int. Colloq. on Group-Theor. Methods in Physics (1984), (World Scientific, Singapore, 1985).
4. G. S. Agarwal, E. Wolf, Phys. Rev.**D2** (1970), 2161, 2187, 2207.
5. C. M. Bender, G. V. Dunne, Polynomials and operator ordering, J. Math. Phys. **29**(8) (1988), 1727-1731.
6. M. García-Bulle, W. Lassner, K. B. Wolf, The metaplectic group within the Heisenberg-Weyl ring, J. Math. Phys. **27**(4) (1986), 29-36.
7. J. Apel, U. Klaus, *Implementational aspects for non-commutative domains,* (in this volume).
8. J. Dixmier, *Algèbres enveloppantes,* (Gauthiers-Villars, Paris 1974).
9. L. Abellanas, L. Martinez Alonso, *Quantization from the algebraic view point,* J. Math. Phys. 17(8) (1976), 1363-1367.
10. U. Petermann, J. Apel, *A program for algebraic computations in quotient skew fields of enveloping algebras of Lie algebras - An application of LOGLAN 82 to symbolic computation,*(Univ. Leipzig, KMU-NTZ-88-02, 1988).
11. V. S. Varadarjan, *Lie groups, Lie algebras and their representation,* (Prentice-Hall, Englewood-Cliffs, 1974).
12. L. Stannarius, *Algorithmen zur Berechnung der Baker-Campbell-Hausdorff-Reihe,* Diplom-Thesis (Univ. Leipzig 1986).
13. K. Zies, *Erzeugung von Twistproduktoperatoren für Elemente aus Enveloping-Algebren,* Diplom-Thesis (Univ. Leipzig 1988).
14. A. C. Hearn, *REDUCE User's Manual.Version 3.3.*(The Rand Corporation, Santa Monica,1987).

CONSTRUCTION OF A LIE ALGEBRA BY A SUBSET OF GENERATORS AND COMMUTATION RELATIONS

I.R.Akselrod, V.N.Robuk
Phys.-Techn. Inst. of Low Temperature,Kharkov,USSR.

V.P.Gerdt
Joint Institute for Nuclear Research
Head Post Office P.O.Box 79, Moscow, USSR

V.E.Kovtun
Kharkov State University,Kharkov,USSR.

Abstract

The problem of constructing the quotient algebra for a free Lie algebra over an ideal given by a subset of generators and commutation relations is investigated. The method proposed to solve this problem can be applied in particular for finding a L-A pair for nonlinear evolution equations. The algorithm is based on the concept of Hall basis for a free Lie algebra and is implemented in the computer algebra system REDUCE.

§1 Introduction

In the present paper we describe approach to investigating the problem of solvability of a given system of nonlinear algebraic equations in the class of Lie algebras over field C of complex numbers. We use the concept of the Hall basis for a free Lie algebra [1,2].

The indicated problem is very actual, in particular, for Walhquist-Estabrook (W.-E.) method in the theory of nonlinear partial differential equations [3]. In fact, numerous successful attempts, including computer-aided ones [4], to realize W.-E. scheme in Lie algebras were based on *a priori* additional assumptions on a class of Lie algebra for the solution. It could lead to complete or partial loss of solutions unlike our approach which is directed to obtaining all the solutions of an initial algebraic system. It should be noted that in finite-dimensional case the effective algorithms are developed for the problem of construction of matrix representations for an algebra generated by a finite set of polynomials [5].

The method is described below for regenerating Lie algebras can be applied also for infinite-dimensional case and has an algorithmic form as well. We present our algorithm and its realization in REDUCE [6].

§2 Main notations and definitions.

Let L be Lie algebra over the field C and $[.,.]$ denotes Lie product. The set $X=\{x_1,...,x_k\}$ of *generators* ("alphabet") is a set of elements of L from which any other element may be constructed by Lie product, addition and multiplication by complex numbers from C. A *basis* $R(X)$ of L is a minimal set of elements such that any other element is a linear combination of them. A *Lie monomial* ("word") $m(X)$ is any element from L constructed of the generators x_i by Lie products. *Lie polynomial* $P(X,C)$ is a sum of Lie monomials with a set of coefficients $C(c_1,...c_k) \in \mathbb{C}$.

Determining relations ("phrases") are Lie polynomial equalities of the form $P_i(X,C)=0$. In the general case a set of determining relations leads to a system of algebraic equations in unknowns $c_i \in \mathbb{C}$, $x_k \in L$. If the values c_i are determined a Lie algebra is given by factorization a free Lie algebra according to an ideal given by a set of generators and determining relations.

The algorithm proposed below for solving this problem is based on the concept of "Hall structure" [1] of a Lie monomial. A *Hall family of bases* of a free Lie algebra with an alphabet X is a set R of Lie monomials with a linear ordering $\leq$ such that the following conditions are satisfied:

1. $X \subseteq R$.
2. An element m of the form $[m_1,m_2] \in R$ iff
   (a) $m_1, m_2 \in R$,
   (b) $m_1 < m_2$,
   (c) if $m_2 = [m_3,m_4]$ then $m_1 \geq m_3$.
3. $[m_1,m_2] > m_1$.

for basis elements in form of Lie monomials of the structure given above.

Some of the Lie monomials from R may remain free when the free Lie algebra will be factorized. These Lie monomials will play the role of basis elements for the quotient Lie algebra. In addition to it each of these monomials is either a Lie product of a pair of other basis elements or a separate generator which can be treated as a pair as well. This face follows from the definition given above. Therefore we shall call basis monomials free pairs. R^f will denote a set of such pairs. Correspondingly R^b will denote the set of pairs $[m_1,m_2]$, such that $m_1,m_2 \in R^f$, $m_1<m_2$ and $[m_1,m_2] \notin R^f$. The elements of R^b are called bound ones. In the process of computation the bound pairs are expressed in terms of linear combinations of basis elements with coefficients

from C. Expressions of such a kind will be denoted phrases. Thus $R^f \cup R^b$ is a commutant of the desired Lie algebra together with generators.

The concept of weights $V(x_i)$ for generators is useful for subsequent descriptions. That means to each generator a certain natural number is assigned. The weight of a Lie monomial is defined as the sum of weights of its constituents (letters), for example:

If $V(x_1)=1$, $V(x_3)=2$, then $V([x_1,[x_1,[[x_1,x_3]x_3]]])=7$.

§3 Algorithm description

Let R_k be a subset of $R^f \cup R^b$, such that $[u,v] \in R$, if $u,v \in R^f$ and $V(u)+V(v)= k$ $(k \in \mathbb{Z})$. Then $R^f = \bigcup_k R^f_k$; $R^b = \bigcup_k R^b_k$; $R^f \cup R^b = \bigcup_k R_k$. The problem is to construct $n+1$-th row R_{n+1} for given n rows, i.e. for $\bigcup_{k=1}^{n} R_k$.

An algorithm for solving this problem starts with a ordering by numeration of the words of Lie algebra L. The search for a given word in whole list of words is performed by its number. For abbreviation, we shall denote a word m by a number in parenthesis (i), $i \in N$, omitting the symbol m. Then any basis element of the form $[m_i,m_j] \in R$ will have the form of a pair of number (i,j). We call this pair free if it is a basis element and therefore can be found by its number in the set R. A bound pair from L does not have its own number. Jacobi identities in L have to be fulfilled.

The ordering has to be a Hall ordering. Therefore it is necessary to know also subwords of a word represented by a number pair. Let us describe the procedure of construction of R_{n+1}, the row with weight $n+1$ by given rows with weight $1,2,\dots,n$.

Step 1. Generate a word with weight $n+1$ from free words. This can be done, for example, by adding a word with weight $k+1$ to a word with weight $n-k$. Since the number of the first word is smaller than the number of the second one, the rule *(a)* in the definition of the Hall basis family (§2) is satisfied.

Step 2. Assign to pairs, obtained at the previous step, numbers in accordance to the consecutive numbering of the basis elements. A number is assigned to a word if the two other rules *(b)* and *(c)* in the Hall basis definition are satisfied. If these rules are not satisfied one has to make substitutions of the subwords which are obtained from Jacobi identities (Lie differentiation):

$[a,[b,c]] := [[a,b],c] - [[a,c],b]\ ,\ a,b,c \in N$.

The appearing pairs $[a,b]$ and $[a,c]$ are replaced by a word having a number.

Step 3. Construct a word with weight $n+1$ from free pairs with weight $k+1$ and bound pairs with weight $n-k$. Add from the left words with a number from the initial data in such a way that the resulting weight is still equal to $n+1$. Differentiate the word with weight $n+1$ in the same way as it has been done at step 2. On other hand the same word can be generated from a Lie product by means of adding a free pair, i.e a word with a number, to the right side of a bound pair. Everywhere substitute words with numbers instead of number pairs. To solve the obtained equation with respect to the word with leading number. This equation is called "key phrase". There occurs a branching of the problem into parts with different solutions. If the leading word has a number less than $n+1$ the process is interrupted. It means that the problem must be solved considering the new phrase that has been obtained. In other words it is necessary to start with new initial data. Other solutions, leading essentially to new tasks, will appear from an analysis of the letter coefficients of a key phrase.

Step 4. Initial data are supplemented by the addition of the row with weight $n+1$ and then the next rows with weights $n+2,n+3,...$ are computed. The process of computation is finished when, beginning with some row, there are no free words or when we obtain enough data to construct the Lie algebra by induction.

Now we can present our algorithm in a REDUCE-like form:

Comment: input of initial data, contents of first i rows R_i and their lengths $l_i = |R_i|\ ,\ i = 0,1,...,n$.;

Input : $R_i\ ,\ i = 0,1,...,n\ ;\ l_i = |R_i|$
Output : $R_{i+1}\ ,\ l_{i+1} = |R_{i+1}|$

```
for k := 1 : n do                          % start of steps 1 and 2;
    for j1 := 1 : l_{k+1} do               % generation of words with weight= n+1;
        for j2 := 1 : l_{n-k} do
        if j1 < j2 then                    % verification of Hall conditions;
        if x_{j1} ∈ R^f_{k+1} and
        x_{j2} ∈ R_{i-k} then
        if x_{j1} > y_1 and y_2 = 0        % y_1 := car x_{j2} , y_2 := cadr x_{j2} ;
        then
```

```
        [x_j1 ,x_j2] := Σ_{s=1}^{i} l_s + N    % assignment of ordinal numbers to pairs;
        else
        [x_j1,[y_1,y_2]] := [[x_j1,y_1],y_2]-[[x_j1,y_2],y_1]
end;
```

Comment: We obtain all possible pairs with weight $n+1$ as a result of steps 1 and 2;

```
for k := 1 : n do                 % generation of words, weight= n+1 ;
    for j1 := 1 : l_{k+1} do
        for j2 := 1 : l_{n-k} do
        if j1 < j2 then           % selection of pairs, satisfying condition:;
        if x_j1 ∈ R^f and         % first subword is basis element, second ;
        x_j1 < y_1 and            % subword does not satisfy to the Hall ;
        x_j2 ∈ R^b                % conditions ;
        then                      % y_1 := car x_j2 ;
        { 0 := [x_j1,x_j2]-[x_j1,[y_1,y_2]] }
end;
```

Comment: At step 3 we obtain equations in coefficients c_i and basis elements of L. Their analysis and solving are carried out in the block *{...}* . It is realized in the following way. The basic element with the leading word is selected and then it is expressed in terms of the remaining elements of the phrase. It could not be done if all monomials in the equation contain coefficients c_i or they do not belong to the row with weight $n+1$.;

Output 1: R_{n+1} , $l_{n+1} = |R_{n+1}|$ % row with weight $n+1$ and its length ;

Output 2: "key phrase"

Comment: Further analysis of the key phrase and rows are done "by hand" ;
end of algorithm.

We have implemented the above algorithm in the REDUCE computer algebra system. An interactive regime should be used. The user of the program will obtain relevant information from the process of row generations and the key phrases analysis in order to reconstruct the Lie algebra from the given initial data.

As an illustration let us consider one of the problem has been investigated in [7]. The determining relations for that problem are

$$[x_1,[x_2,x_1]] - c_1 [x_2,x_1] + c_3x_2 = 0$$
$$[x_2,[x_2,x_1]] - [x_1,x_3] - c_4x_2 = 0$$
$$[x_2,x_3] - c_2[x_2,x_1] - c_5x_2 = 0$$

There is four different solutions. One of them is $c_3=c_4=0$ (other c_i are arbitrary). Lie algebra basis: x_2,x_3, $ad^k x_2(x_1),\{k\}_0^\infty$. Let $y_k = ad^k x_2(x_1)$. Then the Lie algebra has infinite dimension and the following structure:

$$[x_2,x_3] = c_1y_1 + c_5x_2,\ [x_2,y_k] = y_{k+1},\ [y_0,y_n] = c_1y_n,$$
$$[y_0,x_3] = y_2,\ [y_n,x_3] = y_{n+2} + (nc_5 - c_1c_2)y_n,$$
$$[y_p,y_n] = 0,\quad \{n\}_1^\infty,\ \{p\}_1^\infty.$$

To obtain this solution, the knowledge of twelve rows is necessary. The following table gives computing time on an IBM PC AT (12 Mhz):

row number	time in secs	comment
4 5 6	10 17 25	before the key phrase is produced
4 5 6 7 8 9 10 11 12	10 13 18 25 32 40 51 62 77	after substituting the conditions $c_3 = c_4 = 0$ from solution 3 in initial data

§4 Conclusion

The algorithm does not pretend to be complete and effective for any problem of Lie algebra reconstructing by a given subset of generators and determining relations. The reasons for such a conclusion are the following:

1) The key phrases, which are obtained in the computation process, may lead to complicated systems of nonlinear algebraic in scalar variables. To solve them the Groebner basis technique [8] has to be used. It is remarkable that one can apply to these problems the same computer algebra methods as in the problems of investigating integrability of polynomial-nonlinear evolution

equations with arbitrary scalar coefficients [9].

2) Our algorithm (§3) uses only a part of facilities given by the Hall basis definition (§2). The third point of that definition has been enhanced by the condition $[m_1,m_2] > m$, for $V(m_1)+V(m_2) > V(m)$. That may restrict the possibility of optimizing the computational process.

3) Possible fast growing of the number of basis elements makes serious difficulties in analysis of the results of computation by induction.

In future we hope to weaken the restrictions 1)-3) by means of an extension of the algorithm with a special tool for analyzing and solving algebraic equations similar to that presented in [9].

References

[1] Hall M. The Theory of Groups. The Macmillan Co.,New York,1959.

[2] Bakturin Yu.A. Identities in Lie Algebras.Nauka.M.1985(in Russian).

[3] Walhquist H.D.,Estabrook F.B. Prolongation Structures of Nonlinear Evolution Equations. *J.Math.Phys.*16,1,1-7,1975.

[4] Reiman A. Computer-Aided Closure of the Lie Algebra Associated with a Nonlinear PDE. Proc.of the 1989 MACSYMA Users Conference. MIT, Cambridge,Mass.,1979,385-396.

[5] Labonte G. An Algorithm for the Construction of Matrix Representations for Finitely Presented Non-commutative Algebras. *J.Symb.Comp.*9,1, 27-38,1990.

[6] Hearn A.C. REDUCE User's Manual.Version 3.3. The Rand Corporation, Santa Monica,1987.

[7] Robuk V.N. On Classification of Second Order Evolution Equations. In: Operators in Functional Spaces and Problems in Theory of Functions. (V.A.Marchenko,ed.). Kiev,Naukova Dumka,1987,58-66 (in Russian).

[8] Buchberger B. Groebner Bases: an Algorithmic Method in Polynomial Ideal Theory. In:(Bose N.K.,ed.) Recent Trends in Multidimensional System Theory, Reidel,1985.

[9] Gerdt V.P.,Khutornoy N.V.,Zharkov A.Yu. Solving Algebraic Systems which Arise as Necessary Integrability Conditions for Polynomial-Nonlinear Evolution Equations. In this volume.

A PACKAGE FOR COMPUTATIONS IN SIMPLE LIE ALGEBRA REPRESENTATIONS

A.A.Zolotykh
Moscow State University, Moscow

Let L be a simple finite-dimensional Lie algebra over an algebraically closed field of characteristic zero (for example, over the field of complex numbers). It is well known that any finite-dimensional irreducible representation of L is determined by its highest weight.

The aim of this paper is to announce the package of applied programs that was worked out by the author.

The package means to be used for calculation of some integer-valued characteristics of simple Lie algebras of rank not more than 12 and their irreducible finite-dimensional representations.

Input and output information is in terms of highest weights. This package gives us possibilities to calculate the following characteristics:

- integer-valued characteristics of Lie algebras (dimensions of Lie algebras, Dynkin diagrams, Kartan matrices, dimensions of fundamental representations, orders of Weyl groups, highest roots of Lie algebras, lists of all positive roots);
- lists of minimal representations of Lie algebras, dimensions and characters of minimal representations (all representations are ordered by their dimensions);
- dimensions of Lie algebra representations;
- characters of Lie algebra representations;
- decompositions of tensor products of representations into irreducible components.

The methods of computation are grounded on the well known Weyl formula for dimensions, on the recursion formula for characters[1] and on the Klimyk method for tensor products[2]. More detailed description of tensor product algorithm was published by the author in preprint[3].

Dynkin diagrams for simple Lie algebras with used root-numbering conventions are given in table 1. The shaded circles denote short

Table 1

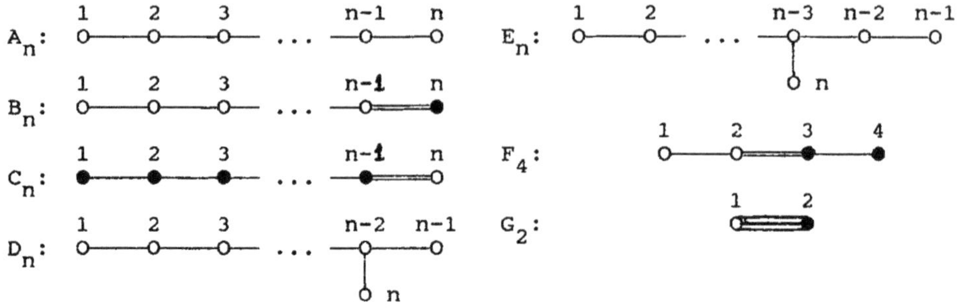

Table 2

Type of Lie algebra	Number of basic representation	Dimension	Time of character computation (seconds)	Time of tensor square computation (seconds)
E_6	1	27	2	2
	2	351	2	2
	3	2925	2	3
	4	351	2	2
	5	27	2	2
	6	78	2	2
E_7	1	56	2	2
	2	1539	2	4
	3	27664	2	7
	4	365750	2	21
	5	8645	2	5
	6	133	2	2
	7	912	2	3
E_8	1	248	2	3
	2	30380	2	9
	3	2450240	2	92
	4	146325270	3	668
	5	6899079264	4	3181
	6	6696000	2	148
	7	3875	2	4
	8	147250	2	22
F_4	1	52	2	2
	2	1274	2	3
	3	273	2	2
	4	26	2	2
G_2	1	14	2	2
	2	7	2	2

Table 3

Type of Lie algebra	Number of basic representation	Dimension	Time of character computation (seconds)	Time of tensor square computation (seconds)
A_{12}	1	13	2	3
	2	78	2	3
	3	286	2	3
	4	715	2	4
	5	1287	2	4
	6	1716	2	5
	7	1716	2	5
	8	1287	2	4
	9	715	2	4
	10	286	2	3
	11	78	2	3
	12	13	2	3
B_{12}	1	25	2	3
	2	300	2	4
	3	2300	2	6
	4	12650	3	12
	5	53130	3	36
	6	17710	3	94
	7	48070	3	198
	8	10815	3	332
	9	20429	3	452
	10	32687	3	525
	11	44574	3	551
	12	4096	2	7
C_{12}	1	24	2	3
	2	275	2	4
	3	2000	2	5
	4	10350	2	10
	5	40480	2	28
	6	12397	2	67
	7	30360	2	129
	8	60087	2	195
	9	96140	3	244
	10	12257	3	264
	11	11886	3	268
	12	74290	3	270
D_{12}	1	24	2	3
	2	276	2	3
	3	2024	2	5
	4	10626	2	10
	5	42504	2	29
	6	13459	2	70
	7	34610	2	135
	8	73547	2	205
	9	13075	3	256
	10	19612	3	278
	11	2048	2	5
	12	2048	2	5

roots.

The package is written in C language and contains more than 4300 operators. It can be used on computers of types IBM PC.

Times of character computations and tensor square computations for the fundamental representations for IBM PC AT (processor 286) are given in table 2 (fundamental representations of exceptional Lie algebras) and in table 3 (fundamental representations of Lie algebras of rank 12). These tables contain also the dimensions of fundamental representations.

The author plans in the nearest future:

- to extend the package for semisimple Lie algebras;
- to add the algebra-subalgebra reduction program connecting representations of Lie algebras and their subalgebras;
- to add the program of tensor degrees, inner degrees and symmetric degrees computation.

More detailed information can be received in the laboratory of computational methods of the mechanical-mathematical department of Moscow State University.

References

1. Moody R.V. and Patera J., Fast Recursion Formula for Weight Multiplicities, Bull. Amer. Math. Soc., 7 (1982), 237 - 242.

2. Klimyk A.U., Decomposition of a Tensor Product of Irreducible Representations of a Semisimple Lie Algebras into a Direct Sum of Irreducible Representations, Ukrainsk. Matem. Zhurn., 18 (1966), N 5, 19 - 27 (Amer. Math. Soc. Transl., Ser. 2, 76 (1968), 63 - 73).

3. Zolotykh A.A., Decomposition of tensor products of representations of Lie algebra E_8, Preprint of VINITI, Moscow State Univ., 1988, Published 28.11.88, N 8360-V88.

AN ALGORITHM FOR REDUCING A BILINEAR ALGEBRAIC SYSTEM INTO TRIANGULAR FORM

Z.T.Kostova, R.M.Yamaleev
Joint Institute for Nuclear Research, Dubna
Head Post Office P.O.Box 79, Moscow, USSR

Abstract

In [1] the method of matrix linearization was proposed as a method of finding general solution of nonlinear algebraic system in the matrix representation was proposed. One of the principal applications of this method is the separation of unknowns in the nonlinear algebraic system.

The main goal of this paper is to show, that matrix linearization method, in principle, is applicable to the algebraic systems, which contain the polynomials of arbitrary form. Here we propose a new formulation of this method, using the vector representation of coefficients of the initial algebraic system.

This representation is very simple and convenient for the realization of the computer algebra methods in solving the nonlinear algebraic systems.

<u>The basic elements of the matrix linearization method.</u>

This method is described in details in [2,3]. We recall the essential part of this method, which is convenient for our purposes .

<u>Proposition 1</u>

There is an algorithm, which for any given M-dimensional nonlinear polynomial algebraic equation of degree N

$$f_1(x_1 , x_2 , \dots , x_m) = 0 \tag{1}$$

associates some bilinear equations with matrix coefficients:

$$[\sum_{m=1}^{M} x_m A_{ml}^{(1j)} + A_{0l}^{(1j)}] \Psi_{(j)l} = 0 \tag{2}$$

such that

$$\mathrm{Det} [\sum_{m=1}^{M} x_m A_{ml}^{(1j)} + A_{0l}^{(1j)}] = f_l \tag{3}$$

The systems (1) and (2) are equivalent, but (2) is more convenient for our purposes. The representation (2) allows us to separate the unknowns if the matrices $A_{mm}^{(1j)}$ are invertible. The second condition imposes that the vectors $\Psi_{(j)l}$ must consist of at least most one nontrivial element or, satisfy the normalization condition. If these two conditions are fulfilled ,then the process of separation of unknowns may be realized with the help of usual methods of matrix algebra.

In our case,for the purposes of computer algebra application, another representation of system (2) is more convenient i.e.the representation with vector coefficients:

$$\sum_{m=1}^{M} x_m X_{ml}^{(l)} + X_{0l}^{(l)} = 0 \tag{4}$$

where $X_{ml}^{(l)} = A_{ml}^{(lj)} \Psi_{(j)l}$

Actually, the system (4) is already in a suitable form to realize the procedure of separation of unknowns. We will attempt to solve the system (4) in the vector form with components $\{x_1, x_2, \ldots, x_m\}$.(One can note, that for the coefficients $X_{ml}^{(l)}$ we have only one operation - the tensor product operation).

Proposition 2

Using the generalization of the Crammer rule [4] for the system with noncommutative elements we can transform the system (4) to the system of the following form

$$\kappa_m \Phi = F_m , \tag{5}$$

where Φ and F_m are vectors with L^M elements.

Further it is essential for us to introduce the notion of determinant of the matrix which contains the elements like $X_{ml}^{(l)}$.

Definition

Let matrix A be defined by the following elements, which are vectors:

$$A := v;X^{(i)}_{ml}\varphi\ ,\ (\ m,l = 1,2,\ldots,\ M\ )\ ,\ (\ i = 1,2,\ldots,\ L\ )].$$

Then we will define Det A by

$$\mathrm{Det}\ A = \sum_{p} (-1)^{t(p)}\ X_{1j_1} \otimes X_{2j_2} \otimes \ldots \otimes X_{nj_n} \qquad (6)$$

(function t(p) is already defined in [4]).z'

This expression can be rewritten in another form

$$\mathrm{Det}\ A = \sum_{m=1}^{M} X_{1m}\ A_{m1}\ ,,$$

where X_{1m} (m = 1,...,M) are the elements of the first line, A_{kl} - is the cofactor of X_{kl} .

One can verify, that matrix $A^{v} = v\ A_{kl}\ v$ is the adjoint matrix, of the matrix A so that

$$A\ A^{v} = A^{v}A = \mathrm{Det}\ A \qquad (7)$$

Using these formulas we can transform the system (4) to the following form:

$$x_m \mathrm{Det}\ v;X_{kl}\ \varphi = -\sum_{l=1}^{M} X_{ol} \otimes A_{lm} \qquad (8)$$

The comparison of (8) with (5) gives:

$$\Phi = \mathrm{Det}\ v;X_{kl}\ \varphi\ ,\ F_m = -\sum_{l=1}^{M} X_{ol} \otimes A_{lm}$$

According to (6), if the vector $X^{(1)}_{kl}$ has an order L, then the order of vectors Φ and F_m is equal to L^{M}.

In the next step we can find the general basis Π, which satisfies the following equations:

$$\Phi = \hat{C}\ \Pi\ ,\ F_m = \hat{B}_m \Pi \qquad (9)$$

where $\hat{C}$ and $\hat{B}_m$ are matrices of order L^{M}.

<u>Proposition 3</u>

Vector Π , which obeys the conditions (9), is defined by the following tensor product

$$\Pi = \Psi_1 \otimes \Psi_2 \otimes \ldots \otimes \Psi_M \qquad (10)$$

As the vectors Ψ_1, Ψ_2,.... ,Ψ_M are normalized hence Π is also normalized.

Let us consider Π as unknown and solve the system (5). Then we have

$$F_m(x_m) := \det(x_m \hat{C} - \hat{B}_m) = 0 \tag{11}$$

$$\Pi^{(p)} = \Pi^{(p)}(x_m) \tag{12}$$

Note that the equation (11) is a one-dimensional polynomial equation and eq.(12) defines the dependence of the components of the vector Π on the x_m.

Solving (11) and using (12) we can obtain the complete information about the all solutions of the system (1).

Example

$$x^2 + 3xy - 5xz + 2yz + x - 3z + 1 = 0$$

$$y^2 + 7xy + 8xz - 14yz + 9x + 13y - 12z - 12 = 0$$

$$z^2 + 9xy + 8xz + 7yz + 6x + 3y - 14z - 20 = 0$$

The one-dimensional polynomial which corresponds to this system is:

$$f(z) := -775877/4\ z^8 + 4639387/2\ z^7 + 316744371/2\ z^6 + 51218771/2\ z^5 - 1644304407/4\ z^4 - 120767319/2\ z^3 + 257195327\ z^2 + 47880637/2\ z + 8433641/2 = 0$$

References

1. Yamaleev R.M. JINR, R11-85-815, Dubna,1985 (in Russian).
2. Yamaleev R.M. JINR, R15-86-250, Dubna,1986;
   R5-86-833,Dubna,1986 (in Russian)
3. Semerdjiev Kh.I.,Yamaleev R.M. SERDIKA,v.13,p.272,1987.
4. Lancaster P. Theory of matrices,Academic Press,New York,1969.

SOLVING ALGEBRAIC SYSTEMS WHICH ARISE AS NECESSARY INTEGRABILITY CONDITIONS FOR POLYNOMIAL-NONLINEAR EVOLUTION EQUATIONS

V.P. Gerdt , N.V. Khutornoy
Laboratory of Computing Techniques and Automation
Joint Institute for Nuclear Research
Head Post Office, P.O. Box 79, Moscow, USSR
A.Yu. Zharkov
Saratov State University, Astrakhanskaya 83, Saratov, USSR

Abstract

We solve, by using computer algebra, some systems of algebraic equations which arise in the investigation of integrability of evolution equations with polynomial-nonlinearity. These systems are equivalent to the first necessary integrability conditions for three multiparametric types of evolution equations with uniform rank: a seventh order family of KdV-like scalar equations, a seventh-order family of MKdV-like scalar equations, and a third order family of coupled KdV-like systems. The corresponding algebraic systems have infinitely many solutions. These solutions have been found with the help of the computer algebra system REDUCE by applying the Groebner basis method.

1. Introduction

The investigation of the problem of integrability of nonlinear evolution equations (NLEEs) can often be reduced to the problem of finding the exact solution of a complicated systems of nonlinear algebraic equations. Such systems arise in particular in the problem of classification [1] of integrable NLEEs of the form

$$\begin{aligned}&u_t = \Delta u_N + F(u, u_1, ..u_{N-1}; \lambda_1, ..\lambda_K),\\ &u = u(x,t) = (u^1, ..u^M), \; u_i = \mathbf{D}^i(u), \; \mathbf{D} = d/dx, \; F = (F^1, ...F^M),\\ &\Delta = diag(\delta_1, \delta_2, ...\delta_M), \; \lambda_i, \delta_j \in \mathbf{C}, \; \delta_i \neq 0, \; \delta_i \neq \delta_j \; (i \neq j)\end{aligned} \tag{1}$$

where F is a polynomial in its arguments. In order to be integrable, (1) must satisfy the necessary integrability conditions which follow from the existence of higher infinitesimal symmetries and which have the form of conservation laws

$$\frac{d}{dt}(R(i,j)) \in Im\mathbf{D}, \; i \in \mathbf{N}, \; j = 1, 2, ..., M \tag{2}$$

The densities $R(i,j)$ depend on a finite number of dynamical variables taken from an infinite set $u, u_1, u_2, \ldots$. Any of the conditions (2) means that the l.h.s. is a total derivative of some other function of a finite number of dynamical variables with respect to x. If arbitrary numerical parameters (λ_i and/or δ_j) are present in (1) the conditions (2) lead to a system of algebraic equations in these parameters. The densities $R(i,j)$ can be calculated, and, in the absence of arbitrary parameters, the conditions (2) can be verified completely automatically by using special computer algebra packages [2, 3]. In the presence of arbitrary parameters, in order to select from (1) all integrable NLEEs, we have to find all the solutions of the algebraic system which is equivalent to (2).

Finding the solution of a system of algebraic equations is one of the most interesting area of application of computer algebra. The main results in this field have been achieved by using the fundamental results obtained by Buchberger on the Groebner basis method [4] and in a number of other papers, for example, [5]-[9].

Here we apply this method to obtain all (infinitely many) solutions of the systems of algebraic equations which are obtained by the above procedure for the classification of the three different multiparametric families of NLEEs: the seventh order scalar KdV-like equations, the seventh order MKdV-like equations, and the third order coupled KdV-like systems. These equations have certain homogeneity properties with respect to the scale transformation $x \Rightarrow \alpha x$, $u \Rightarrow \beta u$ and form an important subclass of (1). They are called NLEEs with uniform rank [10].

We would like to emphasize that all our computations have been carried out by using the computer algebra system REDUCE (version 3.2) on an IBM PC AT-like computer.

2. Systems of Algebraic Equations from Classification Problems

In this section we give an explicit form of three different systems of algebraic equations which arise in classification problems [11]. These systems and others which follow from the necessary integrability conditions (2) are generated completely by computer [3]. In general, each condition (2) generates several algebraic equations, some of them may be identical with others which follow from a different condition (2). In order to simplify the generated system, it is sufficient to reduce the input polynomials according to the prescription of the Buchberger algorithm [4].

I.The seventh order scalar NLEEs of KdV type (M=1,N=7 in (1))

$$u_t = u_7 + \lambda_1 u u_5 + \lambda_2 u_1 u_4 + \lambda_3 u_2 u_3 + \lambda_4 u^2 u_3 + \lambda_5 u u_1 u_2 + \lambda_6 u_1^3 + \lambda_7 u^3 u_1. \quad (3)$$

The first seven conditions (2) $(0 \le i \le 6)$ lead to a system of thirteen algebraic equations in seven variables

$$\begin{aligned}
&\lambda_1(\lambda_4-\lambda_5/2+\lambda_6)=(2/7\lambda_1^2-\lambda_4)(-10\lambda_1+5\lambda_2-\lambda_3)=0,\\
&(2/7\lambda_1^2-\lambda_4)(3\lambda_4-\lambda_5+\lambda_6)=0,\\
&a_1(-3\lambda_1+2\lambda_2)+21a_2=a_1(2\lambda_4-2\lambda_5)+a_2(-45\lambda_1+15\lambda_2-3\lambda_3)=0,\\
&2a_1\lambda_7+a_2(12\lambda_4-3\lambda_5+2\lambda_6)=b_1(2\lambda_2-\lambda_1)+7b_2=b_1\lambda_3+7b_2=0,\\
&b_1(-2\lambda_4-2\lambda_5)+b_2(2\lambda_2-8\lambda_1)+84b_3=0,\\
&b_1(8/3\lambda_5+6\lambda_6)+b_2(11\lambda_1-17/3\lambda_2+5/3\lambda_3)-168b_3=0,\\
&15b_1\lambda_7+b_2(5\lambda_4-2\lambda_5)+b_3(-120\lambda_1+30\lambda_2-6\lambda_3)=0,\\
&-3b_1\lambda_7+b_2(-\lambda_4/2+\lambda_5/4-\lambda_6/2)+b_3(24\lambda_1-6\lambda_2)=0,\\
&3b_2\lambda_7+b_3(40\lambda_4-8\lambda_5+4\lambda_6)=0,
\end{aligned}\tag{4}$$

where

$$\begin{aligned}
&a_1=-2\lambda_1^2+\lambda_1\lambda_2+2\lambda_1\lambda_3-\lambda_2^2-7\lambda_5+21\lambda_6,\quad a_2=7\lambda_7-2\lambda_1\lambda_4+3/7\lambda_1^3,\\
&b_1=\lambda_1(5\lambda_1-3\lambda_2+\lambda_3),\quad b_2=\lambda_1(2\lambda_6-4\lambda_4),\quad b_3=\lambda_1\lambda_7/2,
\end{aligned}$$

define the structure of conservation law densities in (2) (with $R_{i+1}=R(i,1)$, $i=0,1,\ldots$).

$$R_1=\lambda_1 u,\quad R_3=(2/7\lambda_1^2-\lambda_4)u^2,\quad R_5=a_1u_1^2+a_2u^3,\quad R_7=b_1u_2^2+b_2uu_1^2+b_3u^4.$$

Densities R_i $(i=2,4,..)$ do not produce any restrictions on λ_j.

II.The seventh order scalar NLEEs of a modified KdV (MKdV) type (M=1,N=7 in (1))

$$\begin{aligned}
u_t=\;&u_7+\lambda_1(u_1u_5+3u_2u_4+2u_3^2)+\lambda_2(u^2u_5+6uu_1u_4+10uu_2u_3+\\
&6u_1^2u_3+7u_1u_2^2)+\lambda_3(2u_1u_2^2+u_1^2u_3)+\lambda_4(3u^2u_1u_3+12uu_1^2u_2+\\
&3u^2u_2^2+2u_1^4)+\lambda_5(u^4u_3+8u^3u_1u_2+6u^2u_1^3)+\lambda_6u^6u_1.
\end{aligned}\tag{5}$$

The computation of the densities R_i $(i=1,3,5)$ and the verification of (2) gives a system of nine algebraic equations in six variables

$$\begin{aligned}
&a_1\lambda_1=a_1\lambda_2+14a_2=a_1\lambda_4=a_1(6\lambda_2+2\lambda_3+3\lambda_4)+168a_2=0,\\
&a_1\lambda_5+5a_2\lambda_2=5b_1\lambda_1+21b_2=10b_1\lambda_2+14b_3=0,\\
&105b_4-5b_1\lambda_5-b_3\lambda_2=5b_1\lambda_4+2b_2\lambda_2=0,
\end{aligned}\tag{6}$$

where

$$\begin{aligned}
&a_1=-4\lambda_2+\lambda_3-2/7\lambda_1^2,a_2=\lambda_5-2/7\lambda_2^2,\\
&b_1=7\lambda_2,b_2=6\lambda_1\lambda_2-2\lambda_1\lambda_3-7\lambda_4+3/7\lambda_1^3,\\
&b_3=-42\lambda_5-6\lambda_1\lambda_4-2\lambda_2\lambda_3+9/7\lambda_1^2\lambda_2+16\lambda_2^2,b_4=-2\lambda_2\lambda_5+7\lambda_6+3/7\lambda_2^3.
\end{aligned}$$

As in the previous case, we have from the structure of the conservation law densities

$$R_1 = \lambda_2 u^2, \quad R_3 = a_1 u_1^2 + a_2 u^4, \quad R_5 = b_1 u_2^2 + b_2 u_1^3 + b_3 u^2 u_1^2 + b_4 u^6.$$

III. The system of two coupled nonlinear equations of KdV type (M=2, N=3 in (1)) which satisfies the conditions (10) for $i = 0, 1, 2, 3$ [11]

$$\begin{aligned} u_t &= \lambda_1 u_3 + \lambda_5 u u_1, \\ v_t &= (\lambda_1 - 1) v_3 + \lambda_2 u u_1 + \lambda_3 v u_1 + \lambda_4 u v_1. \end{aligned} \tag{7}$$

The conditions (2) for $i = 4, 5, 6;\ j = 1, 2$ do not depend on λ_2 and generate a system of four algebraic equations in four variables

$$\begin{aligned} &-2\lambda_4^3\lambda_1 + (3\lambda_4^2\lambda_1 - 2\lambda_4^2 - 6\lambda_4\lambda_3\lambda_1 + 6\lambda_4\lambda_3 + 6\lambda_3^2\lambda_1 - 6\lambda_3^2)\lambda_5 - \lambda_4\lambda_1\lambda_5^2 = 0, \\ &18\lambda_4^3\lambda_1^2 - 9\lambda_4^3\lambda_1 - 18\lambda_4^2\lambda_3\lambda_1^2 + 18\lambda_4^2\lambda_3\lambda_1 + 18\lambda_4\lambda_3^2\lambda_1^2 - 18\lambda_4\lambda_3^2\lambda_1 + \\ &(-27\lambda_4^2\lambda_1^2 + 24\lambda_4^2\lambda_1 - 5\lambda_4^2 + 63\lambda_4\lambda_3\lambda_1^2 - 78\lambda_4\lambda_3\lambda_1 + 15\lambda_4\lambda_3 - 63\lambda_3^2\lambda_1^2 + \\ &78\lambda_3^2\lambda_1 - 15\lambda_3^2)\lambda_5 + 9\lambda_4\lambda_1^2\lambda_5^2 = 0, \\ &-8\lambda_4^4\lambda_1 + (6\lambda_4^3\lambda_1 - 6\lambda_4^3 - 12\lambda_4^2\lambda_3\lambda_1 + 12\lambda_4^2\lambda_3 + 12\lambda_4\lambda_3^2\lambda_1 - 12\lambda_4\lambda_3^2)\lambda_5 + \\ &(5\lambda_4^2\lambda_1 - 4\lambda_4^2 - 18\lambda_4\lambda_3\lambda_1 + 18\lambda_4\lambda_3 + 18\lambda_3^2\lambda_1 - 18\lambda_3^2)\lambda_5^2 - 3\lambda_4\lambda_1\lambda_5^3 = 0, \\ &(3\lambda_1 - 5)\lambda_4^2\lambda_3 - 15(\lambda_1 - 1)\lambda_4\lambda_3^2 + 10(\lambda_1 - 1)\lambda_3^3 + \\ &(\lambda_4\lambda_3 + 3\lambda_3^2\lambda_1 - 3\lambda_3^2)\lambda_5 - \lambda_3\lambda_1\lambda_5^2 = 0. \end{aligned} \tag{8}$$

3. The Complete Set of Solutions

The systems (4),(6),(8), as well as higher necessary integrability conditions (2) for NLEEs (3),(5),(7), always have, evidently, the trivial solution λ_i=0 which corresponds to (integrable) linear evolution equations. If there is a nontrivial solution, then it follows from the structure of the multiparametric families (3),(5),(7) that there are infinitely many solutions. Indeed, these families are invariant under the scale transformation

$$x \Rightarrow \alpha x, \quad t \Rightarrow \alpha^N t, \quad u \Rightarrow \rho u, \quad v \Rightarrow \sigma v, \quad \alpha, \rho \in C, \quad \alpha \neq 0, \rho \neq 0, \sigma \neq 0, \tag{9}$$

where N is the order of NLEEs (see (1)). It is obvious that the transformation (9) does not change the property of integrability. Thus, the set of solutions is invariant under the corresponding scale transformation of λ_i.

Because of the fact that the computer algebra system REDUCE 3.2 (in particular on IBM PC, and unlike REDUCE 3.3), has no built-in package for computation of Groebner basis, we have written our own program in Rlisp in order to solve systems of algebraic equations using Buchberger's algorithm [4]. To make the program efficient we have used the distributive form for the internal representation of polynomials together with multivariate factorization.

In order to obtain the (infinitely many) solutions, we construct a lexicographic Groebner basis, then we compute, according to [12], the dimension and independent sets of variables for the ideal which is generated by the input system. Thereafter, we consider each set of variables as free parameters and compute a Groebner basis leaving the order of the others unchanged. As a result we obtain a set of Groebner bases with a simple structure, and the solution can be found in an easy way. From all the sets of solutions found by this procedure, we choose a complete independent subset. This is not difficult, because we have obtained previously all solutions in explicit form. In the worst case the problem is reduced to quadratic equations.

To find a good ordering we have started the process by using an heuristic procedure given in [5]. It turned out that for the systems (6),(8) this is the best choice. For the system (4), however, we did not succeed in constructing a Groebner basis in this way on an IBM PC because of lack of memory . Using a different ordering (see below), we were able to overcome this problem.

It should be noted that our system (4) had been obtained already in [13] and was included in a list of examples for Groebner basis computations [5]. It had been analyzed also in [9, 12]. It seems, however, that no solution has been published as yet. The system (6) has been obtained in [14].

We now give a complete list of solutions for the problems of Sect.2., as obtained by the above procedure. The total computing time is given for an IBM PC AT-like computer (10 Mhz) and for the non- compiled program

I. Ordering: $\lambda_7 > \lambda_6 > \lambda_5 > \lambda_4 > \lambda_3 > \lambda_2 > \lambda_1$, time (min): 48.

Free variables	Solutions
λ_1	$\lambda_2 = 7/2\lambda_1, \lambda_3 = 6\lambda_1, \lambda_4 = 2/7\lambda_1^2, \lambda_5 = 9/7\lambda_1^2, \lambda_6 = 5/14\lambda_1^2, \lambda_7 = 4/147\lambda_1^3.$
λ_1	$\lambda_2 = 3\lambda_1, \lambda_3 = 5\lambda_1, \lambda_4 = 5/14\lambda_1^2, \lambda_5 = 10/7\lambda_1^2, \lambda_6 = 5/14\lambda_1^2, \lambda_7 = 5/98\lambda_1^3.$
λ_1	$\lambda_2 = 2\lambda_1, \lambda_3 = 3\lambda_1, \lambda_4 = 2/7\lambda_1^2, \lambda_5 = 6/7\lambda_1^2, \lambda_6 = 1/7\lambda_1^2, \lambda_7 = 4/147\lambda_1^3.$
λ_2, λ_4	$\lambda_1 = 0, \lambda_3 = 5\lambda_2, \lambda_5 = 1/14\lambda_2^2 + 9/2\lambda_4, \lambda_6 = 1/14\lambda_2^2 + 3/2\lambda_4, \lambda_7 = 0.$
λ_3, λ_6	$\lambda_1 = 0, \lambda_2 = r_1, \lambda_4 = 0, \lambda_5 = -2/161 r_1\lambda_3 + 40/23\lambda_6, \lambda_7 = 10/25921 r_1\lambda_3^2 - 74/483 r_1\lambda_6 - 16/3703\lambda_3\lambda_6,$ *where* $r_1 = (5\lambda_3 \pm \sqrt{25\lambda_3^2 - 5152\lambda_6})/46.$
$\lambda_2, \lambda_3, \lambda_5$	$\lambda_1 = 0, \lambda_4 = 0, \lambda_6 = 1/21\lambda_2^2 + 1/3\lambda_5, \lambda_7 = 0.$

II. Ordering: $\lambda_6 > \lambda_5 > \lambda_4 > \lambda_3 > \lambda_2 > \lambda_1$, time (s): 24.

Free variables	Solutions
λ_1	$\lambda_2 = -1/7\lambda_1^2, \lambda_3 = -2/7\lambda_1^2, \lambda_4 = -2/147\lambda_1^3, \lambda_5 = 2/343\lambda_1^4, \lambda_6 = -4/50421\lambda_1^6.$
λ_2	$\lambda_1 = 0, \lambda_3 = 3\lambda_2, \lambda_4 = 0, \lambda_5 = 5/14\lambda_2^2, \lambda_6 = 5/98\lambda_2^3.$

III.Ordering: $\lambda_3 > \lambda_4 > \lambda_1 > \lambda_5$, time (min): 11.

Free variables	Solutions
λ_3	$\lambda_1 = 1, \lambda_4 = 0, \lambda_5 = 0.$
λ_4	$\lambda_1 = 0, \lambda_3 = \lambda_4, \lambda_5 = 0.$
λ_4	$\lambda_1 = 0, \lambda_3 = \lambda_4/2, \lambda_5 = 0.$
λ_4	$\lambda_1 = 1/3, \lambda_3 = \lambda_4, \lambda_5 = -\lambda_4.$
λ_4	$\lambda_1 = -1/3, \lambda_3 = \lambda_4/2, \lambda_5 = \lambda_4.$
λ_3, λ_6	$\lambda_3 = 0, \lambda_4 = 0.$

4. Conclusion

Our analysis shows that the Groebner basis method allows us to obtain the complete set of exact solutions for systems of nonlinear algebraic equations which are the necessary integrability conditions for NLEEs and therefore to select all integrable evolution equations. It turns out from practice that if the first integrability conditions (2) are fulfilled, then often all the others are fulfilled as well. In other words, the fact that some necessary conditions are fulfilled is often sufficient for integrability. In particular, this is the case for the problems I and II. All the solutions correspond to integrable NLEEs from the families (3),(5) which are the higher infinitesimal symmetries of well-known integrable equations of lower order [13]-[14].

In the case of the problem III we have a more complicated situation. The first three solutions do not satisfy the restrictions on the eigenvalues of the matrix Δ in (1) ($\delta_1 = \lambda_1 \neq 0, \delta_2 = 1 - \lambda_1 \neq 0$) and should therefore be omitted in view of the integrability of (1). The solutions 4-6 in the above list lead to integrable NLEEs of the form (7) only in the case λ_2=0 [15]. Note that λ_2 is not present in the system (8), and the same is probably true also for higher integrability conditions. On the other hand, higher infinitesimal symmetries (which means integrability [1]) exist only for λ_2=0.

It is clear that the solvability of the above systems (with infinitely many solutions) and of even more complicated ones (see, for example [15]) is closely connected with the property of integrability. In addition to their importance in the theory and application of NLEEs, such systems are very useful for testing different computer algebra algorithms.

References

[1] Mikhailov A.V., Shabat A.B., Yamilov R.I. Symmetry approach to classification of nonlinear equations.The complete list of integrable systems. *Usp.Mat.Nauk.*42, 3-53, 1987 (in Russian).

[2] Gerdt V.P., Shabat A.B., Svinolupov S.I., Zharkov A.Yu. Computer algebra application for investigating integrability of nonlinear evolution systems. In: "EUROCAL'87", Lecture Notes in Computer Science 378, 81-92, Springer-Verlag, Berlin, 1989.

[3] Gerdt V.P., Zharkov A.Yu. Computer generation of necessary integrability conditions for polynomial-nonlinear evolution equations. In: "ISSAC'90" (International Symposium on Symbolic and Algebraic Computation), ACM Press, Addison-Wesley Publ.Co., 1990, pp.250-254. Tokyo, Japan.

[4] Buchberger B. Groebner basis: a method in symbolic mathematics. In: (Bose N.K., ed.) Multidimensional Systems Theory, 184-232, Dordrecht: Reidel, 1985.

[5] Boege W., Gebauer R., Kredel H. Some examples for solving systems of algebraic equations by calculating Groebner basis, *J.Symb.Comp.*2, 83-98, 1986.

[6] Kobayashi H., Fujise T., Furukawa A. Solving systems of algebraic equations by a general elimination method, *J.Symb.Comp.*5, 303-320, 1988.

[7] Kobayashi H., Moritzugu S., Hogan R. Solving systems of algebraic equations.In: "ISSAC'88", Lecture Notes in Computer Science 358, 139- 149, Springer-Verlag, Berlin, 1989.

[8] Czapor S.R. Solving algebraic equations: combining Buchberger's algorithm with multivariate fectorization, *J.Symb.Comp.*7, 49-53, 1989.

[9] Melenk H., Moller H.M., Neun W. On Groebner bases computation on a supercomputer using REDUCE,Konrad-Zuse-Zentrum für Informationstechnik Berlin, Preprint SC 88-2, 1988.

[10] Ito M., Kako F. A REDUCE program for finding conserved densities of partial differential equations with uniform rank, *Comp.Phys.Comm.*, 38, 415-419, 1985. Ito M. A REDUCE program for finding symmetries of nonlinear evolution equations with uniform rank, *Comp.Phys.Comm.*42, 351-375, 1986.

[11] Gerdt V.P., Kostov N.A., Zharkov A.Yu. Nonlinear evolution equations and solving of algebraic systems: the importance of computer algebra. In: "Solitons and Applications". World Scientific Publ.Co., Singapore,1990,pp.120-128.

[12] Kredel H., Weispfenning V. Computing dimension and independent sets for polynomial ideals, *J.Symb.Comp.*6, 231-247, 1988.

[13] Gerdt V.P., Shvachka A.B., Zharkov A.Yu. Classification of integrable high-order KdV-like equations, Preprint JINR R5-84-489, Dubna, 1984.

[14] Gerdt V.P., Zharkov A.Yu. Computer classification of integrable seventh-order MKdV-like equations.In: "EUROCAL'87", Lecture Notes in Computer Science 378, 93-94, Springer-Verlag, Berlin, 1989.

[15] Gerdt V.P., Zharkov A.Yu. Computer classification of integrable coupled KdV-like systems, Preprint JINR E5-89-232, Dubna,1989. To be published in *J.Symb.Comp.*

DETERMINATION OF DISTRIBUTION PARAMETERS FOR ω_n^2 AND ω_n^3 TEST STATISTICS BY THE COMPUTER ALGEBRA APPLICATION

P.V. ZRELOV, V.V. IVANOV

Joint Institute for Nuclear Research, Dubna, USSR

ABSTRACT

Among so called nonparametric goodness- of- fit criteria a special place hold the criteria, which allow one to analize nonhystogrammed experimental data. The omega- squared criterion (ω_n^2), as well as, proposed by the authors the omega- cube criterion (ω_n^3), relate to the indicated class of criteria, in particular. An effectivity of employed criterion is determined by its power function. The computer algebra systems REDUCE and muMATH were used to get exact parameters of the power function for ω_n^2 criterion, then those obtained by D. Chapman approximately. Moreover, the corresponding parameters obtained for ω_n^3 statistics, as well, which allowed one to determine the power function for new criterion.

Introduction

In this paper we shall consider the criteria for testing the null - hypothesis H_0 : $F=F_0$ against the right - sided alternative hypothesis $G<F_0$. The criteria ω_n^2 [1] and ω_n^3 [2] are based on the statistics

$$\omega_n^2 = n \int_{-\infty}^{\infty} [S_n(x) - F(x)]^2 dF(x), \omega_n^3 = n^{3/2} \int_{-\infty}^{\infty} [S_n(x) - F(x)]^3 dF(x),$$

where $S_n(x)$ - empirical distribution function for the sample $x_1, x_2, \ldots, x_n$, relating to F(x) distribution, n - the sample size.

In the event that the sample belongs to G(x) distribution, the corresponding statistics take the form :

$$\tilde{\omega}_n^2 = n \int_{-\infty}^{\infty} [\ \tilde{S}_n(x) - F(x)\ ]^2\ dF(x) = n \int_0^1 [\ G_n(u) - u\ ]^2\ du, \tag{1}$$

$$\tilde{\omega}_n^3 = n^{3/2} \int_{-\infty}^{\infty} [\ \tilde{S}_n(x) - F(x)\ ]^3\ dF(x) = n^{3/2} \int_0^1 [G_n(u) - u\ ]^3 du, \tag{2}$$

where $\tilde{S}_n(x)$ - empirical distribution function for the sample $\tilde{x}_1, \tilde{x}_2, \ldots, \tilde{x}_n$ from $G(x)$ distribution, n- the sample size, $G_n^1(u)$ - empirical function corresponding to the sample $u_1 = F(\tilde{x}_1)$, $u_2 = F(\tilde{x}_2), \ldots, u_n = F(\tilde{x}_n)$.

D. Chapman showed [3] that the power of the ω_n^2 criterion for large values of n is determined from the equality:

$$P_{\omega_n^2} \simeq 1 - \Phi \left\{ \frac{Z_\alpha - E(\tilde{\omega}_n^2)}{\sqrt{D(\tilde{\omega}_n^2)}} \right\}, \tag{3}$$

where Φ - the distribution function of the density $N(0,1)$, Z_α - critical point of the ω_n^2 criterion for the significance level α, E and D denotes respectively the mean value and variance.

It can be shown that for the right- sided ω_n^3 criterion the power function is determined from the equality :

$$P_{\omega_n^3} \simeq \Phi \left\{ \frac{Z_\alpha^* - E(\tilde{\omega}_n^3)}{\sqrt{D(\tilde{\omega}_n^3)}} \right\}, \tag{4}$$

where Z_α^* - critical point of the right- sided ω_n^3 criterion for the significance level α [4].

It has been shown by D.Chapman that among all one - sided alternative hypothesis (for definiteness, assume $G<F_0$) for which $\sup_{-\infty<x<\infty} [\ F(x) - G(x)\ ] = \Delta\ (0 < \Delta < 1)$ is fixed it can be chosen one , that minimize (G_{mu_0}) or maximize (G_M) the power P : $P_{min}(\Delta) = \inf P(G_{mu_0})$, $P_{max}(\Delta) = P(G_M)$, where G_{mu_0} and G_M are determined so that

$$G_{mu_0} = \begin{cases} 0, & u < 0, \\ u, & 0 \le u < u_0, \\ u_0, & u_0 \le u < u_0 + \Delta, \\ u, & u_0 + \Delta \le u < 1, \\ 1, & 1 \le u, \end{cases} \qquad G_M = \begin{cases} 0, & u < \Delta, \\ u - \Delta, & \Delta \le u < 1, \\ 1, & 1 \le u, \end{cases} \tag{5}$$

where $0 \le u_0 \le 1 - \Delta$

Exact distribution parameter values of statistics (Eq.1) and (Eq.2) for the expressions (Eq.3) and (Eq.4) are very awkward and it is almost impossible to obtain them by standard techniques. For ω_n^2 criterion D.Chapman received the E and D parameters approximately.

1. Calculation method

The scheme of calculations is illustrated for the ω^3_n criterion case. The mean value and variance are found from

$$E(\tilde{\omega}^3_n) = n^{3/2} \int_0^1 E\{ [G_n(u) - u]^3 \} du, \tag{6}$$

$$D(\tilde{\omega}^3_n) = E[(\tilde{\omega}^3_n)^2] - [E(\tilde{\omega}^3_n)]^2,$$

$$E[(\tilde{\omega}^3_n)^2] = 2n^3 \int\int_{u_2 > u_1} E\{ [G_n(u_1) - u_1]^3 [G_n(u_2) - u_2]^3 \} du_1 du_2, \tag{7}$$

where E - the operator of mathematical expectation, D - the variance. The empirical function $G_n(u)$ is written with the help of indicator ξ_i :

$$G_n(u) = 1/n \sum_{i=1}^{n} \xi_i, \text{ where } \xi_i = \begin{cases} 1, & u_i < u, \\ 0, & u_i > u, \quad i = 1, 2, \dots, n, \end{cases} \tag{8}$$

and u_i - an element of the random sample $u_1, u_2, \dots, u_n$, since $E\xi_i = G(u)$.

To calculate (Eq.7), $G_n(u_1)$ and $G_n(u_2)$ are determined in a similar way :

$$G_n(u_1) = 1/n \sum_{i=1}^{n} \xi^I_i, \qquad G_n(u_2) = 1/n \sum_{j=1}^{n} \xi^{II}_j,$$

where ξ^I_i, ξ^{II}_j are presented in the same way as (Eq.8), $\xi^I_i = G(u_1)$, $E\xi^{II}_j = G(u_2)$ and for $u_2 > u_1$ $\xi^I_i \xi^{II}_i = \xi^I_i$.

It is convenient to present the expression for $\tilde{\omega}^3_n$(Eq.2) as

$$\tilde{\omega}^3_n = n^{3/2} \sum_{l=0}^{3} C^l_3 \int_0^1 du\, [G_n(u) - G(u)]^{3-l} g_l(u),$$

where $g_l(u) = (G(u) - u)^l$ and we suppose $0^0 = 1$, C^l_3 - binomial coefficients.

Thus , the problem on $E(\tilde{\omega}^3_n)$ and $D(\tilde{\omega}^3_n)$ determination reduces to the integrals calculation :

$$E(\tilde{\omega}^3_n) = n^{3/2} \sum_{l=0}^{3} C^l_3 \int_0^1 du\, g_l(u) \frac{1}{n^{3-l}} E\Big\{ \underbrace{\sum_{i=1}^{n} \dots \sum_{j=1}^{n}}_{3-l} (\xi_i - E\xi_i) \dots (\xi_j - E\xi_j) \Big\}, \tag{9}$$

$$E[(\tilde{\omega}^3_n)^2] = 2n^3 \sum_{l=0}^{3} \sum_{k=0}^{3} C^l_3 C^k_3 \frac{1}{n^{6-l-k}} \int\int_{0<u_1<u_2<1} du_1 du_2 g_l(u_1) g_k(u_2) \times$$

$$\times E\Big\{ \underbrace{\sum_{i=1}^{n} \dots \sum_{j=1}^{n}}_{3-l} \underbrace{\sum_{m=1}^{n} \dots \sum_{p=1}^{n}}_{3-k} (\xi^I_i - E\xi^I_i) \dots (\xi^I_j - E\xi^I_j)(\xi^{II}_m - E\xi^{II}_m) \dots \times$$

$$\times\ (\xi_p^{II} - E\xi_p^{II})\Big\}, \qquad (10)$$

The calculation of $E(\tilde{\omega}_n^3)$ and $E\,[(\tilde{\omega}_n^3)^2]$ is evidently cumbersame problem. So, to find the value $E[(\tilde{\omega}_n^3)^2]$ 96 multiple integrals for $G=G_{mu_0}$ and 48 integrals for $G=G_M$ should be calculated.

2. Calculation results

The muMATH system was mainly used to calculate multiple integrals and for their summation. The integrand functions were calculated using the REDUCE system.

1). Parameters for $\tilde{\omega}_n^2$ statistics :

a) $G(u) = G_M(u)$

$$E(\tilde{\omega}_n^2) = n\Delta^2(1 - 2/3\Delta) + 1/6\ (1 - 3\Delta^2 + 2\Delta^3), \qquad (11)$$

$$D(\tilde{\omega}_n^2) = 1/n\ [\Delta^2(-\ 1/6 + 4/3\ \Delta - 11/4\ \Delta^2 + 34/15\ \Delta^3 - - 2/3\ \Delta^4) - 1/60\ ] + + [\Delta^3(-\ 16/9 + 5\Delta - 24/5\ \Delta^2 + 14/9\ \Delta^3) + 1/45] + + n\ [\Delta^2(1/3 - 2\Delta^2 + 8/3\ \Delta^3 - \Delta^4\ )\ ] \qquad (12)$$

б) $G(u) = G_{mu_0}(u)$

$$E(\tilde{\omega}_n^2) = n\Delta^3/3\ (1 + 1/n) + [\ 1/6 + \Delta^2(u_0 - 1/2)], \qquad (13)$$

$$D(\tilde{\omega}_n^2) = 1/n\ [\ u_0\Delta(\ -1/3 + 6\Delta - 11\Delta^2 + 34/3\ \Delta^3 - 4\Delta^4 + 4u_0 - - 29u_0\Delta + 22u_0\Delta^2 - 10u_0\Delta^3 - 11u_0^2 + 124/3\ u_0^2\Delta - 12u_0^2\Delta^2 + + 34/3\ u_0^3 - 18\ u_0^3\Delta - 4u_0^4) + \Delta^2(\ -\ 1/6 + 4/3\ \Delta - 11/4\ \Delta^2 + + 34/15\ \Delta^3 - 2/3\ \Delta^4) - 1/60\ ] + [u_0\Delta\ (\ -\ 1 - 5/3\ \Delta + 5\Delta^2 - - 23/3\ \Delta^3 + 8/3\ \Delta^4 - u_0 + 14u_0\Delta - 11u_0\Delta^2 + 8u_0\Delta^3 + 5u_0^2 - - 64/3\ u_0^2\Delta + 6u_0^2\Delta^2 - 17/3\ u_0^3 + 9u_0^3\Delta + 2u_0^4) + \Delta^3(\ -\ 4/9 + + \Delta - 4/5\ \Delta^2 + 2/9\ \Delta^3) + 1/45\ ] + n\ [u_0\Delta^4(\ 1 - u_0)\ ]. \qquad (14)$$

2). Parameters for $\tilde{\omega}_n^3$ statistics :

a) $G(u) = G_M(u)$

$$E(\tilde{\omega}_n^3) = n^{3/2}[\ \Delta^2/2n^2\ (\ 1 - \Delta\ )^2 + \Delta/2n\ (-\ 2\Delta^3 + 3\Delta^2 - 1) + + \Delta^3/4\ (\ 3\Delta - 4\ )\ ], \qquad (15)$$

$$D(\tilde{\omega}_n^3) = 23/560 + 1/n^2[\ \Delta^3(\ 8/3 - 73/4\ \Delta + 46\Delta^2 - 111/2\ \Delta^3 +$$

$$
\begin{aligned}
&+ 228/7\ \Delta^4 - 15/2\ \Delta^5) + 1/84\ ] + 1/n\ [\ \Delta^2(-1/4 - 3/2\ \Delta + \\
&+ 111/4\ \Delta^2 - 454/5\ \Delta^3 + 503/4\ \Delta^4 - 1131/14\ \Delta^5 + 159/8\ \Delta^6) - \\
&- 11/280\ ] + \Delta^2/2\ (- 1/2 - 5/2\ \Delta - 75/4\ \Delta^2 + 577/5\ \Delta^3 - \\
&- 407/2\ \Delta^4 + 2109/14\ \Delta^5 - 327/8\ \Delta^6) + n\Delta^2\ (1/2 - 3/4\ \Delta^2 - \\
&- 13\Delta^3 + 36\Delta^4 - 33\Delta^5 + 41/4\ \Delta^6) + 3n^2\Delta^4/4\ (1 - 6\Delta^2 + \\
&+ 8\Delta^3 - 3\Delta^4\ ), \qquad (16)
\end{aligned}
$$

б) $G(u) = G_{mu_0}(u)$

$$
\begin{aligned}
E(\tilde{\omega}^3_n) &= n^{3/2}\ [\ \Delta^2/2n^2\ (- 1 + 6u_0 - 6u_0^2 + 2\Delta - 4\Delta u_0 - \Delta^2) + \\
&+ 3\Delta^2 u_0/2n\ (u_0 - 1\ ) - \Delta^4/4\ ], \qquad (17)
\end{aligned}
$$

$$
\begin{aligned}
D(\tilde{\omega}^3_n) &= 23/560 + 1/n^2\ [\Delta^2 u_0(7 - 75u_0 + 230u_0^2 - 270u_0^3 + \\
&+ 108u_0^4 - 66n + 318\Delta u_0 - 504\Delta u_0^2 + 252\Delta u_0^3 + 222\Delta^2 - \\
&- 711\Delta^2 u_0 + 792\Delta^2 u_0^2 - 270\Delta^2 u_0^3 - 330\Delta^3 + 648\Delta^3 u_0 - 360\Delta^3 u_0^2 + \\
&+ 228\Delta^4 - 210\Delta^4 u_0 - 60\Delta^5) + \Delta^3(8/3 - 73/4\ \Delta + 46\Delta^2 - \\
&- 111/2\ \Delta^3 + 228/7\ \Delta^4 - 15/2\ \Delta^5) + 1/84\ ] + \\
&+ 1/n\ [\ \Delta^2 u_0(- 3 + 147/2\ u_0 - 259u_0^2 + 315u_0^3 - 126u_0^4 + \\
&+ 60\Delta - 333\Delta u_0 + 546\Delta u_0^2 - 273\Delta u_0^3 - 471/2\ \Delta^2 + 1689/2\ \Delta^2 u_0 - \\
&- 1029\Delta^2 u_0^2 + 783/2\ \Delta^2 u_0^3 + 357\Delta^3 - 756\Delta^3 u_0 + 450\Delta^3 u_0^2 - \\
&- 489/2\ \Delta^4 + 477/2\ \Delta^4 u_0 + 63\Delta^5) + \Delta^2(- 1/4 - \Delta + \\
&+ 63/4\ \Delta^2 - 229/5\ \Delta^3 + 231/4\ \Delta^4 - 240/7\ \Delta^5 + 63/8\ \Delta^6) - \\
&- 11/280\ ] + [\ \Delta^2 u_0(- 9/4 - 15/4\ u_0 + 75/2\ u_0^2 - \\
&- 105/2\ u_0^3 + 21u_0^4 - 3/2\ \Delta + 32\Delta u_0 - 61\Delta u_0^2 + 61/2\ \Delta u_0^3 + \\
&+ 111/4\ \Delta^2 - 657/4\ \Delta^2 u_0 + 531/2\ \Delta^2 u_0^2 - 513/4\ \Delta^2 u_0^3 - 38\Delta^3 + \\
&+ 108\ \Delta^3 u_0 - 81\Delta^3 u_0^2 + 91/4\ \Delta^4 - 111/4\ \Delta^4 u_0 - 9/2\ \Delta^5) + \\
&+ \Delta^3(- 1/2 - 3/8\ \Delta + 21/5\ \Delta^2 - 13/2\ \Delta^3 + 57/14\ \Delta^4 - \\
&- 15/16\ \Delta^5)\ ] + n\ [\ \Delta^3 u_0/2(1 - 2u_0^2 + u_0^3 + 9\Delta u_0 - 18\Delta u_0^2 + \\
&+ 9\Delta u_0^3 - 12\Delta^2 + 42\Delta^2 u_0 - 30\Delta^2 u_0^2 + 8\Delta^3 - 12\Delta^3 u_0 - 3\Delta^4)\ ] + \\
&+ n^2\ [\Delta^6 u_0(1 - u_0)\ ]. \qquad (18)
\end{aligned}
$$

3. Control tests

Clearly for $\Delta = 0$ the alternative hypothesis $G(u)$ coincides with the null- hypothesis H_0 and hence the $\tilde{\omega}^2_n$ and $\tilde{\omega}^3_n$ statistics parameters must be of the same form as the ω^2_n (see, for instance, [1]) and ω^3_n [2,4] :

$$
E(\omega^2_n) = 1/6\ , \qquad D(\omega^2_n) = (4n - 3\ )/180n\ ,
$$

$$E(\omega_n^3) = 0\ , \qquad D(\omega_n^3) = 23/560 + 1/84n^2 + 11/280n\ .$$

It is easy to verify that this condition is satisfied substituting $\Delta = 0$ in (Eqs. 11-18).

That is more, if the expression for $D(\tilde{\omega}_n^2)$ is reduced only to terms with coefficient n, the results will agree with those obtained by D. Chapman approximately :

$$D(\tilde{\omega}_n^2) \cong n\ (\ \Delta^2/3 - 2\Delta^4 + 8/3\ \Delta^5 - \Delta^6) \qquad \text{for } G(u) = G_M(u),$$

$$D(\tilde{\omega}_n^2) \cong n\ u_0\ (\ 1 - u_0)\Delta^4 \qquad \text{for } G(u) = G_{mu_0}(u).$$

The mean values of $\tilde{\omega}_n^2$ statistics calculated by D.Chapman for both hypothesis kinds have the same form as in (Eq. 11) and (Eq. 12).

Conclusion

The parameters of $\tilde{\omega}_n^2$ and $\tilde{\omega}_n^3$ statistics distributions (Eq. 11-18) obtained with the help of REDUCE and muMATH systems were used to find the corresponding criteria powers from (Eq. 3) and (Eq. 4). This permitted to compare the ω_n^2 and ω_n^3 criteria for both alternative hypothesis. The results are distinct from those obtained by means of approximate expressions for $\tilde{\omega}_n^2$ and $\tilde{\omega}_n^3$ parameters.

References

1. G.V. Martynov , *Omega- squared criteria*, (Nauka, Moscow, 1978).
2. P.V. Zrelov, V.V. Ivanov, *JINR communication*, P10-88-321, (Dubna, 1988).
3. D.G. Chapman, *Ann. Math. Statist*. **v.29** (1958), p.655.
4. P.V. Zrelov, V.V. Ivanov, *JINR communication*, P10-89-577, (Dubna, 1989).

ON CA APPLICATIONS IN SOLVING SOME STATISTICAL DYNAMICAL PROBLEMS

V. V. MALANIN
Department of Mech. and Math., the Perm State Univ., st. Bukireva, 15, Perm, 614600, U.S.S.R.

I. E. POLOSKOV
Department of Mech. and Math., the Perm State Univ., st. Bukireva, 15, Perm, 614600, U.S.S.R.

ABSTRACT

Nonlinear dynamical systems disturbed by a stationary vector Markov process or random parameters are studied by means of FORMAC and REDUCE programs be making use of symbolic manipulations and differentiations, deriving of corresponding equations e.a.

Computer algebra (CA) systems and separate symbolic manipulation programs are currently a most important tool to implement methods and algorithms of the nonlinear objects random vibration study. CA (REDUCE, MACSYMA, FORMAC et al.) scope in statistical dynamics is: symbolic transformations[1], iterative procedures design[2,3], automatic derivation of systems of ordinary differential equations (ODE) for subsequent computations[3,4] et al. The present paper considers a number of problems of nonlinear probability objects analysis which have to be solved by means of CA systems.

Some analitical methods for Markov random processes require the Fokker-Planck-Kolmogorov equation (FPK Eq.)[5]

$$\frac{\partial p(\bar{x},t)}{\partial t} = L_t p(\bar{x},t), \quad p(\bar{x},t_o)=p_o(\bar{x}), \tag{1}$$

$$L_t u \equiv -\sum_{i=1}^{n} \frac{\partial}{\partial x_i}\left\{a_i(\bar{x},t)u\right\} + \frac{1}{2}\sum_{i,j=1}^{n} \frac{\partial^2}{\partial x_i \partial x_j}\left\{b_{ij}(\bar{x},t)u\right\} \equiv$$

$$\equiv \frac{1}{2}\sum_{i,j=1}^{n} A_{ij}\frac{\partial^2 u}{\partial x_i \partial x_j} + \sum_{i=1}^{n} B_i \frac{\partial u}{\partial x_i} + C\,u,$$

$$a_i = f_i + \frac{1}{2}\sum_{j=1}^{n}\sum_{k=1}^{m}\frac{\partial g_{ik}}{\partial x_j} g_{jk}, \qquad b_{ij} = \sum_{k=1}^{m} g_{ik}\, g_{jk},$$

to be solved for the probability density $p(\bar{x},t)$ of the phase vector x of a nonlinear dynamic system described by stochastic differential equations of the following form

$$dx_i = f_i(\bar{x}(t),t)\,dt + \sum_{i=1}^{m} g_{ij}(\bar{x}(t),t)\,dw_j(t), \qquad (2)$$
$$i = 1,2,\dots,n$$

where $w_j(t)$ are independent Wiener's random processes, t is time, $p_o(\bar{x})$, $f_i(\cdot,\cdot)$, $g_{ij}(\cdot,\cdot)$ are given sufficiently smooth functions.

One of the procedures to solve the Eq. 1 is the following. Let us write Eq. 1 in the form:

$$p = p_o + \int_0^t L_\tau p\, d\tau \equiv p_o + \mathbb{L}p.$$

The iterative operator method* is to construct the functions

$$p_k = \mathbb{L}p_{k-1}, \qquad p_o = p_o(\bar{x}) \qquad (3)$$

and to present density approximation as

$$p_M^* = \sum_{k=0}^{M} p_k.$$

Using a FORMAC program the above method has been applied to investigate various systems and in particular to approximate probability density in the problem of a rigid body rotating around the fixed point and excited by the external random moments.

In addition to the functions $p(\bar{x},t)$ Eq. 1 is also satisfied by the transition probability density $\pi(\bar{x},\bar{y},t)$ (for time homogeneous Markov systems) the knowledge of which allows such statistical characteristics as a correlation function, the number of process intersections of a given level et al. to be found. But the Dirac multidelta functions being present in the initial condition

$$\pi(\bar{x},\bar{y},t) = \delta(x-y),$$

make it difficult to determine $\pi(\bar{x},\bar{y},t)$.

One can overcome the arising difficulties by writing the transition density π in the following form

$$\pi(\bar{x},\bar{y},t) = \sum_{\alpha=0}^{\infty} \pi_\alpha(\bar{y},t)\,\delta^{(\alpha)}(x-y).$$

Assuming the FPK Eq. coefficients to be presented as

$$A_{ij} = \sum_{\beta=0}^{\infty} A_{ij\beta}\, z^\beta/\beta!, \qquad B_i = \sum_{\beta=0}^{\infty} B_{i\beta}\, z^\beta/\beta!, \qquad C = \sum_{\beta=0}^{\infty} C_\beta\, z^\beta/\beta!$$

($\bar{z} = \bar{x} - \bar{y}$) the following infinite ODE system has been formed to determine the coefficients $\pi_\alpha(\bar{y},t)$, vector $\bar{y}$ being a parameter

$$\frac{\partial \pi_\alpha}{\partial t} = V_\alpha(\ \pi\ ), \qquad \alpha \geq 0, \qquad \pi_\alpha(\bar{y},0) = \delta_{\alpha o}, \tag{4}$$

where

$$V_\alpha = \sum_{\beta=o}^{\infty} (-1)^{|\beta|} C_\nu^\beta \left\{ \frac{1}{2} \sum_{i=1}^{n} A_{ii\beta}\, \pi_{\nu-2e_i} + \sum_{i=1}^{n-1} \sum_{j=i+1}^{n} A_{ij\beta}\, \pi_{\nu-e_i-e_j} + \right.$$
$$\left. + \sum_{i=1}^{n} B_{i\beta}\, \pi_{\nu-e_i} + C_\beta\, \pi_\nu \right\}, \qquad \nu = \alpha + \beta.$$

The ODE system 4 can be solved in the following way. Let us represent π_α as

$$\pi_\alpha = \sum_{k=o}^{\infty\alpha} \pi_{k\alpha}(\bar{y})\ u_k(t), \qquad u_k(t) = \frac{t^k}{k!}\, e^{-t}.$$

Then $\pi_{k+1,\alpha} = \pi_{k\alpha} + V_\alpha(\pi_k)$.

Using the above method and CA system FORMAC program the transition density $\pi(x,y,t)$ for the system

$$\dot{x} + c\ x + \mu\ x^3 = \sqrt{g}\ \xi \tag{5}$$

has been approximated, where $c>0, \mu>0, g>0$ are constants, $\xi(t)$ is the white noise. Applying the relation

$$\tilde{x}(t) = \tilde{x}_o - \int_{-\infty}^{\infty} \pi_1(y,t)\ p_o(y)\ dy,$$

the mean value $\tilde{x}$ of displacement x has been computed.

The transition processes in the system 2 having attenuated, it can change its state to the one specified by stationary probability density $p_{st}(\bar{x})$. There arises the problem of finding the density.

Consider the following computation algorithm $p_{st}(\bar{x})$. Let the nonlinear object be described by the following equations

$$d\bar{x}(t) = \Big[\ A\ \bar{x}(t) + \bar{f}(\bar{x}(t))\ \Big]\ dt + B\ d\bar{w}(t), \tag{6}$$

where A and B are constant matrices, f are the polynomials. Find the approximation $p^*_{st}(\bar{x})$ of the stationary FPK Eq. solution

$$L\ p_{st}(\bar{x}) = 0$$

as follows

$$p^*_{st}(\bar{x}) = \sum_{k=o}^{M} c_k\ h_k(\bar{x}), \tag{7}$$

with a linear independent functions system $\{h_k(\bar{x})\}$ constructed through iterations

$$h_k(\bar{x}) = L\, h_{k-1}(\bar{x}) \qquad (k>0) \tag{8}$$

in terms of the function $h_0(\bar{x})$ which is steady-state density of the corresponding linear system known to be a multidimensional Gaussian one. The coefficients c_k are found as solutions of a system of linear algebraic equations of the Galerkin method:

$$\sum_{s=0}^{M} c_s \int_{R^n} h_l(\bar{x})\, h_{s+1}(\bar{x})\, d\bar{x} = 0, \qquad l=1,2,\ldots,M \tag{9}$$

($c_0=1$ due to the density normalization condition).

Using FORMAC and FORTRAN-IV programs which implement the above method the stationary density approximation $p^*_{st}(x)$ of the system 5 displacement x has been formed. If $c=\mu=1$, $g=0.125$, $M=10$ the difference between the exact and approximate solutions is less then 0.0005 . To analyse the system with two degrees of freedom of the following form

$$x_1'' + 2\,\varepsilon_1 x_1' + \omega_1^2 x_1 + \mu_1 x_1^3 = k_1 x_2 + \sqrt{g_1}\;\xi_1,$$

$$x_2'' + 2\,\varepsilon_2 x_2' + \omega_2^2 x_2 + \mu_2 x_2^3 = k_2 x_1 + \sqrt{g_2}\;\xi_2$$

where ε_i, ω_i, μ_i, g_i are positive constants $(i=1,2)$, k_1, k_2 = const the algorithm 7-9 had been implemented in terms of REDUCE and FORTRAN77 on IBM PC/AT computer.

To examine the performance of complex engineering systems one has to estimate the effect of random design parameters and technological errors on such systems dynamics. The exact and complete solution of this problem connected with the probability density computation is often impossible because computer resources are limited. Therefore one has to use simple but more efficient here methods of the theory of sensitivity[7].

Let the phase vector $\bar{x} \in R^n$ be described by Eqs.

$$\dot{\bar{x}} = f(\bar{x},\bar{p},t) \tag{10}$$

where $\bar{p} \in R^m$ is a random parameters vector, $\bar{p}_0$ is the mean value of vector $\bar{p}$, $R = \{r_{ij}(t)\}$ is a sensitivity matrix

$$r_{ij}(t) = \partial x_i / \partial p_j \,\Big|_{\bar{p}=\bar{p}_0}. \tag{11}$$

Using the relation 11 and Eq. 10 the following ODE system can be constructed to determine the functions r_{ij}:

$$\dot{R} = \partial f(\bar{x},\bar{p}_o,t)/\partial\bar{x}^T\, R + \partial f(\bar{x},\bar{p}_o,t)/\partial\bar{p}^T.$$

If the object has many degrees of freedom and is affected by many random parameters the above algorithm is impossible to implement without computer processing.

To overcome these difficulties a FORMAC program has been designed allowing partial derivatives of the right-hand sides (r.h.s.) of a sequence of algebraic equalities to be found, the equalities written in FORTRAN-like manner (particularly, it may be a FORTRAN-IV program source text to compute the r.h.s. of the system of Eqs.10) in terms of a list of independent variables and to present the differentiation result as a FORTRAN-IV program complex. Among the equalities (formula statements) there may be subroutine call instructions (CALL), unmanipulated conditional statements, commentaries and others.

The program algorithm is to read the next text element in a sequence (up to its end) from a set of data,to find the elements type and the presence of independent variables and aggregates (left-hand side identifiers mentioned before), to differentiate in terms of the arguments list in the case of algebraic equalities, to enter left-hand sides-input parameters relations data (including the relations through aggregates) in information tables.

It should be noted again in the conclusion that the above results could hardly be obtained without using CA systems.

References

1. M.L.Rehak, F.L.Dimaggio, H.Benaroya, I.Elishakov, *Computer Meth. in Appl.Mech.* **61** (1987) 61.
2. V.I.Lumpov,V.V.Malanin, in *Dynamics of Controlled Mechanical systems* (IPI, Irkutsk, 1982) 158 (in Russian).
3. I.E.Poloskov, in *Proc.of 3d Int. Conf. on Computer Algebra in Phys. Research* (JUNR, Dubna, 1985) 289 (in Russian).
4. V.S.Pugachev, I.N.Sinitsyn, V.I.Shin, *Automatics and Telemechanics* (1987) No 2, 62 (in Russian).
5. M.F.Dimentberg, *Nonlinear Stochastic Problems of Mechanical Vibrations* (Nauka, Moscow, 1980) (in Russian).
6. V.V.Malanin, G.A.Zhdanov, in *Problems of Mech. of Controlled Motion. Non-Linear Dyn. Syst.* (PGU, Perm, 1982) 103 (in Russian).
7. R.Tomovic, *Sensitivity Analysis of Dynamic Systems* (Belgrad, 1963).

SYMBOLIC COMPUTATION OF ALTERNATIVE LAGRANGIANS

S.G.Shorokhov

Friendship of Nations University, Moscow

The problem of existence and determination of alternative variational formulations for different evolution systems has been intensively investigated over the last years. The importance of the problem to a certain extent follows from its close connections with different aspects of quantization [1] and integrability [2].

In particle dynamics the problem of alternative Lagrangians is stated as follows. Given the system of second order PDE's

$$\frac{d}{dt}\left(\frac{\partial L}{\partial \dot{q}^k}\right) - \frac{\partial L}{\partial q^k} = 0, \quad k=\overline{1,n} \tag{1}$$

with some regular Lagrangian $L(t,q,\dot{q})$, find out if the system (1) admits any other Lagrangian representation with a Lagrangian $L'(t,q,\dot{q})$, such, that the relation

$$L'(t,q,\dot{q}) = C\,L(t,q,\dot{q}) + \frac{d}{dt}\varphi(t,q), \quad C = \text{const} \tag{2}$$

doesn't hold true for any constant C and any function $\varphi(t,q)$. For the determination of the totality of Lagrangians for (1) this system is to be solved with respect to its second order derivatives:

$$\ddot{q}^k = f^k(t,q,\dot{q}), \quad k = 1,n \tag{3}$$

Now the problem of alternative Lagrangians for (1) reduces to the well-known inverse problem of the calculus of variations for (3), that is to determine the totality of matrix variational multipliers $\alpha = (\alpha_{ij}(t,q,\dot{q}))$, $\det(\alpha) \neq 0$, such, that equations (3), multiplied by α, admit a Lagrangian form.

For the solution of the inverse problem one has to integrate the following system of linear first order PDE's and linear algebraic equations with respect to elements of a nonsingular $n \times n$ - matrix α [3]:

$$\alpha_{ij} = \alpha_{ji}, \tag{4}$$

$$\frac{\partial \alpha_{ik}}{\partial \dot{q}^j} = \frac{\partial \alpha_{jk}}{\partial \dot{q}^i}, \tag{5}$$

$$\frac{D\alpha_{ij}}{Dt} = \alpha_{ik} A_{kj} + A_{ki} \alpha_{kj}, \tag{6}$$

$$\alpha_{ie} \Phi_{ej} = \alpha_{je} \Phi_{ei}, \tag{7}$$

where $\frac{D}{Dt} = \frac{\partial}{\partial t} + \dot{q}^k \frac{\partial}{\partial q^k} + f^k(t,q,\dot{q}) \frac{\partial}{\partial \dot{q}^k}$, $A_{ij} = -\frac{1}{2} \frac{\partial f^i}{\partial \dot{q}^j}$,

$\Phi_{ij} = B_{ij} - A_{ik} A_{kj} - \frac{D}{Dt} A_{ij}$, $B_{ij} = -\frac{\partial f^i}{\partial q^j}$

(here and later on the summation convention is adopted).

For $n > 1$ the system (4)-(7) is an overdetermined system of PDE's and integration of this system in general leads to unpredictable tedious calculations. Thus the investigation of the problem of alternative Lagrangians for any sapid dynamical system is practically imposssible without symbolic computations.

The methods of general theory of compatibility of differential systems [4] can be applied to the system (4)-(7), but the specific structure of this system essentially facilitates its solution, based on the algorithm, following from the results of [3,5,6].

The algorithms starts with the computation of the elements of $n \times n$ -matrixes A, B, Φ by known right-hand sides $f^k(t,q,\dot{q})$ of (3). After that by means of different procedures a number of new linear algebraic equations should be constructed and added to (7). Differentiating (7) with respect to $\dot{q}^k$ and adding terms, obtained through cyclic rearrangement of indexes i, j, k, one arrives to $n(n-1)(n-2)/6$ linear algebraic equations [6] $(i < j < k)$

$$\alpha_{il}\left[\frac{\partial\Phi_{lj}}{\partial\dot{q}^k} - \frac{\partial\Phi_{lk}}{\partial\dot{q}^j}\right] + \alpha_{jl}\left[\frac{\partial\Phi_{lk}}{\partial\dot{q}^i} - \frac{\partial\Phi_{li}}{\partial\dot{q}^k}\right] + \alpha_{kl}\left[\frac{\partial\Phi_{li}}{\partial\dot{q}^j} - \frac{\partial\Phi_{lj}}{\partial\dot{q}^i}\right]=0 \quad (8)$$

Using notation β_k, $k = \overline{1,m}$, $m = n(n+1)/2$ for unknown elements α_{ij}, $1 \leqslant i \leqslant j \leqslant n$, one can rewrite linear algebraic equations (7) and (8) as

$$b_{ij}(t,q,\dot{q})\ \beta_j = 0,\ i = \overline{1,r}. \quad (9)$$

Here $r = (n+1)n(n-1)/6$ and system (9) in general contains some superfluous linear equations. After computation of the rang of matrix (b_{ij}) these superfluous equations can be easily excluded, so without loss of generality one can consider the equations of (9) to be independent and the rang of the matrix (b_{ij}) to be equal to r.

Differentiating (9) with respect to t and taking into account (6), one obtains new linear algebraic equations for β_k, $k = \overline{1,m}$ of the form

$$\frac{D}{Dt} b_{ij}\ \beta_j + b_{ij}\ c_{jk}\ \beta_k = 0 \quad (10)$$

with $m \times m$ -matrix (c_{jk}) being the matrix in the right side of (6), expressed in terms of β_k. If system (10) contains some equations, independent of (9), then these independent equations should be added to (9) and be differentiated with respect to t due to (6), generating new equations of the form (10). The process stops when each equation of (10) is a linear combination of the equations (9).

If the rang r of the matrix (b_{ij}), corresponding to the final composition of the system (9), is equal to m, then the inverse problem of the calculus of variations for (3) has no solution. This situation does not arise when one is looking for alternative Lagrangians since at least one solution of the system (4)-(7) does exist.

Taking into account equations (9) one can express β_i, $i = \overline{1,m}$ in terms of new unknowns σ_j, $j = \overline{1,m-r}$ by means of linear equalities

$$\beta_i = d_{ij}(t,q,\dot{q})\,\sigma_j \tag{11}$$

with vectors $d_i = (d_{ij})$ being the basis in **Ker b, b** - linear map determined by matrix (b_{ij}). In particular, if $r = m-1$, then $\beta = \sigma_1\,[b_1,b_2,\dots,b_{m-1}]$, $b_i = (b_{ij})$, where square brackets stand for m-dimensional vector product.

After substitution of (11) into (5) and (6) one obtains a system of linear homogeneous first order PDE's with respect to unknowns σ_j, $j = \overline{1,m-r}$. Suppose that some equations (or some linear combinations of equations) of this system form a subsystem of PDE's with respect to a single unknown σ_l, then one can apply the method of Jacobi's brackets and obtain the general solution of this subsystem. In case when one can not share out any subsystems with respect to a single unknown, it is extremely difficult to integrate the whole system [3].

Futher progress in the problem of symbolic computation of alternative Lagrangians can be worked out for two particular cases. If the original system (1) is linear with respect to $q,\dot{q},\ddot{q}$ with coefficients, depending on t, the search for matrix va- riational multipliers α can be limited by matrixes $\alpha = \alpha(t)$. Then (5) trivially holds true and direct substitution of (11) in (6) gives a linear homogeneous system of ODE's with respect to σ_j, $j = \overline{1,m-r}$. Thus the problem under study can be solved for linear systems of ODE's by means of merely matrix calculations [7]. The case n = 2 was studied in [3] in details, including cases when the system of first order PDE's can not be separated into subsystems with respect to a single unknown. The main steps of the algorithm, described in [3], can also be used for symbolic computation of alternative Lagrangians for two-dimensional systems.

References

1. Додонов В.В., Манько В.И., Скаржинский В.Д. // Тр. ФИАН СССР.- 1983.- Т.152.- С.37-89.

2. Crampin M. // Differential Geometry and Its Applications. Proceed. Conf., Aug. 24-30, 1986, Brno. - Dordrecht, 1987. - P.1-12.

3. Douglas J. // Trans.Amer.Math.Soc.- 1941.- V.50, № 1.- P.71-128.

4. Pommaret J.F. Systems of partial differential equations and Lie pseudogroups. - N.Y., 1978. - 420 P.

5. Sarlet W. // J.Phys.A.- 1982.- V.15, № 5.- P.1503-1517.

6. Sarlet W. // Acad.Analecta.- 1987.- V.49, № 1.- P.27-57.

7. Sarlet W., Engels E., Bahar L.Y. // Int.J.Eng.Sci.- 1982. - V.20, № 1.- P.55-66.

CALCULATION OF BEAM DYNAMICS IN MAGNETIC FIELD WITH PLANE SYMMETRY USING THE METHOD OF IMMERSION IN THE MOMENT PHASE SPACE

E.L. Ivanov, Z.T. Kostova, E.A. Perelstein

Joint Institute for Nuclear Research

Post Office P.O.Box 79, Moscow, USSR

ABSTRACT

In the present paper we study the charged beam dynamics in magnetic field with plane symmetry using the method of immersion in the moment phase space. The system of nonlinear differential equations which describes the particle dynamics in the phase space [x, y,x',y'] approximately can be rewritten as a system of linear differential equations in the space of phase moments of n-th degree. We obtain the exact analytical solution of this linear system using computer algebra REDUCE 3 on ES-1061 (IBM 370) computer. Based on this solution the Fortran program is used to study the particle dynamics in the isochronous cyclotron. As an example the results of computations for the cyclotron LNR JINR IC-100 are given.

The motion of a particle with charge Ze and momentum $\vec{P}$ in static magnetic field $\vec{B}$ in Cartesian coordinate system is described by the following system of nonlinear differential equations:

$$\ddot{X} = \frac{Ze}{M}(\dot{Y}Bs-\dot{S}By)$$
$$\ddot{Y} = \frac{Ze}{M}(\dot{S}Bx-\dot{X}Bs) \qquad (1)$$
$$\ddot{S} = \frac{Ze}{M}(\dot{X}By-\dot{Y}Bx)$$

where, M is ion mass and by ($\dot{}$) we denote the differentiation with respect to time.

Introducing new independent variable S and using

$$\dot{X}=\dot{S}X' \;;\; \dot{Y}=\dot{S}Y' \;;\; \ddot{X}=X''(\dot{S})^2+X'\ddot{S} \;;\; \ddot{Y}=Y''(\dot{S})^2+Y'\ddot{S},$$

we obtain the following system of ordinary differential equations

$$X''=\frac{Ze}{M}[(X')^2+(Y')^2+1]^{1/2}\{X'Y'Bx-[1+(X')^2]By+Y'Bs\}$$
$$Y''=\frac{Ze}{M}[(X')^2+(Y')^2+1]^{1/2}\{[1+(Y')^2]Bx-X'Y'By-X'Bs\} \qquad (2)$$

The static magnetic field is described by Maxwell equations:

$$\text{rot}\ \vec{H}=\frac{4\pi}{c}\vec{J}$$
$$\text{div}\ \vec{B}=0 \qquad (3)$$

where $\vec{H}$, $\vec{B}$ и $\vec{J}$ are magnetic field strength, magnetic field induction and electric current density, respectively.

Also the following relations for the magnetic field components with plane symmetry hold on

$$Bx(X,-Y,S)=-Bx(X,Y,S)\ ;\ By(X,-Y,S)=By(X,Y,S)$$
$$Bs(X,-Y,S)=-Bs(X,Y,S)\ ;\ Bx(X,0,S)=0\ ;\ Bs(X,0,S)=0 \qquad (5)$$

Here Y-coordinate is normal to median plane, which is the symmetry plane.

Using (4) and (5) and the Tailor expansions of the magnetic field components we have:

$$By(X,Y,S)=-By(Xo,0,So)+\frac{\partial By}{\partial X}(X-Xo)+\frac{\partial By}{\partial S}(S-So)+$$
$$\frac{1}{2}\{\frac{\partial^2 By}{\partial S^2}[(S-So)^2-2Y^2]+(\frac{\partial^2 By}{\partial X^2}[(X-Xo)^2-2Y^2]\}+\frac{\partial^2 By}{\partial X\partial S}(X-Xo)(S-So)+\dots$$
$$Bx(X,Y,S)=\frac{\partial By}{\partial X}Y+\frac{\partial^2 By}{\partial X^2}(X-Xo)Y+\frac{\partial^2 By}{\partial X\partial S}(S-So)Y+\ \dots \qquad (6)$$
$$Bs(X,Y,S)=\frac{\partial By}{\partial S}Y+\frac{\partial^2 By}{\partial S^2}(S-So)Y+\frac{\partial^2 By}{\partial X\partial S}(X-Xo)Y\ +\ \dots$$

Inserting $[(X')^2+(Y')^2+1]^{1/2}=1+\frac{(X')^2}{2}+\frac{(Y')^2}{2}+O(4)$ and expansions (6) in (2) we obtain the following nonlinear system of differential equations which describe the motion of charged particles in magnetic field with plane symmetry with account to the second order terms in phase coordinates - X,X',Y,Y'

$$X''=-2AX-FxX^2-FyY^2-Be[3(X')^2+(Y')^2+2]+EYY'$$
$$Y''=2AY+2FxXY-2BeX'Y'-EX'Y \qquad (7)$$

where

$$A\ =\ \frac{Ze}{2P}[\frac{\partial By}{\partial X}\ -\ \frac{\partial^2 By}{\partial X^2}Xo+\frac{\partial^2 By}{\partial X\partial S}(S-So)]$$
$$Be=\ \frac{Ze}{2P}[By(Xo,0,So)+\frac{\partial By}{\partial S}(S-So)-\frac{\partial By}{\partial X}Xo+\frac{1}{2}[\frac{\partial^2 By}{\partial S^2}(S-So)^2+\frac{\partial^2 By}{\partial X^2}Xo^2]-$$
$$-\frac{\partial^2 By}{\partial X\partial S}(S-So)Xo$$
$$Fx\ =\ \frac{Ze}{2P}\ \frac{\partial^2 By}{\partial X^2}$$
$$Fy\ =\ \frac{Ze}{2P}(\frac{\partial^2 By}{\partial X^2}\ +\ \frac{\partial^2 By}{\partial S^2})$$
$$E\ =\ \frac{Ze}{P}[\frac{\partial By}{\partial S}+\frac{\partial^2 By}{\partial S^2}(S-So)-\frac{\partial^2 By}{\partial X\partial S}Xo]$$

Let us introduce the following vectors:

$$Z^1_1(S)=|X,X',Y,Y'|;Z^2_1(S)=|X^2,XY,Y^2,X'^2,X'Y,X'Y',Y'^2,X'X,Y'X|$$

$$\vec{Z}(S)=\begin{vmatrix}\vec{Z}^{11}\\ \vec{Z}^{21}\end{vmatrix} \;;\; F(S)=|0,2Be,0,0,0,0,0,0,0,0,0,0,0,0| \qquad (8)$$

Using these notations the system (7) can be approximately written as a system of linear differential equations [1]

$$\vec{Z}^{i}(S)=P^{ij}(S)\vec{Z}^{j}(S)+\vec{F}^{i}(S) \qquad i\le j \quad i,j=1,2 \qquad (9)$$

The solution of system (9) is expressed in terms of matrizant $H^{ij}(S/So)$ of homogeneous system by

$$\vec{Z}(S)=H(S/So)[\vec{Z}(0)+\int_{So}^{S} H(\tau)^{-1}\vec{F}(\tau)d\tau]$$

where $\vec{Z}(0)$ is the vector fixed by initial conditions and the vector $\vec{F}(S)$ of the free terms is defined in (8).

The solution of homogeneous system can be written in the following form

$$\vec{Z}^{i}(S)= H^{ij}(S/So)\vec{Z}^{j}(So),$$

where we use the system (9) with $F^{i}(S)=0 \quad i=1,\ldots,14$.

Hence the solution of nonhomogeneous system (9) can be written as

$$\vec{Z}^{i}(S)= H^{ij}(S/So)\vec{Z}^{j}(So)+\vec{V}^{i}(S)$$

where the matrizant $H^{ij}(S/So)$ is defined by

- for diagonal block matrices $H^{jj}(S/So)$ j=1....N

$$[H^{jj}]_{kl}=\sum_{i=1}^{j}\{[H^{11}]_{k_1 l_1}\cdots\cdots\cdots\cdots[H^{11}]_{k_j l_{i+j-1}}\}$$

when the multi-index l have any different components, and

$$[H^{jj}]_{kl}=[H^{11}]_{k_1 l_1}\cdots\cdots\cdots\cdots[H^{11}]_{k_n l_1}$$

when the multi-index l have only coincided components,

- for nondyagonal matrices $H^{ij}(S/So)$ i<j

$$H^{ij}(S/So)= \sum_{l=i+1}^{j}\int_{So}^{S} H^{il}(S/\tau)P^{il}(\tau)H^{lj}(\tau/So)d\tau$$

All analytical calculations are performed on ES-1061 (IBM 370) computer using computer algebra REDUCE 3.2. Using our program we may obtain the approximate analytical solution of any system of the following type

$$\vec{X}'=f(X,t)$$

where the right hand side $\vec{f}$ fulfills the following requirements

$$fm(X,t) = \sum_{|k|=1}^{m} \frac{X_1^{k_1} \cdots X_n^{k_n}}{k_1! \cdots k_n!} F_{,k}(O,t) = \sum_{k=1}^{m} \Pi_{1k}(t) Z_k .$$

The natural requirements are also related to the degree of the phase vector $\vec{Z}_k(S)$ and depend on the computer storage and speed. For example to obtain the general solution of the nonhomogeneous system (7) when the general solution of the homogeneous system $H^{ij}(S/So)$ is known (see the formula (7)) we compute the determinants of (14x14) matrices. When we study the equations of third degree in terms of phase coordinates we compute the determinants of (34x34) matrices. From another point a view the computation of large determinants is not useful because the general form of the vector $\vec{V}^i(S)$ in (9) is independent from the degree of the phase vector $\vec{Z}^i(S)$ in (8).Then if we consider the system of equations up to the third order and higher degree in phase coordinates the time of computation increase only due to increasing the number of integrals.

Based on the analytical solution (7) we developed the program which is called CYCLOTRON for the computation of beam dynamics of isochronic cyclotron. The input program parameters are fixed by the geometry of central region and accelerating structure of accelerator. At every step we detect the particle position to account the accelerating electric field. The influence of electric field is essential in the central region of the accelerator. By the rule form this region we obtain the conserved amplitudes of radial and vertical betatron oscillations.

For finding electric field in the free gap of the puler and in accelerating regions of the dee at the present time we have used the following methods

- the electric field distribution is used as homogeneous in the accelerating regions,
- the exact solutions of Poisson equation for the given geometry of accelerating structure are searched,
- the electric field distribution in the central domain is obtained by modeling in electrolytic and magnetic container.

-use the analytical approximation of real electric field distribution.

The first approach is a very bad approximation. The second one give the exact distribution but is very time consuming. The third one give the electric field distribution but is related to computation of great number of experimental data. Because of the problems with computational time we haven't possibility to make the quick comparison between the different variants of the geometry of the central region of the accelerator.Then when we compute the ion motion in the accelerator the program CYCLOTRON uses the Gauss distribution of the electric field [2]. The horizontal component of the electric field in the dee free distance and in the source puler are obtained from

$$Exs=\frac{U_g}{(2\pi)^{1/2}C_n}\exp\left(-\frac{\Delta L^2}{2.C_n^2}\right)\cos[n\Omega(t-to)]$$

$$Exs=\frac{U_g}{(\pi/2)^{1/2}C_p}\exp\left(-\frac{\Delta L^2}{2\ C_p^2}\right)\cos[n\Omega(t-to)],$$

where $C_n=0.2HDD+0.4GD$, $C_p=0.2HD+0.8G$, HDD is the dee aperture,GD is the distance dee-ground, G is the distance source-puler, HD is the vertical aperture of the puler and Ug is the potential of the accelerating field. Here in first expression by ΔL we denote the modulus of the distance normal to the particle position at the center of the accelerating gap and in the second expression the distance in the normal direction to the position of the ion source.

For the vertical component of the electric field we have

$$Ey=\frac{\Delta\ LY_i}{C_n}Exs,$$

where Y_i is the vertical coordinate of the particle.

The results of the numerical computations using CYCLOTRON shows that we have a successful simulation of the beam dynamics during the full circle of the acceleration of the isochronous cyclotron.In particular this program is used to find the equilibrium orbits, to compute the beam energy dispersion due to distribution of the initial phases and phase motion of the particles during the acceleration.

REFERENCES

1. A.D.Dymnicov,E.A.Perelstein *Moment method in dynamics of charged particle beams*,Nuclear Instruments and Methods,148,1978,p.567.

2.J.M.Neuwland,N.Hazewindus *Some aspects of the design of a cyclotron central region*. Philips Res.Repts-29,1974,p.528-559

INTEGRATION OF SOME CLASSES OF LINEAR ORDINARY DIFFERENTIAL EQUATIONS

L.M.Berkovich
Dept. of Algebra & Geometry,Kulbyshev State Univ.
Kulbyshev,USSR

V.P.Gerdt, Z.T.Kostova
Joint Institute for Nuclear Research
Post Office P.O.Box 79, Moscow, USSR

M.L.Nechaevsky
Dpt. of Higher Mathematics, Kulbyshev Aviation Inst.
Kulbyshev, USSR

ABSTRACT

Two algorithms with their mathematical foundation concerning 2nd order OLDE are briefly described. One is intended to find a fundamental system of solutions and factorization for OLDE reducible to ones with constant coefficients by a variable transformation. Another allows to generate four-parameter families of OLDE with coefficients of preassigned (in the scope of the procedure) structures integrable in terms of a given equation. The algorithms have been implemented in the REDUCE computer algebra system.

One of the most effective means for investigating and solving differential equations is the method of factorization and transformations, the main ideas and results of which are developed in [1]. This method is constructive in its character and capable of algorithmization of its techniques and computer algebra application.

Among the ordinary linear differential equations (OLDE) those play a special role which are reduced to equations with constant coefficients by means of the local point transformation

$$y = v(x)\ z\ ,\ dt = u(x)\ dx\ ,\ u,v \in C^2 \tag{1}$$

to be called the Kummer - Liouville transformation.

Our further considerations of reducible equations are based on the following theorem.

Theorem. For OLDE of the form

$$y'' + a_1(x)y' + a_0(x)y = 0,\ a_k(x) \in C^k,\ (')=d/dx \tag{2}$$

to be reduced to the equation with constant coefficients

$$\ddot{z} + b_1\dot{z} + b_0z = 0,\quad b_1,b_0 = \text{const},\ (\dot{})=d/dt \tag{3}$$

by means of transformation (1), it is necessary and sufficient to fulfil one of these equivalent conditions:

(i) OLDE (2) admits factorization of the form

$$(D - \alpha_2(x))(D - \alpha_1(x))y = 0\ ,\ D=d/dx,$$

$$\alpha_2(x) = v'/v + u'/u + r_2u,\ \alpha_1(x) = v'/v + r_1u, \tag{4}$$

where $r_{1,2}$ are the characteristic roots of equation (3).

(ii) The transformation functions u(x) and v(x) satisfy therelations:

$$(1/2)(u''/u) - (3/4)(u'/u)^2 - (1/4)\delta u^2 = A_0(x),\ \delta = b_1^2 - 4b_0,$$

$$A_0 = a_0 - (1/2)a' - (1/4)a_1^2\ , \tag{5}$$

$$v = |u|^{-1/2}\exp((-1/2)\textstyle\int a_1dx + (1/2)b_1\int u\ dx). \tag{6}$$

The general solution of Eq.(5) can be expressed in terms of the linearly independent solutions $y_1(x)$ and $y_2(x)$ of OLDE (2):

$$u(x) = \exp(\textstyle\int a_1dx)(Ay_2^2 + By_1y_2 + Cy_1^2)^{-1},\ B^2-4AC = \delta. \tag{7}$$

Supposing u(x) known the function v(x) satisfies the OLDE

$$v'' + a_1(x)v' + [a_0(x) - b_0u^2(x)]v = 0. \tag{8}$$

There are some ways for solving OLDE in the scope of this approach. One is to search directly for a transformation of type (1), which reduces a given OLDE to the form (3), using an ansatz "powerful" enough to cover wide classes of OLDE. Such an explicit form of the transformation is presented in [2]:

$$y = |P|^{-1/4}\exp((-1/2)\int a_1 dx + (1/2)b_1\int |P|^{1/2}dx)\ z,$$

$$dt = |P|^{1/2}dx, \tag{9}$$

where the function P(x) is chosen, particularly, from the additive parts (summands) of the semi-invariant $A_0(x)$. The choice is correct if a value of the constant δ (discriminant) exists such that Eq.(5) is fulfilled identically in x.Then the factorization coefficients and fundamental system of solutions of the OLDE under consideration are obtained by:

$$\alpha_2 = (-1/2)(a_1 - (1/2)P'/P + (\delta|P|)^{1/2}),$$

$$\alpha_1 = (-1/2)(a_1 + (1/2)P'/P - (\delta|P|)^{1/2}); \tag{10}$$

$$y_{1,2} = |P|^{-1/4}\exp((-1/2)\int a_1 dx \pm (1/2)\int(\delta|P|)^{1/2}dx),\ \delta\neq 0;$$

$$y_1 = |P|^{-1/4}\exp((-1/2)\int a_1 dx),\ y_2 = y_1\int |P|^{1/2}dx),\ \delta=0. \tag{11}$$

The corresponding algorithm is briefly described below, more details are given in [2,3].

Algorithm SOLDE.

Given a 2-nd order OLDE (2) and functions $p_1(x)$ and $p_0(x)$ (the user's "advice", if any).

S1. $A_0 := a_0 - (1/2)a' - (1/4)a_1^2$;

If A_0= const then << $\delta := -4A_0$; go to S5 >>

else << $q_1 := p_0 - (1/2)a_1 p_1 - (1/2)p_1' + (1/4)p_1^2$;

$q_2 := a_0$; $q_3 := A_0$ >>.

S2. Determine the total number N of the summands q_i, i = 4:N+3, in the expressions for A_0 and q_1 (after grouping in each).

i:=1

S3. If q_i=const then i:=i+1

else << P:= q_i; d:= $(1/P)(P''/P^2 - (5/4)P'^2/P^2 - 4A_0)$;

if d=const then << δ:= d; go to S5 >> else i:=i+1 >>.

S4. If $i \le N+3$ then go to S3

else return "no transformation has been found".

S5. Determine $\alpha_2(x)$, $\alpha_1(x)$ and $y_1(x)$, $y_2(x)$ according to Eqs.(10) and (11) respectively and return them.

Another way of treating OLDE is the reverse of opposite to the previous one. Knowing the structure of a given semi-invariant $A_0(x)$ we can try to arrive at it by direct modification of those coefficients of OLDE with known solutions. The procedure is based on the Kummer-Liouville (K-L) transformation (1) and is called by their names.

The idea is to consider the accompanying OLDE (8) for v(x) as a modification of the initial OLDE.

Let us consider the OLDE:

$$y'' + a_{(1)}(x)y = 0 \qquad (a_{(1)})$$

with a known fundamental system of solutions $y_{1,2}(x)$. We can determine an appropriate transformation function u(x) by Eq.(7).Then redenoting the dependent variable in the corresponding OLDE for v(x) we have a new equation:

$$y'' + a_{(2)}(x)y = 0,$$

$$a_{(2)}(x) = a_{(1)}(x) - b_{0(1)}u^2_{(1)}(x),\ b_{0(1)} = (1/4)(b^2_{1(1)} - \delta_{(1)}) \qquad (a_{(2)})$$

with modified but integrable "potential" and known solutions (see expression (6)). The process can be continued up to reaching the desirable structure of $a_{(n)}(x)$. Moreover, because of the presence of four arbitrary parameters (b_1, δ, A or B, and C) 4- parametric OLDE families arise on each step of the K-L procedure. The number of these families grows according to geometric progression with ratio 8.

The scheme of this process is convenient to be considered as a tree (graph), where the root is the initial (generating) OLDE and the nodes on each level are the members of the next generation of equation families.

The connection between the solutions of OLDE of two neighboring generations is presented in the table where the subscript (n) is omitted for more obviousness.

More details and examples the reader can find in [3,4].

Table

Transformation function $u_{(n)}(x)$	Fundamental system of solutions $y_{(n+1)1,2}(x)$		
	1) $b_1 = \mathrm{Re}\, b \neq 0$	2) $b_1 = \mathrm{Im}\, b = iw$	3) $b_1 = 0$
1. $\delta = (\alpha_1\beta_2 - \alpha_2\beta_1)^2 > 0$ $u_{(n)}(x) = (1)$	(1,1)	(1,2)	(1,3)
2. $\delta = B^2 - 4AC < 0$ $u_{(n)}(x) = (2)$	(2,1)	(2,2)	(2,3)
3. $\delta = 0$ $u_{(n)}(x) = (3)$	(3,1)	(3,2)	(3,3)

Notation:

$$E(x) = ((\alpha_1 y_2 + \beta_1 y_1)(\alpha_2 y_2 + \beta_2 y_1))^{1/2}$$

$$P(x) = |(\alpha_1 y_2 + \beta_1 y_1)/(\alpha_2 y_2 + \beta_2 y_1)|$$

$$F(x) = (Ay_2^2 + By_1 y_2 + Cy_1^2)^{1/2}$$

$$Q(x) = \mathrm{atan}((2Ay_2 + By_1)/((-\delta)^{1/2} y_1))$$

$$R(x) = y_1/(\alpha y_2 + \beta y_1).$$

1. $\delta = (\alpha_1 \beta_2 - \alpha_2 \beta_1)^2 > 0,$

$$u_{(n)}(x) = (\alpha_1 y_2 + \beta_1 y_1)^{-1} (\alpha_2 y_2 + \beta_2 y_1)^{-1}; \qquad (1)$$

1) $b_1 = \mathrm{Re}\, b_1 \neq 0,$

$$y_{(n+1)1,2}(x) = E(x)[P(x)]^{(\pm b_1/(2\delta^{1/2}))}; \qquad (1,1)$$

2) $b_1 = \mathrm{Im}\, b_1 = iw,$

$$y_{(n+1)1,2}(x) = E(x) \begin{matrix}\cos\\ \sin\end{matrix} \{(w/(2\delta^{1/2}))\log P(x)\}; \qquad (1,2)$$

3) $b_1 = 0$,

$$y_{(n+1)1,2}(x) = \left\{ \begin{matrix} E(x) \\ E(x)\log P(x) \end{matrix} \right\}. \qquad (1,3)$$

2. $\delta = B^2 - 4AC < 0$,

$$u_{(n)}(x) = (Ay_2^2 + By_1y_2 + Cy_1^2)^{-1}; \qquad (2)$$

1) $b_1 = \mathrm{Re}\ b_1 \neq 0$,

$$y_{(n+1)1,2}(x) = F(x) \begin{matrix} \cosh \\ \sinh \end{matrix} \{(b_1/(-\delta)^{1/2})Q(x)\}; \qquad (2,1)$$

2) $b_1 = \mathrm{Im}\ b_1 = iw$,

$$y_{(n+1)1,2}(x) = F(x) \begin{matrix} \cos \\ \sin \end{matrix} \{(w/(-\delta)^{1/2})Q(x)\}; \qquad (2,2)$$

3) $b_1 = 0$,

$$y_{(n+1)1,2}(x) = \left\{ \begin{matrix} F(x) \\ F(x)Q(x) \end{matrix} \right\}. \qquad (2,3)$$

3. $\delta = 0$,

$$u_{(n)}(x) = (\alpha y_2 + \beta y_1)^{-2}; \qquad (3)$$

1) $b_1 = \mathrm{Re}\ b_1 \neq 0$,

$$y_{(n+1)1,2}(x) = (\alpha y_2 + \beta y_1) \begin{matrix} \cosh \\ \sinh \end{matrix} \{(b_1/(2\alpha))R(x)\}; \qquad (3,1)$$

2) $b_1 = \mathrm{Im}\ b_1 = iw$,

$$y_{(n+1)1,2}(x) = (\alpha y_2 + \beta y_1) \begin{matrix} \cos \\ \sin \end{matrix} \{(w/(2\alpha))R(x)\}; \qquad (3,2)$$

3) $b_1 = 0$,

$$a_{(n+1)} = a_{(n)}, \quad y_{(n+1)} = y_{(n)} . \qquad (3,3)$$

<u>Algorithm GOLDE</u>

Given a 2-nd order OLDE $y''+a(x)y = 0$ with known fundamental system of solutions $y_{1,2}(x)$ and two arrays $\{r_i\},\{c_i\}$, $i=1:n$, where n is the number of the desirable generations (not counting the initial one). The sequence of the pairs (r_i,c_i), $i=1:n$, determines the path along the tree leading to the needed OLDE $(a_{(n)})$:

$$y'' + a_{(n)}(x)y = 0,$$

$$a_{(n)} = a_{(1)} - (1/4)\sum_{l=1}^{n-1}(b^2_{1(l)} - \delta_{(l)})u^2_{(l)}.$$

G1. $a_{(1)} := a(x)$; $y_{(1)1} := y_1(x)$; $y_{(1)2} := y_2(x)$.

i:=1

G2. Determine $u_{(i)}(x)$ using the expression in the r_i-th row of the table with $y_{(i)1}(x)$ and $y_{(i)2}(x)$.

G3. $a_{(i+1)} := a_{(i)} - (1/4)(b^2_{1(i)} - \delta_{(i)})\, u^2_{(i)}$.

G4. Determine $y_{(i+1)1}(x)$ and $y_{(i+1)2}(x)$ by the formulas atthe intersection of the r_i-th row and c_i-th column of the table with $y_{(i)1}(x)$ and $y_{(i)2}(x)$.

G5. i:=i+1;

If i ≤ n then go to G2

else return $a_{(n+1)}(x)$, $y_{(n+1)1}(x)$ and $y_{(n+1)2}(x)$.

The algorithms SOLDE and GOLDE have been implemented in the REDUCE computer algebra system.

REFERENCES

1. Berkovich L.M., *Factorization and transformations of ordinary differential equations*, (Saratov University Publishers, Saratov,1989), in Russian.

2. Berkovich L.M., Gerdt V.P., Kostova Z.T., and Nechaevsky M.L., *Second order reducible linear differential equations*, (Preprint JINR E5-89-141, Dubna , 1989).

3. Berkovich L.M. and Nechaevsky M.L., in: *Modern group analysis*, (Elma Publishers, Baku, 1989) , pp.53-63, in Russian.

4. Berkovich L.M., Gerdt V.P., Kostova Z.T., and Nechaevsky M.L., *Computer algebra generating related 2nd order linear differential equations*, (To be published in JINR, Dubna ,1990).

AUTOMATION OF NUMERICAL SOLUTION OF PDE SYSTEMS

R. Liska
Faculty of Nuclear Science and Physical Engineering
Technical University of Prague
Břehová 7, 115 19 Prague 1, Czechoslovakia

Abstract

The article deals with automation of the process of numerical solving of partial differential equations systems (PDES) by means of computer algebra. For solving PDES the finite difference method is applied. The computer algebra system REDUCE and the numerical programming language FORTRAN are used in the methodology presented, its main aim being to speed up the process of preparing numerical programs for solving PDES. Quite often, especially for complicated systems, this process is a tedious and time consuming task.

In the process several stages can be found in which computer algebra can be used for performing routine analytical calculations, namely: transformation of differential equations into different coordinate systems, discretisation of differential equations, analysis of difference schemes, and generation of numerical programs. The presented FIDE package automate some of these tasks.

1 Introduction

Mathematical models of many problems in almost all branches of science are described by PDES which can be solved only numerically. One of the most frequent methods for solving PDES is the finite difference method, in which derivatives are replaced by quotients of discrete values differences. Classical numerical programs for automatic solution of PDES are limited only to some classes of PDES. Real problems are usually complicated and require a special method of solution, which a general program cannot ensure. During the solution of a real complicated problem by the finite difference method one has to perform all the steps of the method. In the individual steps many mechanical, analytical computations, which can be accomplished by a computer algebra system, have to be performed. Using computer algebra speeds up the process of designing numerical programs and decreases the probability of bugs in these programs. We intend to automate the mechanical analytical computation only. By automating the whole process we would reach the same conclusion as classical programs, i.e. a limited set of solvable problems. In order to allow the user to influence the process of solution, the control over this process should remain in his hands.

We present here the FIDE package (FInite difference method for solving partial Differential Equations systems), implemented in computer algebra system REDUCE [1]. The package automates mechanical analytical computations performed during designing a numerical program for solving PDES by the finite difference method. The individual modules of the FIDE package are designed for experts to help them in mechanical tasks and allow them to concentrate on the concepts of solution. The package can be used for any PDES and is designed especially for complicated systems, for which its gain increases.

A similar approach was presented in [2], where the MACSYMA package performing such tasks was implemented. The other PhD. thesis [3] deals with designing numerical programs for a magnetohydrodynamic model of axisymmetric plasma by using the MACSYMA computer algebra system. Several works of Wang et al., e.g. [4], describe development of the FINGER package, written under MACSYMA, and performing automation of the finite element method for solving PDE. Another group of papers of Steinberg and Roache, e.g. [5], uses MACSYMA in preparing numerical programs for solving simple elliptic differential equations in complex 2-D or 3-D regions.

2 Transformation of differential equations into orthogonal coordinate systems

The EXPRES module performs transformation of scalar, vector, or tensor differential equations containing vectors, tensors, and vector differential operators as e.g. gradient into the component form in the given 2-D or 3-D orthogonal coordinate system. The coordinate system is defined by expressing Cartesian coordinates in terms of new coordinates. The transformed vector differential operators are computed with the help of metric coefficients and the Christoffel symbols of the given coordinate system. Complete vector algebra in component form is implemented in the module. All arithmetic operators, differential operators (gradient, divergence, Laplacian, curl, directional derivative, usual derivative), and products (scalar, vector, outer, double scalar) can be applied to any scalar, vector, or tensor quantities to which it makes sense. Vector and tensor variables can be used.

There also exist several programs [2,6] capable of performing such transformations. The EXPRES module remains in the FIDE package because it is compatible with the IIMET module. The EXPRES module has been implemented according to the similar module from [2] written in MACSYMA.

3 Discretization by integro-interpolation method

The present version of the FIDE package deals only with orthogonal grids, which can be uniform or nonuniform. In every coordinate the second grid with grid points halfway between the original ones is defined in the integro-interpolation method [7]. This method was chosen because it is quite universal, it can be applied to nonlinear PDES, and in linear cases it produces conservative difference equations. The basic principle of the integro-interpolation method is in integrating the given PDES over the whole cell of a grid, from which the difference scheme results. If direct integration is not possible, some kind of interpolation is used. Every quantity appearing in a PDES can be discretized in every coordinate on the original grid or on the second one. Every equation of the system can be integrated over the cell of the original grid or of the second one.

The decision on which grids the individual quantities (looked for functions of coordinates) are to be defined and over which grid cells the individual equations are to be integrated is made so as to minimize the number of interpolations used during integrating the whole PDES. Such a decision is really not easy, especially for large systems.

The IIMET module for discretization by the integro-interpolation method accomplishes this decision by either heuristically driven, or full search and from input PDES the module produces the system of difference equations. The discretization of individual basic types of terms appearing in PDES can be defined, and the discretization can be controlled in several manners, so that almost any discretization, according to the user's wish, can be achieved by this module. Input PDES can also include vector or tensor equations, in which case these equations are converted into scalar ones by the EXPRES module before the discretization begins.

4 Analysis of difference scheme: approximation and stability

In order to result in a successful numerical algorithm, a difference scheme ought to have certain properties. Such main properties of a difference scheme are approximation and stability [8]. The order of approximation is expressed in powers of grid steps in individual coordinates. It is obtained by expanding each discrete function value in the difference scheme into the Taylor series. After such expansion and rearranging, the actual PDES solved by the difference scheme is obtained by neglecting all grid steps and the order of approximation is obtained by the lowest powers of grid steps appearing in the rearranged equations. The APPROX module determines the actual PDES solved by a difference scheme and the order of approximation of the scheme. The module enables one to define the center of Taylor expansion and the number of its terms in individual coordinates.

One of the most important properties of a difference scheme with a time coordinate is its stability. Very briefly speaking, the difference scheme is stable if any numerical error in its solution does not grow as the solution continues with increasing the time coordinate. The instability usually causes growing oscillations in the numerical solution. The most frequent stability analysis approach is the Fourier method. It performs Fourier expansion of discrete function values in a difference scheme in all non-time coordinates, and for every Fourier component, it constructs the so called amplifications matrix for transition from one time layer to another. The von Neumann stability condition then states that all eigenvalues of the amplification matrix, i.e. roots of its characteristic polynomial, must lie in the unit circle in the complex plane. The CHARPOL module computes the amplification matrix of a difference scheme and its characteristic polynomial. The difference scheme has to be linear and two-step in time coordinate. A nonlinear scheme must be linearized prior to stability analysis. A multi-step difference scheme can be transformed to an equivalent two-step one. For a detailed description of the algorithm and the module see [9].

For systems of more than two difference equations, i.e. characteristic polynomials of degree higher than 2, checking the von Neumann condition is not easy. After excluding the roots $\lambda = 1$, the characteristic polynomial $P(\lambda)$ is transformed by conformal mapping $\lambda = (z+1)/(z-1)$ to the polynomial $Q(z) = (z-1)^k P(\lambda)$ (k is the degree of polynomial P). The following proposition holds: all roots λ_j of polynomial $P(\lambda)$ lie in the unit circle $|\lambda_j| < 1$, iff all roots z_j of the polynomial $Q(z)$ have a negative real part $\mathrm{Re}(z_j) < 0$. By this proposition we transform the von Neumann condition to the Routh-Hurwitz problem [10], for which there exist several algorithms of solution [11] based on the coefficients of polynomial $Q(z)$. In [9] the Routh method is used. The HURWP module implements the Hurwitz method utilising the Lienard-Chipart stability criterion. As a result, from the characteristic polynomial of the amplification matrix of a difference scheme one obtains several inequalities connected by logical operators AND and OR, which have to hold for the scheme to be stable. Earlier implementation of the Hurwitz method can be found in [12].

5 Generation of numerical programs

After obtaining and analyzing a difference scheme, the method of its solution has to be chosen and implemented in a numerical program. Working with schemes in the REDUCE computer algebra system, the very helpful package GENTRAN [13] can be used for this purpose. It seems to be very useful to work with numerical program libraries containing many subprograms for solving systems of equations, i.e. our difference scheme. A computer algebra system can generate for the user a block of numerical program, which performs filling up the parameters of a subprogram, calling the subprogram and returning the computed solution to determined arrays or variables. Solving a system of linear difference equations we often need to solve a system of linear algebraic equations with a band matrix (after suitable ordering of variables). For solving such a system with a band matrix there exist fast and reliable algorithms implemented in many numerical libraries. The LINBAND module generates automatically the block of FORTRAN code for solving such a system of equations utilising LEQT1B routine from the IMSL library.

6 Conclusion

The basic principles of methods and algorithms implemented in the FIDE package have been shown. The package will be included in the REDUCE program library managed by Anthony C. Hearn, where its full documentation can be found. It is written mainly in the symbolic mode of REDUCE system and contains more than 4000 lines. On two real physical problems the applicability of the package has been shown. Its main advantages are speeding up the process of designing numerical programs for solving PDES and decreasing the probability of bugs appearing in them. Our experience with the package shows another advantage, namely the possibility of very quick modification of the problem solved or its difference scheme, e.g. adding some earlier neglected term to the equations or other type of discretization of some terms.

The FIDE package has been applied to two real physical problems [14]. The first one is the nonlinear Schrödinger equation, which describes several physical phenomena, e.g. the Langmuir waves in plasma. The second one is the Fokker-Planck equation in diffusive approximation, which is used to describe the kinetics of electrons in plasma, including interactions of electrons with a laser beam through inverse bremsstrahlung. The numerical programs have been tested and compared with similar calculations published so far.

References

[1] A. C. Hearn. *REDUCE User's Manual, Version 3.3.* Technical Report CP 78 (Rev. 7/87), The RAND Corporation, Santa Monica, 1987.

[2] M. C. Wirth. *On the Automation of Computational Physics.* PhD thesis, Lawrence Livermore National Laboratory, Livermore, 1980. UCRL-52996.

[3] G. O. Cook Jr. *Development of a Magnetohydrodynamic Code for Axisymmetric High-B Plasmas with Complex Magnetic Fields.* PhD thesis, Lawrence Livermore National Laboratory, Livermore, 1982. UCRL-53324.

[4] P. S. Wang. FINGER: a symbolic system for automatic generation of numerical programs in finite element analysis. *J. Symb. Comp.*, 2(3):305–316, 1986.

[5] S. Steinberg and P. J. Roache. Symbolic manipulation and computational fluid dynamics. *J. Comp. Phys.*, 57:251–284, 1985.

[6] D. Harper. Vector33: A REDUCE program for vector algebra and calculus in orthogonal curvilinear coordinates. *Comp. Phys. Comm.*, 54:295–305, 1989.

[7] V. Vemuri and W. J. Karplus. *Digital Computer Treatment of Partial Differential Equations.* Prentice-Hall, Englewood Cliffs, 1981.

[8] R. D. Richtmyer and K. W. Morton. *Difference Methods for Initial Value Problems.* John Wiley & Sons, New York, 1967.

[9] V. Gansha and R. Liska. Application of the REDUCE computer algebra system to stability analysis of difference schemes. In E. Kaltofen and S. M. Watt, editors, *Computers and Mathematics '89*, pages 119–129, Springer-Verlag, New York, 1989.

[10] V. G. Gansha and E. V. Vorozhtsov. The stability analysis of difference schemes by numerical solution of the generalized Routh-Hurwitz problem. *Comp. Phys. Comm.*, 43:209–216, 1987.

[11] F. R. Gantmacher. *Applications of the Theory of Matrices.* Interscience Publishers, New York, 1959.

[12] S. I. Mazurik. *Algoritmy rešenija zadači o lokalizacii kornej simvolnogo mnogočlena, ich realizacija na EVM i primenenie.* Technical Report 24-85, Institut Teoretičeskoj i prikladnoj mechaniky, SO AN SSSR, Novosibirsk, 1985.

[13] B. L. Gates. *GENTRAN User's Manual, REDUCE Version.* Technical Report, The RAND Corporation, Santa Monica, 1987.

[14] R. Liska and L. Drska. FIDE: A REDUCE package for automation of FInite difference method for solving pDE. In M. Nagata, editor, *ISSAC-90* , ACM, New York, 1990, (to be published).

APPLICATION OF COMPUTER ALGEBRA TO INVESTIGATION OF STABILITY OF DIFFERENCE SCHEMES

S. I. Serdyukova

Joint Institute for Nuclear Research

Dubna , USSR

ABSTRACT

The state of affairs and perspectives of application of computer algebra to the investigation of the difference schemes stability are discussed.

Extensive application of numerical methods for solving differential equations in applied problems has stimulated the difference schemes stability theory. Far from all theoretical results in this field have found practical use . A rather complicated mathematical apparatus is used to prove the difference schemes stability criteria. Without this knowledge the algorithm of stability verification is difficult to understand . Besides , verification of difference schemes stability criteria , as a rule , involves labour consuming and often unrealizable by hand calculations. The progress of computer algebra gives real possibilities to develop standard programs for difference schemes stability verification . This brings up some particular problems which can be solved by special methods . Here we discuss the application of computer algebra to investigation of stability of difference initial-boundary value problems of the form :

$$v_{\nu}^{n+1}=\sum_{l=-r1}^{r2} A_l v_{\nu+l}^{n} \;,\; n\geq 0 \;,\; \nu\geq 1 \;, \qquad (1)$$

$$v_{\nu}^{0} = f_{\nu} \;,$$

$$v_{m}^{n}= \sum_{l=1}^{s1} C_{lm} v_{l}^{n} \;,\; m=-r1+1,\dots,0. \qquad (2)$$

Here v_{ν}^{n} are k-dimensional vectors, A_l, C_{lm} are constant matrices, A_{-r1}, A_{r2} are non-singular, $r1\geq 1$. Such model problems are usually used to study instabilities observed near boundaries . Full theoretical investigation of numerical algorithms for solving complicated applied problems , as a rule , is as unreal as solving them without computer . However , when the instability cause is understood, we usually succeed finding an efficient stable algorithm of solving an original applied problem . The problem (1),(2) can be written in the operator form

$$v^{n+1}=Gv^{n}.$$

Problem (1),(2) is stable in space C, if $\|G^{n}\| \leq c$ for all $n\geq 0$; here G is an operator of transition from layer to layer. If there is no stability in space C the grows rate of $\|G^{n}\|$ gives the instability degree of problem (1),(2). As the system (1) is explicit the corresponding Cauchy problem stability in space C^{1} is necessary for stability in space C of the mixed problem (1),(2). Additionally, the spectrum of G must lie inside a unit circle $|z|\leq 1$. Otherwise, $\|G^{n}\|$ grows exponentially with growing n . To get necessary and sufficient stability conditions , we write G^{n} as an integral of the resolvent

$$G^{n}= - \frac{1}{2\pi i} \oint_{\Gamma} (G-zI)^{-1} z^{n}\, dz.$$

The contour Γ includes the spectrum of G in its interior. The G spectrum points are singularities of the resolvent $(G-zI)^{-1}$. We need the explicit form of the resolvent in the vicinity of a unit circle $|z|=1$. The problem reduces [2] to the solution of

$$\sum_{l=-r1}^{r2} A_l u_{\nu+l} - zu_\nu = f_\nu, \quad \nu\geq 1,$$
$$u_m = \sum_{l=1}^{s1} C_{lm} u_l, \quad m=-r1+1,\dots,-1,0. \tag{3}$$

Here $\{f_\nu\}$ is a given sequence, $\{u_\nu\}$ is an unknown one. After the change of variable

$$Y_\nu = (u_{\nu+r2-1},\dots,u_\nu,\dots,u_{\nu-r1})^*$$

the system (3) takes the form

$$Y_{\nu+1} = M(z)Y_\nu + F_\nu, \quad \nu\geq 1,$$
$$[0,I]Y_1 = \sum_{l=r1}^{s1} [0,B_l]Y_{l+1}, \tag{4}$$

$$B_{r1} = \begin{bmatrix} C_{r1,0} & \dots C_{1,0} \\ \vdots & \vdots \\ C_{r1,-r1+1} & \dots C_{1,-r1+1} \end{bmatrix}, \qquad B_l = \begin{bmatrix} C_{l,0} & 0\dots 0 \\ \vdots & \vdots \quad \vdots \\ C_{l,-r1+1} & 0\dots 0 \end{bmatrix}, \quad r1+1 \leq l \leq s1$$

$M(z)$ - a resolvent matrix, $F_\nu = (A_{r2}^{-1} f_\nu, 0,\dots,0)^*$:

$$M(z) = \begin{bmatrix} -A_{r2}^{-1} A_{r2-1} & -A_{r2}^{-1} A_{r2-2} & \dots & -A_{r2}^{-1}(A_0 - zI) & \dots & -A_{r2}^{-1} A_{-r1} \\ I & 0 & \dots & \cdot & \cdot\cdot 0 & 0 \\ 0 & I & \dots & \cdot & \cdot\cdot 0 & 0 \\ \cdot & \cdot & \dots & \cdot & \dots & \cdot \\ 0 & 0 & \dots & 0 & \dots I & 0 \end{bmatrix}.$$

To construct the resolvent, we have to factor M, which has a high dimension $(r1+r2)k$. The problem can be solved in the general form due to the following. The eigenvalues (e.v.) κ of the resolvent matrix and e.v. λ of the characteristic matrix $D = \sum_{l=-r1}^{r2} A_l e^{il\phi}$ are one-to-one valued algebraic functions. If we know the structure of λ we can define the structure of κ. Moreover, if the system (1) is hyperbolic the characteristic and resolvent matrices have the same quasi-jordan normal form. In the construction of the quasi-jordan normal form the resolvent matrix factors automatically. It follows from von Neumann condition: $|\lambda(e^{i\phi})|\le 1$, $0\le\phi\le 2\pi$, that $|\kappa(z)|\ne 1$ for $|z|>1$. From this and from $\kappa(z)$ continuity it follows[2] that for $|z|>1$ the spectrum of $M(z)$ divides into two non-intersecting sets: the set of κ with $|\kappa(z)|<1$ for $|z|>1$ and the set of κ with $|\kappa(z)|>1$ for $|z|>1$. With changing z outside the unit circle κ can not go from one set to the other. Asymptotic behaviour at infinity shows that exactly r1*k e.v. κ belong to the first set and exactly r2*k e.v. κ belong to the second one. When z approaches the unit circle, κ can have limit values equal to the unit in absolute value. The point $e^{i\psi_0}$ is called defining if in this point at least one κ is equal to the unit in absolute value, $\kappa(e^{i\psi_0})=e^{i\phi_0}$. In this case $e^{i\phi_0}$ is a defining point of the characteristic matrix : for some λ $\lambda(e^{i\phi_0})=e^{i\psi_0}$. The converse statement is true . If the Cauchy problem is stable in space C then e.v. of the characteristic matrix in vicinity of defining points have the expansions of the form

$$\lambda(e^{i\phi})=\exp\{i\psi_0+i\gamma(\phi-\phi_0)+(i\alpha-\beta)(\phi-\phi_0)^{2\mu}+O((\phi-\phi_0)^{2\mu+\theta})\}.$$

Here $\phi_0, \psi_0, \gamma, \alpha$ are real, $\beta>0$, $\theta>0$, μ - integer. It is well known[1] that the difference Green function is located near the "difference characteristics " $\nu+\gamma n=0$. Further we suppose that the system (1) is hyperbolic: all λ for all defining points have $\gamma\ne 0$. Then the only κ correspond to every λ :

$$\kappa(e^{i\psi})=\exp\{i\phi_0+i\frac{(\psi-\psi_0)}{\gamma} + \frac{(\beta-i\alpha)}{\gamma^{2\mu+1}}(\psi-\psi_0)^{2\mu}+ \dots \} .$$

From this it follows that λ with outgoing characteristics($\gamma<0$) generate κ with $|\kappa(z)|<1$ for $|z|>1$ and λ with ingoing ($\gamma>0$) characteristics generate κ with $|\kappa(z)|>1$ for $|z|>1$. Denote the principal part of κ as $\bar{\kappa}$:

$$\bar{\kappa}(e^{i\psi}) = \{i\phi_0+i\frac{(\psi-\psi_0)}{\gamma} + \frac{(\beta-i\alpha)}{\gamma^{2\mu+1}}(\psi-\psi_0)^{2\mu} \}.$$

For every defining point $e^{i\psi_0}$ e.v. of M split into classes $K_0, K_1, \dots, K_s, K_\infty$, $s=s(\psi_0)$. K_0 includes κ with e.v. $|\kappa(e^{i\psi_0})|<1$, K_∞ includes κ with $|\kappa(e^{i\psi_0})|>1$. The main e.v. $\kappa_i, \kappa_j \subset K_\xi$, $1\le\xi\le s$, if $\bar{\kappa}_i\equiv\bar{\kappa}_j$.

Theorem 1. *If the Cauchy problem is stable in space C and the system (1) is hyperbolic, then for every defining point* $e^{i\psi_0}$ *there exist*[3] *a constant* ρ *and analytic non-singular in* $|\psi-\psi_0|<\rho$ *matrix* $T(\psi)$ *that reduces* $M(e^{i\psi})$ *to the quasi-jordan normal form*

$$TMT^{-1}=\begin{bmatrix} M_{11} & 0 \\ 0 & M_{22} \end{bmatrix}=\begin{bmatrix} M_1 & 0 & 0 & 0 \\ 0 & A & 0 & 0 \\ 0 & 0 & C & 0 \\ 0 & 0 & 0 & M_2 \end{bmatrix}.$$

All matrices here have no fractional powers in their expansions. K_0 *and* K_∞ *correspond to* M_1 *and* M_2 *respectively. The matrices* A, C *have a block-diagonal structure. The classes* K_ξ *of e.v.* κ *with* $|\kappa(z)|<1$ *for* $|z|>1$ *correspond to simple quasi-jordan blocks on a diagonal of* A. *The classes* K_ξ *of e.v.* κ *with* $|\kappa(z)|>1$ *for* $|z|>1$ *correspond to quasi-jordan blocks on a diagonal of* C. *The matrices* M_{11}, M_{22} *have orders* r1*k, r2*k *respectively;* A *has an order* l_1, C *has an order* l_2. *So,* M(z) *is factored.*

The explicit solution of system (4) is

$$w_\nu^1 = M_{11}^{\nu-1}w_1^1 + \sum_{\xi=1}^{\nu-1} M_{11}^{\nu-1-\xi}(TF_\xi)^1, \quad \nu\ge2,$$

$$w_\nu^2 = -\sum_{\xi=\nu}^{\infty} M_{22}^{\nu-1-\xi}(TF_\xi)^2, \quad \nu\ge1,$$

$$K_1w_1^1 = K_2w_1^2 + \sum_{\xi=1}^{s1} B_\xi^1(TF_\xi)^1 + B_\xi^2(TF_\xi)^2.$$

As w_ν^1, w_ν^2 we denote vectors of upper r1*k and lower r2*k components of $w_\nu = Ty_\nu$; K_1 , K_2 are analytic matrices:

$$K_1 = T^-_{21} - \sum_{\xi=r1}^{s1} B_\xi T^-_{21} M^\xi_{11}, \qquad K_2 = -T^-_{22} + \sum_{\xi=r1}^{s1} B_\xi T^-_{22} M^\xi_{22},$$

$$T^{-1} = \begin{bmatrix} T^-_{11} & T^-_{12} \\ T^-_{21} & T^-_{22} \end{bmatrix} \begin{matrix} r2*k \\ r1*k \end{matrix} .$$
$$\qquad\quad r1*k \quad r2*k$$

If the matrix $K_1(z)$ is singular in $z=z_0, |z_0|\geq 1$,then z_0 is [2] a spectrum point of G. Note that at the points z^*_0 , $|z^*_0|=1$,$K_1^{-1}(z)$ can have singularity only of the integer order. As K(z) is an analytic, G has only a finite number of spectrum points on the unit circle $|z|=1$. Let $l_2(z^*_0)$ be the number of κ, $|\kappa(z^*_0)|=1$, greater than unit in absolute value for $|z|>1$.

Theorem 2. *The difference initial-boundary value problem* (1),(2) *is stable in space* C *iff* [3] *conditions* 1-3 are fulfilled :

1. *The corresponding Cauchy problem is stable* [1] *in space* C.
2. G *has no spectrum points* z_0, $|z_0|>1$.
3. *At the spectrum points* $z^*_0, |z^*_0|=1$, *the elements of* $K_1^{-1}(z)$ *can have poles of the first order only, elements of the first* $l_2(z^*_0)$ *columns of the matrix* $K_1^{-1}K_2(z)$ *have no singularities on the unit circle* $|z|=1$. *If conditions* 1,2 *are fulfilled and condition* 3 *is not then* $\|G^n\|$ *grows as* n^s, *where* s *is a maximum deviation of addmissible singularity orders over all* z^*_0.

It is reasonable to ask why we need it and how we can use it in real life. As a rule, only von Neumann conditions are verified. In the case of the difference systems the Cauchy problem is stable in space L_2 with additional restrictions on outdiagonal elements of characteristic matrix triangle form (see [4,5]). The stability in space C takes place with additional restrictions on λ behaviour in the vicinity of the defining points. Why do we need the stability in space C ? It is well known that in numerical modelling the Gibbs phenomenon is observed in the vicinity of discontinuity. It was proved theoretically [6] that in the case of C stability weak exponentially decreasing oscillations arise in the vicinity of discontinuity. If there is no stability in space C, there arise oscillations essentially perturbing the solution. We met the discontinuity solutions in the numerical simulations of fluxons dynamics[7]. The fluxons dynamics in the systems with mycrohomogeneity is described by the equations

$$u_{tt}=u_{xx}-(1-\mu\delta(x-x_0))\sin(u)-au_t\ ,\quad t\ge 0\ ,-1\le x\le 1\ ,$$

$$u_x(-1)=u_x(1)=0,\quad u(x,0)=f(x),\quad u_t(x,0)=0.$$

In $x=x_0$ the component of the magnetic field u_x , in general , has discontinuity : $u_x(x_0+0)-u_x(x_0-0)=-\mu\sin(u(x_0))$. We used [7] the Rusanov scheme showing a good performance in computation of the discontinuity solutions of gas dynamics equations[8]. In the vicinity of $\phi=0$ the characteristic function (we mean the Rusanov scheme for $u_t=u_x$) has the expansion [9]

$$f(e^{i\phi}) = \exp\{\ i\sigma\phi - \frac{\omega-4\sigma^2+\sigma^4}{24}\phi^4 + i\left[\frac{\sigma(4-\sigma^2+\sigma^4)}{24} + \frac{\sigma(\sigma^2-1)(4-\sigma^2)}{120}\right]\phi^5$$

$$+\frac{(1+3\sigma^2)(\omega-4\sigma^2+\sigma^4)+\sigma(\sigma^2-1)(4-\sigma^2)}{144}\phi^6\ \ldots\}\ .$$

With $4\sigma^2-\sigma^4<\omega<3\ ,0<\sigma\le 1$ the coefficients with the lowest nonlinear term (of even order) is negative, $\phi=0$ is the only defining point. So there is stability in space C . For such σ , ω the scheme has the third order accuracy. On the boundary of the stability domain , for $\omega=4\sigma^2-\sigma^4, 0<\sigma<1$, the scheme has the fourth order accuracy , but there is no stability in space C . In the construction of the expansion the use of computer is advisable. Calculations are hard , but we can work by hand. The matter is much worse in the investigation of the boundary conditions effects. The five-point Rusanov scheme in computation on the finite interval requires additional boundary conditions . With initially chosen boundary conditions string oscillations were observed to spread rapidly from the boundaries. Analytical investigation with the use of the system REDUCE showed [10] that the spectrum point $z_0=-1.063...$ causes instability . As a model we chose the wave equation with the same oscillations in computation . The spectrum is described by the system of polynomial equations:

$$\mathrm{Det}\, R(z,\kappa_1,\kappa_2,\kappa_3,\kappa_4)=0\ , \qquad P_8(\kappa_i,z)=0\ ,\ i=1,\dots,4\ .$$

Here $P_8(\kappa,z)$ - a polynomial of the eighth order to κ . The problem is reduced to the solution of the polynomial equation of one variable by using the symmetry of zeros of $P_8(\kappa,z)$. Pass to new variables $x=\kappa_3^{-1}+\kappa_4^{-1}$, $y=\kappa_3^{-1} * \kappa_4^{-1}$ and z. It follows from the Vieta theorem that

$$x=\frac{(1-31y)}{(1-4y^2)}\ , \qquad z=\frac{1}{48}\,(4y+x(1-x)+21+y^{-1})\ .$$

After a number of elementary transformations we have [10]

$$\mathrm{DetR}=z(\kappa_1\kappa_2\kappa_3\kappa_4)^2(\kappa_1^{-1}-\kappa_2^{-1})(\kappa_3^{-1}-\kappa_4^{-1})\mathrm{DetR}'(x,y,z),$$

where $r'_{11}=-48z+2(31/4-xy)+60-4y$, $r'_{12}=-x-14-1/(2y)$ and so on . Using the system REDUCE we calculate successively :

$$\mathrm{Det}\ R'(x,y,z)\ ,\ \mathrm{Det}\ R'(x,y)\ ,\ \mathrm{Det}\ R'(z)\ ,$$

$$\mathrm{Det}\ R'(y)=P(y)/Q(y)\ , \quad P(y)=\sum_{i=0}^{18} a_i\ y^i\ ,$$

where $a_{18}=-8978432$, $a_{17}=330637312$ and so on [10] ($\omega=2$, $\sigma=1/2$). It is impossible to calculate such a polynomial by hand . As result , we have the spectrum point $z_0=-1.063...$. For the wave equation "good" additional conditions were found and we used them in numerical simulation of fluxons dynamics .

In N.E.Mazepa's dissertation for a practically important class of difference initial-boundary value problems the algorithm was developed[11] reducing the complicated problem of G spectrum determination to the solution of polynomial equation of one of the variables :

$$z\ ,\quad x=\sum_{i=1}^{r1*k}\kappa_i,\quad y=\prod_{i=1}^{r1*k}\kappa_i,\quad (|\kappa_i(z)|<1 \text{ for } |z|>1\ ).$$

The algorithm was realized in computer EC 1061 by using CAS REDUCE as program SPECTR [12]. The efficiency of this program was demonstrated by a great number of problems. We had the polynomials of high order (up to 50) with huge coefficients (integer numbers of orders up to 25). Calculations of such polynomials without computer is simply impossible. The polynomials zeros were found numerically. If the resulting polynomial depends on x or y then from x_i, y_i the "spectrum points" z_i are calculated. The variables x,y,z are connected by a set of equations which follow from the Vieta relations. This system is satisfied by any

$\sum_{i=1}^{r1*k}\kappa_i$, $\prod_{i=1}^{r1*k}\kappa_i$, z and not only by those containing κ_i, $|\kappa_i(z)|<1$ for $|z|>1$. The false spectrum points appear inevitably . The "spectrum points" z_i determined numerically do not satisfy the determinant equation identically. The determined z_i are checked and made more precise.

Remarkable results of difference schemes properties investigation by applying CAS have been obtained in SD of Acad.Sci.USSR by mathematicians of N.N.Janenko's school. These results will be reported at this conference. A wide range questions has been discussed , initial-boundary value problems are not considered. A program stability investigation IBSTAB has been developed by the swedish mathematician M.Thune. [13]

The program SPECTR has greater capabilities. A unique algorithm forms the basis of this program which allows one to reduce the spectrum calculation problem to the solution of one polynomial equation . It is interesting to extend the range of problems to which this algorithm can be applied. This is an algebraic problem. It is also the problem of great interest to automate investigation of spectrum points on the unit circle $|z|=1$. At last, it is interesting to automate the stability verification in space C of the Cauchy problem for linear difference equations systems. Here another interesting problem arises to develop suitable for realization in computer by using computer algebra algorithm for constructing the quasi-jordan normal form of analytic matrices whose powers generate a bounded semi-group.

REFERENCES

1. S. I. Serdyukova, *Zh. Vychisl. Mat. i Mat. Fiz.* (1967) 7, No. 3, p. 497.
2. H.-O. Kreiss, *Math. Comp.* (1968)22, No. 104, p. 703 .
3. S. I. Serdyukova, *Sbornik "Vychislitelnye protsessy i sistemy"* (to be published).
4. H.-O. Kreiss, *Math. Scand.* (1959)7, p. 71.
5. V. Ja. Urm, *Dokl Akad. Nauk. SSSR*(1961)139, No. 1, p. 40.
6. S. I. Serdyukova, *Zh. Vychisl. Mat. i Mat. Fiz.* (1971)11, No. 2, p. 411.
7. G. S. Kazacha, S. I. Serdyukova, A. T. Filippov, JINR, P11-8560, (Dubna, 1985).
8. V. V. Rusanov, *Fluid. Dynam. Trans.* (1969)4, p. 285.
9. S. I. Serdyukova, JINR, P5-10708, (Dubna, 1977).
10. N. E. Mazepa, S. I. Serdyukova, *Izv. vuzov. Matematika*, (1985)10, p. 55.
11. N. E. Mazepa, JINR, R11-89-382, (Dubna, 1989).
12. N. E. Mazepa, JINR, R11-89-383, (Dubna, 1989).
13. M. Thune, Uppsala University, Department of Comp. Sciences, (1984) Report No. 93.

BASES OF SYMBOLIC TRANSFORMATIONS FOR MATHEMATICAL GEODESY PROBLEMS

I. Trenkov
Central Laboratory of High Geodesy,
Bulgarian Academy of Sciences

M. Spiridonova
Institute of Mathematics with Computer Center,
Bulgarian Academy of Sciences

M. Daskalova
Faculty of Geodesy,
High Institute of Architecture and Construction

Abstract

An approach for solving some mathematical geodesy problems using the computer algebra system REDUCE and bases of symbolic transformations is presented. Computation of the geographical coordinates of a point on the earthly ellipsoid when the azimuths of other points with known coordinates are measured is considered as an example of geodesy problem.

In solving many mathematical geodesy problems different groups of formulae have to be used and tedious algebraic computations have to be done. The use of computer algebra systems for solving such problems might be quite usefull.

An experimental application of the computer algebra system REDUCE [6] for solving some mathematical geodesy problems has been done.The capabilities of a program package for manipulation with bases of symbolic transformations have been used. This package was developed in the Artificial Intelligence Department of the Institute of Mathematics with Computer Center at the Bulgarian Academy of Sciences [3].

One specific problem will be formulated. An algorithm for solving it is suggested in this paper.

The problem is:

On the earthly ellipsoid two different points I and K are given. It meens that their geographical coordinates (the geographical density B_I and the geographical longitude L_I of point I and respectively B_k and L_k of point K) are known. The azimuths in I and K to another point P are measured (A_{Ip} in point I and A_{kp} in point K). The geographical coordinates B_p and L_p of point P are unknown and they are to be calculated.

The suggested algorithm consists in following steps:

1. The geographical coordinates of the point P are presented as Taylor series about the point I andabout the point K [1,5]. The variables are the destinations s_{Ip} (between the points I and P) and s_{kp} (between the points K and P) These series are as follows:

$$B_p=B_I+\Delta B_{Ip}=B_I+\sum_{j=0}^{\infty}\frac{1}{j!}\left(\frac{d^jB}{ds_{Ip}^j}\right)_I s_{Ip}^j$$

$$L_p=L_I+\Delta L_{Ip}=L_I+\sum_{j=0}^{\infty}\frac{1}{j!}\left(\frac{d^jL}{ds_{Ip}^j}\right)_I s_{Ip}^j$$

$$B_p=B_k+\Delta B_{kp}=B_k+\sum_{j=0}^{\infty}\frac{1}{j!}\left(\frac{d^jB}{ds_{kp}^j}\right)_K s_{kp}^j$$

$$L_p=L_k+\Delta L_{kp}=L_k+\sum_{j=0}^{\infty}\frac{1}{j!}\left(\frac{d^jL}{ds_{kp}^j}\right)_K s_{kp}^j$$

The coefficients of these series are functions of the main quantities of mathematical geodesy and their derivatives.

2. The subexpressions only for ΔB_{Ip}, ΔL_{Ip}, respectively ΔB_{kp}, ΔL_{kp} will be considered. The first two of them are functions of s_{Ip}, the others are functions of s_{kp}.

By the formulae:

$$\Delta B_{Ik}= \Delta B_{kp}-\Delta B_{Ip}$$
$$\Delta L_{Ik}= \Delta L_{kp}\ \Delta L_{Ip}$$

power series of two variables (s_{ip} and s_{kp}) are obtained.

It should be noted that everywhere in this work the operations on power series are done on truncated power series.

3. The reversed power series of the obtained for ΔBD_{lk} and ΔL_{ik} power series are calculated. The algorithm for doing it is briefly described bellow. The obtained expresions for s_{ip} and s_{kp} are functions of B_{lk} and L_{lk}.

4. A substitution of the expressions for s_{lp} and s_{kp} is made in ΔB_{ip}, ΔL_{lp}, ΔB_{kp}, ΔL_{kp}.

5. The geographical coordinates of the point P are calculated using the formulae:

$$B_p = B_l + \Delta B_{lp} \qquad B_p = B_k + \Delta B_{kp}$$

$$L_p = L_l + \Delta L_{lp} \qquad L_p = L_k + \Delta L_{kp}$$

In such way "intersection" of the calculated geographical coordinates using two given points is done. They could be different as a result of errors in measurements of the azimuths.

6. In case of using two or more given points with measured azimuths, one more step to the algorithm has to be added for doing equalization by the least squares method.

Some steps and features of the described algorithm will be described briefly.

A. Computation of reversed power series of two variables.

Given the series:

$$u = \sum_{l,k=0}^{\infty} a_{lk} x^l y^k, \qquad a_{00}=0$$

$$v = \sum_{l,k=0}^{\infty} \bar{a}_{lk} x^l y^k, \qquad \bar{a}_{00}=0$$

The series:

$$x = \sum_{l,k=0}^{\infty} b_{lk} u^l v^k, \qquad b_{00}=0$$

$$y = \sum_{l,k=0}^{\infty} \bar{b}_{lk} u^l v^k, \qquad \bar{b}_{00}=0$$

are to be obtained.

The coefficients b_{ik} and $\bar{b}_{ik}$ are calculated using the formulae[2,4].

$$-b_{ik} = \sum_{\substack{1 < l+m \\ \leq i+k}} K^{ik}_{lm} D_{lm} \qquad -\bar{b}_{ik} = \sum_{\substack{1 < l+m \\ \leq i+k}} K^{ik}_{lm} \bar{D}_{lm}$$

In those formulae D_{lm} and $\bar{D}_{lm}$ are the following determinants:

$$D_{lm} = \begin{vmatrix} a_{lm} & a_{01} \\ \bar{a}_{lm} & \bar{a}_{01} \end{vmatrix} . D^{-1} \qquad \bar{D}_{lm} = \begin{vmatrix} a_{10} & a_{lm} \\ \bar{a}_{10} & \bar{a}_{lm} \end{vmatrix} . D^{-1}$$

$$D = \begin{vmatrix} a_{10} & a_{01} \\ \bar{a}_{10} & \bar{a}_{01} \end{vmatrix}$$

The coefficients K^{ik}_{lm} are calculated by the formula :

$$K^{ik}_{lm} = \sum_{\substack{\Sigma r_\lambda + \Sigma t_\mu = i \\ \Sigma s_\lambda + \Sigma u_\mu = k}} \prod_{\lambda,\mu}^{l,m} b_{r_\lambda s_\lambda} \bar{b}_{t_\mu u_\mu}$$

The high speed of calculations by this program is due to the efficient checking of the combinatorial conditions in the above formulae.

B. Use of bases of symbolic transformations.

On different steps of the described algorithm several groups of formulae are used. These formulae include the main geodesic quantities and their derivatives, some geodesic constants etc. Several groups of procedures are used as well. The mentioned above package for manipulation with bases of symbolic transformations (BST) is suitable for solvig such problems. A new version of the early developed packade with appropriate structure of the bases and tools for their building, maintenance and application is now developed.

In accordance with the accepted structure of BST, the geodesic formulae are presented by groups of substitution rules of REDUCE system. The algorithms for geodesic calculations are defined by a set of procedures . The substitution rules and procedures are stored as LISP expressions. The use of such BST allows flexibility in solving the described problem, usage of separete parts of the algorithm, convenient extension of the problem, etc.

C. Use of numeric calculations.

As a final result the geographical coordinates of points on theearthly ellipsoid are to be calculated numerically. The calculations are to be done by real numbers with arbitrary precision. If a large number of calculations using many given points are to be performed, a data base could be created. In this work a data base with simple structure was used.

The algorithm, formula bases, procedures, etc. described in this paper are the main part of a program package for solving mathematical geodesy problems being now under development.

References

1. Daskalova M., High Geodesy - Mathematical Geodesy, Technika, Sofia, 1980 (in Bulgarian)

2. Kenig P., K. Veize, Mathematical Foundations of High Geodesy and Cartography, Inostrannaia Litereratura, Moscow,1954 (in Russian)

3. Spiridonova, M.N., Organization and Application of Bases of Symbolic Transformations Using REDUCE System, Proc. of the Intern. Conf. on Computer Algebra and their Applications in Theoretical Physics, Dubna, 1985 , p. 28 (in Russian)

4. Trenkov I., M. Spiridonova, High Geodesy Problem Solving Using Computer Algebra Systems, Proc. of the First Intern. Seminar on Problems and Applications of Artificial Intelligence, Varna, 1987, p. 206 (in Russian)

5. Hristov V. Mathematical Geodesy, Technica, Sofia, 1987 (in Bulgarian)

6. Hearn A. C., REDUCE User's Manual, Version 3.3, RAND Publication CP/78 (Rev. 7/87)

EXTENDED POSSIBILITIES OF SOME COMPUTER ALGEBRA ALGORITHMS FOR SOLVING LINEAR DIFFERENTIAL AND DIFFERENCE EQUATIONS

S.A. ABRAMOV
USSR ACAD. OF SCI., COMPUTING CENTER, MOSCOW, USSR

Many computer algebra systems contain programs for solving certain type differential or difference equations. The linear equations with variable coefficients are quite interesting in this context.

Let the coefficients of equations be in some function field $\mathbb{P}$ closed with respect to an operation d/dx (or Δ in the case of difference equations). Let algorithm $\mathbb{A}_L$ for finding all solutions of the equations in some linear space L be known and L be not closed with respect to an integration (or summation). (For example, L is the rational or algebraic function field, the quasirational function space, etc.) Then one may extend the possibilities of $\mathbb{A}_L$ due to construction of all the solutions with their higher order derivatives (or differences) in L. The lecture demonstrates two algorithms of this construction. The role of this result for equations with polynomial coefficients is discussed.

Computer Algebraic Generation of Three-Point Difference Schemes with Arbitrary Order of Accuracy for Second-Order Nonlinear Ordinary Differential Equations with Piece-Smooth Coefficients

A. A. SAMARSKII
Institute of Applied Mathematics, Moscow, USSR

and

V. L. MAKAROV, L. D. GREKOV, V. M. KRASNIKOV
Department of Cybernetics, Kiev State University, Kiev, USSR

ABSTRACT

The present work deals with algorithm construction of the three-point difference schemes with arbitrary order of accuracy for linear and nonlinear second-order ordinary differential equations with piece-smooth coefficients. For an approximate calculation of coefficients of the t.e.d.s. scheme with a given order of accuracy, which are expressed by the solutions of axiliary Caushi's problems, the Taylor-series expansion method is used. The derivatives of the coefficients and the right part of equations are calculated with recurrence formulas.

1. Introduction

Based on the exact three-point difference schemes theory (e.t.d.s.) [1-5], three-point difference scheme (t.d.s.) of the n-th order of accuracy was constructed[6,7] for the boundary value problem in the form of

$$L^{(k,q)}u \equiv \frac{d}{dx}[k(x)\frac{du}{dx}] - q(x)u(x) = -f(x), \quad x \in (0,1) \qquad (1)$$

$$u(0)=A, \; u(1)=B, \; 0< c_1 \le k(x) \le c_2, \; |q(x)| \le c_2,$$
$$k(x) \in Q^{n+1}[0,1], \; q(x), \; f(x) \in Q^n[0,1], \qquad (2)$$

where $Q^r[0,1]$ is the class of piece-smooth functions with a finite number of the discontinuity points , having r continuous derivatives on any interval continuity for any n. Note that the procedure of constructing t.d.s. of any order f accuracy for sufficiently smooth functions k, q, f was shown in[9,10] and in[11] this procedure was modified for piece-smooth k, q, f with discontinuity points in nodes of the grid. The schemes suggested in[11] are inhomogeneous (as well as that of schemes[9,10]) and their construction requires the solution of an auxiliary algebraic system which order depends on the given accuracy of a difference scheme.

For determining the coefficients of the scheme suggested in[6,7] and its right side in an arbitrary node x_j of the grid $\hat{\omega}_h$ it is necessary to solve four auxiliary initial value problems: two of them on the interval $[x_{j-1},x_j]$ (forwards) and another two problems on the interval $[x_j,x_{j+1}]$ (backwards), both intervals having no discontinuity points as they coincide with nodes of the grid. Each initial value problem is solved in a single step by any one-step method (Taylor-series expansion, Runge-Kutta method) which order of accuracy $\bar{n}=2[(n+1)/2]$ is consistent with smoothness of k, q, f .

As a result a homogeneous t.d.s. of the n-th order of accuracy was obtained. The advantage of the given scheme over those used in[11] for piece-smooth k, q, f, besides homogeneity, is that the three-point approximation of stream $k(x)\frac{du(x)}{dx}$ is constructed in the grid nodes without essential computational efforts. The accuracy of the above approximation in the discrete Chebysheff norm coincides with an accuracy of approximation u(x) (in[11] the error of approximation $u_{\bar{x}}$ is evaluated).

In[8] the existence of t.t.d.s. was first proved for the nonlinear boundary value problem in the form of

$$\frac{d}{dx}[k(x)\frac{du}{dx}] = -f(x,u(x)),\ x \in (0,1),\ u(0)=A,\ u(1)=B, \quad (3)$$

$$0< c_1 \le k(x),\quad k(x) \in C^1[0,\eta] \cup C^1[\eta,1],$$

$$|f(x,u)-f(x,v)| \le L|u-v|,\ \forall\ x \in [0,1],\ u,v \in \Omega_u, \quad (4)$$

$$f_u(x) \equiv f(x,u) \in C[0,\eta] \cup C[\eta,1],\ \eta \in (0,1),\ \forall\ u \in \Omega_u,$$

where Ω_u is the certain neighbourhood of solving problem (1), (2), η is the discontinuity point in which the conditions of consistency should be satisfied

$$u(\eta-0) = u(\eta+0),\ k(\eta-0)\ \frac{du(\eta-0)}{dx} = k(\eta+0)\ \frac{du(\eta+0)}{dx}\ .$$

Proceeding from e.t.d.s. and developing the ideas of [7], t.d.s. of the n-th order of accuracy both in respect to the approximation of function u(x) and its stream in the nodes of grid ω_h were constructed and substantiated[8] for any $n \in \mathbb{N}$ on condition that

$$k(x) \in C^{n+1}[0,\eta] \cup C^{n+1}[\eta,1],\ f_u(x) \in C^n[0,\eta] \cup C^n[\eta,1].$$

For their construction in each node $x_j \in \omega_h$ one should solve two nonlinear initial value problems on the intervals $[x_{j-1}, x_j]$ (forwards) and $[x_j, x_{j+1}]$ (backward), which is performed using a certain one-step method in one step (η is the grid node). Note that an attempt to construct t.d.s. of the 4-th order of accuracy for (3) was undertaken in[12], but substantiation was given only for $k(x) = e^{bx/a}$. The technique suggested in[12] cannot be used even for this special case to construct t.d.s. of higher order of accuracy ($n > 4$).

The purpose of the present work is the computer-analytical generation of t.d.s. of the arbitrary order of accuracy n for problems (1)-(2), (3)-(4). Inasmuch as the construction of the Runge-Kutta methods of high orders of accuracy is rather complicated and the highest order actually achieved for explicitly constructed methods is equal to ten [13], we'll base on the Taylor series expansion method .

For the purpose of computer determination of Taylor series expansion coefficients of solutions of the appropriate initial value problems (skeleton functions) indispensable for constructing t.d.s., one should use recurrence formulas which are also employed in the present work . In this case it is supposed that the functions $k(x)$, $q(x)$, $f(x)$, $f(x,u)$ belong to the class of partially recurrence functions and, therefore, can be represented as a result of using the finite number of superposition, branching and cycling operations over a set Ω of the simplest base functions $o(x), x$, $s(x)$, $x \pm y$, $x \cdot y$, x/y, x^{α}, e^{x}, $an(x)$, $\sin x$, $\cos x$ and predicates $x < y$, $x \le y$ (here $o(x)$ is the zero function, $s(x)$ is the successor function)[14]. Note that the above set of base functions is overdetermined, but from the point of view of practical realization it is more preferable.

2. T.d.s. of any order of accuracy for the second order linear ordinary differential equations

Let us introduce the nonuniform grid

$$\hat{\omega}_h = \left\{ x_j \in (0,1), \ j = \overline{1, N-1}, \quad x_j - x_{j-1} = h_j > 0, \quad \sum_{j=1}^{N} h_j = 1 \right\}$$

so that the discontinuity points of $k(x)$, $q(x)$ and $f(x)$ functions coincide with the grid nodes ω_h. We denote a set of all discontinuity points by ρ and take N such that $\rho \subseteq \omega_h$. Let the solution of problem (1), (2) satisfy discontinuity conditions

$$u(x_i - 0) = u(x_i + 0), \ k(x)\frac{du(x)}{dx}\bigg|_{x=x_i-0} = k(x)\frac{du(x)}{dx}\bigg|_{x=x_i+0}, \qquad (5)$$

We define skeleton functions $v_{\alpha}^{j}(x)$, $w_{\alpha}^{j}(x)$, $\alpha = 1,2$, as the solutions of initial value problems

$$L^{(k,q)}v_\alpha^j(x) = 0, \ x \in (x_{j-1}, x_{j+1}),$$

$$v_\alpha^j(x_{j+(-1)^\alpha}) = 0, \ \left[k(x) \frac{dv_\alpha^j(x)}{dx}\right]_{x=x_{j+(-1)^\alpha}} = (-1)^{\alpha+1}, \quad (6)$$

$\alpha=1,2, \ j=\overline{1,N-1}$

$$L^{(k,q)}w_\alpha^j(x) = -f(x), \ x \in (x_{j-1}, x_{j+1}),$$

$$w_\alpha^j(x_{j+(-1)^\alpha}) = 0, \ \left[k(x) \frac{dw_\alpha^j(x)}{dx}\right]_{x=x_{j+(-1)^\alpha}} = 0, \quad (7)$$

$\alpha=1,2, \ j=\overline{1,N-1}$.

In[7] the following result is proved

Lemma 1. Let $k(x) \in Q^1[0,1]$, $q(x)$, $f(x) \in Q[0,1]$, then at $h= \max\limits_{1\le i\le N} h_i \le h_o = \frac{1}{2}\sqrt{\frac{c_1}{c_2}}$ and supposing that homogeneous boundary value problem (1), (2), (5) has only a trivial solution, then there exist a unique e.t.d.s. for (1), (2), (5)

$$\Lambda y \equiv (a y_{\bar{x}})_x - d y = -\varphi(x), \ x \in \hat{\omega}_h, \ y(0) = A, \ y(1) = B \quad (8)$$

having a unique solution $y(x)\equiv u(x)$, $\forall \ x \in \bar{\hat{\omega}}_h$, where

$$a(x_j)=\left[\frac{1}{\hbar_j} v_1^j(x_j) \right]^{-1},$$

$$d(x_j)=\hbar_j^{-1} \sum_{\alpha=1}^{2} \left[v_\alpha^j(x_j) \right]^{-1} \left[(-1)^{\alpha+1} m_\alpha^j(x_j)-1\right], \quad (9)$$

$$\varphi(x_j)=\hbar_j^{-1} \sum_{\alpha=1}^{2} (-1)^\alpha \left[l_\alpha^j(x_j) - m_\alpha^j(x_j) \frac{w_\alpha^j(x_j)}{v_\alpha^j(x_j)} \right],$$

$$m_\alpha^j(x)=k(x) \frac{dv_\alpha^j(x)}{dx}, \ l_\alpha^j(x)=k(x) \frac{dw_\alpha^j(x)}{dx}, \ \alpha=1,2.$$

From (9) it is seen that for determining the coefficients $a(x_j)$, $d(x_j)$ and the right side $\varphi(x_j)$ of t.d.s. $\forall \ x_j \in \omega_h$ one should solve four initial value problems with smooth coefficients (6), (7) at $\alpha=1$ on the interval $[x_{j-1}, x_j]$ and (6), (7) at $\alpha=2$ on the interval $[x_j, x_{j+1}]$. Each of these problems is solved approximately in one step by the Taylor series expansion method of the $\bar{n}=2[(n+1)/2]$-th order of accuracy. To find the coefficients of a Taylor series, the recurrence formulas from[13] are used. Thus, for example, for calculating the functions $v_\alpha^j(x)$, $m_\alpha^j(x)$, $\alpha=1,2$ we have

$$v_\alpha^j(x_j)=\sum_{i=0}^{n}\left[(-1)^{\alpha+1}h\right]^i V_{\alpha,i}\,,\quad V_{\alpha,i}=\frac{1}{i!}\frac{d^i v_\alpha^j(x_{j+(-1)^\alpha})}{dx^i},\quad i=\overline{0,n},$$

$$m_\alpha^j(x_j)=\sum_{i=0}^{\bar{n}}\left[(-1)^{\alpha+1}h\right]^i M_{\alpha,i}\,,\quad M_{\alpha,i}=\frac{1}{i!}\frac{d^i m_\alpha^j(x_{j+(-1)^\alpha})}{dx^i},\quad i=\overline{0,\bar{n}},\tag{10}$$

$$V_{\alpha,i+1}=\frac{1}{i}\sum_{s=0}^{i}K_{\alpha,s}M_{\alpha,i-s}\,,\quad i=\overline{0,n-1},\quad V_{\alpha,0}=0,\quad \alpha=1,2,$$

$$M_{\alpha,i+1}=\frac{1}{i}\sum_{s=0}^{i}Q_{\alpha,s}V_{\alpha,i-s}\,,\quad i=\overline{0,\bar{n}-1},\quad M_{\alpha,0}=(-1)^{\alpha+1},\quad \alpha=1,2,\tag{11}$$

where

$$K_{\alpha,i}=\frac{1}{i!}\frac{d^i\left(\frac{1}{k(x)}\right)}{dx^i}\,x_{j+(-1)^\alpha},\quad Q_{\alpha,i}=\frac{1}{i!}\frac{d^i(q(x))}{dx^i}\,x_{j+(-1)^\alpha},\tag{12}$$

$i=\overline{0,\bar{n}-1}$.

Therefore, to find $v_\alpha^j(x_j)$, $m_\alpha^j(x_j)$ according to formulas (10), it is necessary to determine the coefficients of Taylor series $K_{\alpha,i}$, $Q_{\alpha,i}$, $\alpha=1,2$, $i=\overline{0,\bar{n}-1}$ for the functions $[k(x)]^{-1}$, $q(x)$, respectively, in the neighbourhood of point $x_{j+(-1)^\alpha}$. By virtue of suggestion relative to functions $k(x)$, $q(x)$, (see Introduction), we can automatically gain an idea of each of them, which makes it possible to determine consequently the requisite number of coefficients of a Taylor series based on the recurrence formulas. On expanding the functions $w_\alpha^j(x)$, $l_\alpha^j(x)$ by their Taylor series expansion, the coefficients of a Taylor series expansion $F_{\alpha,i}$, $\alpha=1,2$, $i=\overline{0,\bar{n}-1}$ for $f(x)$ are also required, for which calculation the same technique is used.

The approximations obtained are denoted by an additional upper index n:

$$v_\alpha^{(n)j}(x_j),\ m_\alpha^{(n)j}(x_j),\ w_\alpha^{(n)j}(x_j),\ l_\alpha^{(n)j}(x_j).$$

Considering the foregoing, we have the following algorithm to realize scheme (8), (9):

1. Perform the lexical analysis of input information ($k(x)$, $q(x)$, $f(x)$) and, as a result of which, find the representation of each function as the application of a finite number of superposition, branching and cycling operations of the set Ω.
2. Using the information on smoothness of functions $k(x)$, $q(x)$, $f(x)$ and the information obtained at previous steps, we find the solutions of auxiliary initial value problems in the grid nodes ω_h: $v_\alpha^{(n)j}(x_j)$, $m_\alpha^{(n)j}(x_j)$, $w_\alpha^{(n)j}(x_j)$, $l_\alpha^{(n)j}(x_j)$.

3. Construct t.d.s. in the form of

$$(a^{(n)}y^{(n)}_{\bar{x}})_{\hat{x}}-d^{(n)}y^{(n)} = -\varphi^{(n)}(x), \quad x \in \hat{\omega}_h, \qquad (13)$$
$$y^{(n)}(0) = A, \quad y^{(n)}(1) = B$$

where

$$a^{(n)}(x_j)=\left[\frac{1}{\hbar_j} v_1^{(n)j}(x_j)\right]^{-1},$$

$$d^{(n)}(x_j)=\hbar_j^{-1}\sum_{\alpha=1}^{2}\left[v_\alpha^{(n)j}(x_j)\right]^{-1}\left[(-1)^{\alpha+1}m_\alpha^{(\bar{n})j}(x_j)-1\right], \qquad (14)$$

$$\varphi^{(n)}(x_j)=\hbar_j^{-1}\sum_{\alpha=1}^{2}(-1)^{\alpha}\left[l_\alpha^{(n)j}(x_j) - m_\alpha^{(\bar{n})j}(x_j)\frac{w_\alpha^{(\bar{n})j}(x_j)}{v_\alpha^{(n)j}(x_j)}\right].$$

4. We solve system (13) by the elimination method[15].

An accuracy of the approximate solution obtained as a result of the operation of the above algorithm is characterized by the following statement.

Theorem 1. Let the conditions of lemma 1 be fulfilled and besides $k(x) \in Q^{n+1}[0,1]$, $q(x)$, $f(x) \in Q^{n}[0,1]$, then for the error of scheme (13), (14) at sufficiently small h we have

$$\| u-y^{(n)} \|^{*}_{1,\infty,\hat{\omega}_h} =$$
$$= \max\left\{\|u-y^{(n)}\|_{0,\infty,\hat{\omega}_h},\ \left\|k\frac{du}{dx} - k\frac{dy^{(n)}}{dx}\right\|_{0,\infty,\hat{\omega}_h}\right\} \le Mh^{n}, \qquad (15)$$

where
$$k(x_j)\frac{dy^{(n)}(x_j)}{dx}=\sum_{\alpha=1}^{2}\left\{m_\alpha^{(\bar{n})j}(x_j)\, y^{(n)}(x_{j+(-1)^{\alpha}}) + (-1)^{\alpha+1}\times\right.$$
$$\times\left[m_1^{(\bar{n})j}(x_j)m_2^{(\bar{n})j}(x_j)w_\alpha^{(\bar{n})j}(x_j)+m_\alpha^{(\bar{n})j}(x_j)v_{3-\alpha}^{(n)j}(x_j)l_{3-\alpha}^{(n)j}(x_j)\right]\Big\}\times$$
$$\times\left[\sum_{\alpha=1}^{2}(-1)^{\alpha}v_\alpha^{(n)j}(x_j)\, m_{3-\alpha}^{(\bar{n})j}(x_j)\right]^{-1}. \qquad (16)$$

3. T.d.s. of any order of accuracy for nonlinear second order ODE

For simplicity, we confine ourselves to the uniform grid $\omega_h = \{ x_j=jh,\ j=\overline{1,N-1},\ h=1/N \}$ and one discontinuity point η coinciding with i-th node of the grid x_i. Let us introduse functions $w_\alpha^j(x,u)$, $l_\alpha^j(x,u)$, $\alpha=1,2$, being the solutions of two initial value problems

$$\frac{d}{dx}w_{\alpha}^{j}(x,u) = \frac{l_{\alpha}^{j}(x,u)}{k(x)}, \qquad x_{j-2+\alpha} < x < x_{j-1+\alpha},$$

$$\frac{d}{dx}l_{\alpha}^{j}(x,u) = -f\left[x,\ \hat{u}(x) - \frac{w_{\alpha}^{j}(x_j,u)}{v_{\alpha}^{j}(x_j)}\,v_{\alpha}^{j}(x) + w_{\alpha}^{j}(x,u)\right], \qquad (17)$$

$$w_{\alpha}^{j}(x_{j+(-1)^{\alpha}},\ u) = l_{\alpha}^{j}(x_{j+(-1)^{\alpha}},\ u) = 0,\quad \alpha=1,2,$$

where

$$\hat{u}(x)=[u(x_j)v_1^j(x)+u(x_{j-1})v_2^{j-1}(x)][v_1^j(x_j)]^{-1},\ x \in [x_{j-1},x_j],$$

$$j=\overline{1,N},\ v_1^j(x) = \int_{x_{j-1}}^{x} \frac{dt}{k(t)},\ v_2^j(x) = \int_{x}^{x_{j+1}} \frac{dt}{k(t)},\ j=\overline{1,N-1}.$$

Lemma 2.[8] Let the problem (3), (4) has a unique solution $u(x)$ and $\max\{ h_o/c_1,\ Lh_o(3+2h_o/c_1) \} = 1$, then $\forall\, h \in (0,h_o)$ problem (17) has a unique solution and

$$u(x) = \hat{u}(x) - \frac{w_{\alpha}^{j}(x_j,u)}{v_{\alpha}^{j}(x_j)}\,v_{\alpha}^{j}(x) + w_{\alpha}^{j}(x,u),\ x \in [x_{j-2+\alpha},x_{j-1+\alpha}],$$

$\alpha=1,2$ and for problem (1), (2) $\exists$ t.t.d.s.

$$(a u_{\bar{x}})_x=-\varphi(x,u),\ x \in \omega_h,\ u(0)=A,\ u(1)=B \qquad (18)$$

where

$$a(x_j)=\left[\frac{1}{h}v_1^j(x_j)\right]^{-1},\ \varphi(x_j,u)=\sum_{\alpha=1}^{2}\left[(-1)^{\alpha}l_{\alpha}^{j}(x_j)+\frac{w_{\alpha}^{j}(x_j,u)}{v_{\alpha}^{j}(x_j)}\right]. \qquad (19)$$

Thus, for constructing t.t.d.s.(18), (19) it is necessary to solve two initial value problems (17): one ($\alpha=1$) on the interval $[x_{j-1},x_j]$ forwards, and the second ($\alpha=2$) on the interval $[x_j,x_{j+1}]$ backwards, both having smooth coefficients. To solve the problem, as well as in the linear case, one can use the Taylor series expansion method of the $2[(n+1)/2]$-th order of accuracy. Considering the fact that the right part of (17) depends on $w_{\alpha}^{j}(x_j,u)$, the algorithm can be constructed as follows

$$w_{\alpha}^{(s)j}(x_j,u) = -\frac{h^2}{2}\,\frac{f_{j+(-1)^{\alpha}}}{k_{j+(-1)^{\alpha}}} +$$

$$+\sum_{p=3}^{s}\frac{[(-1)^{\alpha+1}h]^p}{p!}\ \frac{d^p w_{\alpha}^{j}(x_{j+(-1)^{\alpha}},u,w_{\alpha}^{(s+2-p)j}(x_j,u))}{dx^p},\ s=\overline{2,\bar{n}}, \qquad (20)$$

$$\sum_{p=3}^{2}=0,\quad w_{\alpha}^{(1)j}(x_j,u)=0,$$

$$l_{\alpha}^{(s)j}(x_j,u) = (-1)^{\alpha}\, h\, f_{j+(-1)^{\alpha}} +$$

$$+\sum_{p=2}^{s}\frac{[(-1)^{\alpha+1}h]^p}{p!}\;\frac{d^p l_{\alpha}^{j}(x_{j+(-1)^{\alpha}},u,w_{\alpha}^{(s+1-p)j}(x_j,u))}{dx^p},\ s=\overline{1,n}, \quad (21)$$

$$\sum_{p=2}^{1}=0, \quad l_{\alpha}^{(0)j}(x_j,u)=0,$$

the functions $w_{\alpha}^{j}(x,u,w_{\alpha}^{(s)j}(x_j,u))$, $l_{\alpha}^{j}(x,u,w_{\alpha}^{(s)j}(x_j,u))$ are the solution of problem (17), where in the second argument f the number $w_{\alpha}^{j}(x_j,u)$ is substituted for $w_{\alpha}^{(s)j}(x_j,u)$. To find the derivatives occuring in (20), (21) one can use the numerical and analytical approach described to fit the linear problem. The solution of the resulting non-linear t.d.s. of the form

$$(a^{(n)}y_{\bar{x}}^{(n)})_x=-\varphi^{(n)}(x,y^{(n)}),\ x\in\omega_h,\ y^{(n)}(0)=A,\ y^{(n)}(1)=B \quad (22)$$

can be obtained using the successive approximation method.

Theorem 2.[8] Let the conditions of lemma 2, $0< c_1\le k(x)$, $k(x)\in W_{\infty}^{n+1}(0,x_1)\cup W_{\infty}^{n+1}(x_1,1)$, $f(x,u)\in W_{\infty}^{n}((0,x_1)\times\Omega_u)\cup W_{\infty}^{n}((x_1,1)\times\Omega_u)$ be fulfilled, where $L/c_1<1$.

Then, $\exists\, h_1\in(0,h_0)$ that $\forall\, h\in(0,h_1]$ the successive approximation method

$$\begin{gathered}(a^{(n)}y_{\bar{x}}^{(n,m)})_x = -\varphi^{(n)}(x,y^{(n,m-1)}), \quad x\in\omega_h,\\ y^{(n,m)}(0)=A, \quad y^{(n,m)}(1)=B,\end{gathered} \quad (23)$$

$$m=1,2,\dots, \quad y^{(n,0)}(x)=A(1-x)+Bx$$

converges and error estimate is fulfilled

$$\|u-y^{(n,m)}\|^{*}_{1,\infty,\omega_h}\le M(h^n+q^m), \quad (24)$$

where M is the constant independent of h, n, m, and the value $q<1$, $q\approx L/c_1$.

4. Numerical experiments

Program realization of algorithms given in §§ 2, 3 were approved by numerous examples. We give two of them.

Example 1 [11].

$$k(x)=e^{x-0.5H(0.75-x)},\quad q(x)=[1+H(x-0.75)]\,/\,(1+x),$$

$$u(x)=\begin{cases}e^{x}, & 0\le x\le 0.75,\\ e^{x-0.5}+e^{0.75}-e^{0.25}, & 0.75\le x\le 1,\end{cases}$$

where H(x) is the Heaviside function.

Example 2.

$$k(x) = [1+H(x-0.5)] / (1+x),$$

$$f(x,u)=\begin{cases} -\alpha u^3(x)+\alpha e^{6x}-\frac{2}{1+x}e^{2x}\left[2-\frac{1}{1+x}\right], & 0 \le x \le 0.5, \\ -\alpha u^3(x)+\alpha e^{3x+1.5}-\frac{2}{1+x}e^{x+0.5}\left[1-\frac{1}{1+x}\right], & 0.5 \le x \le 1, \end{cases}$$

$$u(x) = \begin{cases} e^{2x}, & 0 \le x \le 0.5 , \\ e^{x+0.5}, & 0.5 \le x \le 1 , \end{cases}$$

The calculations were carried out using computer IBM PC/AT with double precision. For practical estimation of the rate of convergence the value

$$p = \log_2 \frac{\|u-y^{(n)}\|^*_{1,\infty,\omega_h}}{\|u-y^{(n)}\|^*_{1,\infty,\omega_{h/2}}}$$

was used. At h=0.5 and n=14, p=14.5 and 14.8 were obtained for both examples, respectively. A further increase in the order of accuracy of the method is restricted by the length of mantissa in the computer used.

References

1. Tichonov A.N. and Samarskii A.A., *Dokl. Akad. Nauk SSSR* **122** (1958), pp. 562-565 (Russian).
2. Tichonov A.N. and Samarskii A.A. *Ž. Vyčisl. Mat. i Mat. Fiz.* **1** (1961), pp. 5-63 (Russian) .
3. Tichonov A.N. and Samarskii A.A. *Ž. Vyčisl. Mat. i Mat. Fiz.* **1** (1961), pp. 425-440 (Russian).
4. Godev K.N., Lazarov R.D., Makarov V.L. and Samarskii A.A. *Math. USSR Sb.* **10** (1986) (Russian).
5. Samarskii A.A., Lazarov R.D. and Makarov V.L., *Difference Schemes for Differential Equations with weak solutions*, (Wyshaja Shcola, Moscow, 1987), p.296 (Russian).
6. Samarskii A.A. and Makarov V.L., *Dokl. Akad. Nauk SSSR* **312** (1990), pp. 538-543 (Russian).
7. Samarskii A.A. and Makarov V.L., *Differential Equations* **26** (1990), pp.1254-1265 (Russian).
8. Makarov V.L. and Samarskii A.A., *Dokl. Akad. Nauk SSSR* **312** (1990), pp. 795-800 (Russian).
9. Lynch R.E. and Rice J.R., *Math. Comp.* **34** (1980), pp.333-372.
10. Doedel E.J. *SIAM J. Numer. Anal.* **15** (1978) pp.450-465.
11. Gartland E.C.Jr. *IMA J. Numer. Anal.* **9** (1989), pp.243-260.
12. Iyengar S.R.K. and Pillai A.C.R. *Appl. Math. Modelling* **13** (1989), p.58-62.
13. Hairer E., Nørsett S.P. and Wanner G. *Solving ordinary differential equations I (nonstiff problems)*

(Springer-Verlag, Berlin Heidelberg New-York London Paris Tokyo, 1987).
14. Bui D.V. and Redko V.N. *Dokl. Akad. Nauk Ukrain. SSR Ser. A* (1985), pp. (Russian).
15. Samarskii A.A. and Nikolaev E.S. *Numerical Methods for Grid Equations. Volume I. Direct Methods.* (Birkhäuser Verlag, Basel-Boston-Berlin, 1989).

AUTOMATIC GENERATION OF FINITE-VOLUME AND FINITE-DIFFERENCE CODE

S. Steinberg and P.J. Roache

University of New Mexico and Ecodynamics Research Associates, Inc.

Introduction

Part of a project to build modeling programs for porous media flow in a complex three-dimensional underground region required the generation of FORTRAN code for solving a boundary value problem (BVP) for a linear, second-order, symmetric, elliptic, partial differential equation (PDE) with Robin (mixed) boundary conditions (BCs). Numerical grid-generation techniques are used to handle the complex geometry. Because of the nature of the porous media, the coefficients of the second derivatives in the PDE form a general symmetric matrix with discontinuous entries. The problem that was modeled required a mixture of Dirichlet and Neumann boundary conditions, so to cover all cases, full Robin boundary conditions were implemented.

The numerical algorithms are derived by first transforming the BVP to general boundary-conforming coordinates and then discretizing both the PDE and the BCs. Two ideas are used to derive difference schemes: 1) a finite-difference approach based on nodal positions for the unknowns; and 2) a finite-volume approach based on cell-centered positions for the unknowns. Approach 2) is the most difficult to program, so this paper only considers that approach. The paper/1/ gives a detailed derivation of these algorithms. In the papers/2,3,4,5/, related work on the use of symbol manipulators to generate FORTRAN is presented.

The main parts of the FORTRAN code are the I/O interfaces, the grid generator, the stencil loader, and a linear equations solver. The stencil loader is the code that computes the coefficients of the difference equations. A current generation multigrid algorithm is used to solve the linear equations. This paper describes the use of the symbol manipulator MACSYMA/6/ to generate FORTRAN code for the stencil loader. The stencil loader has code for: 1) evaluating metrics; 2) transforming the coefficients of the PDE and BCs; 3) averaging of coefficients; 4) calculating stencils; 5) calculating the residue; and 6) evaluating fluxes. The FORTRAN code generated by MACSYMA is combined with existing software to create the complete program. The symbol programs are capable of writing FORTRAN code for any spatial dimension; the FORTRAN code for three dimensions is twenty-seven pages long.

This project required a substantial symbol-manipulator programming effort. However, this approach was very successful, producing essentially error-free FORTRAN code with a reasonable programming effort. (This will be discussed further in the conclusions section.)

A MACSYMA/6/ reference manual will be very helpful for reading the remainder of this paper.

FORTRAN Samples

In this section, some samples of the three-dimensional FORTRAN code generated by MACSYMA are presented; the samples are presented in a special font, essentially as MACSYMA produced them (some editing has been done for compactness of printing). The techniques used to generate the code are discussed in the next section. The three-dimensional code has 125 declarations: 32 real; 75 array; and 14 integer (all variables are declared). Also, the code contains 39 loops. The generation of code this size is, clearly, a complex task.

There are a number of useful things that MACSYMA can do besides compute formulas. First, certain parts of comment statements can be computed along with the formulas. This is very helpful for human readers of the code. Secondly, it is easy to keep track of all variables introduced by MACSYMA and then use these in the declaration statements. When this is combined with the default variable type being *undeclared*, then errors are much easier to detect. Also, MACSYMA computes the upper and lower values for array dimensions, and then it builds array declarations using this information. The following FORTRAN fragments illustrate these points. For example, the upper and lower limits on the i index have been computed by MACSYMA.

```
c     The arrays of grid points.
c         Stored at corners: il-1 <= i <= il+2, etc.
      real x(i1d:ild, j1d:jld, k1d:kld)
```

A numerical grid gives a correspondence between points in physical space and the indices that lie in a box in a space which has the same number of dimensions as the physical space. In the finite-volume approach, physical space is divided into cells and the coordinates of the transformation are tabulated at the corners of the cells. The solution values are tabulated at the centers of cells in the interior of the region, while the fluxes are tabulated at cell face centers.

A box has one interior, six faces, twelve edges, and eight corners. In the interior of the region, loops must be constructed for calculations over cell corners, cell centers, and cell face centers. Similar loops must be generated for the faces, edges, and corners of the box. Samples of such loops follow:

```
c         Loop is over cell centers.
      do  1   k  =  kl - 1  , kl + 1
      do  1   j  =  jl - 1  , jl + 1
      do  1   i  =  il - 1  , il + 1

c         Loop is over face centers.
      do 21 k = kl-1,kl+1
      do 21 j = jl-1,jl+1
      do 21 i = il,il+1

c         Loop is over face centers.
      i  =  il
      do  611 j  =  jl ,jl
      do  611 k  =  kl ,kl

c         Loop is over edge centers.
      j  =  jl - 1
      k  =  kl - 1
      do  9111 i  =  il, il
```

Formulas for derivatives of the transformation, computing metrics, averaging, evaluating the stencils, etc. must be produced. A few samples are listed below:

```
c The derivatives of the coordinate change.
      x1 = ( x(i+1,j+1,k+1)+x(i+1,j+1,k)+x(i+1,j,k+1)+x(i+1,j,k)-
     1        x(i,j+1,k+1)-x(i,j+1,k)-x(i,j,k+1)-x(i,j,k)  )/h1/4.0

c Compute the metric quantities.
      vk11 = y2*z3-y3*z2
      vjac = vk13*x3+vk12*x2+vk11*x1

c     The transformed coefficients.
      a11(i,j,k) = ( vk31**2*vsig33+2*vk21*vk31*vsig23+vk21**2*vsig22
     1    +2*vk11*vk31*vsig13+2*vk11*vk21*vsig12+vk11**2*vsig11)/vjac
```

```
c The stencil formulas.
c The coefficient of  f(i - 1, j, k)
      stn011(i,j,k) = ah11(i,j,k)/h1**2
c The coefficient of  f(i - 1, j - 1, k)
      stn001(i,j,k) = (a12(i,j-1,k)+a12(i-1,j,k))/(h1*h2)/4.0
c The coefficient of  f(i, j, k)
      stn111(i,j,k) = -stn211(i,j,k)-stn121(i,j,k)-stn112(i,j,k)
     1                -stn110(i,j,k)-stn101(i,j,k)-stn011(i,j,k)
```

Basic Symbol Tools

There is no symbol manipulator which provides all of the basic tools needed for FORTRAN code generation and analysis. In particular, tools are needed for manipulating finite difference equations (FDEs) in both the continuum and discrete settings. Simple tools were built for: 1) basic finite difference formulas; 2) transforming the PDE and BCs; 3) converting a PDE to a continuum FDE; 4) converting a continuum FDE to a discrete FDE; 5) collecting coefficients and building stencil formulas; and 6) writing FORTRAN code. Here is an example of a central difference formula. The lines labelled with c_ are MACSYMA input lines, those labelled with d_n are MACSYMA output lines.

```
(c_7) cd(f(x),x,1,1/2);
```

$$(d_7) \qquad \frac{f\left(x+\frac{dx}{2}\right)-f\left(x-\frac{dx}{2}\right)}{dx}$$

Tools are needed for both algorithm generation and algorithm checking. The most elementary check is a truncation error analysis.

The code generated by MACSYMA was checked by this method; it took about six cpu hours on a Sun 3 workstation to check the three-dimensional code The Taylor series computation, used in the check, requires the use of the *newdiff* option from the Partial Differentiation Package/6/. An example of a truncation error calculation follows:

```
(c_11) expression : cd(f(x),x,1,1/2)-diff(f(x),x,1);
```

$$(d_{11}) \qquad \frac{f\left(x+\frac{dx}{2}\right)-f\left(\left(x-\frac{dx}{2}\right)\right)}{dx}-\frac{\partial f}{\partial x}(x)$$

```
(c_12) taylor(expression,dx,0,4);
```

$$(d_{12}) \qquad \frac{\frac{\partial^3 f}{\partial x^3}(x)\,dx^2}{24}+\frac{\frac{\partial^5 f}{\partial x^5}(x)\,dx^4}{1920}+\cdots$$

One of basic tools needed for stencil generation is a list of all of the nearest neighbors of a point in n-dimensions. The function *displace* generates a list of the differential displacements of the nearest neighbors of a point.

```
(c_1) displace([x,y],1);
```

$$(d_1) \qquad [[-dx,-dy],[-dx,0],[-dx,dy],[0,-dy],[0,0],[0,dy],[dx,-dy],[dx,0],[dx,dy]]$$

It is particularly useful to develop algorithms in continuum notation (as is used in the previous example) and then convert this notation to an index notation. The function *index_note* was written to do this.

```
(c_14) expression : expand(ca(cd(f(x,y),x,1,1),y,1,1));
```

$$(d_{14}) \quad \frac{f(x+dx,y+dy)}{4\,dx}+\frac{f(x+dx,y-dy)}{4\,dx}-\frac{f(x-dx,y+dy)}{4\,dx}-\frac{f(x-dx,y-dy)}{4\,dx}$$

```
(c_15) index_note(expression,[x,y],[i,j],1);
```

$$(d_{15}) \quad \frac{f(i+1,j+1)}{4\,h1}+\frac{f(i+1,j-1)}{4\,h1}-\frac{f(i-1,j+1)}{4\,h1}-\frac{f(i-1,j-1)}{4\,h1}$$

The *fortran* function can be used to convert the previous expression to FORTRAN syntax. A more compact form of the expression is obtained by using the *factor* or *ratsimp* functions.

The Symbol Programs

The symbol manipulation programs follow the theoretical development given in the paper/1/. Thus, once the basic tools are understood, the main symbol manipulation programs are easily understood. In this section, several basic utility programs are listed and described. (The comments in the symbol programs have been deleted and replaced by descriptions.) The symbol manipulation programs use three sets of variables: `physical` $= [x, y, z]$, `logical` $= [\xi, \eta, \zeta]$, and `index` $= [i, j, k]$. It is assumed that the PDE is transformed from physical space to a unit interval, square, or cube in logical space, and then logical space is divided into identical cells which are located by using the index variables. Because the index variables i,j,and k are used in the FORTRAN code, the symbol code uses the variables `ii`, `jj`, and so forth, to index loops.

The block of code listed below is from the *transformation* function, which is used to generate FORTRAN code for the evaluation of derivatives of the coordinate transformation. Here *cd* is a central difference, *ca* is a central average, and *lt* is a left translate. All other functions are standard MACSYMA utilities.

```
real : [],
formulas : [],
for ii thru nn do for jj thru nn do (
   temp1 : concat(physical[ii], jj),
   real : endcons(temp1, real),
   temp2 : cd(transformation[ii], logical[jj], 1, 1/2),
   for kk thru nn do if kk # jj then
      temp2 : ca(temp2, logical[kk], 1, 1/2),
   for kk thru nn do
      temp2 : lt(temp2, logical[kk], 1/2),
               /* Simplify the derivatives. */
      temp2 : factor(temp2),
   formulas : endcons(temp1 = temp2, formulas),
end_ii_jj_do),
```

The `formulas` list is used to hold the formulas that are later converted to FORTRAN. To understand the first loop, assume that `ii` $= 1$, `jj` $= 1$, and `physical` $= [x, y]$. Then `temp1` $= x1$, `real` $= [x1]$, and `temp2` has as its value the central-difference formula (see above) for the derivative of the x coordinate with respect to its first argument. Because of the nature of the finite-volume formulation, the derivative must be averaged in all variables, except the one in which it is differenced, and then all variables must be translated by one-half step. This is accomplished in the two **kk** loops. After *transformations* is called, the list of formulas in `formulas` is translated to index notation, then to FORTRAN notation, and finally the code is printed into a file. This produces the formula for `x1` given in a previous example.

MACSYMA is used to convert the differential equation to a difference equation by substituting differences for derivatives. In the finite-volume formulation, different differences are used for the diagonal and off-diagonal terms in the PDE, so this is not totally trivial. In our programs, the solution of the PDE is called f, so the previous operations produce an expression (in one dimension) in $f(\xi - d\xi)$, $f(\xi)$, and $f(\xi + d\xi)$. It is the job of the *make_stencils* function to collect the coefficients of the various f values. These are the formulas for the stencils. After translation to index notation and to FORTRAN, formulas such as those for **stn011**, **stn001**, and **stn111**, as shown above, are produced. A listing of part of the *make_stencil* program follows:

```
/* Create a list of all possible differential displacements. */
   list : displace(variables, width),
/* Collect the coefficients of the differences of function in the expression. */
   expression : expand(expression),
   for jj thru length(list) do (
      place : function,
      for ii thru length(variables) do
         place : subst(variables[ii]+list[jj][ii], variables[ii], place),
      parts : bothcoeff(expression, place),
      full : name,
      for ii thru length(variables) do (
         temp : width + coeff(list[jj][ii], concat(d, variables[ii])),
         full : concat(full, temp),
      end_ii_do),
      full : apply(full, variables),
      parts[1] : ratsimp(parts[1]),
      if not(parts[1] = 0) then (
         real : endcons(full , real),
         stencils : endcons([full, parts[1], place], stencils)
      ) else (
         stn_zero : endcons([full, place], stn_zero) ),
   expression : expand(parts[2]),
   end_jj_do),
```

This is a computation in logical space, therefore, in two dimensions **variables** $= [\xi, \eta]$ and for a nearest neighbor stencil **width** $= 1$. The action of *displace* was given above; a typical entry in `list` is a list of two displacements. So, assume `list[jj]` $= [-d\xi, -d\xi]$ and `function` $= f(\xi, \eta)$. After the first pass through the `ii` loop **place** $= f(\xi - d\xi, \eta - d\eta)$. Then **parts** is a list containing the coefficient of **place** in **expression** and the rest of **expression**.

Next, assume **full** = **stn**. After the `ii` loop **full** = **stn00**, that is, the name for the lower left stencil. Next **full** = **stn00**(ξ, η). The conditional checks to see if the coefficient is zero; if not, add **full** to the list of real variables and, if COEFFICIENT is the coefficient just computed, then add

$$[stn00(\xi, \eta), \text{COEFFICIENT}, f(\xi - d\xi, \eta - d\eta)]$$

to the list of stencils. This list is converted to index notation and then to FORTRAN, finally resulting in the following comment and code being inserted into the FORTRAN code.

```
c The coefficient of  f(i - 1, j - 1)
      stn00(i,j) = (a12(i,j-1)+a12(i-1,j))/(h1*h2)/4.0
```

Conclusions

It is our conclusion that the use of a symbol manipulation program to write FORTRAN code is effective and efficient.

The first symbol program we wrote was used to write FORTRAN code based on the finite-difference formulation. Most of the symbol code was developed for the n-dimensional case, but it was tested for increasing dimension. At the point when the correct two-dimensional FORTRAN was finally written, substantially more time was required via the symbol manipulation route than would be required by hand coding the FORTRAN. However, now much of the work for the three-dimensional case was done, by the time the three-dimensional FORTRAN was running, the symbol manipulation route was as efficient as hand coding. The next project was to write a finite-volume formulation code. Much of the symbol code that was written before was reusable, so the symbol manipulation route was substantially more efficient than hand coding. The final symbol code is about one and one-half times as large as the three-dimensional finite-volume FORTRAN code.

Certainly, accuracy is more important that efficiency. An important aspect of using a symbol manipulator to write FORTRAN is that errors in the symbol code typically produce FORTRAN code that is obviously incorrect, and thus, errors are easy to detect. For example, a sign error in the symbol code will typically appear in many places in the FORTRAN code, making it far easier to detect that a single typo. Thus, when one believes that the FORTRAN code is correct, there are very few errors left.

Of course all of the codes were rigorously tested numerically[7] and some additional errors were found. The testing is done one dimension at a time. For the finite-volume approach, during the numerical testing, no errors were found in the one dimensional code, three errors were found in the two dimensional code, and one error was found in the three dimensional code. All errors were easy to find and of the type where a FORTRAN statement is not in the correct loop or a loop index has an incorrect limit.

Because the FORTRAN code is written symbolically, then symbolically tested using a truncation error analysis, and then tested numerically, we believe that the probability of any remaining errors is extremely small.

Acknowledgments

This work was partially supported by the Office of Naval Research, Sandia National Laboratories, the Army Research Office, and the Air Force Weapons Laboratory.

Closely related material has been presented at: The Second International Conference on Expert Systems for Numerical Computing, Purdue University, West Lafayette, April 1990; and The Symposium on Symbolic Computations and Their Impact on Mechanics at the ASME Annual Winter Meeting, Dallas, November 1990.

References

1. S. Steinberg and P.J. Roache, In preparation.
2. S. Steinberg and P.J. Roache, MACSYMA Newsletter, **II-2** (1985), 10-12. Also, Journal of Symbolic Computation, **2** (1986), 213-216.
3. S. Steinberg and P.J. Roache, AIAA Paper No. 86-0241, Proceedings of the AIAA 24th Aerospace Sciences Meeting, Reno, Nevada, 1986. Also, Transactions of the Fourth Army Conference on Applied Mathematics and Computation, Ithaca, NY, 1986. Also, Proceedings of the Coupling Symbolic and Numeric Computing in Knowledge Based Systems Workshop, Boeing Computer Services, Seattle, WA, 1987.
4. M. Florence et al. Proceedings of the Sixth International Conference on Mathematical Modelling, St. Louis, Missouri, 1987.
5. S. Steinberg and P.J. Roache, Symbolic Computation in Fluid Mechanics and Heat Transfer, (H.H. Bau, T. Herbert, M.M. Yovanovich, eds.), HTD-Vol. 105, AMD-Vol. 97, the Heat Transfer Division and the Applied Mechanics Division, ASME ASME Winter Annual Meeting, Chicago, IL, Nov. 1988.
6. MACSYMA Reference Manual (version13), Symbolics, Inc. 1988.
7. P.J. Roache, P.M. Knupp, S. Steinberg, Forum on Benchmark Test Case for Computational Fluid Dynamics, ASME Fluids Engineering Division Spring Conference, Ontario, Canada, June, 1990.

QUALITATIVE INVESTIGATIONS OF SYSTEMS WITH THE AID OF COMPUTER ALGEBRA

A. V. BANSHCHIKOV, L. A. BURLAKHOVA, V. D. IRTEGOV

Irkutsk Computing Center, Siberian Branch, USSR Acad. Sci., Lermontov str., 134, Irkutsk, 664033, USSR

ABSTRACT

The algorithms of constructing of the first integrals and Lyapunov functions for the package of symbolic calculations MECHANIC are described on investigation of stability of system motion of interconnected rigid and deformable bodies.

In 1,2 the operational component of the package for symbolic calculations MECHANIC, intended for constructing differential equations of motion of systems of interconnected rigid and deformable bodies and their qualitative investigation is described. Let us consider expansion of possibilities on investigation of stability. Differential equations of motion can be obtained in Lagrange 2--type form for coordinates q and velocities $\dot{q}$ or in Euler-Lagrange form in quasivariables π. Fore these equations one may automatize computations of generalized integral of Jacobi-type energy and cyclic integrals 3 . Other types of integrals (quadratic and linear of velocity) we obtain using relations from Euler--Lagrange equations for the special choice of quasivelocities:

$$\Omega_i = \sum a_{ij}(q,t)\dot{q} + a_{ik+1}(q,t) \qquad (i,j=1,\dots,k). \tag{1}$$

If quasivelocities for interconnected bodies are chosen so, that they are equal to projections on connected axes of absolute body-bearer angle velocity and to the linear velocity of the pole O_1 of a bearer, and motion of carried bodies is described in relative velocities (quasivelocities), then the correlations hold:

$$\frac{d}{dt}((\partial_v T)^T(\partial_v T)) = 2(\partial_v T)^T P^{(1)}, \tag{2}$$

$$\frac{d}{dt}((\partial_\omega T)^T(\partial_\omega T)) + (\check{v}\,\partial_v T)^T(\partial_\omega T) = 2M^T(\partial_\omega T), \tag{3}$$

$$\frac{d}{dt}((\partial_\omega T)^T(\partial_v T)) = (\partial_\omega T)^T P^{(1)} + (\partial_v T)^T M, \tag{4}$$

$$\frac{d}{dt}(\gamma^T \partial_\omega T + \tilde{\gamma}^T \partial_v T) + \gamma^T \check{v}\,\partial_v T = \gamma^T M + \tilde{\gamma}^T P^{(1)}. \tag{5}$$

Here T -kinetic energy; $\omega = (\omega_1, \omega_2, \omega_3)$- matrix of quasivelocities Ω_1, Ω_2, Ω_3 (projections of absolute angle velocity of bearer),

$v = (v_1, v_2, v_3)$- matrix of quasivelocities on π_4, π_5, π_6 (projections of pole velocities); $P^{(1)}$- matrix of generalized forces on π_4, π_5, π_6; M- matrix of generalized forces π_1, π_2, π_3; $\gamma, \tilde{\gamma}$- matrix of direct cosines of bearer axises with two constant vectors $\xi, \tilde{\xi}$; $\check{\omega}$ - matrix of angular velocity ω; $\check{v}$-is constructed similarly $\check{\omega}$.

In (2)-(5) under definite conditions one can obtain the integrals of Kotelnikov, Volterra, Brun or their generalizations. If a system of bodies is described by Lagrange equations, the relations (2)-(5) can be refined on account of kinematic equations (1).

Further usage of given invariant relations is bound up with theinvestigation of stability of some motions by the method of Lyapunov function for the systems with the first integrals - through the algorithm of the Routh-Lyapunov theorem **4**.

For the system of ordinary differential equations

$$\dot{x} = X(x), \qquad x \in R^n, \tag{6}$$

assuming m+1 the first integrals $V_0(x) = c_0, \ldots, V_m(x) = c_m$, we form linear space of first integrals for R : $K_J = \sum \lambda_{i_j} V_{i_j}(x)$, $J = (i_0, \ldots, i_k) \subset (0, \ldots, m)$, λ_{i_j}=const. Let's investigete the system:

$$\partial_{x_i} K_J = f_i(x_1, \ldots, x_n, \lambda_{i_1}, \ldots, \lambda_{i_k}) = 0, \tag{7}$$

$$V_{i_1}(x) = c_{i_1}, \ldots, V_{i_k}(x) = c_{i_k}. \tag{8}$$

If the Jacobian (7) is nonzero, the solution (7) is the solution of (6), and this is the steady motion. If (7) may be represented in the form: $\partial_{x_i} K_J = \sum \varphi_l(x,\lambda) a_{li}(x,\lambda) = 0$, and rank $\| a_{li}(x,\lambda) \| = k < n$ on $\varphi_l(x,\lambda) = 0$ $(l = 1, \ldots, k)$ for $x_i \in U \subset R^n$, $\lambda_{i_j} \in \Lambda \subset R^m$, then $\varphi_l(x,\lambda) = 0$ $(l = 1, \ldots, k)$ is an invariant manyfold (6) (degenerated invariant manyfold of steady motions -IMSM). If in (7) J does not contain all numbers from $(0, \ldots, m)$, then we may investigate the problem of existence and stability of the corresponding K_J IMSM under perturbations, which preserve (remain invariant) integrals from bundle K_J, while destroying the rest of integrals.

When investigating stability of IMSM by the Routh-Lyapunov method it is necessary to analyse the conditions of (conditional) signdefiniteness of the expansion K_J in the vicinity of IMSM, which begins from the members not lower than the 2nd-order infinitesimal. For the systems assuming (generalized) integral of energy and cyclic integrals, it is possible automate completely the process of construction of equations (7) and sufficient stability conditions for found steady motions 3. If differential equations of motions do not assume the first integrals, then for construction the Lyapunov functions for investigation of stability or instability the need motion one may try to use differential consequences of equations.

When expanding the members in series with respect to degrees of deviation the equilibrium, the differential equations of a mechanical system may be reduced to the form:

$$A\ddot{x} + B\dot{x} + G\dot{x} + Cx + Px = Q(\dot{x}, x), \tag{9}$$

where $x, \dot{x}, \ddot{x}$ -matrices of generalized coordinates, velocities, accelerations (in deviations); A-positive definite symmetric matrix; B,C - symmetric matrices of dissipative and potential forces; G,P -skew-symmetric matrix of gyro- and nonconservative forces; $Q(x,\dot{x})$- matrix of additives nonlinear with respect to $x, \dot{x}$. It is known, for wide class of nonlinear forces, decomposition of the positional forces into potential $Q^P=\nabla U(x)$ and nonconservative $R(x)$ (satisfying the condition $R^T x=0$) forces, and the forces depended from velocities into dissipative $D=\nabla F(\dot{x})$ and gyro-forces Γ ($\Gamma^T \dot{x}=0$) is possible.

Let us consider two groups of invariant relations for the system (9) which is possible to obtain by the left-hand side multiplication (9) on the matrices $(L_1\dot{x})$ and (Lx) with subsequent transformations. By choosing various L_1 and L one can obtain, e.g., the following relations:

$$(\dot{x}^T A\dot{x} + x^T Cx + 2x^T\tilde{P}x - U(x))^{\cdot}/2 = -\dot{x}^T B\dot{x} + 2\dot{x}^T\tilde{P}x - x^T(\partial_x R)\dot{x} \quad (10)$$

$$(\text{ if } Q(x,\dot{x}) = \nabla U(x) + R(x) + \Gamma(\dot{x})),$$

$$(\dot{x}^T AA\dot{x} + x^T(AP+ AC)^c x)^{\cdot}/2 = \dot{x}^T AQ - \dot{x}^T(AB + AC)^c\dot{x} - \\ - \dot{x}^T(AP + AC)^s x, \quad (11)$$

here $\tilde{P}$ - triangular matrix obtained from P by replacement of elements below the diagonal by zeroes; $()^s$ - skew-symmetric part of the matrix, $()^c$ - symmetric part of the matrix. For any symmetric matrix L_1= M and for any L we have:

$$(\dot{x}^T M\dot{x} + x^T(MA^{-1}(C + P))^c x)^{\cdot}/2 = -\dot{x}^T(MA^{-1}(B + G))^c\dot{x} - \\ - \dot{x}^T(MA^{-1}(C + P))^s x + \dot{x}^T MA^{-1}Q, \quad (12)$$

$$(x^T L^T A\dot{x} + 1/2\, x^T(L^T B + L^T G)^c x)^{\cdot} = x^T L^T Q + \dot{x}^T(L^T A)^c\dot{x} - \\ - x^T(L^T B + L^T G)^s\dot{x} - x^T(L^T P + L^T C)^c x, \quad (13)$$

$$(\dot{x}^T P^T A\dot{x} + 1/2\, \dot{x}^T(P^T B + P^T G)^c x)^{\cdot} = x^T P^T Q + \dot{x}^T(P^T A)^c\dot{x} - \\ - x^T(P^T B + P^T G)^s\dot{x} - x^T(P^T P + P^T C)^c x. \quad (14)$$

If $Q(\dot{x}) = \nabla F(\dot{x})$, it is possible to obtain the relation for (9):

$$\frac{d}{dt}F(\dot{x}) = -(\nabla F(\dot{x}))^T A^{-1}(B\dot{x} + G\dot{x} + Cx + Px) + \\ + (\nabla F(\dot{x}))^T A^{-1}\nabla F(\dot{x}). \quad (15)$$

In more common case $Q(x,\dot{x})$:

$$\frac{d}{dt}\int_0^{\dot{x}} Q(0,\dot{x})d\dot{x} = -(Q(0,\dot{x}))^T A^{-1}(B\dot{x} + G\dot{x} + Cx + Px) + \\ + (Q(0,\dot{x}))^T A^{-1}Q(x,\dot{x}). \quad (16)$$

(10)-(16) give a rich set of functions V_i, satisfying some differential relations due to (9): $dV_i/dt = W_i$. If some V_i (or bandles from V_i) with their derivative W_i satisfy either of Lyapunov, Chetajev, Krasovsky or oth. theorems, we obtain sufficient conditions of stability or instability (depending on signdefiniteness of V_i and W_i) of equilibrium position for the system (9). Such a way it is possible to prove all known theorems on effect of potential, giroscopic, dissipative and nonconservative positional forces on stability of the equilibrium position. In concrete problems it is possible to obtain inequalities for system parameters with respect to these properties. Such a problem was solved for a system with 11 degrees of freedom. The relation (10) is employed in the "energetic" method. Examples of usage of correlations (15)-(16) may be found in 5. When constructing signdefinite and signvariable Lyapunov functions V with the aid of (10)-(16), we can obtain the possibility of defining (i) the domain of variation of system parameters corresponding to either stability or instability of the solution $x=\dot{x}=0$, (ii) the attraction domain or its estimate. In problems of this kind application of computer algebra (CA) side by side with computer graphic aids in considered perspective.

E x a m p l e. When the attraction domain of an attractor is noncompact, for the purpose of its estimation it is possible to use a theorem with a signvariable Lyapunov function **4**: let for the systems $\dot{x}=Ax+F(x)$, $x\in R^n$, $F(x)$ be the polynomial, there be the polynomial function $V(x)$, for which the derivative due to the system is $\dot{V}=aV(1+f(V))$, where $f(V)$ is the polynomial of V, and 1) $a=SpA=\sum\lambda_i<0$; 2) the set $M\{x|V(x)=0\}$ contains the point $x=0$ uniformely asymptotically stable on the whole for the equations defined on M by the initial system; 3) $V(x)$ is monotonously increasing with the growth of $\rho(x,V(x)=0)$, then hypersurface $V=\mu_{min}$ (if it devides the phase space R^n), where μ_{min} is a modulo-minimal real root of the equation $(1+f(V))=0$, will define the attraction domain of the solution $x=0$.

E.g., for well-known system considered in 6:

$$\dot{x} = -x + 2x^2y, \qquad \dot{y} = -y \tag{17}$$

Let us choose in accordance with (13) the function $V=xy$, for which $\dot{V}=-2V(1-V)$. Having defined $\rho=\min(|x|,|y|)$, we make sure that the conditions of the theorem are satisfied. Hence the curve $1-xy=0$ defines the boundary of attraction domain of the origin. Let us demonstrate another approach to defining the attraction domain by the same example. From (10) and (13) is follows for (17) that:

$$\dot{V}_1=1/2d(x^2+y^2)/dt=-x^2-y^2+x^3y \quad \text{and} \quad \dot{V}_2= d(xy)/dt=-2xy+2x^2y^2.$$

In the small neighbourhood of zero $V_1>>0$ has $\dot{V}_1<<0$, i.e. $x=y=0$ is asymptotically stable. The equation $\dot{V}_2=-2V_2(1-V_2)$ has only two steady-state solutions $V_2=0$, $V_2=1$. The first value ($V_2=0$) correspond to the following solutions: 1) $x=y=0$ (asymptotically stable), 2) $x=0$, y is arbitrary, for it the vector field is $\dot{y}=-y$; 3) $y=0$, x- is arbitrary, and for it the vector field is $\dot{x}=-x$. Hence, the solutions beginning on coordinate axises do not leave them and tend to the origin. The integral curve $xy=1$ corresponds to the solution $V_2=1$. The set $M(x,y|xy=1)$ is unstable. Consequently, the whole domain between the hyperbola branches xy=1 is the attraction domain of the origin. Moreover, the qualitative mode of behaviour of phase trajectories is determined. The example confirms the efficiency of employment of a few Lyapunov functions defining various properties of motion (stability and instability).

The usage of the proposed Lyapunov functions for linear systems is expedient in critical cases , when there is no asymptotic stability, or when the analysis of Routh-Hurwitz conditions is somehow complicated. When there are two zero roots of the characteristic equations with the help of Lyapunov functions it is possible, e.g., to determine the structure of Jordan cells of the canonical matrix, which correspond to these roots, without any normalizing transformation. To this end, it is sufficient to determine the fact of either stability or instability of the solution x=x=0. Relations (10)-(16) yield the conditions of existence a definite type of the integrals.

The discussed problems of qualitative investigations can be solved by computers,if the system CA offers the following possibilities: partial differentiation; Taylor expension of functions, matrix multiplication; separation of main diagonal minors; computation of determinants; explication of (i) Sylvester criterion of signdefiniteness of quadratic forms and (ii) the generalized Sylvester criterion of conditional signdefiniteness of quadratic forms on linear manifolds; substitutions; opening brackets; reduction of similars; trigonometric transformations, etc. Note, when verifying the conditions of existence of first integrals of account of (2)-(6) and the Jacoby integral, in trigonometric transformations not only substatutions $\sin^2x+\cos^2x=1$ are productive but also the following ones: $\sin^2x=1-\cos^2x$; $\cos^2x=1-\sin^2x$. In the package MECHANIC the system CA possesses the abovementioned possibilities.

References

1. A.Banshchikov,L.Burlakhova, V.Irtegov and oths, *in International conference on system of analitical computing and their applications in theoretical physics*,(JINR,Dubna,D11-85-71,1985),65-70.

2. A.Banshchikov and oths., *in book PPP.Results and applications*, (Nauka, Novosibirsk, 1986), 96-105.

3. L.Burlakhova, V.Irtegov, M.Pochtarenko, *in book A theory of stability and its applications*,(Nauka, Novosibirsk, 1979),247-254.

4. V.D.Irtegov, *Invariant manifold of stationar motions and their stability*,(Nauka, Novosibirsk, 1985),p.141.

5. E.Barbashin, *Lyapunov functions*,(Nauka, Moskow, 1970),p.239.

6. V.Zubov. *Stability of motion*,(High scool,Moskow, 1973),p.220.

USE OF SYMBOLIC CALCULATIONS IN PROBLEMS OF MECHANICS

V.M. RUDENKO
Institute for Problems of mechanics, Academy of Sciences of USSR, 117526, Prospect Vernadskogo 101, Moscow, USSR

and

V.V. LEONOV, A.I. ZHUROV
Institute for Problems of mechanics, Academy of Sciences of USSR, 117526, Prospect Vernadskogo 101, Moscow, USSR

ABSTRACT

The paper presents the features of a program package "Polymech-symbol" helping to solve some laborious mechanical problems. The package was written by means of the REDUCE system and contains several algorithms in a form of REDUCE procedures.

1. Introduction

In the Institute for Problems of Mechanics, a program package "Polymech-symbol" written by means of the symbolic computation system REDUCE has been created. The package in particular includes procedures realizing algorithms of some laborious mechanical problems such as deducing the kinetic and the potential energies of mechanical systems, obtaining the equations of motion in Lagrange's form, reducing the equations to the Caushy normal form, linearizing the equations near a partial solution, constructing equation solutions in a form of Taylor series with respect to an independent variable.

The paper aims to present the features of the "Polymech-symbol" package and in particular the latest ones, i.e., deriving the kinetic and the potential energies of a mechanical system. However, because of the paper volume limitations we have to say only some words about the method of deducing the kinetic energy and just a bit about the other topics.

2. Short description

Consider a system of rigid bodies with holonomic constraints. Let's assign a number n to each body of the mechanical system and choose a coordinate system $P_n x_n y_n z_n$ which we assume fixed in body n. Point P_n for each body may be taken arbitrary. Let's mentally make a kinetic chain of the bodies, i.e., define body j for each body i that is "attached" (physically or logically) to the one. "Attaching" means giving characteristics of the body i motion with respect to body j, i.e., generalized coordinates and parameters completely describing the kinematics of the body i motion with respect to body j.

We consider the motion of the whole mechanical system relative to an inertial coordinate system $P_0x_0y_0z_0$ which is taken as body 0.

To describe the body i motion relative to body j one should define the translation motion of point T_i fixed in body i and the rotational motion of the $P_ix_iy_iz_i$ with respect to the $P_jx_jy_jz_j$.

In the simplest case when constraints don't restrict the body i motion, the point T_i cartesian coordinates in the $P_ix_iy_iz_i$ and the Euler-Krylov (or Euler) angles of the $P_ix_iy_iz_i$ rotation relative to the $P_jx_jy_jz_j$, should be taken as the body i generalized coordinates. More complex cases should be described with the help of additional kinematic parameters. We have no opportunity to discuss it in detail.

Procedure $KINLINE$ is intended to describe the generalized coordinates and the kinematic chain. There are additional procedures for defining the kinematic parameters, the major central moments of inertia, the body i mass etc. As a result of $KINLINE$ working, the body i kinetic energy value is assigned to operator $TT(I)$.

Now we'll say some words about the other stages of research into the dynamics of a mechanical system.

In order to construct the equations of motion of a mechanical system, one may utilize procedure $EQLG$[2]. The procedure uses the technique of obtaining Lagrange's equations following the well known formula

$$\frac{d}{dt}\frac{\partial L}{\partial \dot{q}_i} - \frac{\partial L}{\partial q_i} = 0, \qquad i = 1, \ldots, N,$$

where $L(t, q_1, \ldots, q_N, \dot{q}_1, \ldots, \dot{q}_N)$ is the Lagrangian function of the mechanical system, q_i are the generalized coordinates.

For further research of the equations of motion, one usually transforms them to the Caushy normal form, i.e., rewrites them to solve for the highest derivatives. Since the highest derivatives come into Lagrange's equations as linear members it is sufficient to reverse the matrix of coefficients by the highest derivatives. Procedure NFK[2] realizes this algorithm.

Usually, one linearizes the equations of motion near a partial solution for researching stability of the equilibrium condition and for other aims. Procedure $LINEU$[2] solves this problem.

Procedure $PUAS1$[2] constructs equation solutions in a form of Taylor series with respect to an independent variable. The procedure uses Lie's group technique[3].

3. Example

To illustrate how the procedures involved work let's consider, as an example, a gyro with a gimbal suspension whose rotation axes of the inside and the outside rings aren't perpendicular to each other[4].

Unfortunately, we'll not be able to present the explicit form of the results to be obtained in the example, since these are too large to be included into the paper. Therefore, we will write out the results only as functions depending on parameters which we'll write explicitly.

3.1. Deducing the kinetic energy

Consider the gyro as a system of three bodies, i.e., the inside ring (body 1), the outside ring (body 2), the rotor (body 3).

Introduce a coordinate system $Px_iy_iz_i$ for body i (i=1,2,3; point P is the gyro center of inertia) and an inertial coordinate system $Px_0y_0z_0$ (body 0). Let axes Py_1 and Pz_1 be directed along the rotation axes of the inside ring and the rotor respectively, axis Px_2 be directed along the outside ring rotation axis, and axis Py_2 be in the outside ring plane. There is an angle $\delta \neq 0$ between axes Py_1 and Py_2 because the rotation axes of the inside and the outside rings aren't perpendicular to each other. Let at an initial moment of time the coordinate system $Px_1y_1z_1$ coincides with the $Px_3y_3z_3$, $Px_0y_0z_0$ coincides with the $Px_2y_2z_2$, and the plane Px_1y_1 coincides with the Px_2y_2. Thus we have the following kinematic chain

$$\textit{body}\,0 \;\rightarrow\; \textit{body}\,2 \;\rightarrow\; \textit{body}\,1 \;\rightarrow\; \textit{body}\,3$$

Let's define the inertial parameters of the outside ring (A_2, B_2, C_2 are the major central moments of inertia, M_2 is the ring mass):

$INERTIA(2, A2, B2, C2, M2);$

Compute the kinetic energy of the outside ring:

$KINLINE(0, 2, 0, 0, 0, 0, AL, 0, 0);$

Here AL is a rotation angle (α) of the outside ring around axis Px_0. Let's write the same for the inside ring:

$ANGLESO(1, 0, 0, 0, DE);$
$INERTIA(1, A1, B1, C1, M1);$
$KINLINE(2, 1, 0, 0, 0, 0, 0, BE, 0);$

Here $ANGLESO$ is a procedure defining the angle DE (δ) between the rotation axes of the rings, BE is a rotation angle (β) of the inside ring around its axis.

For the rotor we can write:

$INERTIA(3, A, A, C, M);$
$KINLINE(1, 3, 0, 0, 0, 0, 0, 0, GA);$

Here GA is a rotation angle (γ) of the rotor around its axis.

As a result, the kinetic energies of the bodies 1, 2, 3 are assigned to $TT(I)$, $I = 1, 2, 3$.

3.2. Obtaining Lagrange's equations.

Find the Lagrangian function:

$LG(1) := FOR\ I := 1 : 3\ SUM\ TT(I);$

It has the following form: $LG(1) = F(\beta, \dot{\alpha}, \dot{\beta}, \dot{\gamma};\ \delta, A, C, A_1, B_1, C_1, A_2)$. Derive Lagrange's equations:

$LET\ DF(AL, T) = YT(1),\ \ DF(BE, T) = YT(2),\ \ DF(GA, T) = YT(3);$
$Y(1) := AL; \quad Y(2) := BE; \quad Y(3) := GA;$

```
EQLG(3, LG, BU, T, 1);
CLEAR  DF(AL,T),  DF(BE,T),  DF(GA,T);
```

The left parts of the obtained equations are assigned to $BU(I)$, $I = 1,2,3$.

3.3. Reducing the equations to the Caushy normal form.

Using the system first integral $H = C(\dot{\gamma} + \dot{\alpha} \cdot sin\beta \cdot cos\delta) = const$, we exclude $\dot{\gamma}$ from the equations:

```
LET  DF(GA,T) = H/C - DF(AL,T) * SIN(BE) * COS(DE);
ARRAY  U1(1,2);     U1(1,1) := BU(1);     U1(1,2) := BU(2);
CLEAR  DF(GA,T);
```

Introduce new variables $u = \dot{\alpha}, w = \dot{\beta}$:

```
Y(1) := AL;     Y(2) := BE;     Y(3) := U;     Y(4) := W;
```

Reduce the equations to the Caushy normal form:

```
NFK(1, 2, U1, U2, 1, T);
```

The right parts of the obtained equations are assigned to $U2(1,I)$, $I = 1,2,3,4$ ($U2$ is an array). The equations have the following form

$$\dot{\alpha} = u, \qquad \dot{\beta} = w,$$

$$\dot{u} = f_1(\beta, u, w;\ \delta, A, H, A_1, C_1, A_2),$$

$$\dot{w} = f_2(\beta, u, w;\ \delta, A, H, A_1, B_1, C_1, A_2).$$

3.4. Linearizing the equations near a partial solution.

Consider the following partial solution of the above equations:

$$\alpha = 0, \qquad \beta = \beta_0, \qquad u = 0, \qquad w = 0.$$

The equations are defined as follows:

```
ARRAY  FL(4),  FP(4),  NU(4,0);
LET  AL = Y(1,T),  BE = Y(2,T),  U = Y(3,T),  W = Y(4,T);
FOR  J := 1 : 4  DO  <<
     FL(J) := DF(Y(J,T),T);
     FP(J) := U2(1,J)  >>;
NU(1,0) := 0;        NU(2,0) := BE0;
NU(3,0) := 0;        NU(4,0) := 0;
```

Now we linearize the equations:

```
LINEU(4, 2, EP, T);
```

The linear parts the obtained equations are assigned to $RRML(I)$, and the nonlinear parts are assigned to $RRMP(I)$, $I = 1,2,3,4$ ($RRML$, $RRMP$ are arrays).

3.5. Constructing the solution in a form of Taylor series with respect to independent variable t.

Let's take the following initial conditions ($t = 0$)

$$\alpha = 0, \qquad \beta = 0, \qquad u = \dot{\alpha}_0, \qquad w = 0.$$

The equations to be solved are defined in the following way:

$ARRAY\ \ PA(4),\ \ PU(4);$
$FOR\ \ ALL\ \ J\ \ LET\ \ Y(J,T) = Y(J);$
$EP := 1;$
$FOR\ \ J := 1:4\ \ DO\ \ PA(J) := Y(J);$
$FOR\ \ J := 1:4\ \ DO\ \ PU(J) := RRMP(J) - RRML(J) + DF(Y(J,T),T);$

The solution is obtained with the help of

$PUAS1(4,2,4,PA,PU,UU,T,Y);$
$AL := 0; \qquad BE := BE0; \qquad U := AT0; \qquad W := 0;$

The right parts of the obtained expressions are assigned to $UU(I)$, $I = 1,2$ (UU is an array). The solution has the following form:

$$\alpha(t) = a_0 + a_1 t + a_2 t^2 + \ldots$$
$$\beta(t) = b_0 + b_1 t + b_2 t^2 + \ldots$$

Here a_i, b_i, $(i = 0,1,2,\ldots)$ are functions of parameters δ, A, H, A_1, B_1, C_1, A_2, β_0, $\dot{\alpha}_0$.

4. Conclusion

The given example shows what a deep analysis can be made with the help of the REDUCE system and the "Polymech-symbol" package. It should be noted that the package has much more features than those of described here. It includes procedures allowing the user to carry out deeper analysis of a mechanical system at different stages. There are opportunities, for example, to find solutions for linear parts of differential equations, to analyze them, to rewrite differential equations as those in normal coordinates[2,3], to analyze a mechanical system characteristic determinant, to investigate transient motion for nonlinear oscillating systems etc.

5. References

1. A.C.Hearn, *Reduce User's Manual. Version 3.0* (Santa Monica, CA 90406, 1983).
2. D.M.Klimov and V.M.Rudenko, *Methods of Computer Algebra in Problems of Mechanics,* (Nauka, Moscow, 1989), p. 216 (in Russian).
3. D.M.Klimov and V.F.Zhuravlyov, *Applied Methods of Vibration Theory,* (Nauka, Moscow, 1988), p. 328 (in Russian).
4. D.M.Klimov and S.A.Kharlamov, *Dynamics of Gyroscope with Gimbal Suspension,* (Nauka, Moscow, 1978), p. 208 (in Russian).

THE ROTATIONAL EVOLUTION OF MACLAURIN ELLIPSOIDS WHICH ACCRETE MATTER VIA A DISC: AN APPLICATION OF COMPUTER ALGEBRA

I.L.Andronov
Dept. of Astronomy, Odessa State University,
T.G.Shevchenko Park,Odessa, USSR, 270014

Z.T.Kostova
Joint Institute for Nuclear Research
Head Post Office P.O.Box 79,101000 Moscow, USSR

K.S.Kölbig
Visitor from CERN, European Laboratory for Particle Physics,
CH-1211 Geneva 23, Switzerland

Abstract

The rotation of stars and planets is one of the most significant phenomena which affect the evolution of these astronomical objects (see e.g.,Kippenhahn and Tomas 1981),Packet (1981 a,b), Zahn and references therein). In this paper we study the rotational evolution of incompressible axial-symmetric self-gravitating bodies, onto which matter form outer space falls via an accretion disc. The viscosity forces are assumed to be sufficiently strong to bring the outer and inner layers into synchronous rotation.No other dissipative forces, for example tidal distortion by a companion star, are taken into account. This restricted problem can be solved analytically for the main physical parameters, i.e. the mass m, the equatorial radius (major semi - axis) a, the polar radius (minor semi-axis) c, the eccentricity e, the angular velocity w, and the angular momentum l. With the help of a program package for manipulating power series, written in the computer algebra language REDUCE, we have derived the leading coefficients in the power series for these parameters as functions of both the eccentricity e and the mass m.

With the assumptions made above, the objects under considaration are ellipsoids (Chandrasekhar (1969) and references therein), and the basic equations can be written in the following way (Andronov (1988));

$w(e) = \mathrm{sgn}(e) f_1^{1/2}(e)$ – angular velocity in units $\Omega_0 = (2\pi G\rho)^{1/2}$

$m(e) = a^3(e)(1-e^2)^{1/2}$ – mass in units $M_0 = 4\pi/3\rho A_0^3$

$l(e) = a^5(e) f_3(e)$ – angular momentum in units $L_0 = M_0 A_0 \Omega_0$

$c(e) = a(e)(1-e^2)^{1/2}$ – polar radius in units A_0;

$dl/dm = a^2(e) f_2^{1/2}(e)$;

where G is the gravitational constant, ρ is the density, and where :

$$f_1(e) = \left(\frac{\arcsin(e)}{e} (1 - e^2)^{1/2} (3 - 2e^2) - 3(1 - e^2)\right)/e^2,$$

$$f_2(e) = \left(\frac{\arcsin(e)}{e} (1 - e^2)^{1/2} - (1 - e^2)\right)/e^2,$$

$$f_3(e) = \frac{2}{5} \mathrm{sgn}(e) \left((1 - e^2)^{1/2} f_1(e)\right)^{1/2}.$$

The factor sgn(e) allows us to consider negative values of e (see bellow). At initial time t = 0, which corresponds to e = 0, the values of the mass and the equatorial radius are $m(0) = M_0$ and $a(0) = A_0$ respectively.In dimensioneless units we can set $m(0)= a(0)=c(0)=1$ and $w(0)=l(0)= 0$. Combining the above expressions for m(e), l(e) and dl/dm, we obtain the differential equation

$$da/de = a(e) f_4(e),$$

where

$$f_4(e) = \frac{df_3(e)/de + e(f_2(e)/(1 - e^2))^{1/2}}{3(f_2(e)(1 - e^2))^{1/2} - 5f_3(e)}.$$

Thus

$$a(e) = \exp \int_0^e f_4(e)de.$$

The other functions w(e), m(e), c(e), l(e) can be calculated easily by using a(e) and the above equations. They have been computed numerically by Andronov (1988), and are given in Table 1.

In order to obtain the leading coefficients of the power series for the above functions, we use a program package (Feldmar and Kölbig (1986)) for manipulating generalized power series (gps) of the form:

$$y(x) = \sum_{j=0}^{\infty} a_j x^{\mu+\nu j} \qquad (\mu,\nu \text{ rational})$$

written in the computer algebra language REDUCE. This package handles the multiplication and division of two gps, the raising of a gps to an arbitrary power, the substitution of one gps into another, the reversion of a gps, and the deparametrization of two gps given in terms of a common parameter.

The power series for some of the above functions (of e) as obtained on a personal computer by using this package are presented in Table2. If one introduces a dimensionless parameter

$$q = m - 1 = \dot{M}.(t - t_0)/M_0,$$

where $\dot{M}$ is the acceretion rate (mass of infalling matter accreted per unit time), one can obtain, by reversion of the power series for m(e) = 1 + q(e) and substitution, the leading coefficients of the power series for e(q), a(q), c(q), and l(q) as function of q. These series are given in Table 3.

A plot of the functions e(m), a(m), c(m), l(m) (dimensionless units) is given in Fig 1. We can extend these functions formally to negative values of the eccentricity e(m<1), which correspond to negative values of time t, i.e. to a situation in which the angular velocity of the infalling matter and that of the rigidly rotating accreting object are pointed in opposite directions. The mass m = 0.6677 corresponds to eccentricity e = -0.99, and it appears that there exists a lower limit of mass m $\gtrsim$0.6 corresponding to e = -1, i.e. to a flat disc. As the mass increases, the polar radius c increases until its maximum of 1.0057 at m = 1.0357 (e = 0.1340) and then decreases monotonically to zero; while e tends to 1 as m goes to infinity. The equatorial radius a,on the contrary, decreases until its minimum of 0.9903 at m = 0.9450 (e = -0.2301), and then tends to infinity at e to ±1. Obviusly, the angular momentum l and the eccentricity e increase with m. The angular velocity w (not shown in the fiqure) has its maximum at m = 1.7569 (e = 0.9300) and its minimum at m=0.7362 (e = -0.9300). It is equal to zero at e = -1,0,1.

The absolute value of the equatorial velocity v = wa (which is negative for e>0) increases as |e| increases from zero, and reaches the "first cosmic (or Keplerian) velocity" at e = ±1.

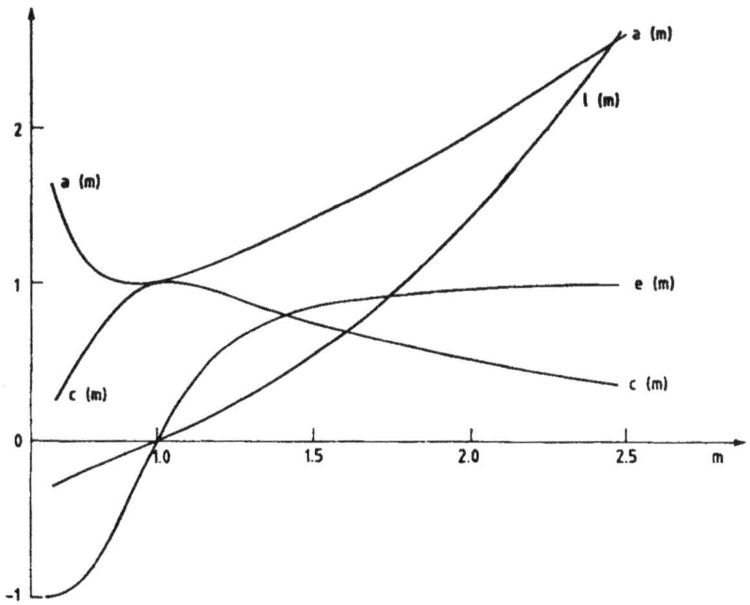

Fig.1 The equatorial radius a(m), the polar radius c(m), the angular momentum l(m) and the eccentricity e(m) as functions of the mass.
(All variables in demensionless units.)

The fact that m tends to infinity as e approaches unity shows that an object which consists of incompressible fluid may accrete an infinite amount of matter, in contrast to the case of a compressible gas.

The "incompressible fluid" approximation corresponds to the value $n = 0$ in the polythropic equation of state $P = K_\gamma \rho^\Gamma$, $\gamma = 1+1/n$, where P and ρ are the local pressure and density respectively, and K_γ is a constant. It might also be possible to apply this approximation to asteroids consisting of rocks and dust, but not to stellar objects for which the effective value of n lies between 1.5 to 3, or when the equation of state is more comlicated. For these cases numerical methods are more appropriate. This will be discussed elsewhere.

Acknowledgement

K.S.Kölbig would like to thank JINR at Dubna for its hospitality.

References

1. Andronov I.L., in:"*The Problems of Astronomy, vol.3*",Preprint Ukr. NIINTI (Ukrainian Sci.Res.Sci.Tech.Inform.,Kiev) No. 2921, 1988, 217-221.
2. Chandrasekhar S., *Elipsoidal Figures of Equilibrium*, New Haven and London, Yale Univ.Press, 1969, 254.
3. Feldmar E.,Kölbig K.S., Computer Phys. Comm. 39, 1986, 267.
   Kippenhahn R., Thomas H.-C.,in, "Fundamental Problems in Theory of Stellar Evolution", eds. D. Sugimito et al.(IAU Symp. 93), Reidel, Dordrecht, 1981, 237.
4. Packet W., Astron.Astrophys. 102, 1981 a) 179, b) 17.
5. Zahn J.P.,in: "*Observational Tests of the Stellar Evolution Theory*", eds. A. Maeder and A. Renzini (IAU Symp.105), Reidel, Dordrecht, 1981, 379.

Table 1

e	a(e)	c(e)	m(e)	l(e)	w(e)
-0.99	1.67904021	0.23685777	0.66774375	-0.29659891	-0.39389233
-0.98	1.51176616	0.30083767	0.68754552	-0.27325735	-0.43475157
-0.97	1.42343703	0.34604454	0.70114609	-0.25809379	-0.45418474
-0.96	1.36486632	0.38216257	0.71191539	-0.24652536	-0.46472175
-0.95	1.32174007	0.41271321	0.72100866	-0.23703071	-0.47044845
-0.90	1.20057494	0.52331848	0.75430089	-0.20410580	-0.46932339
-0.85	1.13892648	0.59996675	0.77824899	-0.18189850	-0.45046307
-0.80	1.09965720	0.65979432	0.79785361	-0.16444897	-0.42612137
-0.75	1.07202466	0.70907766	0.81489820	-0.14971756	-0.39966841
-0.70	1.05149914	0.75092059	0.83025568	-0.13673572	-0.37238510
-0.65	1.03576827	0.78711574	0.84443030	-0.12495786	-0.34483756
-0.60	1.02350408	0.81880326	0.85774603	-0.11404097	-0.31729486
-0.55	1.01388073	0.84675735	0.87042772	-0.10375162	-0.28988680
-0.50	1.00635492	0.87152893	0.88264112	-0.09392064	-0.26267246
-0.45	1.00055447	0.89352371	0.89451485	-0.08441875	-0.23567303
-0.40	0.99621715	0.91304810	0.90615331	-0.07514228	-0.20888857
-0.35	0.99315498	0.93033763	0.91764487	-0.06600432	-0.18230715
-0.30	0.99123242	0.94557546	0.92906734	-0.05692877	-0.15590988
-0.25	0.99035249	0.95890467	0.94049183	-0.04784618	-0.12967391
-0.20	0.99044785	0.97043674	0.95198576	-0.03869054	-0.10357410
-0.15	0.99147495	0.98025740	0.96361516	-0.02939676	-0.07758404
-0.10	0.99341033	0.98843080	0.97544685	-0.01989827	-0.05167665
-0.05	0.99624831	0.99500222	0.98755036	-0.01012480	-0.02582450
0.00	1.00000000	1.00000000	1.00000000	0.00000000	0.00000000
0.05	1.00469313	1.00343648	1.01287710	0.01056125	0.02582450
0.10	1.01037296	1.00530840	1.02627261	0.02165611	0.05167665
0.15	1.01710405	1.00559653	1.04029025	0.03339779	0.07758404
0.20	1.02497329	1.00426463	1.05505054	0.04592081	0.10357410
0.25	1.03409438	1.00125758	1.07069599	0.05938806	0.12967391
0.30	1.04461427	0.99649850	1.08739807	0.07400041	0.15590988
0.35	1.05672250	0.98988448	1.10536682	0.09001026	0.18230715
0.40	1.07066473	0.98128043	1.12486429	0.10774120	0.20888857
0.45	1.08676271	0.97051014	1.14622410	0.12761710	0.23567303
0.50	1.10544455	0.95734306	1.16988054	0.15020693	0.26267246
0.55	1.12729198	0.94147442	1.19641365	0.17629607	0.28988680
0.60	1.15311750	0.92249400	1.22662180	0.20700500	0.31729486
0.65	1.18409622	0.89983522	1.26164443	0.24399758	0.34483756
0.70	1.22200629	0.87268705	1.30318303	0.28987088	0.37238510
0.75	1.26970409	0.83983031	1.35393116	0.34894858	0.39966841
0.80	1.33216999	0.79930199	1.41850276	0.42908446	0.42612137
0.85	1.41919089	0.74760519	1.50575370	0.54645543	0.45046307
0.90	1.55408400	0.67740951	1.63606391	0.74179077	0.46932339
0.95	1.81797831	0.56766355	1.87615365	1.16685752	0.47044845
0.96	1.91204548	0.53537273	1.95727878	1.33015459	0.46472175
0.97	2.04006093	0.49594884	2.06406398	1.56063722	0.45418474
0.98	2.23377506	0.44451562	2.21802229	1.92462415	0.43475157
0.99	2.60302464	0.36720181	2.48806300	2.65616745	0.39389233

Table 2

$$f1(e) = e^{2}\left(\frac{4}{15} + \frac{4}{105}e^{2} - \frac{32}{3465}e^{6} - \frac{512}{45045}e^{8} - \frac{512}{45045}e^{10} - \frac{8192}{765765}e^{12} - \frac{4096}{415701}e^{14}\right.$$

$$\left. - \frac{131072}{14549535}e^{16} - \frac{131072}{15935205}e^{18} - \frac{4194304}{557732175}e^{20} - \frac{1048576}{152108775}e^{22} + \ldots\right)$$

$$f2(e) = \frac{2}{3} - \frac{2}{15}e^{2} - \frac{8}{105}e^{4} - \frac{16}{315}e^{6} - \frac{128}{3465}e^{8} - \frac{256}{9009}e^{10} - \frac{1024}{45045}e^{12}$$

$$- \frac{2048}{109395}e^{14} - \frac{32768}{2078505}e^{16} - \frac{65536}{4849845}e^{18} - \frac{262144}{22309287}e^{20} - \frac{524288}{50702925}e^{22} + \ldots$$

$$f3(e) = \frac{1}{\sqrt{15}}\, e\left(\frac{4}{5} - \frac{12}{35}e^{2} - \frac{32}{245}e^{4} - \frac{3952}{56595}e^{6} - \frac{75168}{1716715}e^{8} - \frac{31072}{1030029}e^{10}\right.$$

$$\left. - \frac{447404032}{20224619415}e^{12} - \frac{106453127552}{6276373558455}e^{14} - \frac{7679227940992}{571149993819405}e^{16} + \ldots\right)$$

$$f4(e) = e\left(\frac{83}{225} + \frac{26977}{70875}e^{2} + \frac{1222349}{3189375}e^{4} + \frac{4240040501}{11051184375}e^{6} + \frac{2478465297181}{6464942859375}e^{8} + \ldots\right)$$

$$+ \frac{1}{\sqrt{10}}\left(\frac{2}{75} + \frac{146}{4725}e^{2} + \frac{223994}{7441875}e^{4} + \frac{106866446}{3683728125}e^{6} + \frac{2958208352434}{105594066703125}e^{8} + \ldots\right)$$

Table 2 (cont.)

$$w(e) = \frac{1}{\sqrt{15}} e \left(2 + \frac{1}{7} e^{2} - \frac{1}{196} e^{4} - \frac{3103}{90552} e^{6} - \frac{2648929}{65921856} e^{8} - \frac{36733}{921984} e^{10} \right.$$
$$\left. - \frac{78008429911}{2071001028096} e^{12} - \frac{135157071912505}{38562039143147 52} e^{14} - \frac{60752933003304757}{1871544299747426304} e^{16} + \ldots \right)$$

$$a(e) = 1 + \frac{47}{250} e^{2} + \frac{8191121}{70875000} e^{4} + \frac{7576346389}{88593750000} e^{6} + \frac{9835966321272353}{143256093750000000} e^{8} + \ldots$$
$$+ \frac{1}{\sqrt{10}} e \left(\frac{4}{15} + \frac{18014}{118125} e^{2} + \frac{339633229}{3100781250} e^{4} + \frac{15498733229293}{179070117187500} e^{6} + \frac{31706881174610534843}{439975277929687500000} e^{8} + \ldots \right)$$

$$m(e) = 1 + \frac{32}{375} e^{2} + \frac{62498}{984375} e^{4} + \frac{811833248}{16611328125} e^{6} + \frac{132760018664653}{3357564697265625} e^{8} + \ldots$$
$$+ \frac{1}{\sqrt{10}} e \left(\frac{4}{5} + \frac{42548}{118125} e^{2} + \frac{122459812}{516796875} e^{4} + \frac{7939938379724}{44767529296875} e^{6} + \frac{58492898451843476 56}{4124768230590820 3125} e^{8} + \ldots \right)$$

$$l(e) = \frac{1}{\sqrt{15}} e \left(\frac{4}{5} + \frac{734}{1575} e^{2} + \frac{120761}{330750} e^{4} + \frac{3532606327}{11460487500} e^{6} + \frac{101248670991067}{375445570500000} e^{8} + \ldots \right)$$
$$+ \sqrt{\frac{2}{3}} e^{2} \left(\frac{8}{75} + \frac{764}{7875} e^{2} + \frac{152743}{1771875} e^{4} + \frac{6647892817}{85953656250} e^{6} + \frac{592807811275289}{8447525336250000} e^{8} + \ldots \right)$$

$$c(e) = - 1 + \frac{39}{125} e^{2} + \frac{916313}{8859375} e^{4} + \frac{322635421}{5537109375} e^{6} + \frac{44048811820799}{1119188232421875} e^{8} + \ldots$$
$$+ \frac{1}{\sqrt{10}} e \left(\frac{4}{15} + \frac{2264}{118125} e^{2} - \frac{79948}{1550390625} e^{4} - \frac{176547619208}{44767529296875} e^{6} - \frac{666806993838718 76}{1374922743530273437 5} e^{8} + \ldots \right)$$

Table 3

$$e(q) = \sqrt{10}\, q \left(\frac{5}{4} - \frac{5}{3} q - \frac{26305}{6048} q^2 + \frac{1773395}{72576} q^3 - \frac{2480659}{108864} q^4 - \frac{2815548935}{15676416} q^5 + \frac{34849141931245}{50697529344} q^6 + \frac{268049207625205}{1216740704256} q^7 - \frac{7133836041437149645}{854151974387712} q^8 + \dots\right)$$

$$a(q) = 1 + \frac{1}{3} q + \frac{359}{144} q^2 - \frac{7795}{1296} q^3 + \frac{6617035}{870912} q^4 + \frac{9826973}{2612736} q^5 - \frac{27361607657}{2633637888} q^6 - \frac{1329349034183}{7900913664} q^7 + \frac{102663043589968135}{116807107608576} q^8 + \dots$$

$$l(q) = \sqrt{\frac{2}{3}}\, q \left(1 + \frac{1}{3} q + \frac{509}{432} q^2 - \frac{4693}{2592} q^3 + \frac{4475287}{4354560} q^4 + \frac{1926667}{559872} q^5 - \frac{124602704183}{18435465216} q^6 - \frac{326386048669}{15801827328} q^7 + \frac{117630507987090559}{1051263968477184} q^8 + \dots\right)$$

$$c(q) = 1 + \frac{1}{3} q - \frac{383}{72} q^2 + \frac{7915}{648} q^3 + \frac{547145}{108864} q^4 - \frac{71299663}{653184} q^5 + \frac{4451693231}{20575296} q^6 + \frac{274864764337}{493807104} q^7 - \frac{6664140895670435}{1825111056384} q^8 + \dots$$

GROUP-CLASSIFIED POLYNOMIALS OF PHASE SPACE AND OPERATIONS AMONG THEM

KURT BERNARDO WOLF and GUILLERMO KRÖTZSCH

INSTITUTO DE INVESTIGACIONES EN MATEMÁTICAS APLICADAS Y EN SISTEMAS
UNIVERSIDAD NACIONAL AUTÓNOMA DE MÉXICO
Национальний Автономический Университет Мексика

Apartado Postal 139-B, 62191 Cuernavaca, México; Куэрнавака, МЕКСИКО

Abstract: We use polynomial functions of classical phase space $(p_1, p_2; q_1, q_2)$, on a reference screen in geometric aberration optics. The polynomials are irreducibly classified by their transformation properties under linear transformations of phase space; for axis-symmetric systems we use the solid spherical (*symplectic*) harmonics in the variables p^2, $\mathbf{p}\cdot\mathbf{q}$ and q^2 . The operations in which the polynomials participate are linear combinations, products, Poisson brackets, and Baker-Campbell-Hausdorff compounding that embodies the aberration group composition law. We handle the polynomials through their coefficients in two bases: the *monomial* basis and the *symplectic* basis, and show that the symplectic basis is more efficient —by some 40%— for Poisson brackets and aberration-group compounding.

1. Introduction

The functions we must handle in higher-order perturbation theory are polynomials in the phase space variables of a system, that are subject to linear combination, multiplication, Poisson brackets, and other nonlinear compounding that represents the concatenation of optical elements of a system, calculated up to a given (typically seventh) aberration order. We present results on the computation complexity for polynomials expressed in the basis of *monomials* and in the *symplectic* basis built from spherical harmonics. The mathematics are applicable to many common Hamiltonian mechanical systems such as perturbed harmonic oscillators.

The phase space coordinates $\mathbf{p} = (p_1, p_2)$, $\mathbf{q} = (q_1, q_2)$ are defined to satifsy the Lie-Poisson bracket relations of the two-dimensional Heisenberg-Weyl algebra,

$$\{q_i, p_j\} = \delta_{i,j}, \qquad \{q_i, q_j\} = 0 = \{p_i, p_j\}, \tag{1.1}$$

where the operation $\{\circ, \circ\}$ is bilinear, antisymmetric, and satisfying the Jacobi identity. The functions

$$p^2 = p_1^2 + p_2^2, \quad \mathbf{p}\cdot\mathbf{q} = p_1 q_1 + p_2 q_2, \quad q^2 = q_1^2 + q_2^2, \quad \mathbf{q}\times\mathbf{p} = q_1 p_2 - q_2 p_1, \tag{1.2}$$

are invariant under joint rotations of the $\mathbf{p}$ and $\mathbf{q}$ planes. Axis-symmetric systems will be described by polynomials of (1.2) taken as variables. The Leibnitz identity for the Poisson bracket, $\{f, gh\} = g\{f,h\} + \{f,g\}h$, finds

$$\{\mathbf{p}\cdot\mathbf{q}, p^2\} = 2p^2, \qquad \{\mathbf{p}\cdot\mathbf{q}, q^2\} = -2q^2, \qquad \{q^2, p^2\} = 4\mathbf{p}\cdot\mathbf{q}, \quad \{\mathbf{q}\times\mathbf{p}, f\} = 0, \tag{1.3}$$

where the last equation holds for any function f of the variables. This characterizes (1.2) as the symplectic *Lie* algebra $\mathrm{sp}(2,\Re)$, and $\mathbf{q}\times\mathbf{p}$ as an invariant, that is is not functionally independent of the others, since its square, the *Petzval* invariant of optics

$$|\mathbf{q}\times\mathbf{p}|^2 = q^2p^2 - (\mathbf{p}\cdot\mathbf{q})^2 \tag{1.4}$$

is the *Casimir* invariant of the algebra.

Aberration optics requires polynomial functions of phase space of degree greater than two. The Lie group exponentiated from the above Lie algebra is $Sp(2,\Re)$. It is used to classify the polynomials by symplectic *spin*, similar to ordinary angular momentum in the three-dimensional harmonic oscillator /1/, /2/. Aberrations are thus classified into irreducible *symplectic multiplets*, which mix only amonst themselves under the linear transformations of paraxial optics.

The monomial basis of functions is reviewed in Section 2 as a multi-level vector with the coefficients of the polynomials. In Section 3 we present the symplectic basis /2/. We give the relative advantages for symbolic computation in the two bases, under the operations of ordinary polynomial multiplication in Section 4, Poisson brackets/3/ in Section 5, and elements of aberration group in Section 6. Nonlinear transformations are thus treated as linear perturbation expansions /4/. We have observed that in multiplication, the monomial basis tables are roughly 40% shorter, while for Poisson brackets, there is a 40% advantage in the choice of the symplectic basis.

Our current aim is to produce a package of functions for Lie optics design of axis-symmetric systems to *seventh* aberration order. Aberration order three may be managed by hand, while orders nine and beyond sometimes end up with expressions too long for a 612 KB PC. Order seven is a compromise that is nontrivial and within desktop capacity.

This conference report is a condensed version of reference /5/; the tables mentioned in the text are being integrated into a .SYS file in muSIMP; the tables themselves, to be used in MACSYMA or FORTRAN, for example, can be obtained from the author through formal request.

2. The monomial basis

We consider *polynomials* $\mathcal{A}(p^2,\mathbf{p}\cdot\mathbf{q},q^2)$ in the variables (1.2), of degrees 1 up to some $k_{\max}$, written as sums of *monomials*

$$M_{k_+,k_0,k_-}=(p^2)^{k_+}(\mathbf{p}\cdot\mathbf{q})^{k_0}(q^2)^{k_-}. \tag{2.1}$$

The polynomial $\mathcal{A}$ may be written then

$$\mathcal{A}=\sum_{1\le k_++k_0+k_-\le k_{\max}} A^{\mathrm{M}}(k_+,k_0,k_-)\,M_{k_+,k_0,k_-}=\sum_{k=1}^{k_{\max}}\sum_{n=1}^{\frac{1}{2}(k+1)(k+2)} A^{\mathrm{M}}[k,n]\,M_{k_+,k_0,k_-}, \tag{2.2}$$

where $k=k_++k_0+k_-$ is the *rank*, and $n=\frac{1}{2}(k_0+k_-)(k_0+k_-+1)+k_-+1$ numerates the $\{k_+,k_0,k_-\}$ of fixed rank to lexicographical order. For example, for $k=2$, the labels $n=1,\ldots,6$ enumerate the monomials 200, 110, 101, 020, 011, and 002, respectively.

In the monomial basis we handle the polynomial $\mathcal{A}$ by a two-level column vector $\mathcal{A}^{\mathrm{M}}_{\mathrm{vec}}$, whose elements $\mathcal{A}^{\mathrm{M}}_{\mathrm{vec}}[k,n]$, contain the polynomial coefficients $A^{\mathrm{M}}(k_+,k_0,k_-)$. In fact we avoid the use of *constant* terms in polynomials so as not to have a $k=0$ subcomponent. For rank $k=2$, second-order perturbation theory, the polynomials are represented as the nine vector components in

$$\mathcal{A}^{\mathrm{M}}_{\mathrm{vec}} = \begin{pmatrix} \begin{pmatrix} A^{\mathrm{M}}[1,1] \\ A^{\mathrm{M}}[1,2] \\ A^{\mathrm{M}}[1,3] \end{pmatrix} \\ \begin{pmatrix} A^{\mathrm{M}}[2,1] \\ A^{\mathrm{M}}[2,2] \\ A^{\mathrm{M}}[2,3] \\ A^{\mathrm{M}}[2,4] \\ A^{\mathrm{M}}[2,5] \\ A^{\mathrm{M}}[2,6] \end{pmatrix} \end{pmatrix} = \begin{pmatrix} \begin{pmatrix} A^{\mathrm{M}}(1,0,0) \\ A^{\mathrm{M}}(0,1,0) \\ A^{\mathrm{M}}(0,0,1) \end{pmatrix} \\ \begin{pmatrix} A^{\mathrm{M}}(2,0,0) \\ A^{\mathrm{M}}(1,1,0) \\ A^{\mathrm{M}}(1,0,1) \\ A^{\mathrm{M}}(0,2,0) \\ A^{\mathrm{M}}(0,1,1) \\ A^{\mathrm{M}}(0,0,2) \end{pmatrix} \end{pmatrix}, \quad \text{coefficient of} \begin{cases} p^2 \\ \mathbf{p}\cdot\mathbf{q} \\ q^2 \\ (p^2)^2 \\ p^2\mathbf{p}\cdot\mathbf{q} \\ p^2q^2 \\ (\mathbf{p}\cdot\mathbf{q})^2 \\ \mathbf{p}\cdot\mathbf{q}\,q^2 \\ (q^2)^2. \end{cases} \tag{2.3}$$

In mechanics, this set corresponds to the three-dimensional harmonic oscillator Cartesian basis with one and two energy quanta, the p and d shells of vector and quadrupole deformation. The number of components in each subvector k is $\frac{1}{2}(k+1)(k+2)$; the sum of these dimensions for $k' = 1, 2, \ldots k$ represents the $\frac{1}{6}k(k^2+6k+11)$ distinguishable summands. For aberration orders $A = 2k-1 = 3$, 5, and 7 (rank or perturbation order $k = 2$, 3, 4) there are thus 9, 19, and 34 terms in the ring of polynomials modulo terms of rank higher than k.

3. The symplectic basis

We use the classification of polynomials as vectors in the *symplectic* basis ${}^k\chi^j_m\,(p^2, \mathbf{p}\cdot\mathbf{q}, q^2)$ /1/, /2/. These functions are defined in terms of the solid spherical harmonics /6/ and the Sp(2,$\Re$) invariant in (1.4). They are

$${}^k\chi^j_m = |\mathbf{p}\times\mathbf{q}|^{k-j}\frac{(j+m)!\,(j-m)!}{2^m(2j-1)!!}\sum_n \frac{1}{2^{2n}}\frac{(p^2)^{m+n}}{(m+n)!}\frac{(\mathbf{p}\cdot\mathbf{q})^{j-m-2n}}{(j-m-2n)!}\frac{(q^2)^n}{n!}. \tag{3.1}$$

We have ${}^k\chi^k_k = (p^2)^k$ with unit coefficient, and all other ${}^k\chi^j_m$'s with *rational* coefficients. We thus avoid the square roots of the spherical harmonic normalization.

The first *triplet* of harmonics are the $k=1$ functions

$${}^1\chi^1_1 = p^2 = M_{100}, \quad {}^1\chi^1_0 = \mathbf{p}\cdot\mathbf{q} = M_{010}, \quad {}^1\chi^1_{-1} = q^2 = M_{001}. \tag{3.2}$$

The six $k=2$ generators of *third*-order aberrations, now break break into a quadrupole $j=2$ quintuplet

$${}^2\chi^2_2 = M_{200}, \quad {}^2\chi^2_1 = M_{110}, \quad {}^2\chi^2_0 = \tfrac{1}{3}M_{101}+\tfrac{2}{3}M_{020}, \quad {}^2\chi^2_{-1} = M_{011}, \quad {}^2\chi^2_{-2} = M_{002}, \tag{3.3a}$$

and a monopole $j=0$ singlet

$${}^2\chi^0_0 = M_{101} - M_{020} \tag{3.3b}$$

that has zero Poisson bracket with every polynomial of the sp(2,$\Re$) generators.

A polynomial $\mathcal{A}$, such as (2.2), can be written thus as

$$\mathcal{A} = \sum_{k=1}^{k_{\max}}\ \sum_{j=k(-2)}^{1\text{ or }0}\ \sum_{m=-j}^{j} A^{\mathrm{S}}(k,j,m)\,{}^k\chi^j_m = \sum_{k=1}^{k_{\max}}\ \sum_{n_j=1}^{[k/2]+1}\ \sum_{n_m=1}^{2j+1} A^{\mathrm{S}}[k,n_j,n_m]\,{}^k\chi^j_m, \tag{3.4}$$

where k, $n_j = \frac{1}{2}(k-j)+1$, and $n_m = j-m+1$, number the element of a three-level vector $\mathcal{A}^{\rm S}_{\rm vec}$, with values $A^{\rm S}(k,j,m)$. For ranks $k=1,2$, the nine vector components are

$$\mathcal{A}^{\rm S}_{\rm vec} = \begin{pmatrix} \begin{pmatrix} \begin{pmatrix} A^{\rm S}[1,1,1] \\ A^{\rm S}[1,1,2] \\ A^{\rm S}[1,1,3] \end{pmatrix} \end{pmatrix} \\ \begin{pmatrix} \begin{pmatrix} A^{\rm S}[2,1,1] \\ A^{\rm S}[2,1,2] \\ A^{\rm S}[2,1,3] \\ A^{\rm S}[2,1,4] \\ A^{\rm S}[2,1,5] \end{pmatrix} \\ \begin{pmatrix} A^{\rm S}[2,2,1] \end{pmatrix} \end{pmatrix} \end{pmatrix} = \begin{pmatrix} \begin{pmatrix} \begin{pmatrix} A^{\rm S}(1,1,1) \\ A^{\rm S}(1,1,0) \\ A^{\rm S}(1,1,-1) \end{pmatrix} \end{pmatrix} \\ \begin{pmatrix} \begin{pmatrix} A^{\rm S}(2,2,2) \\ A^{\rm S}(2,2,1) \\ A^{\rm S}(2,2,0) \\ A^{\rm S}(2,2,-1) \\ A^{\rm S}(2,2,-2) \end{pmatrix} \\ \begin{pmatrix} A^{\rm S}(2,0,0) \end{pmatrix} \end{pmatrix} \end{pmatrix}, \quad \text{coefficient of} \begin{cases} {}^1\chi^1_1 \\ {}^1\chi^1_0 \\ {}^1\chi^1_{-1} \\ {}^2\chi^2_2 \\ {}^2\chi^2_1 \\ {}^2\chi^2_0 \\ {}^2\chi^2_{-1} \\ {}^2\chi^2_{-2} \\ {}^2\chi^0_0 \end{cases} \tag{3.5}$$

To work with polynomials and their vector representations we have developed a number of muSIMP functions to handle polynomials in three variables either

- in explicit form $\mathcal{A}(p^2, \mathbf{p}\cdot\mathbf{q}, q^2)$, or through
- their monomial coefficients $A^{\rm M}(k_+, k_0, k_-)$, or
- their symplectic coefficients $A^{\rm S}(k,j,m)$.

We can convert freely between the two vector representations, to build and encode the explicit form. The formula that inverts (3.1) may be found /1/; for example, it holds that $M_{523} = (p^2)^5(\mathbf{p}\cdot\mathbf{q})^2(q^2)^3 = -\frac{160}{3003}\,{}^{10}\chi^2_2 + \frac{16}{143}\,{}^{10}\chi^4_2 + \frac{14}{17}\,{}^{10}\chi^6_2 + \frac{29}{19}\,{}^{10}\chi^8_2 + {}^{10}\chi^{10}_2$.

4. The two bases in multiplication

Suppose we have two polynomials $\mathcal{A}$ and $\mathcal{B}$, and a binary operation $\circ$, that is generally nonlinear and non-commutative, that gives a third polynomial $\mathcal{C} = \mathcal{A}\circ\mathcal{B}$. Suppose we have introduced a grading by *rank* such that the subpolynomials $\mathcal{A}_k$, $\mathcal{B}_{k'}$ of strict ranks k and k' yield a subpolynomial $\mathcal{C}_{k''}$, also of strict rank k'', such that $k'' = k+k'$ as in multiplication, or $k'' = k+k'-1$ as in Poisson brackets, so that the ring of polynomials (modulo terms of rank higher than a certain $k_{\rm max}$) be nilpotent or solvable. Finally, suppose we have calculated the table of the operation $\circ$ between general polynomials in terms of their coefficients 'once and for all' in some basis, to some rank. We now pose the concrete question: is the monomial or is the symplectic basis more efficient for numeric and symbolic computation? If the operation $\circ$ is 'arbitrary' there probably is no fast answer. So let us examine multiplication first.

The operation of multiplication "$\times$" between polynomials is distributive among its summands and commutes with numerical multiplication, so it may be defined through the basis of monomials as

$$M_{k_+,k_0,k_-} \times M_{k'_+,k'_0,k'_-} = M_{k_++k'_+,k_0+k'_0,k_-+k'_-}. \tag{4.1}$$

A multiplication table in the monomial basis specifies $C^{\rm M}(k_+,k_0,k_-)$ in $\mathcal{C} = \sum C^{\rm M}_{\cdots} M_{\dots}$ for $\mathcal{C} = \mathcal{A}\times\mathcal{B}$, given as sums of bilinear products of the coefficients of $\mathcal{A} = \sum A^{\rm M}_{\cdots} M_{\dots}$ and $\mathcal{B} = \sum B^{\rm M}_{\cdots} M_{\dots}$. For example /5/, $C^{\rm M}_{020} = A^{\rm M}_{010} B^{\rm M}_{010}$ is a brief entry with one term, and a longer one is $C^{\rm M}_{111} = A^{\rm M}_{001}B^{\rm M}_{110} + A^{\rm M}_{010}B^{\rm M}_{101} + A^{\rm M}_{011}B^{\rm M}_{100} + A^{\rm M}_{100}B^{\rm M}_{011} + A^{\rm M}_{101}B^{\rm M}_{010} + A^{\rm M}_{110}B^{\rm M}_{001}$.

Multiplication of polynomials in the symplectic basis requires the product of two of its elements

$$ {}^{k}\chi^{j}_{m} \times {}^{k'}\chi^{j'}_{m'} = \sum_{j''=|j-j'|}^{j+j'} X^{\mathrm{S}}(j,j',m,m';j'')\ {}^{k+k'}\chi^{j''}_{m+m'} \,. \tag{4.2} $$

Ranks sum, and so does the m label; this leaves a single sum over the symplectic spin index values j'' that are allowed by the total rank, $k'' = k + k' \geq j''$, with $k - j''$ even, and $|m + m'| \leq j''$. The coefficients $X^{\mathrm{S}}(j,j',m,m';j'')$, are *rational* numbers that may be obtained from spherical harmonics [reference /6/, Eqs.(4.138) and (4.189)]. We build similarly a multiplication table for the coefficients of $C^{\mathrm{S}}(k,j,m)$ in terms of those of A and B and keep it as a symbolic function or numerical computation routine. As examples of a short and a long entry /5/ we have $C^{\mathrm{S}}_{222} = A^{\mathrm{S}}_{111}B^{\mathrm{S}}_{111}$ and $C^{\mathrm{S}}_{310} = 2/5A^{\mathrm{S}}_{11-1}B^{\mathrm{S}}_{221} + A^{\mathrm{S}}_{110}B^{\mathrm{S}}_{200} - 4/15A^{\mathrm{S}}_{110}B^{\mathrm{S}}_{220} + 2/5A^{\mathrm{S}}_{111}B^{\mathrm{S}}_{22-1} + A^{\mathrm{S}}_{200}B^{\mathrm{S}}_{110}$. Now, let us count the numbers of summands present in the two tables. We regard these as a measure of the computational complexity of the operation.

The number of AB summands present in the whole monomial-basis multiplication table, to fifth rank (or perturbation, *i.e.*, ninth aberration order), are:

RANK k	1	2	3	4	5
NUMBER OF BASIS VECTORS	3	6	10	15	21
CUMULATIVE DIMENSION	3	9	19	34	55
TERMS IN MONOMIAL MULTIPLICATION	0	9	36	96	120
CUMULATIVE LENGTH	0	9	45	141	261

In comparison, the number of summands in the *symplectic*-basis multiplication table are:

RANK k (SPIN j)	1(1)	2(2,0)	3(3,1)	4(4,2,0)	5(5,3,1)
NUMBER OF BASIS VECTORS	3	6(5+1)	10(7+3)	15(9+5+1)	21(11+7+3)
TERMS IN MULTIPLICATION	0	12(9+3)	54(30+24)	156(67+77+12)	222(70+98+54)
CUMULATIVE LENGTH	0	12	66	222	444

The monomial basis is evidently more efficient because the product rule (4.1) has only one summand in the right-hand side. The symplectic basis, however, is not far behind: in the product (4.2), k and m are fixed and the sum extends only over j''. One may show/5/ that the *average* number of summands is $(k+1)(k+2)/2(2k+1)$, asymptotically $\sim \frac{1}{4}k$. The ratios of the cumulative lengths of monomial to symplectic basis multiplication tables is 1, 0.75, 0.68, 0.64, and 0.59 for the first five ranks tabulated above.

The monomial basis has another advantage in multiplication: the *coefficients* of the AB terms are unity. The symplectic basis (for $j < k$) has *fractions* in its multplication table. To compare CPU running times we multiplied polynomials with symbolic coefficients; the monomial to symplectic ratios were 0.43, 0.33, 0.22, and 0.18, for ranks $k = 2,3,4$, and 5. For polynomials with 'arbitrary' simple fraction coefficients, the ratios were 0.58, 0.49, 0.49, and 0.41, with statistical fluctuations depending on the internal state of the system.

5. Poisson brackets

The Poisson bracket /3/ between two polynomial functions $A(p^2, \mathbf{p}\cdot\mathbf{q}, q^2)$ and $B(p^2, \mathbf{p}\cdot\mathbf{q}, q^2)$ can be obtained performing the partial derivatives

$$\{A, B\} = \frac{\partial A}{\partial q_1}\frac{\partial B}{\partial p_1} + \frac{\partial A}{\partial q_2}\frac{\partial B}{\partial p_2} - \frac{\partial A}{\partial p_1}\frac{\partial B}{\partial q_1} - \frac{\partial A}{\partial p_2}\frac{\partial B}{\partial q_2} \tag{5.1a}$$

$$= 2p^2\left(\frac{\partial A}{\partial \mathbf{p}\cdot\mathbf{q}}\frac{\partial B}{\partial p^2} - \frac{\partial A}{\partial p^2}\frac{\partial B}{\partial \mathbf{p}\cdot\mathbf{q}}\right) + 4\mathbf{p}\cdot\mathbf{q}\left(\frac{\partial A}{\partial q^2}\frac{\partial B}{\partial p^2} - \frac{\partial A}{\partial p^2}\frac{\partial B}{\partial q^2}\right)$$
$$+ 2q^2\left(\frac{\partial A}{\partial q^2}\frac{\partial B}{\partial \mathbf{p}\cdot\mathbf{q}} - \frac{\partial A}{\partial \mathbf{p}\cdot\mathbf{q}}\frac{\partial B}{\partial q^2}\right). \tag{5.1b}$$

To handle and calculate Poisson brackets efficiently is of interest for many computational tasks in applications of Lie methods to high-order perturbation problems /7/. If we denote by A_k, B_k, and C_k subpolynomials of *strict* rank k, *i.e.*, homogeneous of degree $2k$ in the basic phase space variables $(\mathbf{p}, \mathbf{q})$, the Poisson bracket of $C = \{A, B\}$ is seen to have the rank grading as

$$\begin{pmatrix} C_1 \\ C_2 \\ C_3 \\ C_4 \\ \vdots \end{pmatrix} = \left\{ \begin{pmatrix} A_1 \\ A_2 \\ A_3 \\ A_4 \\ \vdots \end{pmatrix}, \begin{pmatrix} B_1 \\ B_2 \\ B_3 \\ B_4 \\ \vdots \end{pmatrix} \right\} = \begin{pmatrix} \{A_1, B_1\} \\ \{A_1, B_2\} + \{A_2, B_1\} \\ \{A_1, B_3\} + \{A_2, B_2\} + \{A_3, B_1\} \\ \{A_1, B_4\} + \{A_2, B_3\} + \{A_3, B_2\} + \{A_4, B_1\} \\ \vdots \end{pmatrix}. \tag{5.2}$$

In the monomial basis we have the closed expression

$$\left\{M_{k_+,k_0,k_-}, M_{k'_+,k'_0,k'_-}\right\} = 4(k_-k'_+ - k_+k'_-)M_{k_++k'_+-1,k_0+k'_0+1,k_-+k'_--1}$$
$$+ 2(k_0k'_+ - k_+k'_0 + k_-k'_0 - k_0k'_-)M_{k_++k'_+,k_0+k'_0-1,k_-+k'_-}. \tag{5.3}$$

Each elementary Poisson bracket in the basis thus yields two, one, or no monomials. This tactic provides a shortcut to write recursive symbolic Poisson bracket functions for sparse polynomials. We take again a tabular approach: for $C = \{A, B\}$ we calculate 'once and for all' $C^{\mathrm{M}}(k_+, k_0, k_-)$ in terms of coefficients times products of $A^{\mathrm{M}}(k'_+, k'_0, k'_-)$ and $B^{\mathrm{M}}(k''_+, k''_0, k''_-)$. Thus for example, for $C_2 = \{A_1, B_2\}$ we have $C^{\mathrm{M}}_{020} = 4A^{\mathrm{M}}_{001}B^{\mathrm{M}}_{110} + 4A^{\mathrm{M}}_{100}B^{\mathrm{M}}_{011}$, and for $C_3 = \{A_2, B_2\}$ we have $C^{\mathrm{M}}_{111} = 16A^{\mathrm{M}}_{002}B^{\mathrm{M}}_{200} + 4A^{\mathrm{M}}_{011}B^{\mathrm{M}}_{110} - 4A^{\mathrm{M}}_{110}B^{\mathrm{M}}_{011} - 16A^{\mathrm{M}}_{200}B^{\mathrm{M}}_{002}$. We count the relative complexity through the number of terms in the table, below.

The symplectic basis has a number of Lie-algebraic properties with respect to the Poisson bracket. The fundamental one, that may be verified from the explicit formula (3.1), is

$$\left\{{}^{\kappa}\chi^1_1, {}^{k}\chi^j_m\right\} = 2(m-j)\ {}^{\kappa+k-1}\chi^j_{m+1}, \qquad \text{RAISING} \tag{5.4a}$$

$$\left\{{}^{\kappa}\chi^1_0, {}^{k}\chi^j_m\right\} = 2m\ {}^{\kappa+k-1}\chi^j_m, \qquad \text{WEIGHT} \tag{5.4b}$$

$$\left\{{}^{\kappa}\chi^1_{-1}, {}^{k}\chi^j_m\right\} = 2(m+j)\ {}^{\kappa+k-1}\chi^j_{m-1}, \qquad \text{LOWERING} \tag{5.4c}$$

$$\left\{{}^{\kappa}\chi^0_0, {}^{k}\chi^j_m\right\} = 0. \qquad \text{SCALARS} \tag{5.4d}$$

These brackets are valid for *all* triplets ${}^{\kappa}\chi^1_m = |\mathbf{p}\times\mathbf{q}|^{\kappa-1}\,{}^1\chi^1_m$ for κ odd, all singlets ${}^{\kappa}\chi^0_0$ for κ even, and all k, since $\left\{|\mathbf{p}\times\mathbf{q}|^2, f(p^2,\mathbf{p}\cdot\mathbf{q},q^2)\right\} = 0$.

The general expression for the Poisson brackets between the symplectic basis elements ${}^k\chi^j_m$ can be given [see reference /2/, equations (7.14), (7.21), and (7.52)]. For our purposes it is sufficient to give its overall structure: the label m is again summed and conserved, so there is a *single* sum over symplectic spin in

$$\left\{{}^k\chi^j_m, {}^{k'}\chi^{j'}_{m'}\right\} = \sum_{\substack{j''=|m+m'| \\ k+k'+j''\ \text{odd}}}^{j+j'-1} P^{j,j',j''}_{m,m'}\ {}^{k+k'-1}\chi^{j''}_{m+m'}. \tag{5.5}$$

For example, in the tables for $C_2 = \{A_1, B_2\}$, there appears $C^S_{222} = 4A^S_{110}B^S_{222} - 2A^S_{111}B^S_{221}$, and for $C_3 = \{A_2, B_2\}$, we have $C^S_{310} = 32/5A^S_{22-2}B^S_{222} - 4/5A^S_{22-1}B^S_{221} + 4/5A^S_{221}B^S_{22-1} - 32/5A^S_{222}B^S_{22-2}$.

In the monomial basis, the number of AB terms in the Poisson bracket tables of A_1 with B_1, B_2, B_3, and B_4, are 6, 16, 32, and 52. Similarly, the tables for $C_3 = \{A_2, B_2\}$ and $C_4 = \{A_2, B_3\}$ have 34 and 72 terms, respectively. Now we count totals for the monomial basis Poisson bracket table according to the grading (5.2) for ranks $k = 1,2,3,4$. The rank $k = 1$ table contains only 6 terms. While strict rank $k = 2$ treats 32 terms (16 for $\{A_1, B_2\}$ and 16 for $\{A_2, B_1\}$), $k = 3$ involves $2\times 32+34 = 98$ terms, and $k = 4$, $2\times 52 + 2\times 72 = 248$ terms. These numbers and their cumulative sum appear in the table below.

The *symplectic* basis reduces both the number of independent tables and yields a *single* term between brackets involving an A_1 triplet. Their Poisson bracket tables with B_1, B_2, B_3, and B_4, that are 6, 12, 24, and 36. Similarly, the tables for $C_3 = \{A_2, B_2\}$ and $C_4 = \{A_2, B_3\}$ have 32 and 64 summands, respectively. The cumulative totals appear below:

RANK k	1	2	3	4
MONOMIAL BASIS	6	32	98	248
SYMPLECTIC BASIS	6	24	80	200
MONOMIAL, CUMULATIVE	6	38	136	384
SYMPLECTIC, CUMULATIVE	6	30	110	310

For seventh aberration order ($k = 4$), the use of the symplectic basis has thus about one quarter less complexity. Again, the symplectic basis table coefficients include *fractions*, and this introduces extra CPU computation time. The ratio of monomial to symplectic average running times for symbolic coefficients are 1.397, 1.475, and 1.393. For fractional numeric coefficients they are 1.165, 1.176, and 1.151. For low ranks, thus, the advantage of using the symplectic basis is higher than 40% for symbolic and 15% for numeric calculations.

6. Lie transformations and the optical aberration group

Optical devices in the paraxial approximation produce *linear* transformations of phase space,

$$G(\mathbf{M}) : \begin{pmatrix}\mathbf{p}\\ \mathbf{q}\end{pmatrix} \mapsto \begin{pmatrix}\mathbf{p}'\\ \mathbf{q}'\end{pmatrix} + \mathbf{M}^{-1}\begin{pmatrix}\mathbf{p}\\ \mathbf{q}\end{pmatrix} = \exp\{A_1, \circ\}\begin{pmatrix}\mathbf{p}\\ \mathbf{q}\end{pmatrix}. \tag{6.1}$$

For the transformation to be canonical,[3] $\det \mathbf{M} = 1$. The last member writes the linear transformation as a *Lie transformation* generated by the rank-1 polynomial $\mathcal{A}_1$ of p^2, $\mathbf{p}\cdot\mathbf{q}$, and q^2, with coefficients that may be related to the elements of $\mathbf{M}$. It is built out of the Poisson bracket as the formal series [8]

$$\exp x\{\mathcal{A},\circ\} = \sum_{n=0}^{\infty} \frac{x^n}{n!}(\{\mathcal{A},\circ\})^n = \sum_{n=0}^{\infty} \frac{x^n}{n!}\{\mathcal{A},\{\mathcal{A},\cdots\{\mathcal{A},\circ\}\cdots\}\}, \tag{6.2}$$

that acts on the phase space variables, $(\mathbf{p},\mathbf{q})$, or functions thereof.

Under linear transformations of $(\mathbf{p},\mathbf{q})$, the *coefficients* $\mathcal{B}^{\mathrm{S}}_{\mathrm{vec}}$ of polynomials $\mathcal{B}$ in the symplectic basis, given as in (3.5), transform through

$$G(\mathbf{M}): \begin{pmatrix} (B^{\mathrm{M}}_{11\cdot}) \\ \begin{pmatrix} B^{\mathrm{M}}_{22\cdot} \\ B^{\mathrm{M}}_{20\cdot} \end{pmatrix} \\ \begin{pmatrix} B^{\mathrm{M}}_{33\cdot} \\ B^{\mathrm{M}}_{31\cdot} \end{pmatrix} \end{pmatrix} = \begin{pmatrix} (\tilde{B}^{\mathrm{M}}_{11\cdot}) \\ \begin{pmatrix} \tilde{B}^{\mathrm{M}}_{22\cdot} \\ \tilde{B}^{\mathrm{M}}_{20\cdot} \end{pmatrix} \\ \begin{pmatrix} \tilde{B}^{\mathrm{M}}_{33\cdot} \\ \tilde{B}^{\mathrm{M}}_{31\cdot} \end{pmatrix} \end{pmatrix} = \mathbf{D}(\mathbf{M}^{-1})^{\top} \begin{pmatrix} (B^{\mathrm{M}}_{11\cdot}) \\ \begin{pmatrix} B^{\mathrm{M}}_{22\cdot} \\ B^{\mathrm{M}}_{20\cdot} \end{pmatrix} \\ \begin{pmatrix} B^{\mathrm{M}}_{33\cdot} \\ B^{\mathrm{M}}_{31\cdot} \end{pmatrix} \end{pmatrix}, \tag{6.3a}$$

where the matrix $\mathbf{D}(\mathbf{M}^{-1})^{\top}$ is the transpose, inverse of the block-reduced matrix

$$\mathbf{D}(\mathbf{M}) = \begin{pmatrix} \boxed{\mathbf{D}^{j=1}_{[3\times3]}(\mathbf{M})} & \mathbf{0} & \mathbf{0} \\ \mathbf{0} & \boxed{\begin{matrix} \mathbf{D}^{j=2}_{[5\times5]}(\mathbf{M}) & 0 \\ 0 & \mathbf{D}^{j=0}_{[1\times1]}(\mathbf{M}) \end{matrix}} & \mathbf{0} \\ \mathbf{0} & \mathbf{0} & \boxed{\begin{matrix} \mathbf{D}^{j=3}_{[7\times7]}(\mathbf{M}) & 0 \\ 0 & \mathbf{D}^{j=1}_{[3\times3]}(\mathbf{M}) \end{matrix}} \end{pmatrix}. \tag{6.3}$$

(Were it in the monomial basis, the diagonal submatrices k would not be completely reduced.) In the symplectic basis, j labels the sub-submatrices (and m their rows), each a finite-dimensional *representation*[9] of $\mathrm{Sp}(2,\Re)$ that is *independent* of rank k. This is not a property of the monomial basis.

In the monomial basis there are $d_k^2 = \frac{1}{4}(k+1)^2(k+2)^2 \sim \frac{1}{4}k^4$ matrix elements for each k, in multiplication with the polynomial vector coefficients. The symplectic basis [5] counts are $\frac{1}{6}(k+2)(4k^2+10k+3)$ for k even and $\frac{1}{6}(k+1)(4k^2+14k+9)$ for k odd, and asymptotically $\sim \frac{2}{3}k^3$. For $k = 1,2,3,4,\cdots$ the number of elements in the monomial basis is 9, 36, 100, 225, ..., *versus* 9, 26, 58, 107, ..., in the symplectic basis.

Lie aberration theory describes *nonlinear* transformations of phase space through the exponential Poisson operator (6.1) for ranks $k > 1$. The position coordinate, for example, transforms as $\exp\{\mathcal{A}_k,\circ\} : \mathbf{q} = \mathbf{q} + \{\mathcal{A}_k,\mathbf{q}\} + \frac{1}{2!}\{\mathcal{A}_k,\{\mathcal{A}_k,\mathbf{q}\}\} + \cdots$ with summands of degrees 1, $2k-1$, $4k-3$, $\cdots$ up to the aberration order $2k_{\max}-1$, in $(\mathbf{p},\mathbf{q})$ [10]. Lie transformations of this quotient space are a finite-parameter *aberration* group. Its structure is $A_{k_{\max}} = \mathrm{Sp}(2,\Re) \triangleleft \mathcal{K}_{k_{\max}}$, where the first factor group is semisimple and normal, and the second is the nilpotent group of all *pure* aberrations; it is abelian for $k = 2$, but not beyond: aberrations interact to higher orders[11]. The elements of the aberration group may be parametrized by the coefficients of a polynomial $\mathcal{A}$ as a *factored product* [12]. For rank $k = 3$,

$$\begin{aligned} \gamma(\mathcal{A}) &= \gamma(\mathcal{A}_3,\mathcal{A}_2,\mathcal{A}_1) = \exp\{\mathcal{A}_3,\circ\}\ \exp\{\mathcal{A}_2,\circ\}\ \exp\{\mathcal{A}_1,\circ\} \\ &= \gamma(\mathcal{A}_3,\mathcal{A}_2,0)\ \gamma(0,0,\mathcal{A}_1) = \Gamma(A)\ G(\mathbf{M}). \end{aligned} \tag{6.4}$$

The action of each optical element is thus represented by a polynomial. An optical system composed of more than one optical element is the aberration group *composition* of the representing polynomials /2/, /4/. In particular, $G(\mathbf{M})\,G(\mathbf{N}) = G(\mathbf{MN})$.

The composition of two optical elements, $\gamma(\mathcal{A})\,\gamma(\mathcal{B}) = \gamma(\mathcal{C})$, is an operation $\mathcal{A} \circ \mathcal{B} = \mathcal{C}$ in the space of polynomials of rank up to k_{max}. We may separate the product according to (6.4) as

$$\begin{aligned}\gamma(\mathcal{A})\,\gamma(\mathcal{B}) &= \Gamma(A)\,G(\mathbf{M})\,\Gamma(B)\,G(\mathbf{N}) = \Gamma(A)\,G(\mathbf{M})\,\Gamma(B)\,G(\mathbf{M})^{-1}\,G(\mathbf{M})\,G(\mathbf{N}) \\ &= \Gamma(A)\,\Gamma(G(\mathbf{M}):B)\,G(\mathbf{MN}) = \Gamma(A\sharp[G(\mathbf{M}):B])\,G(\mathbf{MN}).\end{aligned} \tag{6.5}$$

Here "$\sharp$" is the *'gato'* operation of *pure aberration composition*; it may be found explicitly from the Baker-Campbell-Hausdorff formula /8/. Indicating $A = \{\ldots, \mathcal{A}_3, \mathcal{A}_2\}$, $G(\mathbf{M}) : B = \tilde{B} = \{\ldots, \tilde{\mathcal{B}}_3, \tilde{\mathcal{B}}_2\}$ and $C = A\sharp\tilde{B} = \{\ldots, \mathcal{C}_3, \mathcal{C}_2\}$,

$$\mathcal{C}_2 = \mathcal{A}_2 + \tilde{\mathcal{B}}_2, \tag{6.6a}$$

$$\mathcal{C}_3 = \mathcal{A}_3 + \tilde{\mathcal{B}}_3 + \tfrac{1}{2}\left\{\mathcal{A}_2, \tilde{\mathcal{B}}_2\right\}, \tag{6.6b}$$

$$\mathcal{C}_4 = \mathcal{A}_4 + \tilde{\mathcal{B}}_4 + \left\{\mathcal{A}_2, \tilde{\mathcal{B}}_3\right\} + \tfrac{1}{3}\left\{\mathcal{A}_2, \left\{\mathcal{A}_2, \tilde{\mathcal{B}}_2\right\}\right\} - \tfrac{1}{6}\left\{\left\{\mathcal{A}_2, \tilde{\mathcal{B}}_2\right\}, \tilde{\mathcal{B}}_2\right\}, \tag{6.6c}$$

to aberration order seven. The *pure* aberration subgroup of rank $k = 2$ has abelian composition —a simple 6-vector sum of the perturbations of the factors. Fifth-order aberrations (second order perturbations) sum those of the factors, *plus* a bilinear compounding of third-order aberrations. Finally, the seventh-order coefficients (third order perturbation) involve linear, bilinear, and trilinear compounding between the factor elements.

The tables of the $\sharp$ operation in the monomial and in the symplectic bases have been found, the latter appears in Ref. /4/. The following are the counts of terms involving linear, bilinear, and trilinear A and $\tilde{B}$ factors. For ranks $k = 2, 3, 4$, the monomial basis has 12, 54, and 422 terms while the symplectic basis has 12, 52, and 318 terms. Since each of the $\tilde{B}$ terms is generally a linear combination of B's through the $\mathbf{D}^{j}_{[(2j+1)\times(2j+1)]}(\mathbf{M})$ matrices, for $k = 2$ there are 6 summands in both bases, but the 6 terms in $\tilde{B}$ add to 36 B's in the monomial basis while there are 26 in the symplectic one. For $k = 3$, the monomial basis has complexity 314 *versus* 228. Lastly, for $k = 4$, the count /5/ bloats to 6720 terms in the monomial basis, *versus* 3360 terms in the symplectic one. Hence the total cumulative counts for aberration order 7 stand in the monomial basis at 7680 terms, against only 3882 terms in the symplectic basis. So, there should be nearly 100% advantage in using the symplectic basis for symbolic aberration calculations.

More generally, we may say that the symplectic basis is suited for operations involving Poisson brackets between functions of p^2, $\mathbf{p}\cdot\mathbf{q}$, and q^2. The reason is that we use the raising, lowering, and weight form of the defining Lie brackets; they return a single term after operation. For other semisimple Lie algebras built out of $\mathbf{p}$'s and $\mathbf{q}$'s, whose complete list of phase-space realizations by Poisson brackets are known /13/, we may use the finite-dimensional irreducible representations of the group as *eigenbasis* for the operation. Upon this semisimple group [here it was $\mathrm{Sp}(2, \Re)$], we may build the group of *aberrations*, *i.e.* nonlinear transformations, of the same group manifold. In this case we were fortunate that the manifold is quite simple.

References.

1. K.B. Wolf, Symmetry in Lie optics. *Annals of Physics* **172**, 1–25 (1986).

2. A.J. Dragt, E. Forest, and K.B. Wolf, Foundations of a Lie algebraic theory of geometrical optics. In *Lie Methods in Optics*, Lecture Notes in Physics, Vol. 250, Ed. by J. Sánchez Mondragón and K.B. Wolf (Springer Verlag, Heidelberg, 1986).

3. H. Goldstein, *Classical Mechanics* (Addison-Wesley, Reading, Mass, 1963); Chapters 7–8.

4. K.B. Wolf, Nonlinearity in Aberration Optics. In *Symmetries and Nonlinear Phenomena*, Proceedings of the CIF International School on Applied Mathematics (Paipa, Colombia, February 22-26, 1988). Ed. by D. Levi and P. Winternitz (World Scientific, CIF series, Singapore, 1989).

5. K.B. Wolf and G. Krötzsch, Group-classified polynomials of phase space in higher-order aberration expansions. Comunicaciones Técnicas IIMAS, Investigaciones No. 563 (1990), submitted to *Journal of Symbolic Computation.*

6. L.C. Biedenharn and J.D. Louck, *Angular Momentum in Quantum Physics. Theory and Application.* Encyclopedia of Mathematics, Vol. 8 (Academic Press, New York, 1981); Eqs. (3.138) and (3.189).

7. E. Forest and M. Berz, Canonical integration with non-standard analysis; L.M. Healy and A.J. Dragt, Concatenation of Lie algebraic maps. In *Lie Methods in Optics, II Workshop*, Lecture Notes in Physics, Vol. 352, Ed. by K.B. Wolf (Springer Verlag, Heidelberg, 1989).

8. S. Steinberg, Lie series, Lie transformations, and their Applications. In *Lie Methods in Optics*, Lecture Notes in Physics, Vol. 250, *op.cit.*

9. R. Gilmore, *Lie Groups, Lie Algebras, and Some of their Applications* (Wiley Interscience, New York, 1978).

10. M. Navarro-Saad and K.B. Wolf, The group–theoretical treatment of aberrating systems. I. Aligned lens systems in third aberration order *J. Math. Phys.* **27**, 1449–1457 (1986).

11. K.B. Wolf, Symmetry-adapted classification of aberrations. *J. Opt. Soc. Am. A* **5**, 1226–1232 (1988).

12. A.J. Dragt and J. Finn, Lie series and invariant functions for analytic symplectic maps. *J. Math. Phys.* **17**, 2215–2227 (1976); S. Steinberg, Factored product expansions of nonlinear differential equations. *SIAM J. Math. Anal.* **15**, 108–115 (1984).

13. M. Hávliček and W. Lassner, Canonical realizations of the Lie algebra $\mathrm{sp}(2n, \Re)$, *Int. J. Theor. Phys.* **15**, 867–875 (1976); *ib.* On the 'near ot minimal' canonical realizations of the Lie algebra C_n, *ibid.* **15**, 877-887 (1976).

The program of analytical calculation of the normal Birkhoff-Gustavson form

V. Yu. Gonchar,[+)1] *N. A. Chekanov*[+)], *B. L. Markovski, V. A. Rostovtsev,*
S. I. Vinitsky JINR-Dubna

1. Introduction

Transformation of the initial classical Hamiltonian to a more simple form, called normal form, is one of the universal methods of analyzing and integrating classical equations of motion [1)]. In the present paper, we discuss the normal Birkhoff-Gustavson form [2,3)] for the Hamiltonians that can be represented as a polynomial near the equilibrium points, i.e.:

$$H(q,p)= H^{(2)}(q,p) + V(q) \tag{1}$$

where

$$H^{(2)}(q,p)= \sum_{\nu} 1/2\ \omega_{\nu}^2\ (p_{\nu}^2 + q_{\nu}^2), \tag{2}$$

$$V(q) = \sum_{|j|\geq 3} V_{j_1 j_2} q_1^{j_1}\ q_2^{j_2}\ , q=(q_1,q_2), p=(p_1,p_2) \tag{3}$$

Here we consider the system with <u>two degrees of freedom</u>, however, the method can be applied for Hamiltonian systems with arbitrary number of degrees of freedom.

The procedure of reducing to the normal Birkhoff-Gustavson form (NBGF) and its realization depend on the relation between frequencies ω_{ν}. If the frequencies ω_{ν} of Hamiltonian (1) are incommensurable, there exists a canonical transformation $(q,p) \to (\xi,\eta)$ such that in the new variables the Hamiltonian $\Gamma(\xi,\eta)$ will be a function of only the following variables $I_{\nu}=1/2(\xi_{\nu}^2+\eta_{\nu}^2)$

---

[1]Kharkov Institute for Thechnical Physics

$$H(q,p) \Rightarrow \Gamma(\xi,\eta) = \omega_1 I_1 + \omega_2 I_2 + \sum \alpha_{\mu\nu} I_\mu I_\nu + \dots \quad (4)$$

If the frequencies ω_ν are commensurable, i.e. if there exist resonance relations of the type $n\omega_1+m\omega_2 = 0, (n,m \neq 0)$ the Hamiltonian (1) can not be reduced to the NBGFdue to the appearance of zero denominators $n\omega_1+m\omega_2= 0$. Therefore, one should not cancel some terms in the Hamiltonian (1), and the normal form becomes complicated [3] and will contain apart from I_ν other combinations of variables ξ_ν and η_ν as well.For the two-dimensional system with the resonance condition $n\omega_1+m\omega_2= 0$ these combinations are the following $\xi_1\eta_2-\xi_2\eta_1$, $\eta_1\eta_2+ \xi_1\xi_2$. Such an extended normal form is used to be called the Birkhoff- Gustavson form (B-G).

The first rather quick programs in FORTRAN for making cumbersome calculations of the normal form were realized in ref.[3,4]. We shall point out other way of constructing canonical transformations (see formula (5)) suggested by Hori[5] and Deprit [6] whose practical realization is based on the use of Lie transformations [7-9]. The computer program for normalization of Hamiltonian systems in this approach were realized in ref. [10].

In the present paper, we expound two REDUCE programs for analytical construction of the NBGF according to the Gustavson algorithm [3] which is briefly represented in sec.2. The first program GITA gives the normal form of a given order s_{max} if one introduces the initial Hamiltonian (1) and resonance relations between frequencies ω_1, $\omega_2$or frequencies themselves if they are incommensurable. The second program GITN is applicable exceptionally for the resonance case $\omega_1= \omega_2$,i.e. for obtaining a modified normal form whose definition will be given in sec.2,3.

Note that since $I_1(\xi,\eta)$ and $\Gamma_s(\xi,\eta)$ are independent integrals of motion in the S approximation, then expressing ξ and η through the initial coordinates q,p one can obtain an approximate integral of motion $I_1(q,p)$. It can be used in different applications, for instance, in ref.[11] for elucidating the criterion of negative curvature. However, the range of ap- plicability of the normal B-G form goes beyond the limits of classical description of the system: it has successfully been applied for quantizing the classical Hamiltonian as well [12-16].

2. The method of reducing the Hamiltonian to the normal form

2.1. The normal Birkhoff-Gustavson form

As it has been point out above, the Hamiltonian (1) can always be reduced to the normal form making successive canonical transformations

$$\xi_\nu = q_\nu + \frac{\partial W^{(j)}(q,\eta)}{\partial \eta_\nu}; \quad p_\nu = \eta_\nu + \frac{\partial W^{(j)}(q,\eta)}{\partial q_\nu}, \quad \nu = 1,2 \tag{5}$$

with the help of the generating function, or more exactly, the polynomial $W^{(s)}$ of the powers

$$F_2(q,\eta) = q_\nu \eta_\nu + W^{(s)}(q,\eta) \tag{6}$$

The integer numbers means the power of that part of the Hamiltonian which is reduced to the normal form by using the polynomial $W^{(s)}$. In this case, the parts of $H^{(i)}$ with $i < s$ do not change under the canonical transformation (5). Then under the canonical transformation the Hamiltonian H(q,p) turns into

$$H(q,\eta + \frac{\partial W^{(s)}}{\partial q}) = \Gamma(q + \frac{\partial W^{(s)}}{\partial \eta}, \eta), \tag{7}$$

Expanding this expression in the Taylor series near $q=\xi, \eta=p$ one equates the terms of the same power. For i=2 we get the known results that the Hamiltonian (2) is still in a normal form

$$\Gamma^{(2)}(\xi,\eta) = H^{(2)}(q,p); \quad q=\xi, p=\eta \tag{8}$$

For the terms of power i=s=3, by which the procedure is starting the reducion to the normal form, we derive the generic equation in the form

$$D(q,\eta)W^{(s)}(q,\eta) = -H^{(s)}(q,\eta) + \Gamma^{(s)}(q,\eta) \tag{9}$$

where the operator of the normal form

$$D(\xi,\eta) = \sum \omega_\nu (\eta_\nu \frac{\partial}{\partial \xi_\nu} - \xi_\nu \frac{\partial}{\partial \eta_\nu}) \tag{10}$$

is introduced. At each subsequent (S-2) step of reducing the Hamiltonian to the normal form we should solve eq. (9). So we have to find two unknown functions: the generating polynomial $W^{(s)}$ and Hamiltonian $\Gamma^{(s)}$ in the normal form, satisfying condition $D\Gamma^{(s)} = 0$

Before discussing the way of solving eq. (9) we first write down the relations resulting from the higher powers i=s+1, s+2, - that will be necessary for the next step of reducing

$$\Gamma^{(j)}(\xi,\eta) = H^{(j)}(\xi\ \eta) + \sum_{k} \frac{1}{K!} \left\{ (\frac{\partial W^{(s)}}{\partial \xi})^{k} (\frac{\partial^k H^{(l)}}{\partial \eta^k}) - (\frac{\partial W^{(s)}}{\partial \eta})^{k} \frac{\partial^k \Gamma^{(l)}}{\partial \xi^k} \right\} \tag{11}$$

Here the two-dimensional integer-valued vector K satisfies the conditions

$$l-|K| + |K|(s-1) = j,\ 1 \le |K| \le l < j,\ l \ge 2,\ s \ge 3$$

and the notations $K! = k_1!\ k_2!$ and $|K| = k_1 + k_2$ are introduced. Having calculated at the (S-2) step, beginning with S=3, and generating polynomial $W^{(s)}$, we find the next higher power contribu-tions to the Hamiltonian (11) to be used in performing the above procedure of reducing to the normal form. The number of iterations determines the order of approximations s_{max}.

2.2 The method of solving the basic equation

The linear operator D is determined in the functional space of all homogeneous polynomials $\Phi^{(s)}(\xi,\eta)$ of the power s of four variables (ξ,η) . In this space(dim= m=(s+3)!/3!s!) the setof monomials of the type $\xi^k \eta^l \equiv \xi_1^{k_1} \xi_2^{k_2} \eta_1^{l_1} \eta_2^{l_2}$, $(|k|+|l| \equiv k_1+k_2+l_1+l_2 = s)$ define the basis. In our (two-dimensional) case m=20, 35, 56, 84, 120 and 165, respectively, for s = 3, 4, 5, 6, 7, and 8.

To solve the equation (9) the canonical transformation to the complex variables x, y is made

$$q_\nu = 1/\sqrt{2}\,(x_\nu + iy_\nu), \eta_\nu = i/\sqrt{2}\,(x_\nu - iy_\nu) \tag{12}$$

in which the operator of the normal form will be diagonal

$$\tilde{D}(x,y) = i \sum \omega_\nu (x_\nu \frac{\partial}{\partial x_\nu} - y_\nu \frac{\partial}{\partial_\nu y}), \tag{13}$$

and the basic equation becomes

$$\tilde{D}\tilde{W}^{(s)}(x,y) = -\tilde{H}(x,y) + \tilde{\Gamma}^{(s)}(x,y) \tag{14}$$

The basis of functional space consists of monomials of the form

$$\Phi^{(s)} = x_1^{l_1} x_2^{l_2} y_1^{m_1} y_2^{m_2} \tag{15}$$

which, as one can easily verify, are eigenfunctions of the operator D, i.e.

$$\tilde{D}\,\Phi^{(s)} = \left[i \sum \omega_\nu (m_\nu - l_\nu)\right]\Phi^{(s)} \tag{16}$$

Then, the solution of the basic equation can be written as

$$\tilde{W}^{(s)} = \tilde{D}^{-1}[\tilde{\Gamma}^{(s)} - \tilde{H}^{(s)}] \tag{17}$$

where the inverse operator D^{-1} is determined by the relation

$$\tilde{D}\,\Phi^{(s)} = \Phi^{(s)}/[i \sum \omega_\nu (m_\nu - l_\nu)] \tag{18}$$

Since $\Gamma^{(s)}$ is not yet determined, we choose it so that the quantity $\tilde{W}^{(s)}$ would have no poles $\sum \omega_\nu (m_\nu - l_\nu)$. The Hamiltonian $\Gamma^{(s)}$ should be chosen so as to compensate exactly those terms $H^{(s)}$ which lead to the resonance condition $\sum\omega_\nu (m_\nu - l_\nu) = 0$.. But if this condition is fulfilled for some terms of the Hamiltonian $H^{(s)}$ this implies the normal form and consequently $\tilde{D}\tilde{H}^{(s)}=0$. Thus, the quantity $\tilde{H}^{(s)}$ can unambiguously be represented as a sum of polynomials $\tilde{H}^{(s)} = \tilde{N}^{(s)} + \tilde{R}^{(s)}$ with $\tilde{D}\tilde{N}^{(s)} = 0$. Choosing $\tilde{\Gamma}^{(s)} = \tilde{N}^{(s)}$ (the normal form) we see that the terms of the Hamiltonian $\tilde{H}^{(s)}$ leading to the condition $\sum\omega_\nu (m_\nu - l_\nu) = 0$ are canceled by the same terms in the Hamiltonian $\Gamma^{(s)}$, the generating function $W^{(s)}$being finite and determined from the algebraic equation $\tilde{D}W^{(s)} = R^{(s)}$ Moreover, relation $\sum\omega_\nu (m_\nu - l_\nu) =$

0 may holds if $m_\nu = l_\nu$ and ω_ν are incommensurable (the normal Birkhoff form) or if $m_\nu \neq l_\nu$ but there are resonance relations of $\sum n_\nu \omega_\nu = 0$ (n_ν being integers). In the last case, the structure of the normal form Γ becomes more complicated, as it has been mentioned in the Introduction.

Finally going back from the complex variables x and y to variables ξ, η we find the normal Hamiltonianform $\Gamma^{(s)}(\xi,\eta)$ and the generating function $W^{(s)}$ in the s order of approximation in terms of the initial variables.

2.3. The modified normal form of Birkhoff-Gustavson

In this section we present a certain modification [10] of the procedure of reducing the Hamiltonian to NBGF . Before reducing the initial Hamiltonian (1) to the normal form according to the Gustavson algorithm [3], (Sec 2.1 and 2.2) we first do the canonical change $(q,p) \Rightarrow (\mathcal{P}, Q)$ which would transform the operator form Eq. (10) into the diagonal one. For the Hamiltonian (1) with the resonance relation $\omega_1 = \omega_2$ the above transformation becomes:

$$q_1 = i/2(-Q_1 + Q_2 + \mathcal{P}_1 + \mathcal{P}_2), p_1 = 1/2(Q_1 - Q_2 + \mathcal{P}_1 - \mathcal{P}_2),$$
$$q_2 = 1/2(Q_1 + Q_2 + \mathcal{P}_1 + \mathcal{P}_2), p_2 = i/2(Q_1 + Q_2 - \mathcal{P}_1 - \mathcal{P}_2). \tag{20}$$

The Hamiltonian (1) in the new variables looks like

$$K(Q,\mathcal{P}) = K^{(2)}(Q,\mathcal{P}) + \sum_{j>2} K^{(j)}(Q,\mathcal{P}), \tag{21}$$

were

$$K^{(2)}(Q,\mathcal{P}) = i(Q_1\mathcal{P}_1 + Q_2\mathcal{P}_2)$$

and the canonically conjugate coordinates and momenta are $Q=(Q_1, Q_2)$ $\mathcal{P}=(\mathcal{P}_1\ \mathcal{P}_2)$. Note that transformation (20) are canonical with "valence" equal to imaginary unit [17].

Each term $K^{(j)}$ of Hamiltonian (21), which is a homogeneous polynomial in variables Q and $\mathcal{P}$ of power j ,is reduced to the normal form by making a series of canonical transformations $(Q,\mathcal{P}) \Rightarrow (\xi,\eta)$

with the generating function (6). Reducing to the normal form is simplified by a solution of the basic equation

$$DW^{(s)} = \Gamma^{(s)} - K^{(s)} \qquad (22)$$

owing to the diagonality of the operator D. The introduction of $(Q,\mathcal{P})$ variables simplifies the algorithm providing an useful tool for quantization of (1) for $\omega_1 = \omega_2$.

3. Description of the program in REDUCE

The GITA program consists of 6 basic and 4 auxiliary blocks, some of them being created as procedures.

Block 1. Transformation of the coordinates $p_\nu \to \sqrt{\omega_\nu} p_\nu, q_\nu \to q_\nu/\sqrt{\omega_\nu}$ for reducing $H^{(2)}(p,q)$ to the NF (2). The nonlinear part of the Hamiltonian (3) is simultaneously transformed too.

Block 2. The beginning of the cycle $s=3,\ldots s_{max}$ where s_{max} is the maximal order of reducing to the normal form. The procedure - SUPQXY performs transformation to the complex variables (12) of the homogeneous part of the s-order $\tilde{H}^{(s)}(x,y) = H^{(s)}(p,q)$.

Block 3. The SEPA procedure makes division of $\tilde{H}^{(s)}(x,y)$ into monomials (15).

Block 4.*T* The BASIS procedure solves the basic equation (16) and finds the generating function $\tilde{W}^{(s)}(x,y)$ and the *s*th order NF of the Hamiltonian $\tilde{\Gamma}^{(s)}(x,y)$.

Block 5. The inverse transformation is fulfilled from the complex variablex, y to the initial variables q,p.

Block 6. It implies the transformation of the remaining part of the Hamiltonian containing all higher than s orders which are necessary for the next reduction step. At the end of the cycle one goes back to block 2.

Upon completion of the cycle with respect to s one gets a Hamiltonian in the NF up to order and the "abnormal" part of the order higher than s_{max}.

Additional blocks.

Block 7. One transforms the NF $\Gamma^{(s)}(q,p)$ from the Cartesian coordinates to the action-angle variables I, φ. If the case is non-resonant one the NF $\Gamma^{(s)}(I)$ depending only on the action I (i.e., it is angle-independent).

Block 8. The basic results are collected into output file for further convenience.

The auxiliary procedures FAC(N) and CNM(N,M) calculate the factorials and binomial coefficients correspondingly. The GITN program calculating the modified B-G form has the same structure as GITA.

4. Examples

4.1 The NBGF. Resonance case.

The potential function of quadruple vibrations of the surface of a spherical liquid drop of matter can be parameterized as[18-19]

$$V(q_1,q_2) = \frac{1}{2}(q_1^2+q_2^2) + b(q_1^2q_2 - \frac{1}{3}q_2^3) + c(q_1^2+q_2^2)^2 \quad (23)$$

As is known [19-20], the invariant polynomials of the discrete C_{3v} group can be polynomialy expressed in terms of two basis invariants $(q_1^2+q_2^2)$ and $(q_1^2q_2-1/3q_2^3)$. Therefore, the potential function (23) is the most general among the 4-order polynomials obeying symmetry C_{3v} [21]. Note that for c=0 the expression (23) is reducing to the well known Henon-Heiles potential[22].

The fourth order NBGF for the potential (23) , equals to

$$\Gamma_4(\xi,\eta) = 1/2(\xi_1^2+\xi_2^2+\eta_1^2+\eta_2^2) + \quad (24)$$
$$+(-\frac{5b^2}{48}+\frac{3c}{8})(\xi_1^2+\xi_1^2+\eta_1^2+\eta_2^2) + (-\frac{c}{2}+\frac{7b^2}{12})(\xi_1\eta_2-\xi_2\eta_1)^2$$

In the action-angle variables

$$\xi_\nu = \sqrt{2\bar{I}_\nu}\,\cos\bar{\varphi}_\nu\ ,\ \eta_\nu = \sqrt{2\bar{I}_\nu}\,\sin\bar{\varphi}_\nu\ ,\ \nu=1,2 \quad (25)$$

one have

$$\Gamma_4 = \bar{I}_1 + \bar{I}_2 + \left(-\frac{5b^2}{12} + \frac{3c}{2}\right)(\bar{I}_1^2+\bar{I}_2^2) + (1/3b^2+2c)\bar{I}_1\bar{I}_2 + \qquad (26)$$
$$+(-\frac{7b^2}{6} + c)\bar{I}_1\bar{I}_2 cos\, 2(\bar{\varphi}_2 - \bar{\varphi}_1).$$

or in the new canonically conjugate variables

$$2I_1 = \bar{I}_1 + \bar{I}_2,\ 2\varphi_1 = \bar{\varphi} + \bar{\varphi},\ I_2 = \bar{I}_1 - \bar{I}_2,\ \varphi_2 = \bar{\varphi}_1 - \bar{\varphi}_2 \qquad (27)$$

the normal form becomes

$$\Gamma_4 = 2I_1 + (-\frac{5b^2}{4} + \frac{9}{2}c)I_1^2 + (-\frac{7b^2}{24} + \frac{c}{4})I_2^2 + \frac{1}{4}(\frac{7b^2}{6} - c)I_2^2\, cos\varphi_2 \qquad (28)$$

As the variable φ_1 does not explicitly enter the Hamiltonian (28), the corresponding canonically conjugate momentum I_1 is the integral of motion. Another integral of motion is the Hamiltonian Γ_4 itself. Thus, the Hamiltonian Γ_4 describes the integrable system.

4.2 The modified normal form

For comparison we present the modified NF for the same Hamiltonian as in Sec. 4.1 in the 4th approximation obtained by GITN

$$\Gamma_4 = i\{\xi\ \eta + \xi\ \eta\ + (\frac{b^2}{6} + c)(\xi_1^2\eta_1^2 + \xi\ \eta\ ) - (2b^2 - 2c)\xi_1\xi_2\eta_1\eta_2\} \qquad (29)$$

One can easily verify that independent integrals of motion here are $N=\xi_1\eta_1+\xi_2\eta_2$ and $M=\xi_1\eta_1-\xi_2\eta_2$, and the NF equals to

$$\Gamma_4 = i\{N + N^2(-\frac{11b^2}{24} + \frac{5c}{4}) + M^2(\frac{13b^2}{24} - \frac{3}{4})\} \qquad (30)$$

As it is seen, the separation of variables takes place in coordinates N, M unlike I_1 and I_2 though in the 6-th approximat- ion there is no separation.

4.3. The Birkhoff-Gustavson NF- nonresonance case.

As a potential function we choose the function

$$V(q_1,q_2) = 1/2(\omega_1^2q_1^2 + \omega_2^2q_2^2) + \sqrt{2}\, q_2^3 + q_1^2q_2 + 1/4q_2^4 \qquad (31)$$
$$\omega_1 = \sqrt{2(2+\sqrt{2})} \cong 2.613125,\ \omega_2 = 2.$$

The normal form obtained by GITA in the 4th approximation equals

$$\Gamma_4 = \omega_1 I_1 + \omega_2 I_2 - 0.016735\ I_1^2 - 0.375\ I_2^2 - 0.219363\ I_1\ I_2 \qquad (32)$$

References

1. Arnold V.I. *Mathematical methods of classical mechanics.* M. Atomizdat, 1974, p. 431 (in Russian)
2. Birkhoff G. *Dynamical systems.* M.OGIZ,1941, p.384 (in Russian)
3. Gustavson F.G., Astron.J. v. 71, p. 670.
4. Giorgilli A.,Comput. Phys.Commun.1979, v. 6, p. 331.
5. Hori G.I., Astron.Soc. Japan. 1966, v.18, p. 287.
6. Deprit A., Cel. Mech. 1969, v.1, p. 12.
7. Campbell J., Jeffris W.H. Celest. Mech. 1970, v. 2, p. 467.
8. Henrad J., Roels J. Celest. Mech. !974, v.10, 497.
9. Mersman W.A., Celest.Mech. 1970, v. 30, p. 81.
10. Markeev A.P., Sokolsky A.G., *Some calculation algorithms for the normalization Hamiltonian systems.* Preprint IPM AN SSSR,1976, N 31.
11. Bolotin Yu.L. et al Phys.Elem.Part.At.Nucl.,1989, v.20. part 4, p.878.
12. Swimm R.T., Delos J.B.,J.Chem.Phys. 1979,v. 71, p.1706.
13. Jaffe C., Reinhart W.P. J.Chem. Phys. 1982,v.77,p.5191.
14. Robnik M., J.Phys.A: Math.Gen. 1984, v.17, p.109.
15. Uzer T., Marcus R.A., J.Chem.Phys. 1984, v. 81,p. 5013.
16. Chekanov N.A. Jad.Fiz., 1989, v.50. p.344.
17. Gantmaher F.R. *Lectures on analytical mechanics.* M. Nauka,1966 (in Russian)
18. Eisenberg I.,Greiner V. *Model of nuclei. Collective and one particle states.* M. Atomisdat (Russian transl),1975.
19. Bolotin Yu.L.,Vinitsky S.I.,Gonchar V.Yu., Chekanov N.A., Yad. Fiz.1989, 6(12), p.1563.
20 Bolotin Yu.L et al ,Yad. Phys. 1990,v.51,1818.
21. Sleter Dj. *Electron structure of molecules.* Mir,1965.
22. Henon M., Heiles C. Astron.J., v.69, p. 73.

INVESTIGATION OF A SYSTEM OF EQUATIONS TO CALCULATE THE CHARGED PARTICLE BEAM EMITTANCE

L. P. LAPTEVA, A. V. SAMOILOV
Institute for Nuclear Research, USSR Academy of Sciences
Moscow, USSR

and

I. V. MALYOVANY, B. S. VOLKOV
Institute for High Energy Physics, Protvino, USSR

ABSTRACT

The present paper considers a method of determining in the complete 3-dimensional matrix of second momenta of a beam. It is shown that the problem is reduced to the solution of the system of linear algebraic equations. Applications of various CAS's for the solution of the obtained system of equations are discussed.

To determine the charged particle beam emittance, it is necessary to know all the elements of the second momenta matrix (a covariance matrix of the beam, denote it as CM). A CM is a positive definite symmetric matrix. Its diagonal elements are the dispersions of beam phase vector components, whereas the non-diagonal ones are the corresponding pair correlations. A CM is often called a matrix of the second momenta. It is conventional to choose for the measure of the phase volume of a beam (treated as complex of particles) the value of CM determinant. In the given case, the motion of the particles through the beam line does not effect the value of this determinant:

$$\sigma(2) = \overline{RXX^TR} = R\sigma(1)R^T \qquad (1)$$

where X is a vector, whose components are positions, angles and momentum of the particle with respect to specified reference trajectory; R is a square matrix, which described the action of the beam line optics on the particle coordinates; and $\sigma(1)$ and $\sigma(2)$ are initial and final CM, respectively.

Since

$$\det R = 1 \qquad (2)$$

(of the Liouville theorem), then

$$\det \sigma(2) = (\det R)(\det \sigma(1))(\det R^T) = \varepsilon^2 = \text{inv}.$$

The volume of the ellipsoid defined by CM is usually called the mean squadratic emittance of a beam. This volume depends on the value det σ of the following form:

$$V = \pi^{n/2}(\det \sigma)^{1/2} \;/\; \Gamma(n/2+1),$$

where n is dimension of ellipsoid.

It is very important to know the beam emittance for many various purposes, such as beam dynamics, experimental and applied problems. The methods for defining a CM in real time and without special procedures, which may require some changes in beam line regime or additional equipment, are of most interest. Here we consider a method to determine the covariance matrix elements based on using beam profile monitors.

The analysis is carried out within the first-order theory. Particle motion is examined in the fields with median plane of symmetry. When the motion of particles in two transverse (perpendicular) planes can be considered quite independently. For such beam lines, a CM can be presented as

$$\sigma = \overline{X\,X^T} = \begin{bmatrix} H_{[3x3]} & 0 \\ 0 & V_{[2x2]} \end{bmatrix},$$

where $X = (x,\vartheta,\delta,y,\varphi)$; x, ϑ and y, φ are the beam extents and the beam divergences at the planes H and V, respectively, δ means the momentum spread of particles; H stands for a CM for a bending plane; V is a CM for a plane without bending. Here we ignore the sixth component of the vector X related with the part length difference.

For the bending plane, a CM is of the following form

$$\sigma = \begin{bmatrix} \sigma_{11} & \sigma_{12} & \sigma_{13} \\ \sigma_{12} & \sigma_{22} & \sigma_{23} \\ \sigma_{13} & \sigma_{23} & \sigma_{33} \end{bmatrix},$$

where σ_{11}, σ_{22}, σ_{33} are the spatial, angular and momentum beam dispersions, respectively; σ_{12}, σ_{13}, σ_{23} are the relevant correlations:

$$\sigma_{11}=\overline{x^2};\; \sigma_{22}=\overline{\vartheta^2};\; \sigma_{33}=\overline{\delta^2};\; \sigma_{12}=\overline{x\vartheta};\; \sigma_{13}=\overline{x\delta};\; \sigma_{23}=\overline{\vartheta\delta}.$$

In the given case, the matrices of optics between the beam profile monitors are of the form :

$$R = \begin{bmatrix} R_{11} & R_{12} & R_{13} \\ R_{21} & R_{22} & R_{23} \\ 0 & 0 & 1 \end{bmatrix} \qquad (3)$$

Suppose that there are six beam profile monitors in the beam line; then, using Expression 1, we can obtain the equation for the spatial dispersions on these monitors. Hence, the system of equations defining the CM elements takes the form:

$$M \vec{\sigma} = \vec{D} \qquad \text{or}$$

$$\begin{bmatrix} 1 & 0 & 0 & 0 & 0 & 0 \\ A_{11}^2 & 2A_{11}A_{12} & A_{12}^2 & 2A_{11}A_{13} & 2A_{12}A_{13} & A_{13}^2 \\ B_{11}^2 & 2B_{11}B_{12} & B_{12}^2 & 2B_{11}B_{13} & 2B_{12}B_{13} & B_{13}^2 \\ C_{11}^2 & 2C_{11}C_{12} & C_{12}^2 & 2C_{11}C_{13} & 2C_{12}C_{13} & C_{13}^2 \\ D_{11}^2 & 2D_{11}D_{12} & D_{12}^2 & 2D_{11}D_{13} & 2D_{12}D_{13} & D_{13}^2 \\ E_{11}^2 & 2E_{11}E_{12} & E_{12}^2 & 2E_{11}E_{13} & 2E_{12}E_{13} & E_{13}^2 \end{bmatrix} \begin{bmatrix} \sigma_{11} \\ \sigma_{12} \\ \sigma_{22} \\ \sigma_{13} \\ \sigma_{23} \\ \sigma_{33} \end{bmatrix} = \begin{bmatrix} D_1 \\ D_2 \\ D_3 \\ D_4 \\ D_5 \\ D_6 \end{bmatrix}, \quad (4)$$

with A = a, B = ba, C = cba, D = dcba, E = edcba. Here small letters mean the optics between detectors, while capital ones refer to the relevant cumulative optics. In the simplest case when the elements R_{13} and R_{23} of all the matrices a, b, c, d, e are zeros (the optics in each interval between the monitors is dispersion-free), det M = 0. A similar result is obtained when only one of the matrices a, b, c, d or e has the form of Eq. 3.

Let two of the matrices a, b, c, d, e be like Eq. 3. For this case, all possible locations of the optics in five intervals between monitors are shown in table 1, as well as the values of det M.

Thus, if a beam line includes no less than two matrices like Eq. 3 and at least six beam profile monitors, disposed as in schemes of table 1, then all elements of a three-dimensional CM including $\sigma_{33} = \overline{\delta^2}$, can be obtained in real time, the performance depending, in principle, on how quickly the beam profiles are measured, since the matrix M could also be investigated beforehand. For three and more optics of the form of Eq. 3, the expression for the determinants entering the system of Eq. 4 is too cumbersome and therefore is not given here.

In dispersion-free case ($R_{13} = R_{23} = 0$), the matrix σ is of dimension [2x2], and the knowledge of only three beam profile monitors is enough to find the CM. The equations specifying CM can be easily obtained from Eq. 4. One does not need to apply computer algebra systems (CAS) in this case and the methods for determining of the emittance are already known (see[1]).

To solve this problem we have used different CAS such as REDUCE[2], VAXIMA[3] and HECAS[4]. We have use mainly those facilities of the CAS's that that have to do exclusively with operations of matrix algebra, factorization of analytic expressions and various substitutions.

Table 1

	a	b	c	d	e	det M
1	∇	∇				0
2	∇		∇			$-8b_{12}d_{12}e_{12}(ed)_{12}a_{13}^{-1}c_{13}^{-1}(cb)_{13}^{-1}(cba)_{13}^{-1}$
3	∇			∇		$-8b_{12}c_{12}e_{12}(cb)_{12}d_{13}d_{13}^{-1}(ed)_{13}^{-1}(cba)_{13}^{-1}$
4	∇				∇	0
5		∇	∇			$-8a_{12}d_{12}e_{12}(ed)_{12}b_{13}e_{13}^{-1}(cb)_{13}^{-1}(cba)_{13}^{-1}$
6		∇		∇		$-8a_{12}c_{12}e_{12}((dc)_{12}(ba)_{13}^{-1}b_{13}^{-1}(edc)_{13}(edcb)_{13}-c_{12}d_{13}(a_{12}(ed)_{12}(edc)_{12}b_{13}^{-1}b_{13}(dcb)_{13}-e_{12}(ed)_{13}(ba)_{12}(cba)_{12}b_{13}^{-1}(dcb)_{13}-b_{12}(cb)_{12}(ba)_{13}^{-1}(a_{22}(dcb)_{13}^{-1}-a_{12}(dcb)_{23}^{-1})))$
7		∇			∇	$-8a_{12}c_{12}d_{12}(dc)_{12}b_{13}b_{13}^{-1}(cba)_{13}^{-1}$
8			∇	∇		$-8a_{12}b_{12}e_{12}(ba)_{12}c_{13}d_{13}^{-1}(dc)_{13}(edc)_{13}$
9			∇		∇	$-8a_{12}b_{12}d_{12}(ba)_{12}c_{13}e_{13}(dc)_{13}(edc)_{13}$
10				∇	∇	0

The first attempt to calculate det M directly was undertaken by the HECAS and VAXIMA. The fully expanded expression for this determinant consists of more than 10^8 monomials. In the general case the reduction of the matrix elements could not be fulfilled because of the awkwardness of their the corresponding expressions. Even in the cases listed in table 1, considerably difficulties are there in the grouping monomials and replacing the names of the symbolic expressions by their values. So as to overcome these difficulties, special algorithms were created realizing the search for the necessary combinations of monomials(satisfying the relations of Eq.2) and their substitution into the corresponding symbolic expression for a matrix element (gensub, facsub, coefac).

Since the main difficulty consist of reducing the expanded expressions, to solve this problem, one needs CAS's with rich facilities as far as substitutions of matrix elements, monomials, rational expressions are concerned. Polynomial arithmetics calculations of determinants and inversion of matrices have not been frequently applied. To obtain final result REDUCE and VAXIMA were used in parallel. Thus, both these computer algebras are applicable to solve similar problems but for the purpose one must get through knowledge of their facilities and the niceties of their use. HECAS can also applied when it is enough to get result without further reducing and simplifying of expressions.

To conclude the paper, we express our acknowledgements to Dolgov G.A. for quite a number useful remarks, and to Larin V.N. for the assistance in carrying out computation with the help of HECAS.

References

1. E.Kiyokaru, e.a., *NIM* **v.202** (1982), p.403.

2. A.C.Hearn, *REDUCE User's Manual,* (Third Edition,Santa Monica,1983).

3.R.Bogen, e.a., *MACSYMA Reference Manual,* (Project MAC, MIT, Cambridge,1974).

4. S.N.Grudtsin and V.N.Larin, *Preprint IHEP 87-28,* (Serpukhov, 1987, in Russian).

COMPLEX REDUCE-PROGRAMS FOR ANALYTIC SOLUTION OF SOME PROBLEMS OF BEAM TRANSPORT SYSTEMS

A. L. URINTSEV
Joint Institute for Nuclear Research, Dubna, USSR

A. V. SAMOILOV
Institute for Nuclear Research, USSR Academy of Sciences, Moscow, USSR

ABSTRACT

A complex of programs for analytic solution of various beam transport systems is described. The complex is realized by means of the language REDUCE. This complex permits to calculate in high order of the theory of aberrations the systems which are included in any combination of any sequence of the following elements drift length, solenoid, dipole, quadrupole, sextupole, etc. without principal limitation on the number of poles. The magnetic field in these elements may be realized up to 6 - 10 orders with respect to phase space variables (coordinates and their derivatives, trajectory length differences, momentum dispersion) and components of a particle spin. In the case of spin motion the order of aberrations is the same as for the order of charged particle motion. The analytic expressions for the phase space variables in REDUCE or FORTRAN are generated together with formula answers as the result of program complex work. Expressions for the aberration coefficients can be used for the following steps of the optimization calculations.

1. It is convenient to consider separately a case when the element axis is a straight line (drift length, solenoid, quadrupole, etc.), and a case when a particle trajectory corresponding to the particle motion through a bending magnet (dipole) represents approximately an arc of a circle. In the former case the equations of motion have the following form[1]:

$$x'' = \alpha\,[x'y'B_x - (1+x'^2)B_y + y'B_z],$$

$$y'' = \alpha\,[(1+y'^2)B_x - x'y'B_y - x'B_z], \quad \alpha = qp^{-1}(1+x'^2+y'^2)^{1/2}.$$

Here and below we use SI-units; x, y are the space coordinates of the particle with respect to the axis trajectory (horizontal and vertical, accordingly); z is the axis coordinate which coincides with the axial trajectory; (')=d/dz; q is the charge, p is the momentum, $p=p_0(1+\delta)$, where p_0 is the equilibrium axis momentum and $\delta=\Delta p/p_0$. B_x, B_y, B_z are the components of the magnetic induction vector **B**. In the latter case it is more convenient to use a cylindric coordinate system r, φ, y, where r is the distance from the axis of the system (y-axis) as before, y is the value measured from the median plane of the magnet, φ is the value measured with respect to the direction of the particle motion, where (r, φ, y) orts are the left-hand three. The equations of motion for this case are the following:

$$r'' = r + 2r^{-1}r'^{2} + \alpha[r'y'B_r + ry'B_\varphi - (r'^{2} + r^2)B_y],$$

$$z'' = 2r^{-1}r'y' + \alpha\,[(y'^{2} + r^2)B_r - rr'B_\varphi - r'y'B_y],$$

$$\alpha = qp^{-1}[1 + (r'^{2} + y'^{2})r^{-2}]^{1/2}, \quad (') = d/d\varphi,$$

where B_r, B_φ, B_y are the projections of **B** on the coordinate axes. The necessary procedure of transition from Cartesian to cylindric coordinate systems and vice versa is foreseen for the beam transport systems in which elements like dipoles are presented.

The behaviour of the polarization vector **S** = (S_1, S_2, S_3) in magnetic field is given by BMT-equation[2] :

$$\frac{d\mathbf{S}}{dt} = \frac{q}{m_0\gamma}\left[\frac{a\gamma^2}{c^2(1+\gamma)}(\mathbf{v}\cdot\mathbf{B})\,\mathbf{v} - (1 + a\gamma)\,\mathbf{B}\right] \times \mathbf{S},$$

where t is the time, m_0 is the rest mass, a is an anomalous magnetic moment of a particle (1.7928444 for a proton and 0.001159652209 for an electron); c is the velocity of light, **v** is the particle velocity vector, γ and γ_0 are the Lorentz factors for arbitrary and equilibrium particles which are given by the expressions

$$\gamma = \gamma_0[1+(1-\gamma_0^{-2})\delta(2+\delta)]^{1/2}, \quad \gamma_0 = [1+(p_0c/E_0)^2]^{1/2},$$

where $E_0=m_0c^2$ is the particle rest energy. A scalar form of BMT-equation was used in symbolic REDUCE-calculations and the transition was realized from t-variables to z- or φ-variables which are counted along the axis trajectory.

2. It is known that by considering the static magnetic field we can introduce a scalar potential V in such a way that **B** = grad V

and V obey the Laplace equation (by neglecting its own beam current). At the coaxial approximation we can have a general solution of the Laplace equation for a Cartesian system in the following form:

$$V = \frac{m!}{m} \operatorname{Im}(x+iy)^m \sum_{k=0}^{\infty} (-1/4)^k \frac{(x^2+y^2)^k}{k!\,(m+k)!} \frac{d^{2k} f(z)}{dz^{2k}},$$

where f(z) is an arbitrary function of z; Im is an operation of taking the imaginary part; m is the natural number, m=0 corresponds to a solenoid; m=2 is a quadrupole, etc. It should be noted that when calculating a solenoid it is advisable to change the function f(z) by $\int f(z)dz$ (using its arbitrariness). It is easy to see that by cutting an infinite series at $k_{max} = \max\{0, \mu(n-m)\}$, $\mu(j)=j/2$ if j is even and $\mu(j)=(j+1)/2$ if j is odd, we can ensure the accuracy of solving up to the n-th order. In each case f(z) has the real physical sense: for the solenoid f(z) is the value of the field on the axis; for the quadrupole $f(z) = \partial B_x/\partial y = \partial B_y/\partial x$ is the gradient on axis, etc. To obtain more accurate approximation to the field of the real element it is necessary to supply with the index m the quantities f and V in the Laplace equation solution given above and to take into account the real magnetic potential V as a sum $V = \sum_m V_m$. The necessary procedure is foreseen in the complex.

Using the equations rot **B**=0 and div **B**=0 we can have for dipole field components the following formal representation:

$$B_r = \sum_{k=0}^{\infty} \frac{y^{2k+1}}{2k+1} \frac{\partial B_{2k}}{\partial r}, \qquad B_\varphi = \frac{1}{r} \sum_{k=0}^{\infty} \frac{y^{2k+1}}{2k+1} \frac{\partial B_{2k}}{\partial \varphi},$$

$$B_y = \sum_{k=0}^{\infty} y^{2k} B_{2k}, \quad B_{2k} = \frac{(-1)^k}{(2k)!} \left(\frac{\partial^2}{\partial r^2} + \frac{1}{r}\frac{\partial}{\partial r} + \frac{1}{r^2}\frac{\partial^2}{\partial \varphi^2} \right)^k B_0(r,\varphi),$$

which suggests that we need only one quantity $B_0(r,\varphi)$, i.e. the field value on the median plane. It is assumed that this value is given.

3. To solve the equations of motion we have used 3 methods:
   a) the classic Runge-Kutta method of the 4-th order modified for symbolic calculation;
   b) the expansion of the dependent variables in the Taylor series along the axis coordinate (z or φ);
   c) the consecutive approximation method is based on the inversion of a linear part of the equation of motion and subsequent iterations.

It should be noted that the b-method was the most flexible and effective in the general case. The c-method was useful and rapid only when it had no z-coordinate dependence. Obviously it means that in the first-order solution all coefficients are constant. It is true for rectangular field model only. The computer realization of the c-method demands the presence of symbolic integration operations. Taking into account the specific feature of our integrals we have obtained our own subprogram for symbolic integration without using the universal symbolic integrator mounted into the REDUCE-system[3]. It was possible to make calculations based on a- and b-methods when a firm well-founded exclusion was foreseen in REDUCE-procedures. The alphabetic items whose infinitesimal order was higher than the one given by the user were excluded. The principal difficulty in REDUCE-system was due to the unlimited accuracy arithmetics: in the process of step-by-step integration any incoming number was changed by decimal number with a quantity of digits which corresponded to the necessary accuracy. Zeros were put instead of small numbers. This procedure was negligible for the final results within the given accuracy.

4. The size of the complex is about 4800 lines in the REDUCE-3.0 language of the algebraic mode for the ES-computers operating under the control of a virtual machine system. The procedures realizing rational numbers truncation are written in the RLISP-symbolic mode language. It permitted to obtain the results ≈ 30 times faster. This complex has 28 REDUCE-procedures, a part of which may be used not only for our needs but to solve problems of a more general character.

The complex consists of functional modules. Loading the modules in the operating memory depends on the user needs. In this case the operation HELP can be used. The initial data are controlled automatically. Here are the main steps of symbolic computations:

a) loading the initial data;
b) deduction of formulas for the magnetic field components in the necessary order of aberration theory;
c) deduction of the equations of motion (with or without spin);
d) output of intermediate results on a monitor;
e) storage of the final results in files on a magnetic disk (band).

5. This complex of programs was first used to solve some problems related to the design study of the Moscow meson factory. The symbolic analysis has been realized at ES-1055M, ES-1061 and IBM PC/AT computers up to 4-5 orders. In every case where we had a possibility to compare our results with those obtained in the digital mode by using the available programs[4], we had a good agreement with the accuracy required. At the moment the complex is added by the electrostatic fields and RF-field elements.

6. References

1. V.P.Kartashev and V.I.Kotov, *Foundations of Beam Optics,* Moscow, Energoatomizdat, 1984.
2. V.Bargman et al., *Phys. Rev. Lett.* **v.2** (1959), p.435.
3. A.Hearn, *REDUCE User's Manual, vers. 3.0,* (Rand Publication CP(4/83), Rand Corporation, Santa Monica, California, 1983).
4. K.L.Brown et al., *CERN Report 80-04*, (1980); T.Matsuo et al., *Computer Program TRIO*, (Proc. of the 3-d Symposium on Ion Sources and Application Technology, 1979), p.25-28.

USING REDUCE SYSTEM FOR CALCULATION OF RENORMALISED STRESS-ENERGY TENSOR FOR MATTER FIELDS IN CURVED SPACE-TIME

A.V.NESTERUK
Department of Mathematics, Financial and Economical Institute, Leningrad, 198000 USSR

and

S.A.PRITOMANOV
Department of Physics, Chemical and Technological Institute, Gagarin av. 8, Dniepropetrovsk, 320640 USSR

It is well known that a background gravitational field can create particles and cause the vacuum polarization of material massive and massless fields. Usually these quantum effects are described by the renormalized stress-energy tensor for corresponding fields. Stress-energy tensor calculation is a very difficult and complicated task.

Based on the h-wave regularisation method proposed by Zeldovich and Starobinsky[1] we elaborate an algorithm for calculation stress-energy tensor for scalar, spinor and vector fields in an isotropic and anisotropic spaces. The algorithm was realized as a complex of the programs for **REDUCE-3.2** system on personal computer compatible with IBM PC/XT (with 512 RAM). Analytical computations were performed for local and nonlocal parts of stress-energy tensor, conformal anomaly of scalar[2] field in anisotropic homogeneous space-time Bianchi-I type.

1.Task definition and basic formulae.

In the case of homogeneous and anisotropic space-time with the metric

$$ds^2 = a^2(\eta)[d\eta^2 - \sum_{i=1}^{3} \alpha_i^2(\eta)(dx^i)^2]. \tag{1}$$

one can obtain the expression for the set of $T_{\mu\nu}=\langle 0|N_\eta T_{\mu\nu}|0\rangle$ for the scalar field with the equation

$$\nabla_\mu \nabla^\mu \varphi + (m^2 + R/6)\varphi = 0$$

as in the form of spectral expansion

$$T_0^0(\eta)=\frac{1}{(2\pi)^3 a^4}\int d^3k\ \mu\ K_0\left[S(\eta,\mathbf{k}) - \frac{Q}{2\mu^2 K_0^2}\left(S(\eta,\mathbf{k}) + \frac{1}{2}U(\eta,\mathbf{k})\right)\right]$$
$$T_\mu^\mu(\eta)=\frac{1}{(2\pi)^3 a^4}\int d^3k\ \mu\ K_0\left[\frac{m^2 g^2}{K_0^2}\left(S(\eta,\mathbf{k}) + \frac{1}{2}U(\eta,\mathbf{k})\right)\right] \tag{2}$$

where in distinction of[1] we defined

$k = (K_0, \mathbf{k})$, $\mathbf{k} = (k_1,k_2,k_3) = k(\sin(\theta)\cos(\varphi),\sin(\theta)\sin(\varphi),\cos(\theta))$,

$$K_o^2(\eta,\mathbf{k}) = k^2 + m^2 g^2(\eta,\theta,\varphi), \quad g = a(\eta)/\mu(\eta,\theta,\varphi), \quad C_i = \dot{\alpha}_i/\alpha_i,$$

$$\mu^2(\eta,\theta,\varphi) = \frac{\sin^2(\theta)\cos^2(\varphi)}{\alpha_1^2(\eta)} + \frac{\sin^2(\theta)\sin^2(\varphi)}{\alpha_2^2(\eta)} + \frac{\cos^2(\varphi)}{\alpha_3^2(\eta)},$$

$$Q = [(C_1 - C_2)^2 + (C_2 - C_3)^2 + (C_1 - C_3)^2] .$$

The functions $S(\eta,\mathbf{k})$ and $U(\eta,\mathbf{k})$ are the solutions of the system

$$\begin{cases} \dot{S}_k = \dfrac{W}{2} U_k + \dfrac{\bar{W}}{2} V_k \\ \dot{U}_k = W\,(2S_k + 1) - (\bar{W} + 2K_o)\, V_k \\ \dot{V}_k = \bar{W}\,(2S_k + 1) + (\bar{W} + 2K_o)\, U_k \end{cases} \qquad (3)$$

$$W = \dot{\mu}/\mu + \dot{K}_o/K_o, \quad \bar{W} = Q/(\mu K_o)$$

with $S_k = U_k = V_k = 0$ at $\eta = \eta_o$.

When $k \to \infty$ the integrals for $T^{\mu}_{\nu}(\eta)$ have divergences and so, the regularisation is necessary. As we think the h-wave regularisation method is most easier and simpler adopt for the computer algebra system. In this method we find the solutions of system (3) for S,U,V as

$$S = \sum h^{-n} S_n , \qquad U = \sum h^{-n} U_n , \qquad V = \sum h^{-n} V_n$$

at $h \to \infty$ and substract first two member of these series in (2). After that we must calculate integrals for T^{μ}_{ν}. The main difficulties about this is rapidly growing numbers of members over integrating by k, θ and φ.

II. Computational scheme and results.

We propose a computational scheme for calculation of the regularized stress-energy tensor for arbitrary quantum field in homogeneous anisotropic space-time and realize it as the complex of REDUCE-3.2 programs for personal computers. The scheme consist of six steps:

1). In the first step we investigate the asymptotic structure of system for S, U, V and restrict ourselves the forth order by h. Behind this S_n, U_n and V_n considered as the series of K_o^2. For example

$$S_2^{2n} = \frac{(m^2 g^2)^{n-1}}{K_o^{2n}} J_2^{2n}(g,\dot{g}), \qquad S_4^{2n} = \frac{(m^2 g^2)^{n-2}}{K_o^{2n}} J_4^{2n}(g,\dot{g},\ddot{g},\dddot{g}\,\dots).$$

2). After the substitutions these series in (2) and some simplifications we have only five integrals

$$C_n = \int_0^{\infty} \frac{(m^2 g^2)^{n-2}}{K_o^{2n-1}}\, k^2 dk , \qquad n = 3,4,5,6,7.$$

We integrate over k, manipulate the REDUCE flags and receive the expression for integrating over θ and φ.

2). The integrals over θ have the form

$$\int_0^{\pi} \frac{\sin^n(\theta)\cos^m(\theta)d\theta}{(a^2\cos^2(\theta)+b^2\sin^2(\theta))^{p/2}} .$$

There are only 16 different integrals. But the functions which stand before these conclude each other about 60-70 items. These integrals we calculate separately, because the memory of our PC was very restricted (only 512 K). We integrate over θ using the substitution $x = tg(\theta)$.

4). The integrals over φ have the form

$$\int_0^{2\pi} \frac{\cos^{2m}(\varphi)d\varphi}{(a^2\sin^2(\varphi)+b^2\cos^2(\varphi))^{2p}}$$

These integrals is also 16. We integrate over φ using the substitution $y = tg(\varphi)$.

During all these steps we check our results by two way:

a)visual control - all the expressions must be symmetric over α (η);

b)calculated control - we find the isotropic limit and compare it with the well known results[3].

5). Because the result of integration is very big and conclude very many items we make some substitutions and calculate results as the functions of

$$H = \frac{\dot{a}}{a}, \qquad C_i = \frac{\dot{\alpha}_i}{\alpha_i}$$

and it derivatives We find the final results by using the conditions

$$C_1 + C_2 + C_3 = 0,$$

$$\dot{C}_1 + \dot{C}_2 + \dot{C}_3 = 0,$$

etc.

6). We calculate the well known results for the renormalised stress-energy tensor and conformal anomaly[4] in our notation. At the end of the computation process we compare the results of steps 5 and 6. Our results completely coincide with the manual results calculated by other regularisation methods.

For the conformal scalar field in anisotropic space (1) we obtain for stress tensor

$$T_{\mu}^{\nu}(\eta) = - \frac{1}{2880\pi^2}\left[\frac{1}{6}\,{}^{(1)}H_{\mu}^{\nu}(\eta) + {}^{(3)}H_{\mu}^{\nu}(\eta) \right]$$

and conformal anomaly

$$T_{\mu}^{\mu}(\eta) = \frac{1}{2880\pi^2}\left[C_{\mu\nu\alpha\beta}C^{\mu\nu\alpha\beta} + R_{\mu\nu}R^{\mu\nu} - \frac{1}{3}R^2 - \Box R \right].$$

In conclusion we want pointed that the main difficulties in calculation process were connected with the limited memory space for REDUCE-3.2. We have to divide our expressions for several part, simplifier they separately and connected results again by text editor.

We think that the limited value of memory space for REDUCE-3.2 in personal computer restrict application of REDUCE system in quantum field theory.

References

1. J.B.Zeldovich and A.A.Starobinsky, *JETP* **61** (1971) 2161.
2. A.V.Nesteruk and S.A.Pritomanov, *Class.Quantum Grav.* **6**, (1989) 1739.
3. A.A.Grib, S.G.Mamaev, V.M.Mostepanenko, *Vacuum quantum effects in strong fields*, (M., Energoatomizdat, 1988).
4. N.D.Birrel and P.C.W.Davies, *Quantum Fields in Curved Space*, (London, Cambridge University Press, 1982).

NUMERICAL-ANALITICAL CALCULATIONS OF SYMPLECTIC MAPS FOR PARTICLES INSIDE ACCELERATORS

ANDRIANOV S.N.
Dept.of Applied Mathem.-Contr.Processes, Leningrad University
Leningrad, USSR

This paper presents a use of Lie algebraic methods for modelling and synthesis of particles beams control systems. The Pade approximations of symplectic maps is discussed.

An intense development of fundamental and applied researches in the charged particles beams field lead to necessity to improve the existing methods of analysis and to create the new ones.

The research of phase sets evolution allows to describe the tracking of a particles beam as a particles ensemble and to use the powerful tools of mathematics. Some of the most important problems of accelerator physics are the problem of nonlinear effects modelling, the problem of beam distortions compensations, the problem of synthesis of particles beam control systems. The characteristic properties of beam physics allows to use the well known methods of the pertubation theory. The opportunities of traditional number methods application are limited for long time evolution simulation and for the control system synthesis. The last time the Lie algebraic tools are used actively. Among the advantages of these methods we have to mention the increase of analitical (nonnumerical) calculations contribution and the opportunity of the most ful application of known information about the system (for example motion integrals, symmetries). In the base of our approach there lies Poincare sections which are referred to a design curve and symplectic maps $\mathfrak{M}$ which drive one of them to another. Let $z=(q_1,\dots,q_n,p_1,\dots,p_n)\in\mathbb{R}^{2n}$ is the phase vector, then $z_f=\mathfrak{M}z_i$ where z_f,z_i belong to step-by-step section. The basic problems of beam dynamic simulation are constraction of the map $\mathfrak{M}$, researching its properties for control system synthesis purpose. It is clear that the increasing of the simulation effectiveness can be guarantee at the expense of analitical transformations. Let $\mathcal{H}_s$ is a Hamiltonian of particle a beam control system which is written in the coordinate system referred to some design curve (for example, to an equilibrium orbit). In the capacity of controled actions are some characteristics of electromagnetic field, for example ϕ, $\mathbf{A}$ - electric and magnetic potentials correspondigly. In this case there is

$$\mathcal{H}_s=-\frac{q}{c}A_s-\left(1-hx\right)\left\{\left(\frac{\mathcal{H}-q\phi}{c}\right)^2-\left(p_x-\frac{q}{c}A_x\right)^2-\left(p_y-\frac{q}{c}A_y\right)^2-m^2c^2\right\}^{1/2},$$

where $\mathcal{H}$ is an old Hamilton function referred to the laboratory-coordinates system. Φ , $\mathbf{A}=(A_x,A_y,A_s)$- electric and magnetic potentials are functions of the independent variables s and the transverse phase coordinates: x,p_x,y,p_y. Let represent $\mathcal{H}_s$ in the form of power series

$$\mathcal{H}_s=\mathcal{H}_2+\Sigma_{k=1}^{\infty}\ \varepsilon^{k-2}\ \mathcal{H}_k\ ,\ \varepsilon\ll 1,$$

where $\mathcal{H}_k$ are homogeneous polynomials of k order according to phase variables z. In according to the Maxwell equations there are independent functions ,which define the electromagnetic field fully (for example, the values of field and their derivatives calculated on the disign curve). It is known that a symplectic map $\mathcal{M}$ generated by a Hamiltonian phase flow is defined as a Lie transformation (Lie series). In the stationar case this transformation looks as

$$\mathcal{M}=exp\left\{-\left(s-s_o\right) \mathcal{L}_{\mathcal{H}_s}\right\}=\sum_{k=0}^{\infty} \frac{(s-s_o)^k}{k!} \mathcal{L}^k_{\mathcal{H}_s},$$

where $\mathcal{L}_{\mathcal{H}}=[\mathcal{H}_s,\cdot]$ is a associated with $\mathcal{H}_s$ Lie operator which acts on general functions g ($\mathcal{L}_{\mathcal{H}} g=[\mathcal{H}_s,g]$) and $[\cdot,\cdot]$ is the Poisson bracket, $\mathcal{L}^k_{\mathcal{H}}g=\mathcal{L}_{\mathcal{H}}(\mathcal{L}^{k-1}_{\mathcal{H}}g)$, $\mathcal{L}^0_{\mathcal{H}}g=g$. Let $\mathcal{H}$ and g are analitical functions in some closed set $\mathcal{X}\subset\mathbb{R}^{2n}$, then such maps generate analitical in the set $\mathcal{X}\times I_\tau$ function $f(z,t)=\exp\{-t\,\mathcal{L}_{\mathcal{H}}\}\cdot g(z)$, $t\in I_\tau=[s_o,s_o+\tau]$ and $\tau<r_m(1-x_m/r_m)^{2n+1}(1+1/2n)^{2n}/\ (2n\cdot M^m_g)$, $\|z\|\le x_m<r_m$, $|g(z_i)|\le M_i\sum_{k=0}^{\infty}(z/r)^k$, $(z/r)^k=\prod_{i=1}^{2n}(z_i/r_i)^{k_i}$, $\sum_{i=1}^{2n}k_i=k$, $M^m_i=max\ M_i$.Also, we can define an associate Lie algebra with respect to the Poisson bracket. In this algebra their is an specific element - a Hamiltonian function $\mathcal{H}$, which defines an time evolution of observables with the aid of the symplectic map $\mathcal{M}=exp\{-t\mathcal{L}_{\mathcal{H}}\}$.

According to the known factored product expansion theorem[1] we can write

$$\mathcal{M}=\ldots exp\{\mathcal{L}_{f_4}\}\cdot exp\{\mathcal{L}_{f_3}\}\cdot exp\{\mathcal{L}_{f_2}\},$$

where $\mathcal{L}_{f_k}$ is a Lie operator associated with a homogeneous polynomial f_k of order k. In particularly, $f_2=-(s-s_o)\mathcal{H}_2$ and the map $\mathcal{M}_2=exp\{\mathcal{L}_{f_2}\}$ defines the fundamental solution matrix of the linear ordinary differential equations system determinated by $\mathcal{H}_2$. The group property of Lie transformations allows to write

$\mathcal{M}=\mathcal{M}^m\cdot\mathcal{M}^{m-1}\cdot\ldots\cdot\mathcal{M}^2\cdot\mathcal{M}^1$, where $\mathcal{M}^k=exp(t_k\cdot\mathcal{L}_{g^k})$, g^k is a Hamiltonian for some lattice (for example, drift, dipole, quadrupole and so on) and t_k is its lengths. With the aid of Campbell-Hausdorf formula we can rewrite the map $\mathcal{M}$ as a Lie transformation $exp(\mathcal{L}_G)$, where G can be calculated by using the polynomials g^k. Then there is an expression $G=\sum_{k=1}^{\infty}\varepsilon^{k-1}\cdot G_k$ and G_k is a homogeneous polynomial of k-th order. If our system has some symmetry then it symplifies calculations of G_k. The group analysys of the wanted characteristics of the beam allows to find this symmetry.

Obviously, that the concrete realization of maps $\mathcal{M}$ calculation is possible only approximately. Therefore there is the problem of approximation of a symplectic map $\mathcal{M}$ of given order accuracy. For this purpose in this paper is used the diagonal *Pade* approximants $P_m(-t\mathcal{L}_G)/P_m(t\mathcal{L}_G)$, where $P_0(\tau)=1$, $P_1(\tau)=2+\tau$, $P_k(\tau)=2(2k-1)P_{k-1}(\tau)+\tau^2\cdot P_{k-2}(\tau)$.

The valume of all calculations for many-dimensional dynamic systems is too large. That is why these calculations can be maked only on computer algebra codes (for example, on *REDUCE 3.2*). What is more, the use of computer algebra methods increases the effectiveness of our approach.

The circumscribed approach allows the author to solve some practical problems of analysis and synthesis of particle beams systems, in particularly, the proton microzond[2], the slow extract system for JINR synchrophasotron[3].

1.Steinberg St. *Factored Product Expansion of Solutions of nonlinear differential Equations,* // SIAM J.Math. Anal., v.15,№1,1984,108-115.

2.Andrianov S.N., Dymnikov A.D., Osetinsky G.M. *The Sistem of Forming of Micron Size Proton Beams*. Dubna, JINR P9-12873, 1979, pp.24.

3.Andrianov S.N.,Zacharov V.V. Mathematical modelling of the resonanse extract system. Part 1-3, in publ.

List of Poster Reports

Symbolic Manipulation in the Unicalc System
V.E.Dmitriev,A.L.Semenov,I.E.Shvetsov

Computer Algebra System for Personal Computer on FORTH
E.N.Kryuchkova

FORMIS - an Interactive System of Analytical Calculations
A.Yu.Zharkov,B.L.Faifel

Some Aspects of Symbolic Manipulations Concept in the System for Symbol-Analytic Transformation SANTRA-2
I.B.Shchenkov

Preprocessor "ALGEBRA-88" in the Representation Theory
V.V.Borisenko,Yu.H.Razmyslov

Algebras of Pseudodifferential Operators: Constructive Aspect
S.M.Zverev,P.S.Shpak

The Intermodule Interfase Organizing in the General Purpose CAS
N.A.Kalinina

Analytical Computation in EVROLOG Processor
M.V.Pochtarenko,E.A.Sumenkov

Deriving the Analytical form of Conditions for Signdefinition
M.A.Choubarov /USSR/

Search of the Rational Solutions of the Linear Differential Equations in REDUCE
K.Yu.Kvashenko

Matrix Operation CAS REDUCE Extension
M.U.Drozdov,V.V Malanin

Equal-Leveled Approximation of Mathematical Functions by Splines
B.A.Popov

Generation of REDUCE-Programs by the PLANNER system Using the Example of Evaluating the Determinant
V.A.Eltekov

Use SAC "REDUCE" for Classifying the Staeckel Spaces in Theory of Gravity
V.G.Bagrov,K.E.Osetrin

Research of Eigen Frequencies of Mechanical System by Means of REDUCE System
V.M.Rudenko

Symbolic-Numerical Methods for the Computer-Aided Stability